Plant-Microbe Dynamics: Recent Advances for Sustainable Agriculture

Plant-Microbe Dynamics: Recent Advances for Sustainable Agriculture

Edited by

TANVEER BILAL PIRZADAH

BISMA MALIK

KHALID REHMAN HAKEEM

CRC Press
Taylor & Francis Group
Boca Raton London New York

CRC Press is an imprint of the
Taylor & Francis Group, an **informa** business

First edition published 2021
by CRC Press
6000 Broken Sound Parkway NW, Suite 300, Boca Raton, FL 33487-2742

and by CRC Press
2 Park Square, Milton Park, Abingdon, Oxon, OX14 4RN

CRC Press is an imprint of Taylor & Francis Group, LLC

Library of Congress Cataloging-in-Publication Data

Names: Pirzadah, Tanveer Bilal, editor. | Malik, Bisma, editor. | Hakeem, Khalid Rehman, editor.
Title: Plant-microbe dynamics : recent advances for sustainable agriculture / edited by Tanveer Bilal Pirzadah, Bisma Malik, Khalid Rehman Hakeem.
Description: First edition. | Boca Raton, FL : CRC Press, 2021. | Includes bibliographical references and index.
Identifiers: LCCN 2020058434 | ISBN 9780367618384 (hardback) | ISBN 9781003106784 (ebook)
Subjects: LCSH: Plant-microbe relationships--Molecular aspects. | Microbial biotechnology.
Classification: LCC QR351 .P572 2021 | DDC 579/.178--dc23
LC record available at https://lccn.loc.gov/2020058434

ISBN: 9780367618384 (hbk)
ISBN: 9780367618391 (pbk)
ISBN: 9781003106784 (ebk)

Typeset in Times
by KnowledgeWorks Global Ltd.

Dedication

This book is dedicated to our Beloved Parents

"Your Lord had decreed, that you worship none save Him, and (that you show) kindness to parents. If one of them or both attain old age with you, say not "Fie" unto them nor repulse them, but speak unto them a gracious word. And lower unto them the wing of submission through mercy and say: My Lord! Have mercy on them both, as they did care for me when I was young."

[Qur'an 17:23-24]

Contents

Preface

Plants and microbes have co-evolved and interacted with each other in nature. The microbes associated with plants, often called plant microbiota, are an integral part of plant life. Depending on the health effects of the hosts, plant–microbe interactions are either beneficial or harmful. The role of the microbiota in promoting plant growth and protecting against various stresses is well known. Recently, our knowledge of the community composition of the plant microbiome and important driving factors has improved considerably. Thus, the significance of the plant microbiome is a reliable approach towards sustainability in order to meet the future food crises and to rejuvenate the soil health. Profiling of the plant-associated microbiome (genome assemblies of all microbes) is an emerging concept to understand the plant–microbe interactions. Microbiota extends the plant capacity to acclimatize fluctuating environmental conditions through several mechanisms. Thus, unravelling the mystery of plant–microbe dynamics through latest technologies to better understand the role of metabolites and signal pathway mechanisms is an important approach. Recently discovered clustered regularly interspaced short palindromic repeats (CRISPR)/Cas9-mediated genome editing (GE) tools are of great interest to explore plant–microbe interactions. A systematic understanding of plant–microbe interactions will allow the application of GE tools to improve the ability of microbes or plants to improve agronomic traits. Modern technologies, such as next-generation sequencing (NGS), omics approaches (e.g., metagenomics, transcriptomics, proteomics, and metabolomics), and computational tools, enable the understanding of community-level molecular aspects of the plant–microbe interactions governing the plant traits. The proposed book will focus on various aspects such as plant–microbe interactions, synergistic mechanisms, microbial nano-formulations, application of metabolomics, bioprospecting of novel metabolites using metagenomic approach, and on the question that how microbes acts as alleviating agents to impart abiotic stress in plants.

This book is unique and would be an ideal source of scientific information to the postgraduate students, research scholars, faculty, and scientists involved in agriculture, plant sciences, environmental sciences, molecular biology, microbiology, biochemistry, biotechnology, and food technology.

We are highly grateful to all our contributors for readily accepting our invitation for not only sharing their knowledge and research, but for venerably integrating their expertise in dispersed information from diverse fields for composing the chapters and enduring editorial suggestions to finally produce this venture.

We extend our appreciation to CRC Press, Taylor & Francis Group, for their exceptional kind support which made our efforts successful.

Tanveer Bilal Pirzadah
Bisma Malik
Khalid Rehman Hakeem

List of Contributors

Muhammad Afzal
Zhejiang Provincial Key Laboratory of Agricultural Resources and Environment, Zhejiang University, Hangzhou, China

Mukhtar Ahmad
Centre of Excellence in Molecular Biology (CEMB) University of the Punjab Lahore, Pakistan

Ashia Alam
Department of Biotechnology, Quaid-i-Azam University Islamabad, Pakistan

Syed Muhammad Azam
Institute of Environmental Microbiology, College of Resources and Environment, Fujian Agriculture and Forestry University, Fuzhou, China

Mahima Begum
Department of Agronomy, Assam Agricultural University, Jorhat, India

Pingchuan Deng
Institute of Crop Science, College of Agriculture and Biotechnology, Zijingang Campus, Zhejiang University, Hangzhou, China

Ruth Gill
Department of Biotechnology, Chandigarh University, Punjab, India

Alvina Gul
Atta-Ur-Rahman School of Applied Biosciences, National University of Sciences and Technology, Islamabad, Pakistan

Mahendra K. Gupta
Microbiology Research Lab, School of Studies in Botany, Jiwaji University, Gwalior, Madhya Pradesh, India

Huma Habib
Department of Biochemistry, Islamia College of Science and Commerce, Kashmir, India

Md. Mosfeq Ul Hasan
Institute of Crop Science, College of Agriculture and Biotechnology, Zijingang Campus, Zhejiang University, Hangzhou, China

Assistant Controller of Examinations, Hajee Mohammad Danesh Science and Technology University, Dinajpur, Bangladesh

Irum Iqrar
Department of Biotechnology, Quaid-i-Azam University Islamabad, Pakistan

Faisal Islam
Institute of Crop Science, College of Agriculture and Biotechnology, Zijingang Campus, Zhejiang University, Hangzhou, China

Bushra Jan
Department of Botany, University of Kashmir, Hazratbal, Srinagar, Jammu and Kashmir, India

Ishrat Jan
Department of Zoology, Kashmir University, Srinagar, Jammu and Kashmir, India

Mehmood Jan
Institute of Crop Science, College of Agriculture and Biotechnology, Zijingang Campus, Zhejiang University, Hangzhou, China

Gayatri Goswami Kandali
Department of Soil Science, Assam Agricultural University, Jorhat, India

Navneet Kaur
Department of Biotechnology, Chandigarh University, Punjab, India

Rajinder Kaur
Department of Biotechnology, Chandigarh University, Punjab, India

Sukhminderjit Kaur
Department of Biotechnology, Chandigarh University, Punjab, India

Asad Ullah Khan
Institute of Crop Science, College of Agriculture and Biotechnology, Zijingang Campus, Zhejiang University, Hangzhou, China

Milon Jyoti Konwar
Department of Agronomy, Assam Agricultural University, Jorhat, India

Nazir Ahmad Malik
Department of Botany, Dolphin PG College of Science & Agriculture, Chandigarh-Sirhind State Highway 12-A, Chunni Kalan, Punjab, India

F. A. Mohiddin
Sher-i-Kashmir University of Agricultural Sciences and Technology, Jammu and Kashmir, India

Ali Muhammad
Biomass and Bioenergy Research Center, Huazhong Agricultural University, Wuhan, China

Sajid Muhammad
Institute of Crop Science, College of Agriculture and Biotechnology, Zijingang Campus, Zhejiang University, Hangzhou, China

Muhammad Numan
Laboratory of Molecular Biology and Biotechnology, University of North Carolina Greensboro, USA

Sadhna Pandey
Department of Botany, Govt. K.R.G. PG Autonomous College, Gwalior, Madhya Pradesh, India

Deepu Pandita
Government Department of School Education, Jammu, Jammu and Kashmir, India

Anjali Pathak
Microbiology Research Lab, School of Studies in Botany, Jiwaji University, Gwalior, Madhya Pradesh, India

Amna, Sadia Qamar
Proteomics and Bioinformatics Lab, Department of Biotechnology, Jamia Millia Islamia, New Delhi, India

M. Irfan Qureshi
Proteomics and Bioinformatics Lab, Department of Biotechnology, Jamia Millia Islamia, New Delhi, India

Mir Sajad Rabani
Microbiology Research Lab, School of Studies in Botany, Jiwaji University, Gwalior, Madhya Pradesh, India

Nitu Rani
University Institute of Agricultural Sciences, Chandigarh University, Gharuan, Mohali, Punjab, India

Zafar A. Reshi
Department of Botany, University of Kashmir, Hazratbal, Srinagar, Jammu and Kashmir, India

Sheikh Sajad
Department of Botany, University of Kashmir, Hazratbal, Srinagar, Jammu and Kashmir, India

Amrina Shafi
Education Department, Government of J&K, Srinagar, Jammu and Kashmir, India
Department of Biotechnology, School of Biological Sciences, University of Kashmir, Srinagar, Jammu and Kashmir, India

Syed Mehar Ali Shah
Department of Plant Breeding and Genetics, University of Agriculture, Peshawar, Pakistan

Zabta Khan Shinwari
Pakistan Academy of Sciences, Islamabad, Pakistan

Raffia Siddique
Atta-Ur-Rahman School of Applied Biosciences, National University of Sciences and Technology, Islamabad, Pakistan

Younas Rasheed Tantray
Department of Botany, Government Degree College, Tangdhar, Jammu and Kashmir, India

Shivani Tripathi
Microbiology Research Lab, School of Studies in Botany, Jiwaji University, Gwalior, Madhya Pradesh, India

Mohammad Saleem Wani
Department of Botany, Punjabi University, Patiala, India

Naseema Wani
Department of Botany, Punjabi University, Patiala, India

Insha Zahoor
Department of Biotechnology, School of Biological Sciences, University of Kashmir, Srinagar, Jammu and Kashmir, India
Department of Neurology, Henry Ford Hospital, Detroit, Michigan, USA

Muhammad Zeeshan
Institute of Crop Science, College of Agriculture and Biotechnology, Zijingang Campus, Zhejiang University, Hangzhou, China

Editor Bio

Prof. (Dr.) Khalid Rehman Hakeem, PhD is Professor at King Abdulaziz University, Jeddah, Saudi Arabia. After completing his doctorate (Botany; specialization in Plant Eco-physiology and Molecular Biology) from Jamia Hamdard, New Delhi, India, in 2011, he worked as a lecturer at the University of Kashmir, Srinagar, for a short period of time. Later, he joined Universiti Putra Malaysia, Selangor, Malaysia, and worked there as Post Doctorate Fellow in 2012 and Fellow Researcher (Associate Professor) from 2013 to 2016. Dr. Hakeem has more than 10 years of teaching and research experience in plant eco-physiology, biotechnology and molecular biology, medicinal plant research, plant–microbe–soil interactions as well as in environmental studies. He is the recipient of several fellowships at both national and international levels. Prof. Khalid has recently been elected as the Fellow, Royal Society of Biology, UK. He has also served as the visiting scientist at Jinan University, Guangzhou, China. At present, he is involved with several international research projects with different government organizations.

So far, Dr. Hakeem has authored and edited more than 55 books with international publishers, including Springer Nature, Academic Press (Elsevier), and CRC Press. He also has to his credit more than 120 research publications in peer-reviewed international journals and 60 book chapters in edited volumes with international publishers.

Dr. Hakeem serves as an editorial board member and reviewer of several high-impact international scientific journals from Elsevier, Springer Nature, Taylor and Francis, Cambridge, and John Wiley Publishers. He is included in the advisory board of Cambridge Scholars Publishing, UK. He is also a fellow of Plantae group of the American Society of Plant Biologists, member of the World Academy of Sciences, member of the International Society for Development and Sustainability, Japan, and member of Asian Federation of Biotechnology, Korea. Dr. Hakeem has been listed in Marquis Who's Who in the World, since 2014 to 2020. Currently, Dr. Hakeem is engaged in studying the plant processes at ecophysiological as well as molecular levels.

Bisma Malik, PhD is Assistant Professor at University Centre for Research and Development (UCRD); Chandigarh University, Punjab, India. She is the recipient of Women Scientist Fellowship (WOS-A) from Department of Science and Technology, Government of India. She is actively engaged in research and has 9 years of research experience in plant stress physiology, molecular biology, and environmental studies. She has worked on stress biology and genetic diversity of chicory—an untapped bioresource from Himalayas in Kashmir. She has to her credit 30 publications in international journals and 30 book chapters of international repute. At present, Dr. Bisma is engaged in studying plant endophytes as a promising source of lead compounds that have immense potential in the pharmaceutical sector.

Tanveer Bilal Pirzadah, PhD is Assistant Professor at University Centre for Research and Development (UCRD); Chandigarh University, Punjab, India. After completing his doctorate in Bioresource from University of Kashmir, Srinagar, India, he worked as an Assistant Professor at the Cluster University Srinagar. Dr. Tanveer has about 9 years of research and teaching experience in bioresources management, biofuels, plant stress physiology, Nanobiotechnology and plant–metal interactions. He also has more than 30 research publications in peer-reviewed international journals, more than 30 book chapters in edited volumes, and 3 edited books with international publishers to his credit. Recipient of Young Scientist Award in 5th International Scientist Award on Engineering, Science and Medicine (RA–VD00921).He is also member of the World Academy of Sciences and Plantae group of the American Society of Plant Biologists. He is presently engaged in studying plant processes at proteomic, metabolomics, and molecular levels to better understand the dynamic plant–metal interactions.

1

Plant-Microbe Interactions

Raffia Siddique and Alvina Gul
National University of Sciences and Technology, Islamabad, Pakistan

CONTENTS

1.1 Introduction

Soil microorganisms play an important role in recycling key nutrients such as phosphorous and nitrogen as well as in protecting plants from various abiotic and biotic stress (Bender et al. 2016; Lladó et al. 2017). Although modern agricultural practices enhance crop production, they, at the same time, cause deterioration of the physio-chemical properties of soil and make it infertile (Bouwman et al. 2009). Fertilizers application in intensively engaged agricultural soils can help regain soil fertility, whereas tillage disturbs microbiome communities (Johnson et al. 1997). This is mostly evident in view of contemporary agricultural practices, with degradation of half of the available fertile land, where we are facing agricultural challenges linked with the disturbance of phosphorous and nitrogen cycles. Such a situation is likely to get worse under the aspect of climate change (Rockström et al. 2009; Yuan et al. 2018). As a result, the United Nations has recommended re-introducing practices of sustainable land management to protect land from degradation (Sanz et al. 2017). Such practices include carbon sequestration, control of soil erosion, nutrient recycling, use of local species, intercropping, and use of native species (Sanz et al. 2017). These measures help to enhance soil fertility by suppressing soil diseases and sustain diversity of soil microbiota, which ultimately help to rejuvenate the ecosystem (Sergaki et al. 2018).

In the natural environment, plants are inhabited by a large number of microbiota that include fungi, bacteria, actinomycetes, algae, and protozoa. Among these colonized organisms, 95% are bacteria (Glick 2012). Rhizobacteria Plant growth promoting rhizo bacteria or plant growth promoting bacteria (PGPB) live inside the roots of plants or over them, and are extremely beneficial for their growth and development (Figure 1.1). Among the PGPB, some are grouped as biocontrol agents, biofertilizers or biopesticides (Glick 2012). Rhizobacteria work by augmenting plant nutrition with essential nutrients such as phosphate, nitrogen, and micronutrients. Biofertilizers are grouped based on their structure,

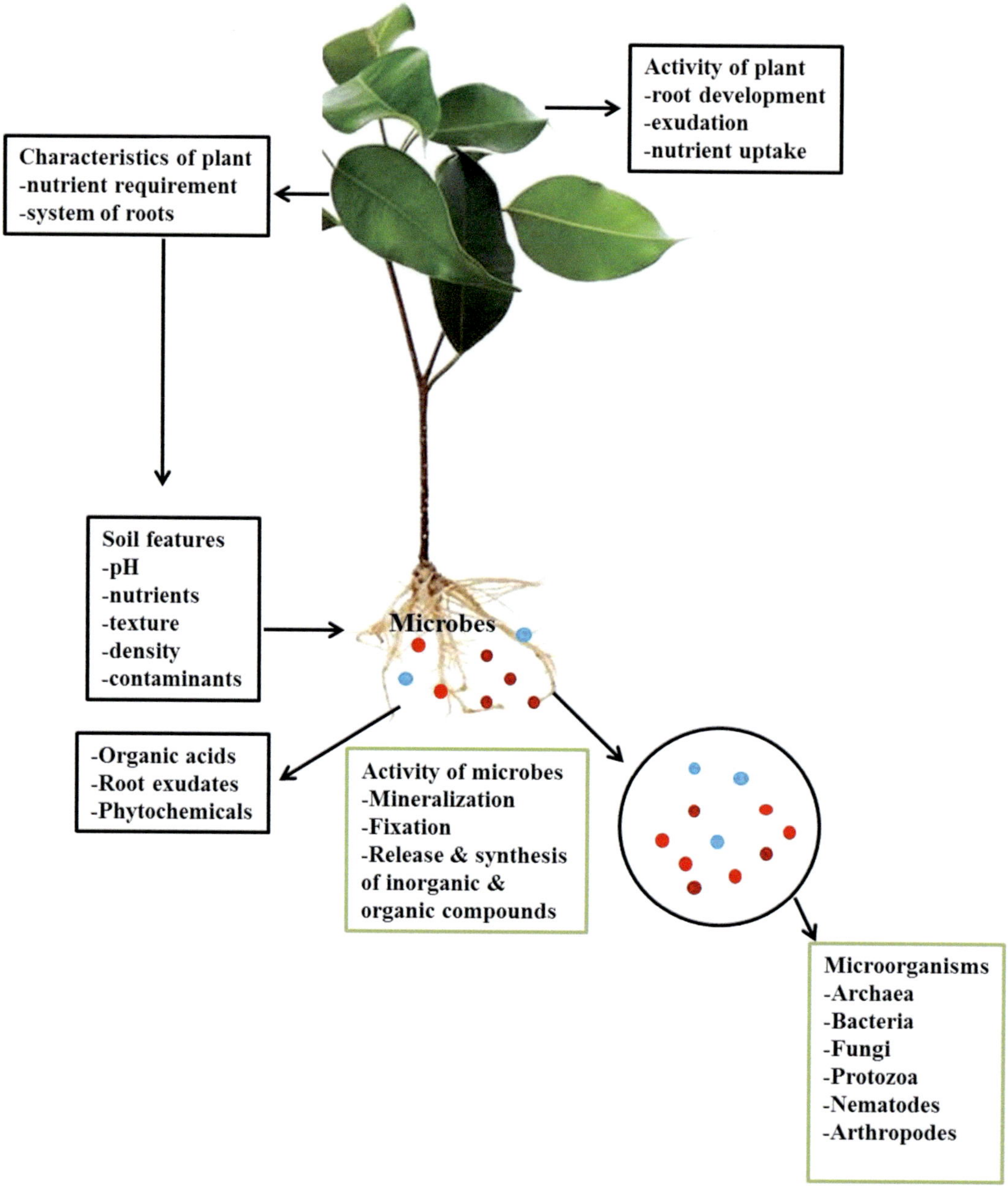

FIGURE 1.1 Mechanism of action of plant microbes with plants.

function, and their ability to fix a specific nutrient such that there are phosphate mobilizing, phosphate solubilizing, and nitrogen fixing bacteria or other biofertilizers for micronutrients. The biopesticides or biocontrol-agents work by controlling or suppressing various disease-causing agents in plants (Glick 2012). However, the plants help these growth promoting microorganisms to proliferate in the nearby areas of roots by a process called roots exudation (Walker et al. 2003). The plant health, genotype, fitness, and developmental stage can determine the properties and composition of root exudates. Root exudation along with cap, border, mucilage, epidermal and cortex cells, in the presence of soil chemicals, leads to redox reactions and change in the pH level of soil that shapes the microbial community near roots

(Kumpf and Nowack 2015). In planted soils, understanding the communication between plant-microbe and microbe-microbe interactions generates a potential to extract beneficial microbiome communities. Here, a question arises as to whether these microbial communities are capable enough to function productively under various agricultural conditions. Though there are differences between PGPR and PGPB organisms, all require similar mechanism directly or indirectly (Glick 1995). These organisms work by providing plants with essential nutrients, hence by improving plant nutrition or by combating diseases. The exact mechanism of action is not well-known, but there are possible explanations that explain hormones production such as gibberellic acid, abscisic acid, auxins, and cytokinins; enzyme production, viz. deaminases to decrease stress conditions in plants by decreasing the concentration of ethylene; contribution toward increase in systemic resistance by producing certain compounds, and biofilm formation by bacteria and rhizobacteria (Aeron et al. 2011; Kim et al. 2012; Timmusk et al. 2015).

1.2 Dynamics and Composition of Microbiome Communities

Microbiome characterization of plant roots' organs has shown a great variety. Studies indicate that bacterial communities are carved out of surrounding environment, as well as host-associated factors such as developmental stage, species, and compartment. Air and soil provide natural reservoir to plant microbiome (Vorholt 2012). The microbiome of plant shoots is influenced by transport processes, while the microbiota of plant roots is influenced by soil type, nutrient content, soil history, and water content (Bogino et al. 2013). In the beginning of growth process, soil interacts and shapes the microbial communities associated with plant both above and below the ground parts (Copeland et al. 2015). The richness of microbial species exists in soil that slowly decreases when it reaches rhizosphere, and in the least quantity around endophytic region, hence making up a selective gradient. Though there is a huge biodiversity among soils, the plant endosphere and rhizosphere are occupied by only four types of bacterial groups: Firmicutes, Actinobacteria, Proteobacteria, and Bacteroidetes (Fierer et al. 2009; Bulgarelli et al. 2012; 2013; Edwards et al. 2015; Rodriguez et al. 2019). Surprisingly, the same bacterial group is present in the human gut microbiome, which suggests that these bacteria are modified to interact with various types of eukaryotes (Ley et al. 2008; Rodriguez et al. 2019). Among these bacterial communities, the constituent bacteria exert an influence on one-another by strong competitive, mutualistic, and antagonistic interactions. The common modes of microbiome interaction include nutrients exchange, competition, and interdependence among microbes (Bais et al. 2006; Rodriguez et al. 2019).

1.3 Driving Factors for Interactions among Plant-Microbe

The microbial habitat below the ground is made of rhizosphere (the soil adjacent to surface of root), rhizoplane (the surface of root), and endosphere (interior part of root). However, the microbial habitat above the ground contains phyllosphere (the above ground area containing the stem and leaves), phylloplane (the one microbial habitat comprising leaves), caulosphere (stem), anthosphere (flowers), spermosphere (seeds), and carposphere (fruits). The plants recruit these microbes from the air and environment actively (Hardoim et al. 2015) (Figure 1.2). Some endophytes (microbes that live inside the tissues of plants and not cause any harm to plants) may follow vertical transmission route (the one involving transfer from seeds of parents), or horizontal transmission route (the one that is directly acquired from the environment) (Frank et al. 2017). The next-generation sequencing (NGS) profiling for composition of microbial communities living above and below the ground habitats of various plants has provided details about the dynamics of microbial community structure such as *Arabidopsis* (Schlaeppi et al. 2014; Bai et al. 2015; Castrillo et al. 2017; Durán et al. 2018; Bodenhausen et al. 2019), agave (Coleman-Derr et al. 2016), cultivated wild barely (Bulgarelli et al. 2015), grapevine (Bokulich et al. 2014; Zarraonaindia et al. 2015), citrus (Xu et al. 2018), petunia (Bodenhausen et al. 2019), maize (Peiffer et al. 2013; Niu et al. 2017; Walters et al. 2018), potato (İnceoğlu et al. 2011), soybean

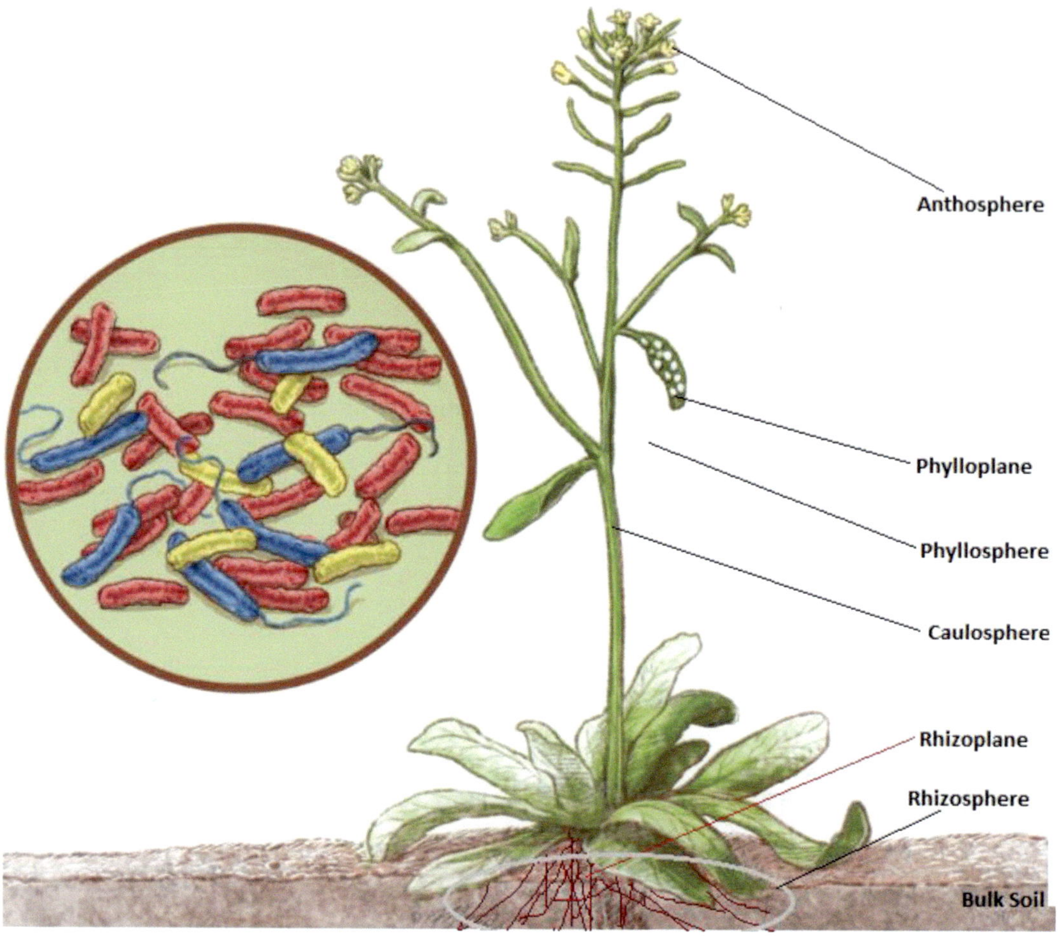

FIGURE 1.2 Microbiome in the various parts of plant body.

(Rascovan et al. 2016), rice (Edwards et al. 2015; Breidenbach et al. 2016), sugarcane (De Souza et al. 2016), sugar beet (Mendes et al. 2011), wheat (Rascovan et al. 2016; Ofek-Lalzar et al. 2014), tomato (Ottesen et al. 2013; Kwak et al. 2018), wild mustard (Wagner et al. 2016), cucumber (Ofek-Lalzar et al. 2014), and lettuce (Cardinale et al. 2015).

Though various metagenomics studies indicate abundant bacterial population in microbiome characterization and composition, other living entities such as protozoa, virus oomycetes, fungi, algae, and nematodes are also fundamental contributors (Table 1.1). By using the various sampling techniques, sequencing technologies, and primers, it is impossible to compare data from such studies (Müller et al. 2016). Many species of microbes are common between root and leaf-associated microbiota in *Arabidopsis thaliana*, grapevine, agave, and mustard (Bai et al. 2015; Bulgarelli et al. 2015; Zarraonaindia et al. 2015; Coleman-Derr et al. 2016). The composition of microbiome of leaf is usually controlled by plant host and various bacterial species that are associated with leaf microbial communities and are shared with microbiome of roots, indicating their attainment from the ground soil (Wagner et al. 2016). On the whole, many reports describe the structure of microbiome is context-related and dependent, and contains taxa having conserved microbes living on a specific plant across multiple species of host plants and environments. Nonetheless, further studies are required to develop the understanding of driving factors of plant-plant, microbe-microbe, and plant-microbe interactions during growth and development of plant.

TABLE 1.1

Various Metagenomics Studies Involve Calculating the Abundance of Only Bacterial Community in the Total Microbiome, the Table below Indicates Some Studies on Plant-Microbiome Structures

Host	Sampling	Key Finding	Reference
Agave	Whole plant, rhizosphere	Plant compartment regulates the microbial composition; however, composition of fungal community is regulated by biogeography of host plant.	(Coleman-Derr et al. 2016)
Arabidopsis	Root, rhizosphere	The composition of rhizospheric bacteria was dependent on environment than host species.	(Schlaeppi et al. 2014)
Arabidopsis	Root, leaf	The genome drafts of over more than 400 isolates indicated an overlap of genome-encoded functions among root and leaf-derived bacteria	(Bai et al. 2015)
Arabidopsis	Root, rhizosphere	The genetic network was explored responsible for controlling response related to phosphate stress that can influence the structure of microbe community of the root.	(Castrillo et al. 2017)
Arabidopsis	Root, rhizosphere	Bacterial microbiome is indispensable for protection and survival of plants against filamentous root fungi.	(Durán et al. 2018)
Maize	Root, rhizosphere	Concerned microbes show variations that are heritable in rhizospheric microbe population.	(Peiffer et al. 2013)
Maize	Root, rhizosphere	Assembled a representative and synthetic and simplified bacterial model having seven dominant strains to study the assembly dynamics of community that was interfering with growth of plant associated pathogen fungus.	(Niu et al. 2017)
Maize	Root, rhizosphere	The composition of microbiome varies with plant age, genotype, and climate events.	(Breidenbach et al. 2016)
Rice	Root, rhizosphere	Rhizosphere, endosphere, and rhizoplane contain a diverse range of microbiota. Cultivation techniques can influence the variety of microbes.	(Edwards et al. 2015)
Rice	Root, rhizosphere	Soil type is a driving cause of structure of microbiome community.	(Breidenbach et al. 2016)
Wheat, soybean	Root, rhizosphere	The properties of soil such as amount of nitrate and pH can influence the structure and composition of microbial community.	(Rascovan et al. 2016)
Sugar beet	Soil	Identify important bacterial genes that can suppress fungal root pathogens.	(Mendes et al. 2011)
Sugarcane	Complete plant, rhizosphere	The microbes enter from local rhizosphere and colonize the organs of plants.	(De Souza et al. 2016)
Tomato	Complete plant, rhizosphere	The microbiome communities here were found closely associated with organs of plants.	(Ottesen et al. 2013)
Tomato	Complete plant, rhizosphere	The microbiota of rhizosphere is responsible for causing diseases to pathogens of soil named as wilt disease.	(Kwak et al. 2018)
Cucumber	Roots	There are differences observed at the level of genes and species among the rhizospheric microbial community from diverse range of plants.	(Ofek-Lalzar et al. 2014)

1.4 Omics Era and Microbiome Research Expansion

Sequencing technologies along with lowering of sequencing costs, and advancement in computational studies have contributed to increase soil microbiome research (Prosser 2015; Fierer 2017). Scientists use three types of sequencing methods: (1) Shotgun sequencing of complete genome or transcriptomic (metatranscriptomatics and metagenomics) information, (2) metataxonomicamplicon primer-based sequencing that emphasizes amplification of genetic markers such as 16SRNA (archaea and bacteria) or region of intergenic spacer (ITS), (3) metabolites detection through nuclear magnetic resonance (NMR) or mass spectrometry (MS) (metabolomics), and (4) detection of fragmented and separated proteins (proteomics) by liquid chromatography and mass spectrometry (LC-MS). The technological limitations of these amplicon sequencing methods and metaomics techniques such as primer biases or sampling errors, the relevant analytical algorithms, and computational power behind them have been elaborated in studies (Hirsch et al. 2010; Pinto and Raskin 2012; Scholz et al. 2012; Tkacz and Poole 2015; Vickers 2017; Sergaki et al. 2018) and will not be explained here independently. There is a major progress achieved in lessening limitations, and obtaining a "perfect" metaomics technology; still, some conceptual limitations exist such as extrapolating correct information from omics-related datasets to draw a purposeful conclusion, and require vigilant experimental designs (Prosser 2015). An example of this includes genes mis-annotation in datasets where data extrapolation from omics datasets is a huge limitation (Lidbury et al. 2014; Fierer 2017; Sergaki et al. 2018). Moreover, in the metaomics field, focus is laid on genes with known functions, while the whole suite of genome information is ignored that provides a way to an extensive research, especially with an emphasis on genes that code ecologically central extracellular proteins (Christie-Oleza et al. 2015).

1.5 Understanding the Relationship between Microbiome-Host Plants with the Help of Syncoms

The complex nature of diverse range of microbes and their interactions in the rhizosphere makes it very difficult to reveal the underlying mechanisms and genetics behind interactions of plant-microbe in natural environment. A strong strategy used to observe the complex nature of plant-microbe interaction in a control setting is the use of syncoms. It starts with collecting isolated cultures of microbes; syncoms are mixed and can be used as inoculants for the observed hosts in gnotobiotic system. With the help of this procedure, one or more members of microbial community that affect plants can be dissected to learn the way host genes affect the composition of microbiome (Bodenhausen et al. 2014). A prominent syncom of maize rhizosphere was employed to study the participation of individual members in the complete community structure of maize plant. It is such that absence of one species in the community causes loss in species richness (Niu et al. 2017) (Table 1.2).

1.6 Role of Plant Microbiota toward Sustainable Agriculture

1.6.1 Beneficial Interactions

The interactions among plant and microbes regulate sequestration of soil carbon by modulating carbon cycle (terrestrial) (Shelake et al. 2019b). The plant-microbiome includes neutral, beneficial, and pathogenic species that help decompose dead organic matter of animals and plant residues. The beneficial microbiome helps plants in many ways and is vital during flowering time, promotes plant growth, and increases the yields of crops (Philippot et al. 2013). Moreover, the plant microbiota helps drive the effect of climate change on agriculture; so, it is of keen interest to employ plant-linked microbiota to minimize the harmful impact of climate change on agriculture and sustainable food production

TABLE 1.2

Collection of Some Microbial Strains Used in Syncom Research

Host	Microbial Kingdom	Strains Number	Tissue Compartment	Microbial Origin	Reference
Arabidopsis thaliana	Bacteria	440	Root (responses to Pi starvation)		(Paredes et al. 2018)
A. thaliana	Bacteria Fungi Oomycete	148 bacteria; 34 fungi; 8 oomycetes	Root, rhizosphere	Cologne agricultural soil (CAS)	(Durán et al. 2018)
Saccharum sp. (sugarcane)	Bacteria	20	Root, rhizosphere, stalks	Greenhouse	(Armanhi et al. 2018)
Trifolium pratense (legume)	Bacteria		Rhizosphere		(Hartman et al. 2017)
Zea mays (maize)	Bacteria	7	Roots	Greenhouse	(Niu et al. 2017)
A. thaliana, Other *Brassicaceae*	Bacteria	35	Roots	North Carolina	(Castrillo et al. 2017)
Solanum lycopersicum (tomato)	Bacteria (Pseudomonas PGPR)	8	Rhizosphere	Nanjing	(Hu et al. 2016)
A. thaliana	Bacteria	218 (leaf); 188 (root and soil)	Leaf, root, and rhizosphere	Cologne, Golm, Widdersdorf, Saint- Evarzec, Roscoff	(Bai et al. 2015)
A. thaliana	Bacteria	38	Roots	North Carolina	(Lebeis et al. 2015)
A. thaliana	Bacteria	7	Leaf	Madrid	(Bodenhausen et al. 2014)

practices (Cavicchioli et al. 2019). A particular plant microbiota has its consequence reliant over its ability to affect plant health and it has a beneficial impact under specific set of conditions, and harmful effect under other conditions. The best example of beneficial plant-microbiota linkage is symbiotic relationship of nitrogen-fixing bacteria in legumes with arbuscular mycorrhizal fungal species that helps the plant in the uptake of phosphorus and nitrogen (Udvardi and Poole 2013; Cao et al. 2017). The symbiotic alliances of plant growth promoting (PGP) microbes help them dominate among the total population of the existing microbes. A great number of PGP microbes affect plant growth by producing phytohormones (such as cytokinin, gibberellins, and auxin) together with beneficial enzymes for plant such as deaminases (Shelake et al. 2019a).

1.6.2 Harmful Interactions

A few plant microbes are harmful causing symptoms of diseases such as *Xanthomonas*, *Ralstonia solanacearum*, *Xylella fastidiosa*, and *Pseudomonas syringae*. The pathogenic plant-microbe interactions infect plants with wounds and natural openings for triggering the immune response and acquiring nutrients (Zoledowska et al. 2019). There are a number of abiotic factors that regulate the effects of interactions among plant-pathogens such as host vulnerability, population size, and climate (Brader et al. 2017). A number of plant-microbiota members are known for increasing plant resistance from phyto-pathogenic microbial species, and hence known as biocontrol agents. While studying the dynamics of plant microbiome, it is a surprising fact that a few non-pathogenic microorganisms start behaving as pathogenic sometimes in some situations such as with an alteration in the host plant or change in size of microbial population (Brader et al. 2017). Hence, modern tools can help understand strange behavior of plant-microbe interactions (Shelake et al. 2019a).

1.7 Beneficial Microbes Functions and Similarities with Pathogens

Amongst beneficial microbiome, the ones that colonize inner surface of root cells (endo-symbionts) are studied extensively because they are capable of promoting stress resistance and plant growth. The best example of most studied endo-symbiosis in plants includes root nodules (Roth and Paszkowski 2017; Keymer and Gutjahr 2018). Nitrogen-fixing bacteria in the root nodules limit fixing nitrogen to one clade of dicots such as Fagales, Fabales, Rosales, and Cucurbitales (legumes) whose root nodules form symbiotic relation with Rhizobium bacteria (Griesmann et al. 2018). However, PGP rhizobacteria are referred to as "plant beneficial bacteria" that are capable of promoting health and growth of plants (Kloepper and Schroth 1981), particularly when a living plant is open to biotic and abiotic stress (Fahad et al. 2015). A number of strains are useful against stress that makes them the best candidates for agricultural applications in various environments. For example, strains of Azospirillum, isolated originally from salty soil in Northern Nigeria, are able to substantially improve yield and growth of durum wheat in areas affected by deprivation of salt such as arid field (Nabti et al. 2010).

1.8 Challenges

It is very challenging to assign a functional group to each microbial group or member based on the microbial structures determined under various environmental conditions. However, species within a specific genus may exhibit different kinds of lifestyles–from disease-causing microbes to mutualistic – based on the environment (Hacquard et al. 2016; Pardee et al. 2016) or because of the horizontal transfer of genes (functional) (Qiu et al. 2009). These differences can cause dramatic shift in microbial phenotypes of desired characteristics such as phosphorous and nitrogen mobilization (Lidbury et al. 2016). More sensitive methods that can distinguish microbial traits beyond their genus are required, together with more clear recognition of individual species that would demand high-throughput technologies and techniques (Schlaeppi and Bulgarelli 2015). The current rate at which the modern technology is making progress is advancing. It is such that transition to metaphenomics from metagenomics is observed, which combines the metagenome (microbial genome containing encoded expressed functions) together with the environment, taking all factors in account that control the dynamics of microbial interactions with community and environment (Jansson and Hofmockel 2018). In this regard, metagenomics is powerful because it provides information about a range of functional and practical capabilities "(such as carbohydrate utilization or production of secondary metabolites) (Bulgarelli et al. 2015) or about particular sets of genes such as for metabolic pathways (Wang et al. 2015; Webb et al. 2017). With the introduction of new analytical methods, we can get more information about specific microbial taxa (carrying essential functional characteristics)(Prosser 2015). The development of commercial deoxyribonucleic acid (DNA) extraction kits has made the process of DNA extraction from a given sample straightforward. However, DNA only shares the functional capabilities of microbial population. Moreover, especially in soils, the huge majority (>90%) of microbial genetic information of biomass is dormant or inactive (Fierer 2017). Even so, this number is going to decrease in the root and rhizosphere area due to step-by-step selection of microbes by plant-associated features (such as more number of microbes are active metabolically in these niches) (Bulgarelli et al. 2013). Efforts are made to extract ribonucleic acid (RNA) from samples of rhizosphere to observe metabolically active microbes and reveal their genetic pathways in response to microbe and plant stimuli (Chaparro et al. 2014; Haichar et al. 2016; Timmins-Schiffman et al. 2017). Hence, studies combine the phenomenon of fortification of ^{13}C-label CO_2 with the field of metatranscriptomics to find out microbes that respond to exudates from plants, to enhance the understanding of microbiome interactions with plants (Haichar et al. 2016; Lueders et al. 2016).

1.9 Conclusion

In contemporary world, the tools and techniques are addressing the composition and properties of microbiome interactions with plants, and how plants recruit the microbial community surrounding them. Moreover, future challenges include engineering and modifying the rhizosphere to help design beneficial microbes – the biggest challenge for coming time.

REFERENCES

Aeron, A., Kumar, S., Pandey, P. and Maheshwari, D. 2011. Emerging role of plant growth promoting rhizobacteria in agrobiology. In: *Bacteria in Agrobiology: Crop Ecosystems.*

Armanhi, J. S. L., De souza, R. S. C., Damasceno, N. D. B., De Araújo, L. M., Imperial, J. and Arruda, P. 2018. A community-based culture collection for targeting novel plant growth-promoting bacteria from the sugarcane microbiome. *Front Plant Sci* 8:2191.

Bai, Y., Müller, D. B., Srinivas, G., Garrido-oter, R., Potthoff, E., Rott, M., Dombrowski, N., Münch, P. C., Spaepen, S. and Remus-emsermann, M. 2015. Functional overlap of the Arabidopsis leaf and root microbiota. *Nature* 528:364–369.

Bais, H. P., Weir, T., Perry, L., Gilroy, S. and Vivanco, J. M. 2006. The role of root exudates in rhizosphere interactions with plants and other organisms. *Annu Rev Plant Biol* 57:233–266.

Bender, S. F., Wagg, C. and Van der Heijden, M. G. 2016. An underground revolution: biodiversity and soil ecological engineering for agricultural sustainability. *Trends Ecol Evol* 31:440–452.

Bodenhausen, N., Bortfeld-Miller, M., Ackermann, M. and Vorholt, J. A. 2014. A synthetic community approach reveals plant genotypes affecting the phyllosphere microbiota. *PLos Genet* 10(4).

Bodenhausen, N., Somerville, V., Desiro, A., Walser, J.-C., Borghi, L., Van der Heijden, M. G. and Schlaeppi, K. 2019. Petunia-and Arabidopsis-specific root microbiota responses to phosphate supplementation. *Phytobiomes J* 3:112–124.

Bogino, P., Abod, A., Nievas, F. and Giordano, W. 2013. Water-limiting conditions alter the structure and biofilm-forming ability of bacterial multispecies communities in the alfalfa rhizosphere. *PLoS One* 8(11).

Bokulich, N. A., Thorngate, J. H., Richardson, P. M. and Mills, D. A. 2014. Microbial biogeography of wine grapes is conditioned by cultivar, vintage, and climate. *Proc Natl Acad Sci* 111:E139–E148.

Bouwman, A., Beusen, A. and Billen, G. 2009. Human alteration of the global nitrogen and phosphorus soil balances for the period 1970–2050. *Global Biogeochem Cy* 23 (12).

Brader, G., Compant, S., Vescio, K., Mitter, B., Trognitz, F., MA, LJ., and Sessitsch, A. 2017. Ecology and genomic insights into plant-pathogenic and plant-nonpathogenic endophytes. *Annu Rev Phytopathol* 55:61–83.

Breidenbach, B., Pump, J. and Dumont, M. G. 2016. Microbial community structure in the rhizosphere of rice plants. *Front Microbiol* 6:1537.

Bulgarelli, D., Garrido-Oter, R., Münch, P. C., Weiman, A., Dröge, J., Pan, Y., Mchardy, A. C. and Schulze-Lefert, P. 2015. Structure and function of the bacterial root microbiota in wild and domesticated barley. *Cell Host Microbe* 17:392–403.

Bulgarelli, D., Rott, M., Schlaeppi, K., Van Themaat, E. V. L., Ahmadinejad, N., Assenza, F., Rauf, P., Huettel, B., Reinhardt, R. and Schmelzer, E. 2012. Revealing structure and assembly cues for Arabidopsis root-inhabiting bacterial microbiota. *Nature* 488:91–95.

Bulgarelli, D., Schlaeppi, K., Spaepen, S., Van Themaat, E. V. L. and Schulze-Lefert, P. 2013. Structure and functions of the bacterial microbiota of plants. *Annu Rev Plant Biol* 64:807–838.

Cao, Y., Halane, M. K., Gassmann, W. and Stacey, G. 2017. The role of plant innate immunity in the legume-rhizobium symbiosis. *Annu Rev Plant Biol* 68:535–561.

Cardinale, M., Grube, M., Erlacher, A., Quehenberger, J. and Berg, G. 2015. Bacterial networks and co-occurrence relationships in the lettuce root microbiota. *Environ Microbiol* 17:239–252.

Castrillo, G., Teixeira, P. J. P. L., Paredes, S. H., Law, T. F., De Lorenzo, L., Feltcher, M. E., Finkel, O. M., Breakfield, N. W., Mieczkowski, P. and Jones, C. D. 2017. Root microbiota drive direct integration of phosphate stress and immunity. *Nature* 543:513–518.

Cavicchioli, R., Ripple, W. J., Timmis, K. N., Azam, F., Bakken, L. R., Baylis, M., Behrenfeld, M. J., Boetius, A., Boyd, P. W. and Classen, A. T. 2019. Scientists' warning to humanity: microorganisms and climate change. *Nat Rev Microbiol* 17:569–586.

Chaparro, J. M., Badri, D. V. and Vivanco, J. M. 2014. Rhizosphere microbiome assemblage is affected by plant development. *ISME J* 8:790–803.

Christie-Oleza, J. A., Armengaud, J., Guerin, P. and Scanlan, D. J. 2015. Functional distinctness in the exoproteomes of marine Synechococcus. *Environ Microbiol* 17:3781–3794.

Coleman-Derr, D., Desgarennes, D., Garcia, C., Gross, S., Clingenpeel, S., Woyke, T., North, G., Visel, A., Martinez, L. P. and Tringe, S. G. 2016. Plant compartment and biogeography affect microbiome composition in cultivated and native Agave species. *New Phytol* 209:798–811.

Copeland, J. K., Yuan, L., Layeghifard, M., Wang, P. W. and Guttman, D. S. 2015. Seasonal community succession of the phyllosphere microbiome. *Mol Plant-Microbe Interact* 28:274–285.

De Souza, R. S. C., Okura, V. K., Armanhi, J. S. L., Jorrín, B., Lozano, N., Da Silva, M. J., González-Guerrero, M., De Araújo, L. M., Verza, N. C. and Bagheri, H. C. 2016. Unlocking the bacterial and fungal communities assemblages of sugarcane microbiome. *Sci Rep* 6:28774.

Durán, P., Thiergart, T., Garrido-Oter, R., Agler, M., Kemen, E., Schulze-Lefert, P. and Hacquard, S. 2018. Microbial interkingdom interactions in roots promote Arabidopsis survival. *Cell* 175:973–983. e14.

Edwards, J., Johnson, C., Santos-Medellín, C., Lurie, E., Podishetty, N. K., Bhatnagar, S., Eisen, J. A. and Sundaresan, V. 2015. Structure, variation, and assembly of the root-associated microbiomes of rice. *Proc Natl Acad Sci* 112:E911–E920.

Fahad, S., Hussain, S., Bano, A., Saud, S., Hassan, S., Shan, D., Khan, F. A., Khan, F., Chen, Y. and Wu, C. 2015. Potential role of phytohormones and plant growth-promoting rhizobacteria in abiotic stresses: consequences for changing environment. *Environ Sci Pollut Res* 22:4907–4921.

Fierer, N. 2017. Embracing the unknown: disentangling the complexities of the soil microbiome. *Nat Rev Microbiol* 15:579.

Fierer, N., Strickland, M. S., Liptzin, D., Bradford, M. A. and Cleveland, C. C. 2009. Global patterns in belowground communities. *Ecol Lett* 12:1238–1249.

Frank, A., Saldierna Guzmán, J. and Shay, J. 2017. Transmission of bacterial endophytes. *Microorganisms* 5:70.

Glick, B. 2012. Plant growth-promoting bacteria: mechanisms and applications. *Scientifica* Article ID 963401, 20 (112).

Glick, B. R. 1995. The enhancement of plant growth by free-living bacteria. *Can J Microbiol* 41:109–117.

Griesmann, M., Chang, Y., Liu, X., Song, Y., Haberer, G., Crook, M. B., Billault-Penneteau, B., Lauressergues, D., Keller, J. and Imanishi, L. 2018. Phylogenomics reveals multiple losses of nitrogen-fixing root nodule symbiosis. *Science* 361(1743).

Hacquard, S., Kracher, B., Hiruma, K., Münch, P. C., Garrido-Oter, R., Thon, M. R., Weimann, A., Damm, U., Dallery, J. F. and Hainaut, M. 2016. Survival trade-offs in plant roots during colonization by closely related beneficial and pathogenic fungi. *Nature Commun* 7:1–13.

Haichar, F., Heulin, T., Guyonnet, J. P. and Achouak, W. 2016. Stable isotope probing of carbon flow in the plant holobiont. *Curr Opin Biotech* 41:9–13.

Hardoim, P. R., Van Overbeek, L. S., Berg, G., Pirttilä, A. M., Compant, S., Campisano, A., Döring, M. and Sessitsch, A. 2015. The hidden world within plants: ecological and evolutionary considerations for defining functioning of microbial endophytes. *Microbiol Mol Biol Rev* 79:293–320.

Hartman, K., Van Der Heijden, M. G., Roussely-Provent, V., Walser, J. C. and Schlaeppi, K. 2017. Deciphering composition and function of the root microbiome of a legume plant. *Microbiome* 5:2.

Hirsch, P. R., Mauchline, T. H. and Clark, I. M. 2010. Culture-independent molecular techniques for soil microbial ecology. *Soil Biol Biochem* 42:878–887.

Hu, J., Wei, Z., Friman, V. P., Gu, S. H., Wang, X.-F., Eisenhauer, N., Yang, T. J., Ma, J., Shen, Q. R. and Xu, Y. C. 2016. Probiotic diversity enhances rhizosphere microbiome function and plant disease suppression. *MBio* 7:e01790–16.

İnceoğlu, Ö, AL-soud, W. A., Salles, J. F., Semenov, A. V. and Van Elsas, J. D. 2011. Comparative analysis of bacterial communities in a potato field as determined by pyrosequencing. *PLoS One* 6(8), e23321

Jansson, J. K. and Hofmockel, K. S. 2018. The soil microbiome—from metagenomics to metaphenomics. *Curr Opin Microbiol* 43:162–168.

Johnson, N. C., Graham, J. and Smith, F.1997. Functioning of mycorrhizal associations along the mutualism–parasitism continuum. *New Phytol* 135:575–585.

Keymer, A. and Gutjahr, C. 2018. Cross-kingdom lipid transfer in arbuscular mycorrhiza symbiosis and beyond. *Curr Opin Plant Biol* 44:137–144.

Kim, Y. C., Glick, B. R., Bashan, Y. and Ryu, C. M. 2012. Enhancement of plant drought tolerance by microbes. *Plant Responses to Drought Stress*. Springer (pp. 383–413). Springer, Berlin, Heidelberg.

Kloepper, J. and Schroth, M. 1981. Relationship of in vitro antibiosis of plant growth-promoting rhizobacteria to plant growth and the displacement of root microflora. *Phytopathol* 71:1020–1024.

Kumpf, R. P. and Nowack, M. K. 2015. The root cap: a short story of life and death. *J Exp Bot* 66:5651–5662.

Kwak, M.-J., Kong, H. G., Choi, K., Kwon, S. K., Song, J. Y., Lee, J., Lee, P. A., Choi, S. Y., Seo, M. and Lee, H. J. 2018. Rhizosphere microbiome structure alters to enable wilt resistance in tomato. *Nat Biotechnol* 36:1100–1109.

Lebeis, S. L., Paredes, S. H., Lundberg, D. S., Breakfield, N., Gehring, J., Mcdonald, M., Malfatti, S., Del Rio, T. G., Jones, C. D. and Tringe, S. G. 2015. Salicylic acid modulates colonization of the root microbiome by specific bacterial taxa. *Science* 349:860–864.

Ley, R. E., Hamady, M., Lozupone, C., Turnbaugh, P. J., Ramey, R. R., Bircher, J. S., Schlegel, M. L., Tucker, T. A., Schrenzel, M. D. and Knight, R. 2008. Evolution of mammals and their gut microbes. *Science* 320:1647–1651.

Lidbury, I., Murrell, J. C. and Chen, Y. 2014. Trimethylamine N-oxide metabolism by abundant marine heterotrophic bacteria. *Proc Natl Acad Sci* 111:2710–2715.

Lidbury, I. D., Murphy, A. R., Scanlan, D. J., Bending, G. D., Jones, A. M., Moore, J. D., Goodall, A., Hammond, J. P. and Wellington, E. M. 2016. Comparative genomic, proteomic and exoproteomic analyses of three pseudomonas strains reveals novel insights into the phosphorus scavenging capabilities of soil bacteria. *Environ Microbiol* 18:3535–3549.

Lladó, S., López-Mondéjar, R. and Baldrian, P. 2017. Forest soil bacteria: diversity, involvement in ecosystem processes, and response to global change. *Microbiol Mol Biol Rev* 81:e00063–16.

Lueders, T., Dumont, M. G., Bradford, L. and Manefield, M. 2016. RNA-stable isotope probing: from carbon flow within key microbiota to targeted transcriptomes. *Curr Opin Biotechnol* 41:83–89.

Mendes, R., Kruijt, M., De Bruijn, I., Dekkers, E., Van Der Voort, M., Schneider, J. H., Piceno, Y. M., Desantis, T. Z., Andersen, G. L. and Bakker, P. A. 2011. Deciphering the rhizosphere microbiome for disease-suppressive bacteria. *Science* 332:1097–1100.

Müller, D. B., Vogel, C., Bai, Y. and Vorholt, J. A. 2016. The plant microbiota: systems-level insights and perspectives. *Annu Rev Genet* 50:211–234.

Nabti, E., Sahnoune, M., Ghoul, M., Fischer, D., Hofmann, A., Rothballer, M., Schmid, M. and Hartmann, A. 2010. Restoration of growth of durum wheat (triticum durum var. Waha) under saline conditions due to inoculation with the rhizosphere bacterium Azospirillum brasilense NH and extracts of the marine alga Ulva lactuca. *J Plant Growth Regul* 29:6–22.

Niu, B., Paulson, J. N., Zheng, X. and Kolter, R. 2017. Simplified and representative bacterial community of maize roots. *Proc Natl Acad Sci* 114:E2450–E2459.

Ofek-Lalzar, M., Sela, N., Goldman-Voronov, M., Green, S. J., Hadar, Y. and Minz, D. 2014. Niche and host-associated functional signatures of the root surface microbiome. *Nat Commun* 5:1–9.

Ottesen, A. R., Peña, A. G., White, J. R., Pettengill, J. B., Li, C., Allard, S., Rideout, S., Allard, M., Hill, T. and Evans, P. 2013. Baseline survey of the anatomical microbial ecology of an important food plant: solanum lycopersicum (tomato). *BMC Microbiol* 13:114.

Pardee, K., Green, A. A., Takahashi, M. K., Braff, D., Lambert, G., Lee, J. W., Ferrante, T., Ma, D., Donghia, N. and Fan, M. 2016. Rapid, low-cost detection of Zika virus using programmable biomolecular components. *Cell* 165:1255–1266.

Paredes, S. H., Gao, T., Law, T. F., Finkel, O. M., Mucyn, T., Teixeira, P. J. P. L., Gonzalez, I. S., Feltcher, M. E., Powers, M. J. and Shank, E. A. 2018. Design of synthetic bacterial communities for predictable plant phenotypes. *PLoS Biol* 16(2):e2003962.

Peiffer, J. A., Spor, A., Koren, O., Jin, Z., Tringe, S. G., Dangl, J. L., Buckler, E. S. and Ley, R. E. 2013. Diversity and heritability of the maize rhizosphere microbiome under field conditions. *Proc Natl Acad Sci* 110:6548–6553.

Philippot, L., Raaijmakers, J. M., Lemanceau, P. and Van der Putten, W. H. 2013. Going back to the roots: the microbial ecology of the rhizosphere. *Nat Rev Microbiol* 11:789–799.

Pinto, A. J. and Raskin, L. 2012. PCR biases distort bacterial and archaeal community structure in pyrosequencing datasets. *PloS One* 7(8):e43093.

Prosser, J. I. 2015. Dispersing misconceptions and identifying opportunities for the use of 'omics' in soil microbial ecology. *Nat Rev Microbiol* 13:439–446.

Qiu, X., Kulasekara, B. and Lory, S. 2009. Role of horizontal gene transfer in the evolution of *Pseudomonas aeruginosa* virulence. *Microbial Pathogenomics*. Karger Publishers.

Rascovan, N., Carbonetto, B., Perrig, D., Díaz, M., Canciani, W., Abalo, M., Alloati, J., González-Anta, G. and Vazquez, M. P. 2016. Integrated analysis of root microbiomes of soybean and wheat from agricultural fields. *Sci Rep* 6:28084.

Rockström, J., Steffen, W., Noone, K., Persson, A., Chapin, F., Lambin, E., Lenton, T., Scheffer, M., Folke, C., Schellnhuber, H. Nykvist, B., De Wit, C. A., Hughes, T., van der Leeuw, S., Rodhe, H., Sorlin S., Snyder, P. K., Constanza, R., Svedin, U., Falkenmark, M., Karlberg, L., Corell, R. W., Fabry, V. J., Hansen, J., Walker, B., Liverman, D., Richardson, K., Crutzen, P., and Foley, J. 2009. Planetary boundaries: exploring the safe operating space for humanity. *Ecol Soc* 14:1–32.

Rodriguez, P. A., Rothballer, M., Chowdhury, S. P., Nussbaumer, T., Gutjahr, C. and Falter-Braun, P. 2019. Systems biology of plant microbiome interactions. *Mol Plant* 12(6):804–821.

Roth, R. and Paszkowski, U. 2017. Plant carbon nourishment of arbuscular mycorrhizal fungi. *Curr Opin Plant Biol* 39:50–56.

Sanz, M., De Vente, J., Chotte, J. L., Bernoux, M., Kust, G., Ruiz, I., Almagro, M., Alloza, J., Vallejo, R. and Castillo, V. 2017. Sustainable land management contribution to successful land-based climate change adaptation and mitigation: a report of the science-policy interface.

Schlaeppi, K. and Bulgarelli, D. 2015. The plant microbiome at work. *Mol Plant-Microbe Interact* 28:212–217.

Schlaeppi, K., Dombrowski, N., Oter, R. G., Van Themaat, E. V. L. and Schulze-Lefert, P. 2014. Quantitative divergence of the bacterial root microbiota in *Arabidopsis thaliana* relatives. *Proc Natl Acad Sci* 111:585–592.

Scholz, M. B., Lo, C.-C. and Chain, P. S. 2012. Next generation sequencing and bioinformatic bottlenecks: the current state of metagenomic data analysis. *Curr Opin Biotechnol* 23:9–15.

Sergaki, C., Lagunas, B., Lidbury, I., Gifford, M. L. and Schäfer, P. 2018. Challenges and approaches in microbiome research: from fundamental to applied. *Front Plant Sci* 9:1205.

Shelake, R. M., Pramanik, D. and Kim, J.-Y. 2019a. Exploration of plant-microbe interactions for sustainable agriculture in CRISPR era. *Microorganisms* 7:269.

Shelake, R. M., Waghunde, R. R., Verma, P. P., Singh, C. and Kim, J.-Y. 2019b. Carbon sequestration for soil fertility management: microbiological perspective. *Soil Fertility Management for Sustainable Development.* Springer. (pp. 25–42). Springer, Singapore.

Timmins-Schiffman, E., May, D. H., Mikan, M., Riffle, M., Frazar, C., Harvey, H., Noble, W. S. and Nunn, B. L. 2017. Critical decisions in metaproteomics: achieving high confidence protein annotations in a sea of unknowns. *ISME J* 11:309–314.

Timmusk, S., Kim, S.B., Nevo, E., Abd EL Daim, I., Ek, B., Bergquist, J. and Behers, L. 2015. Sfp-type PPTase inactivation promotes bacterial biofilm formation and ability to enhance wheat drought tolerance. *Front Microbiol* 6:387.

Tkacz, A. and Poole, P. 2015. Role of root microbiota in plant productivity. *J Exp Bot* 66:2167–2175.

Udvardi, M. and Poole, P. S. 2013. Transport and metabolism in legume-rhizobia symbioses. *Ann Rev Plant Biol* 64:781–805.

Vickers, N. J. 2017. Animal communication: when I'm calling you, will you answer too? *Curr Biol* 27:R713–R715.

Vorholt, J. A. 2012. Microbial life in the phyllosphere. *Nat Rev Microbiol* 10:828–840.

Wagner, M. R., Lundberg, D. S., Tijana, G., Tringe, S. G., Dangl, J. L. and Mitchell-Olds, T. 2016. Host genotype and age shape the leaf and root microbiomes of a wild perennial plant. *Nat Commun* 7:1–15.

Walker, T. S., Bais, H. P., Grotewold, E. and Vivanco, J. M. 2003. Root exudation and rhizosphere biology. *Plant Physiol* 132:44–51.

Walters, W. A., Jin, Z., Youngblut, N., Wallace, J. G., Sutter, J., Zhang, W., González-Peña, A., Peiffer, J., Koren, O. and Shi, Q. 2018. Large-scale replicated field study of maize rhizosphere identifies heritable microbes. *Proc Nat Acad Sci* 115:7368–7373.

Wang, Q., Fish, J. A., Gilman, M., Sun, Y., Brown, C. T., Tiedje, J. M. and Cole, J. R. 2015. Xander: employing a novel method for efficient gene-targeted metagenomic assembly. *Microbiome* 3:32.

Webb, H. E., Brichta-Harhay, D. M., Brashears, M. M., Nightingale, K. K., Arthur, T. M., Bosilevac, J. M., Kalchayanand, N., Schmidt, J. W., Wang, R. and Granier, S. A. 2017. Salmonella in peripheral lymph nodes of healthy cattle at slaughter. *Front Microbiol* 8:2214.

Xu, J., Zhang, Y., Zhang, P., Trivedi, P., Riera, N., Wang, Y., Liu, X., Fan, G., Tang, J. and Coletta-Filho, H. D. 2018. The structure and function of the global citrus rhizosphere microbiome. *Nat Commun* 9:1–10.

Yuan, Z., Jiang, S., Sheng, H., Liu, X., Hua, H., Liu, X. and Zhang, Y. 2018. Human perturbation of the global phosphorus cycle: changes and consequences. *Environ Sci Technol* 52:2438–2450.

Zarraonaindia, I., Owens, S. M., Weisenhorn, P., West, K., Hampton-Marcell, J., Lax, S., Bokulich, N. A., Mills, D. A., Martin, G. and Taghavi, S. 2015. The soil microbiome influences grapevine-associated microbiota. *MBio* 6:e02527–14.

Zoledowska, S., Presta, L., Fondi, M., Decorosi, F., Giovannetti, L., Mengoni, A. and Lojkowska, E. 2019. Metabolic modeling of *Pectobacterium parmentieri* SCC3193 provides insights into metabolic pathways of plant pathogenic bacteria. *Microorganisms* 7:101.

2

Secondary Metabolites: Rapid Evolving Weapon in the Plant-Microbe Warfare

Irum Iqrar[1], Ashia Alam[1], Muhammad Numan[2], and Zabta Khan Shinwari[1,3]
[1] *Quaid-i-Azam University, Islamabad, Pakistan*
[2] *University of North Carolina, Greensboro, North Carolina*
[3] *Pakistan Academy of Sciences, Islamabad, Pakistan*

CONTENTS

2.1 Introduction

Being seized by an enormous number of threats by the antagonists, plants have to suffer a huge loss in productivity and growth every year. The antagonists comprise biotic agents, also known as pathogens (bacteria, fungi), and abiotic factors. Unlike animals that possess significant immunity, plants have developed/evolved themselves with a wide variety of various defense mechanisms by which they can deal with several kinds of biotic and abiotic stresses (Ballhorn et al. 2009) by stimulating the outer surface of the plant and eliciting various enzymes and chemicals to cope up with antagonists (Saeed and Tahira 2019), as shown in Figure 2.1. In the protection against predators and microbial pathogens, plants produce a high diversity of natural compounds/products or secondary metabolites based on the toxic nature of antagonist and repellence to herbivores and microbes; some of them are also involved in defense against abiotic stresses, and also important for the communication of plants with other organisms (Schafer and Wink 2009). The major functions of the secondary metabolites in plants are depicted in Figure 2.2. The secondary metabolites are predominantly used in a variety of biological activities in the form of antiparasitic and antimicrobial agents, antitumor agents, enzyme inhibitors, immunosuppressive agents, etc. (Demain 1999; Thirumurugan et al. 2018). The multifaceted behavior of pathogens interacts with the living plants and tissues by damaging them and minimizing the strength by spreading toxins and disorganizing the genes set-up to utilize nutrients from it. Plant cells resist the penetration of pathogens due to the presence of a cell wall as a first-line defense; important molecules like nuclear binding sites proteins, kinases, and mechanisms elicited in response to the pathogen attack are included in another defense known as inducible defense (Benitez et al. 2004; Chen 2008). Some physical and chemical barriers already present in plants to knock out the pathogens breach come under constitutive defense.

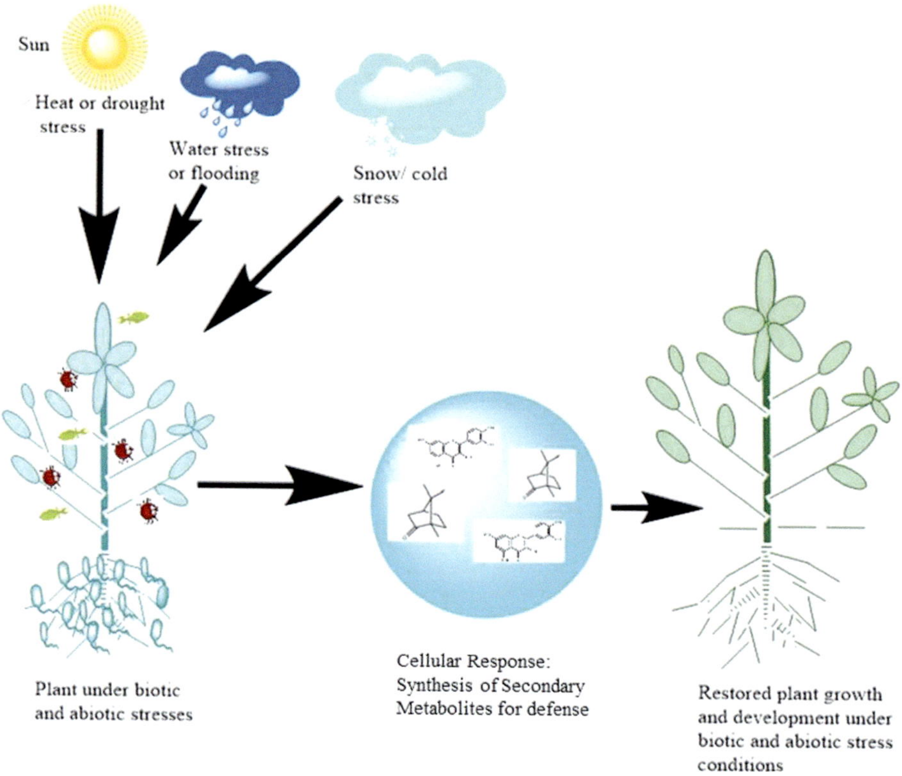

FIGURE 2.1 A general overview of plant responses to biotic and abiotic stresses. Biotic and abiotic stresses severely affect plant growth and development. Under harsh environmental conditions, plants respond by producing certain secondary metabolites to cope with such harsh conditions and to restore normal growth and development.

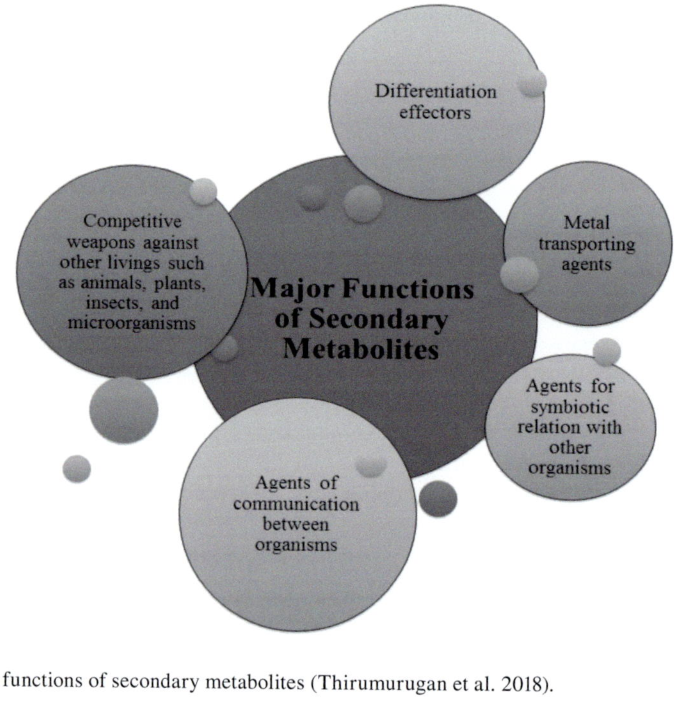

FIGURE 2.2 Major functions of secondary metabolites (Thirumurugan et al. 2018).

2.2 Defense Mechanisms of Plants

Self-protection of plants is considered an important phenomenon as some antimicrobial secondary metabolites have phytotoxic activities (Stergiopoulos et al. 2013). One basic strategy plants adopt for self-protection is the synthesis of specific antimicrobial compounds only when there is a pathogenic attack, and restrict their biosynthesis in the site of infection only (Andersen et al. 2018). Another important protection mechanism is the modification of some biomolecules, which usually act as target sites for the metabolites (Figure 2.3). As an example, the ability of saponins to disrupt a membrane is strictly dependent on the presence of a free 3β-hydroxy group in the membrane (Bednarek and Schulze-Lefert 2018). Another very important mechanism through which plants control the self-phytotoxic activity is to store the active molecules as inactive conjugates or precursors away from the activating enzymes (Bednarek and Schulze-Lefert 2018). Most of the secondary metabolites in the plant cells are accumulated as glycosides in the form of glucose conjugates (Ferrer et al. 2017). Defensive chemicals signal transduction and surveillance are the main responsibilities of inducible defense (Kessler and Baldwin 2002; Ferry et al. 2004), with two broad and significant mechanisms categorized as direct and indirect mechanisms.

2.2.1 Direct Mechanism

The suppression of phytopathogen in direct defense mechanism occurs due to the presence of plant-friendly bacteria and the biochemicals released by the plant tissues in the form of inhibitors and phenolics (such as tannins and dienes), lectins, saponins, and hydrolytic enzymes (Bhatia et al. 2005; Freeman and Beattie 2008; Pundir and Jain 2015). Secretion of these substances through the roots or other surfaces like alpha-terthinyl in marigold, maleic acid in the roots of *Cicerarietinum*, chlorogenic acid in potatoes constrains spore germination, scab bacteria, and the lysis of fungal cell wall that penetrates for nutrient procurement and colonization of roots (Benitez et al. 2004; Haas and Défago 2005; Chen 2008). These friendly bacteria can display biological control activities with proper mechanisms involved in competition, antibiosis, and hyperparasitism (Figure 2.4).

2.2.2 Indirect Mechanism

Plants protect themselves from pathogens using a number of physical and chemical mechanisms. Augmenting plant resistance in response to the stimulus is one of the strategies in which biocontrol agents are not involved in the recognition of pathogens (Pundir and Jain 2015). Due to the presence

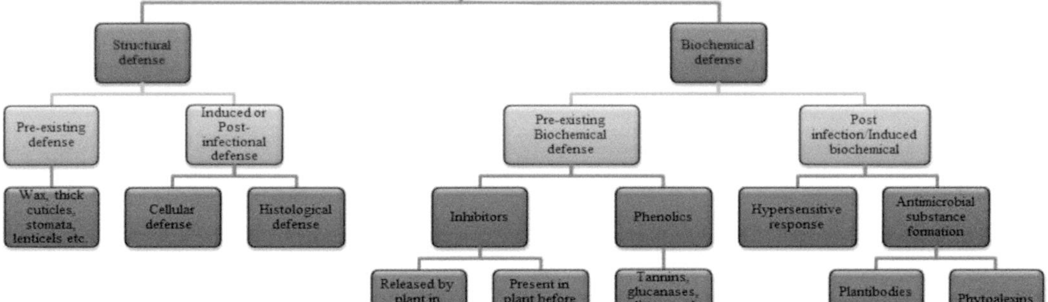

FIGURE 2.3 Summary of interaction of plant and biological control against pathogens.

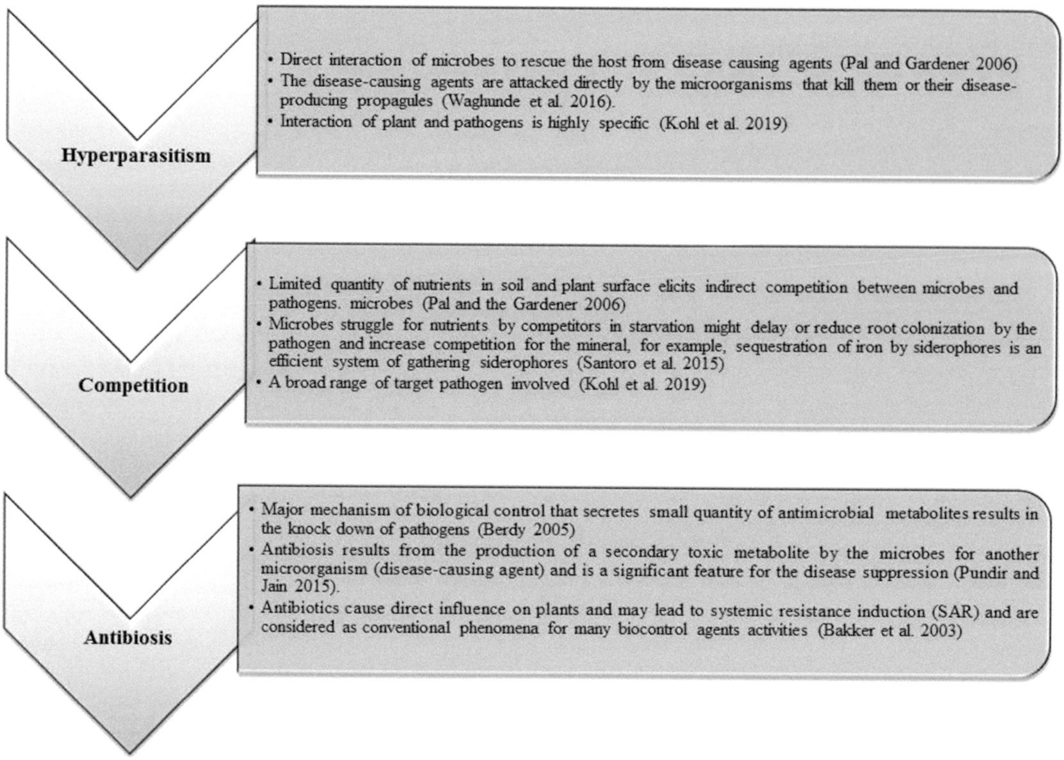

FIGURE 2.4 Schematic diagram of plant defense mechanism against the pathogens.

of multilayer surveillance mechanisms, plants stimulus responds against the breach of the pathogen. Microbe-associated molecular patterns (MAMPs), including specific proteins, cell wall components, and lipopolysaccharides, are typical stimuli that enhance the resistance against the phytopathogens (Hou et al. 2019; Malik et al. 2020). Some fungi, bacteria, viruses, and microscopic worms mimic the defense mechanism to conquer the boundaries of plant tissues, but plants respond in a hypersensitive mode by restricting pathogens to water and nutrients (Freeman and Beattie 2008). Hypersensitive response elicits the systematic acquired resistance (SAR) that necrotizes pathogens, and induced systematic resistance (ISR) enhances the capacity of the entire plant for a wide range of pathogens with the help of beneficial bacteria (Conrath et al. 2015; Singh and Pathak 2015; Shinwari et al. 2019). The SAR pathway triggers the pathogenic war and deliberately produces pathogen-related proteins (PRP) monitored by salicylic acid resulting in the lysis of the pathogenic cell wall and thickening of the cell wall of the plant to resist further infection (Boller and Felix 2009; Waghunde et al. 2017). Certain non-pathogenic substances (ethylene and jasmonic acid) provoke the ISR pathway and result in the production of antibiotics, organic compounds, and siderophores (Prasad et al. 2015), while *Pseudomonas* and *Bacillus* induce the ISR pathway with precursor jasmonic acid and ethylene monitored by salicylic acid (Pieterse et al. 2014; Waghunde et al. 2017).

2.3 Secondary Metabolites: A Rapidly Evolving Weapon in the Plant-Microbe Warfare

Competing with the plethora of pathogens, living plants produce versatile organic compounds while synthesizing primary metabolites (Kim et al. 2010; Isah 2019). Over one million secondary metabolites are known, and they are commonly classified according to their function, vast diversity in structure, and biosynthesis. Among them, about 200,000–250,000 are bioactive metabolites from different sources such as plants, microbes, etc. (Thirumurugan et al. 2018). Some secondary metabolites act as a tool for nitrogen transportation and storage (Chen 2008).

2.3.1 Role of Plant Secondary Metabolites (PSMs) in Plant Defense

Plants with a higher concentration of secondary metabolites are considered more resistant and inversely proportional to plant growth and reproduction with a significant shift in the importance over the decades for their defensive properties against bacteria, fungi, viruses and nematodes in living plants (Mazid et al. 2011). Major chemical compounds involved in the defense of plants include alkaloids, flavonoids, phenolics, non-protein amino acids, cyanogenic glycosides, glucosinolates, and terpenoids categorized as prohibitin, phytoalexins, and phytoanticipins (Mazid et al. 2011; Thirumurugan et al. 2018). Phytoanticipins protect as front-line defenders of plants, and phytoalexins require precursors for their activation. These secondary metabolites are principally divided into three groups based on their composition, i.e., phenolics, terpenes, and sulfur- and nitrogen-containing compounds, and are summarized in Table 2.1. The largest class of secondary compounds, terpenoids with subclasses of mono, di, tri, and sesquiterpenoids, defends living plant and its tissue in multiple ways:

- The essential oil in mono and diterpenoids emits an aroma in plants that keeps fungi and bacteria away from it, e.g. menthol in mint plant and the pungent smell of turpentine in pine tree act as an insect repellent.
- *Gossypium hirsutum* plant produces diterpenes with strong antifungal and antibacterial activities.
- Triterpenoids mimic animal steroids, stimulate insects' mortality, and squirt insecticidal larvae at their developmental stage.
- Sapnonis are triterpenoids present in various plants cell membranes to knock the invading fungi.

TABLE 2.1

Plant Secondary Metabolites and Their Role in Plant Defense System (Mazid et al. 2011; Thirumurugan et al. 2018)

Plant Secondary Metabolite	Secondary Product	Role in the Plant Defense System
Phenolic compounds	Coumarin Flavonoids Isoflavonoids Lignin Furano-coumarins Tannins	Aromatic compound resists insects/ nematode from infection, natural pesticidal property, cell necrotized stimulated by UV light, cell wall thickening protects from harmful UV-B rays
Terpenes	Monoterpenes Sesquiterpenes Diterpenes Triterpenes Polyterpenes	Kills insects, protection against UV-B rays, antiherbivore, keeps mammals away from eating plants, heals wounds of plants affected by insects
Sulfur (S)-containing secondary metabolites	Phytoalexins GSL – Glucosinolates GSH – Glutathione Defensins, thionins, and lectins	Resists microbial infection, herbivore attack, defense against predators, HCN compounds deter snails and insects
Nitrogen (N)-containing secondary metabolites	Non-protein amino acids Cyanogenic glucosides Alkaloids	

Phenolics belong to the second highest class of secondary metabolites formed via pathways of shikimic acid and malonic aid to rescue plants against phytopathogens. The colorful anthocyanin comprises water-soluble pigments to keep plants safe from ultraviolet radiation. Phytoalexins and some flavonoids activate as antibiotic fungicidal upon attack of pathogens. Tannins deter the insects during the nutrition competition by binding with the digestive and salivary proteins, and inactivate them (Ajebli and Eddouks 2019; Laura et al. 2019; Bizuneh 2020). Lignin protects the plant by lignifying the cell wall aiding structurally against the intruders. Similarly, the compounds made up of nitrogen act as a poison to the insects and fungi, and also indulge in the respiration of aerobic microbes (Lee et al. 2019). Sulfur compounds supply antioxidants during stress conditions and enhance the resistance against predators, competitors, and parasites (Mithen 1992; De Vos and Jander 2009). Briefly, PSMs protect living plants and their tissues from bacterial, fungal, viral, and insect's infection by providing structural and biochemical defenses. With the invade of foreign organisms, various pre-existing chemicals, precursors, and inhibitors are activated on the spot and responded as bactericidal, fungicidal, insecticidal, and narcotization of cell components (Wink 2018; Isah 2019; Takshak and Agrawal 2019).

2.3.1.1 Mechanism of Action of PSMs

The complete mechanism of action of the PSMs still needs to be fully explained (Wink 2015). The basic component involved is the biomembrane, which usually isolates the reactive components in a cell. Membrane study/activity of saponin has been studied in detail (Man et al. 2010). Natural as well as artificial lipid bilayer is considered to be disrupted by model metabolite classes such as saponin, avenacine A-1, and α-tomatine (Bednarek and Schulze-Lefert 2018). First, aglycone that is a part of saponin is inserted into the lipid bilayer, followed by the association with the sterol residues in the membrane (Lorent et al. 2014). Second, glucone performs reorganization of sterol within the membrane, which results in pores in the membrane. This membranolytic activity is highly dependent on the presence of sterol in the target membrane having a free 3β-hydroxy group, and also on the presence of sugar moiety in the saponinaglycone at the C-3 position (Piasecka et al. 2015). The budding yeast was used as model eukaryotic cell to understand the complexity of the saponin activity (Coleman et al. 2010). Another class of metabolites known as glycoalkaloids α-tomatine is considered potent antifungal (Morillo et al. 2020). The activity of this metabolite is associated with the presence of tetrasaccharides and the permeabilization of the membrane. Other than the antimicrobial defense of the PSMs, some metabolites are produced for dual function (Sucha and Tomsik 2016), e.g. scopoletin compound performs the antimicrobial activity as well as acts as a buffer to the reactive oxygen species produced as a stressed response to the pathogen (Lattanzio et al. 2006).

2.3.2 Role of Microbial Secondary Metabolites in Plant Defense

Biosynthesis and the supply of secondary metabolites crucially determine the result of plant-microbe interaction. The pathogens often show phytotoxins to interfere with the immunity and metabolism of plants or to kill host cells (Frantzeskakis et al. 2020). Rather, plants produce several antimicrobial secondary metabolites to combat potential invaders. Plant pathogens through specific enzymes encoded by genes or groups of genes in the genomes of plant pathogens can often detoxify host antimicrobial compounds such as the degradation of benzoxazolinones by the *Fusarium pseudograminearum* (Kettle et al. 2015). The appearance of secondary metabolites in plants is often limited to individual phylogenetic lineages such as individual families or genera, signifying that the individual biosynthetic pathways undergo rapid development (Piasecka et al. 2015; Frantzeskakis et al. 2020). Bacteria produce a wide range of secondary metabolites such as soluble and volatile compounds that have important and different ecological functions. The production of secondary metabolites is often triggered by intraspecific and interspecific interactions between bacteria in the soil. Secondary metabolites can be used as warfare agents or as chemicals, with volatile compounds having more potential acting on a larger spatial scale than soluble compounds (Tyc et al. 2017). Some examples are given in Table 2.2.

TABLE 2.2

Bacterial Secondary Metabolites and Their Role in Plant Defense

Bacterial Secondary Metabolite	Source	Function	Reference
Pseudomonine	*Pseudomonas stutzeri* KC	Competitive inhibition of phytopathogens	(Lewis et al. 2000)
Hydrogen cyanide	*P. pseudoalcaligenes* P4	Antifungal	(Ayyadurai et al. 2007)
6-pentyl-α-pyrone (6PP) Harzianolide and Harzianopyridone	*Trichoderma* spp.	Activate plant defense mechanisms and regulate plant growth in pea, tomato, and canola	(Vinale et al. 2009; 2013)
3-octanone and 1-octen-3-ol volatile organic compounds (VOCs)	*Trichoderma*	Induce plant defense genes comparable with the effect of the phytohormones, methyl jasmonate, and methyl salicylate	(Müller et al. 2013; Siddiquee 2014)
Extracellular polysaccharides (EPS) serve as microbe-associated molecular patterns (MAMPs)	*Bacillus cereus* AR156	Induce plant immunity and induced systemic resistance (ISR) in Arabidopsis and take part in the process of AR156/ plant interaction	(Jiang et al. 2016)
Diacetylphloroglucinol (DAPG), Phenazines (PHZ), 2, 4- Pyoluteorin (PLT), Pyrrolnitrin (PRN), Cyclic lipopeptides (CLPs) and VOCs such as hydrogen cyanide (HCN)	*Fluorescent pseudomonads*	Antibacterial, antifungal, antinematicidal antiviral, and antitumor properties	(Mishra and Arora 2018)

2.4 Conclusion and Future Prospect

Plants have developed multiple defense mechanisms against microbial pathogens and various types of environmental stresses. In several previous studies, secondary metabolites (an effective and eco-friendly alternative) can improve plant's ability to cope up with abiotic stresses, and to control a wide variety of plant pathogens that affect commercially (economic and nutritional interest) important crops and agricultural products. Recognizing the importance, the mechanisms that cause systemic induced resistance (SIR) will be a significant milestone for sustainable agricultural production. Therefore, further research in the area of natural pesticide development is needed in the current scenario. In the long term, it will probably be possible to generate genetic cassettes for complete pathways, which could, therefore, be used for the production of valuable secondary defensive metabolites in bioreactors or the metabolic engineering of cultivated plants. This will improve their resistance to herbivores and microbial pathogens as well as various environmental stresses that can potentially lead to the formulation of broad-spectrum biopesticides for sustainable and eco-friendly agriculture.

REFERENCES

Ajebli, M. and Eddouks, M. 2019.The promising role of plant tannins as bioactive antidiabetic agents. *Curr Med Chem* 26(25):4852–4884.

Andersen, E. J., Ali, S., Byamukama, E., Yen, Y. and Nepal, M. P. 2018. Disease resistance mechanisms in plants. *Genes* 9(7):339.

Ayyadurai, N., Naik, P. R. and Sakthivel, N. 2007. Functional characterization of antagonistic fluorescent pseudomonads associated with rhizospheric soil of rice (*Oryza sativa* L.). *J Microbiol Biotechnol* 17(6):919–927.

Ballhorn, D. J., Kautz, S., Heil, M. and Hegeman, A. D. 2009. Analyzing plant defenses in nature. *Plant Signal Behav* 4(8):743–745.

Bednarek, P. and Schulze-Lefert, P. 2018. Role of plant secondary metabolites at the host–pathogen interface. *Annu Plant Rev (Online)* 34:220–260.

Benitez, T., Rincon, A. M., Limon, M. C. and Codon, A. C. 2004. Biocontrol mechanisms of Trichoderma strains. *Int Microbiol* 7(4):249–260.

Bhatia, S., Dubey, R. and Maheshwari, D. 2005. Enhancement of plant growth and suppression of collar rot of sunflower caused by Sclerotium rolfsii through fluorescent Pseudomonas. *Indian Phytopathol* 58(1):17–24.

Bizuneh, G. K. 2020. The chemical diversity and biological activities of phytoalexins. *Adv Tradit Med* 20:1–13.

Boller, T. and Felix, G. 2009. A renaissance of elicitors: perception of microbe-associated molecular patterns and danger signals by pattern-recognition receptors. *Annu Rev Plant Biol* 60:379–406.

Chen, M. S. 2008. Inducible direct plant defense against insect herbivores: a review. *Insect Sci* 15(2):101–114.

Coleman, J. J., Okoli, I., Tegos, G. P., Holson, E. B., Wagner, F. F., Hamblin, M. R. and Mylonakis, E. 2010. Characterization of plant-derived saponin natural products against *Candida albicans*. *ACS Chem Biol* 5(3):321–332.

Conrath, U., Beckers, G. J., Langenbach, C. J. and Jaskiewicz, M. R. 2015. Priming for enhanced defense. *Annu Rev Phytopathol* 53:97–119.

De Vos, M. and Jander, G. 2009. Myzuspersicae (green peach aphid) salivary components induce defence responses in *Arabidopsis thaliana*. *Plant Cell Environ* 32(11):1548–1560.

Demain, A. L. 1999. Pharmaceutically active secondary metabolites of microorganisms. *Appl Microbiol Biotechnol* 52(4):455–463.

Ferrer, A., Altabella, T., Arró, M. and Boronat, A. 2017. Emerging roles for conjugated sterols in plants. *Prog Lipid Res* 67:27–37.

Ferry, N., Edwards, M. G., Gatehouse, J. A. and Gatehouse, A. M. 2004. Plant–insect interactions: molecular approaches to insect resistance. *Curr Opin Biotech* 15(2):155–161.

Frantzeskakis, L., Di Pietro, A., Rep, M., Schirawski, J., Wu, C. H. and Panstruga, R. 2020. Rapid evolution in plant–microbe interactions: a molecular genomics perspective. *New Phytol* 225(3):1134–1142.

Freeman, B. C. and Beattie, G. A. 2008. An overview of plant defenses against pathogens and herbivores. *Plant Health Instr.* doi:10.1094/PHI-I-2008-0226-01.

Haas, D. and Défago, G. 2005. Biological control of soil-borne pathogens by fluorescent pseudomonads. *Nat Rev Microbiol* 3(4):307–319.

Hou, S., Liu, Z., Shen, H. and Wu, D. 2019. Damage-associated molecular pattern-triggered immunity in plants. *Front Plant Sci* 10:646.

Isah, T. 2019. Stress and defense responses in plant secondary metabolites production. *Biol Res 52*(1):39.

Jiang, C. H., Fan, Z. H., Xie, P. and Guo, J. H. 2016. *Bacillus cereus* AR156 extracellular polysaccharides served as a novel micro-associated molecular pattern to induced systemic immunity to PstDC3000 in Arabidopsis. *Front Microbiol* 7:664.

Kessler, A. and Baldwin, I. T. 2002. Plant responses to insect herbivory: the emerging molecular analysis. *Annu Rev Plant Biol* 53(1):299–328.

Kettle, A. J., Batley, J., Benfield, A. H., Manners, J. M., Kazan, K. and Gardiner, D. M. 2015. Degradation of the benzoxazolinone class of phytoalexins is important for virulence of *Fusarium pseudograminearum* towards wheat. *Mol Plant Pathol* 16(9):946–962.

Kim, Y. S., Choi, Y. E. and Sano, H. 2010. Plant vaccination: stimulation of defense system by caffeine production in planta. *Plant Signal Behav* 5(5):489–93.

Lattanzio, V., Lattanzio, V. M. and Cardinali, A. 2006. Role of phenolics in the resistance mechanisms of plants against fungal pathogens and insects. *Phytochem: Adv Res* 661(2):23–67.

Laura, A., Moreno-Escamilla, J. O., Rodrigo-García, J. and Alvarez-Parrilla, E. 2019. Phenolic compounds. In *Postharvest Physiology and Biochemistry of Fruits and Vegetables*. Woodhead Publishing, Elsevier, UK, pp: 253–271.

Lee, M. H., Jeon, H. S., Kim, S. H., Chung, J. H., Roppolo, D., Lee, H. J. and Park, O. K. 2019. Lignin-based barrier restricts pathogens to the infection site and confers resistance in plants. *EMBO J 38*(23):e101948.

Lewis, T. A., Cortese, M. S., Sebat, J. L., Green, T. L., Lee, C. H. and Crawford, R. L. 2000. A *Pseudomonas stutzeri* gene cluster encoding the biosynthesis of the CCl4-dechlorination agent pyridine-2, 6-bis (thiocarboxylic acid). *Environ Microbiol* 2(4):407–416.

Lorent, J., Lins, L., Domenech, O., Quetin-Leclercq, J., Brasseur, R. and Mingeot-Leclercq, M. P. 2014. Domain formation and permeabilization induced by the saponin α-hederin and its aglycone hederagenin in a cholesterol-containing bilayer. *Langmuir* 30(16):4556–4569.

Malik, N. A. A., Kumar, I. S. and Nadarajah, K. 2020. Elicitor and receptor molecules: orchestrators of plant defense and immunity. *Int J Mol Sci* 21(3):963.

Man, S., Gao, W., Zhang, Y., Huang, L. and Liu, C. 2010. Chemical study and medical application of saponins as anti-cancer agents. *Fitoterapia* 81(7):703–714.

Mazid, M., Khan, T. A. and Mohammad, F. 2011. Role of secondary metabolites in defense mechanisms of plants. *Biol Med* 3(2):232–249.

Mishra, J. and Arora, N. K. 2018. Secondary metabolites of fluorescent pseudomonads in biocontrol of phytopathogens for sustainable agriculture. *Appl Soil Ecol* 125:35–45.

Mithen, R. 1992. Leaf glucosinolate profiles and their relationship to pest and disease resistance in oilseed rape. In: *Breeding for Disease Resistance*. Springer, Dordrecht, pp: 71–83.

Morillo, M., Rojas, J., LequartV, L. A. and Martin, P. 2020. Natural and synthetic derivatives of the steroidal glycoalkaloids of solanum genus and biological activity. *Nat Prod Chem Res* 8:371.

Müller, A., Faubert, P., Hagen, M., zu Castell, W., Polle, A., Schnitzler, J. P. and Rosenkranz, M. 2013. Volatile profiles of fungi–chemotyping of species and ecological functions. *Fungal Genet Biol* 54:25–33.

Piasecka, A., Rey, N. and Bednarek, P. 2015. Secondary metabolites in plant innate immunity: Conserved function of divergent chemicals. *New Phytol* 206(3):948–964.

Pieterse, C. M., Zamioudis, C., Berendsen, R. L., Weller, D. M., VanWees, S. C. and Bakker, P. A. 2014. Induced systemic resistance by beneficial microbes. *Annu Rev Phytopathol* 52:347–375.

Prasad, R., Kumar, M. and Varma, A. 2015. Role of PGPR in soil fertility and plant health. In: *Plant-Growth-Promoting Rhizobacteria (PGPR) and Medicinal Plants*. Springer, Cham, pp: 247–260.

Pundir, R. K. and Jain, P. 2015. Mechanism of prevention and control of medicinal plant-associated diseases. In: *Plant-Growth-Promoting Rhizobacteria (PGPR) and Medicinal Plants*. Springer, Cham, pp:231–246.

Saeed, A. and Tahira. 2019. Plant defense mechanisms against bacterial pathogens. *J Genet Genomic Sci* 4(1):3–012.

Schafer, H. and Wink, M. 2009. Medicinally important secondary metabolites in recombinant microorganisms or plants: progress in alkaloid biosynthesis. *Biotechnol J* 4(12):1684–1703.

Shinwari, Z. K., Tanveer, F. and Iqrar, I. 2019. Role of microbes in plant health, disease management, and abiotic stress management. In: *Microbiome in Plant Health and Disease*. Springer, Singapore, pp:231–250.

Siddiquee, S. 2014. Recent advancements on the role and analysis of volatile compounds (VOCs) from Trichoderma. In: *Biotechnology and Biology of trichoderma*. Elsevier, B.V. The Netherlands, pp:139–175.

Singh, S. K. and Pathak, R. 2015. Ecological manipulations of rhizobacteria for curbing medicinal plant diseases. In: *Plant-Growth-Promoting Rhizobacteria (PGPR) and Medicinal Plants*. Springer, Cham, pp:217–230.

Stergiopoulos, I., Collemare, J., Mehrabi, R. and De Wit, P. J. 2013. Phytotoxic secondary metabolites and peptides produced by plant pathogenic Dothideomycete fungi. *FEMS Microbiol Rev* 37(1):67–93.

Sucha, L. and Tomsik, P. 2016. The steroidal glycoalkaloids from Solanaceae: toxic effect, antitumour activity and mechanism of action. *Planta med* 82(5):379–387.

Takshak, S. and Agrawal, S. B. 2019. Defense potential of secondary metabolites in medicinal plants under UV-B stress. *J Photochem Photobiol B* 193:51–88.

Thirumurugan, D., Cholarajan, A., Raja, S. S. and Vijayakumar, R. 2018. An introductory chapter: secondary metabolites. *Second Metab—Sources Appl* 1–21.

Tyc, O., Song, C., Dickschat, J. S., Vos, M. and Garbeva, P. 2017. The ecological role of volatile and soluble secondary metabolites produced by soil bacteria. *Trends Microbiol* 25(4):280–292.

Vinale, F., Ghisalberti, E. L., Sivasithamparam, K., Marra, R., Ritieni, A., Ferracane, R. and Lorito, M. 2009. Factors affecting the production of Trichoderma harzianum secondary metabolites during the interaction with different plant pathogens. *Lett Appl Microbiol* 48(6):705–711.

Vinale, F., Nigro, M., Sivasithamparam, K., Flematti, G., Ghisalberti, E. L., Ruocco, M. and Woo, S. L. 2013. Harzianic acid: a novel siderophore from *Trichoderma harzianum*. *FEMS Microbiol Lett* 347(2):123–129.

Waghunde, R. R., Shelake, R. M., Shinde, M. S. and Hayashi, H. 2017. Endophyte microbes: a weapon for plant health management. In: *Microorganisms for Green Revolution*. Springer, Singapore, pp:303–325.

Wink, M. 2015. Modes of action of herbal medicines and plant secondary metabolites. *Medicines* 2(3):251–286.

Wink, M. 2018. Plant secondary metabolites modulate insect behavior-steps toward addiction? *Front Physiol* 9:364.

3

Unravel the Mystery of Plant-Microbe Interactions: Biotechnological Perspective

Nazir Ahmad Malik[1], Mohammad Saleem Wani[2], Naseema Aqbar Wani[2], and Younas Rasheed Tantray[3]
[1]Dolphin PG College of Science & Agriculture, Punjab, India
[2]Punjabi University, Patiala, India
[3]Government Degree College, Tangdhar, Jammu and Kashmir, India

CONTENTS

3.1 Introduction

The interactions between plants and microorganisms are diversified and flourish underneath or above the ground as well as inside the plants (Vorholt 2012; Bulgarelli et al. 2013). These interactions are complex (Dodds and Rathjen 2010) and may be beneficial, neutral, and disadvantageous, which affect the plant development (Newton et al. 2010). There is a quality of specialization and specificity amongst plant-microbe interactions which contributes to a significant diversity and their expansion over thousands and thousands of years (Galagan et al. 2005). Yet microorganism cooperation with different kinds of flora may be pathogenic and gets contaminated within the flora (Strange and Scott 2005). A commensal and pathogenic relationship requires a specific signaling mechanism for interactions of reciprocal recognition or responses to disease such as strengthening root nodules or rice blast disorder (Riely et al. 2004; 2006; Imam et al. 2014, 2015a; 2015b). Both the interacting partners receive their vitamins from the soil and alter the nature of the soil by using organic cluttering and metabolic activities. Microorganisms directly create an impact on plants by manipulating the hormone signaling and protecting against harmful microbes, and this communication between plants and microorganisms occurs with the release of chemical metabolites through the roots. The major knowledge gaps in understanding the mechanisms of plant-microbe interactions in the rhizosphere are given in Figure 3.1. The studies based on ribonucleic acid (RNA), deoxyribonucleic acid (DNA), statistics of genomics, transcriptomics, metabolomics, metagenomics, next-generation sequencing (NGS) technology, and proteomics strategies have turned out to be precious gear to examine host-microbe interactions and their cooperation. The need of the hour is to understand the interactions between plants and microbes at a molecular level (Madden and

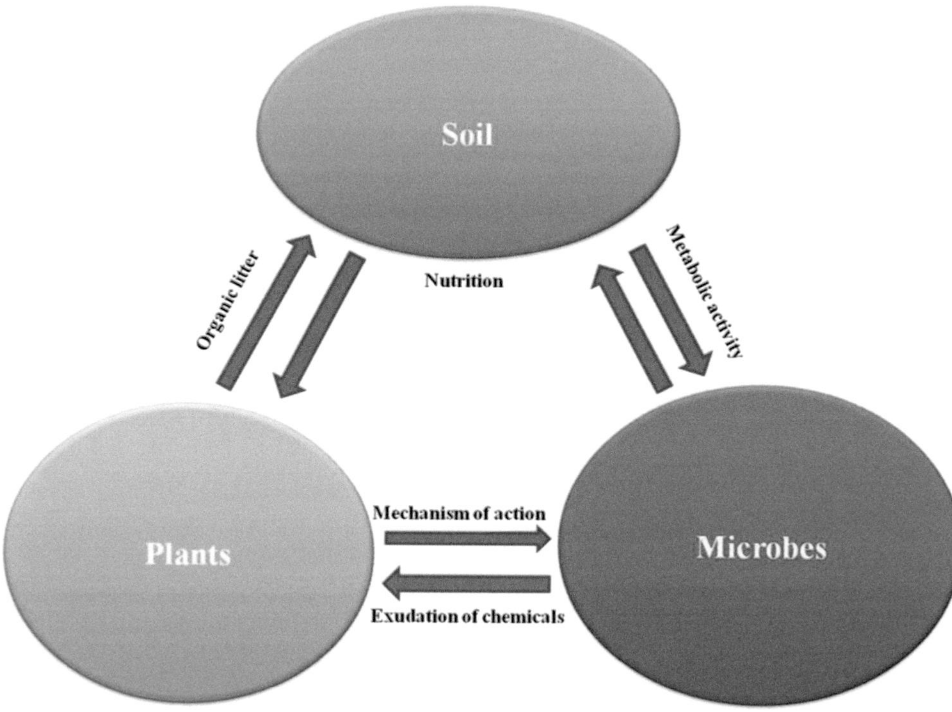

FIGURE 3.1 Interactions among plants, microbes, and soil.

Wheelis 2003). The drastic climatic changes due to global warming have begun to show their destructive consequences on defense structures of plants (Schenk et al. 2008). The methods of transcript profiling offer the study of a large number of genes and examine adjustments in gene expression simultaneously, which might be regulated at transcription level (Wang et al. 2009). The accessibility of the genomic and EST sequences of plants concurrent with the progress in the computational biology tools makes the DNA profiling technique more advanced. It serves in the research of defense system of plants after pathogen harm, in the recognition of pathogenesis-related genes, and likewise in understanding the interactions between specific signaling pathways (Andrew et al. 2002; Libault et al. 2010). A phytopathogenic fungus produces a number of proteins and different metabolites that help to enter and settle inside the plant host to acquire vitamins (Gonzalez-Fernandez et al. 2010). The secretome is a universal class of secreted proteins which are being released into the extracellular space through a cell, tissue, organ, or organism at any given moment and preconditions by known and unknown secretory mechanisms containing essential and controlled secretory organelles (Agrawal et al. 2013). In phytopathogenic fungi, the secretome includes pathogenicity and virulence components, which favor host tissue attack and formation in the sensitive plant (Girard et al. 2013). Much more research has been conducted via the use of both classical methods and modern techniques to understand the plant-microbe interaction mechanisms at the molecular level (Allwood et al. 2008; Tan et al. 2009; Bhadauria et al. 2010; Quirino et al. 2010; Afroz et al. 2011).

3.2 The Relevance of Beneficial Plant-Microorganisms Interactions

The present-day world suffers food crisis to supply the ever-increasing population of the world, and about one billion population is affected by the hunger everyday (Reid 2011). Less productiveness, confined cultivable land, and contaminated irrigation water are the primary reasons for the shortage of meals and grains. Usage of fertilizers, plant breeding, and genetic engineering methods are introduced to enhance the crop harvest, but are costly, moderate, and extremely specific (Reid 2011). Therefore, it is the demand of the time to generate alternative procedures and techniques apart from genetic development of plants against pathogens. A few scientists have contributed to an exclusive aspect and emphasized on the exploitation of plant-related microbes which might be beneficial to plants and have high-quality implements on plant-microbe interactions (Farrar et al. 2014). These useful plant-microbe interactions can be helpful for the growth and development of plants and for increasing their yields to supply them to the world. In useful plant-microbe interactions, plants and microbes have advanced mutualistic and useful interaction, which cooperates in enhancing the host plant protection from an immense variation in stresses such as infections, scarcity, salinity, heavy metals, toxins, nutrient stresses, and excess temperature. This beneficial plant-microbe communication helps in increasing the crop productiveness at cheap-cost expenditures. Currently, this exceedingly new understudied method experience has brought about a fresh desire for the agriculture sector that it can be a part of the modern Green Revolution (Reid 2011). Plants as well as microbes play a significant role in their beneficial association. The well-known example is symbiosis, where both organisms get benefited from each other (Oldroyd et al. 2011). Arbuscular mycorrhizal fungi (AMF), often located in all types of soils, remain in the plant roots and enable phosphate consumption from the soil (Smith and Smith 2011). The inoculation of AMF in the tropical soils decreases the usage of phosphate fertilizers in the association, and increases the crop yield (Bonfante and Genre 2010; Rey and Schornack 2013; Knief 2014; Ramalingam et al. 2015). Quorum sensing and biofilm development allow bacterial populations to stick to plant tissues and contribute to useful plant-microbe interactions (Ramey et al. 2004). Various genes were introduced in microorganisms such as *Bacillus amyloliquefaciens* for biofilm development, root colonization, and plant development (Budiharjo et al. 2014). The bacterial quorum-sensing indicators (QSS) show a thorough impact on plants transcriptome and proteome (Mathesius 2009). Plant production process optimization and bacterial manipulations for higher yield in plants are applied simultaneously using biofilm formation and quorum sensing, which enable the bacterial population to connect the natural surfaces closely with plant tissue, improve cell adhesion among microorganisms, and develop the response of plants to bacterial QSS (Farrar et al. 2014). A few microbial products, in particular the bacterial enzyme, help in the defense mechanism of host plant against biotic and abiotic stresses such as drought and flooding, high salinity and temperature, heavy metals, and pathogens. One of the significant aspects to control drought strain in plants is to develop high trehalose, which permits for maintaining the membranes and enzymes instead of biotechnological engineering to form extra trehalose; it would be further beneficial to use bacteria to provide the surplus trehalose in association with plants. Beneficial endophytes live inside the plants without eliciting any disease reaction and trigger the induced systemic resistance (ISR) in plants when attacked with pathogenic microorganism (Kloepper and Ryu 2006). The advent of NGS technologies and new molecular mechanisms such as entire genome sequencing, metagenomics, transcriptomics, proteomics and fluorescent tagging, and localization applications are of exceptional value in decoding the biological capabilities and useful in plant-microbe interplay research. The particular practice of plant-microbe interaction elicits specific response mechanism, and a special class of plant-bacterium interaction is *Agrobacterium tumefaciens* – a biotrophic interaction in which it does not produce any disease symptoms like pathogenic fungi or bacteria; however, as a substitute, it alters the physiology and morphology of the affected host plant (Pitzschke 2013). Extensive research and massive findings within the past years of *Agrobacterium*-plant interplay proved a top notch paradigm to recognize distinct features of plant-bacterium interactions (Yuan and Williams 2012; Pitzschke 2013; Subramoni et al. 2014). The current traits of *A. tumefaciens* responses to varied plant-derived signaling molecules cooperate in its pathogenicity and the activation of virulence genes, and induce the transmission and assimilation of transfer DNA (T-DNA) from its tumor-inducing (Ti)plasmid into the plant nucleus (Subramoni et al. 2014). Several studies have described the hierarchical activation of virulence because of the response of mixtures of plant-related chemical signals, viz. *Agrobacterium* responses to acidic signals (Winans 1992; Yuan et al. 2008a; Wu et al. 2012), plant-derived phenolic compounds (Brencic et al. 2004; Cho and Winans 2005), and plant-derived sugars in the rhizosphere (He et al. 2009; Hu et al. 2013). Many plant-derived signals and hormones are the characteristic factors that play a pivotal role in modulating *Agrobacterium* virulence, Ti plasmid replica number, and quorum sensing (Yuan et al. 2008a; Subramoni et al. 2014). Mostly, the infestation of *Agrobacterium* in plants results in the conversion of plant cells into a factory to fix the vitamins and keep the genetic integrity of nature (Subramoni et al. 2014). The concern of *A. tumefaciens* contamination and transformation achievements has constantly been a vital feature of studies for scientists. The fulfillment of transformation in large part relies on the host defense mechanisms and recalcitrance to *Agrobacterium*-mediated transformation (Nurnberger et al. 2004;Pitzschke 2013). Many defense components like MAPKs (Mitogen-activated protein kinases), defense gene expression, generation of reactive oxygen species (ROS), and hormonal regulation and its influence via the *Agrobacterium* determine whether the transformation will take place or not (Li et al. 2005; Lee et al. 2009; Zhao et al. 2011). Many biotechnological procedures have been found to get over or enhance the transformation efficiency in recalcitrant plants by the usage of modified *Agrobacterium* strain, change in plant growth factors, proposed manipulation of host plants, and precise information of plant-*Agrobacterium* interactions (Pitzschke 2013). Accordingly, the pathogenicity and high transformation performance of *Agrobacterium* depends on the unique perception, response to and improvement inplant-derived chemical signals, and also response to hormones which lower the defense degree in plants.

3.3 Crop Improvement via Plant-Microbe Interactions

Global warming brings drastic changes in the climatic conditions worldwide due to which crop production and health conditions have been seriously disturbed, and accordingly, the plant production practice needs to be improved for better productivity on the small fertile land. To enhance crop harvest, the utilization of microbes is a stronger solution and could perform a critical role in disease dissemination and management (Reid 2011). The plant-microbe interaction study is mainly focused on three aspects: the symbiotic relation between plants and fungi (Smith and Smith 2011), nitrogen fixation in plants (Oldroyd et al. 2011), and pathogenesis (Dodds and Rathjen 2010; Kachroo and Robin 2013; Wirthmueller et al. 2013). These structures are described and contribute observations into simple and varied signaling network pathways in plant-pathogen interactions. Resistant crop breeding approach involves the use of molecular breeding and genetic engineering method to pass on the resistance genes or QTLs (quantitative trait loci) against pathogens, and is environmentally a good way to offset balance microbial infestation due to the usage of pesticides (Akhon and Machray 2009; Gust et al. 2010). The exploitation of biotic and abiotic scenarios of plants is another eco-friendly way to grow sustainable disease control approaches (Haggag et al. 2015). Nevertheless, the biological technique demands for the in-depth information of how plants and microbes communicate with each other in complex habitat and how those interactions bring about physiological alterations in plants under stress conditions. Moreover, information is required that how plants prepare themselves to combat harsh climatic condition by utilizing their strength and sources for their resistance against the pathogens with respect their own health and improvement (Haggag et al. 2015). Nowadays, proteomics together with bioinformatics and computational biology are extensively used methodologies to analyze plant-pathogen interactions, which might be based on the concept of isolation, characterization, and identification of proteins taking role in the specific conditions at a specific time (Imam et al. 2014). The proteome-level application gives an insight into the original molecules which are held in negotiating particular cellular processes (Kav et al. 2007). However, plant-pathogen proteomics involves the study of proteins that are involved in the interactions between plants and microbes, and it was a very difficult challenge for the scientists to examine and characterize the complex mechanism of these interactions (Mathesius 2009). For the development of crops by the application of biotechnological techniques, the engagement of the key proteins result in a vital interaction and helps in sufficient development and improvement in plants. The proteins of interest are necessary for cellular functions of the plants, which are accomplished with the aid of controlling physiological and biochemical pathways. New studies within the post-genomic generation have shown that both genomics and proteomics are crucial facets for the development of new genes and can be conveniently used in various crop development programs (Baggerman et al. 2005; Scherp et al. 2011; Jayaraman et al. 2012).

3.4 An Immune System of Plants: A Perception into Defense Mechanism

For several microbes such as fungi, bacteria, protists, etc., the richest source of vitamins are plants. Despite the fact that plants lack an appropriate immune mechanism as in animals, they develop specific ways at unique levels such as structural, chemical, and protein-based that altogether knock out the pathogens and provide them resistance for their survival under harsh environmental conditions. Understanding the mechanisms of how plants prevent themselves from pathogens is a requisite for the development of disease-resistant plant. Due to the absence of cellular immune cells and cellular adaptive immune systems, plants particularly rely on intrinsic immunity and active signaling ways (Dangl and Jones 2001; Chisholm et al. 2006). Different pathogens use their own ways of penetration in plant cells: Bacteria penetrate in plant cells through trichomes, lenticels, stomata, and special openings; fungi use a structure referred to as hyphae and formation of penetration peg takes place at the same time; and viruses can simply penetrate in cells of plant by physical injuries (Fox et al. 1971; Getz et al. 1983; Mendgen et al. 1996). Once the pathogen opens the defense barriers, various plant immune responses get elicited such as microbial- (or pathogen-) associated molecular patterns (MAMPs/PAMPs) triggered immunity (MTI/PTI) formerly accepted as basal or horizontal immunity, and effector-triggered immunity (ETI) once called R-gene-based or vertical immunity; the plants also elicit other techniques of immune responses such as systemic acquired response (SAR) and gene silencing (Lu et al. 2003; Durrant and Dong 2004; Padmanabhan et al. 2009; Sahu et al. 2012). The MTI/PTI immune response is elicited during the release of elicitors called MAMPs by pathogens, observed by the pattern recognition receptors (PRRs), a class of plasma-membrane covered by extracellular receptors, which results in the development of active defense responses that break further colonization and proliferation of infection (Hammond-Kosack and Jones 1997; Hammond-Kosack and Parker 2003; Dodds and Rathjen 2010; Beck et al. 2012). The PAMPs mediate the activation of plant defense system. Moreover, structurally related recognition complexes same as animal PAMP receptors have also been recorded in plants, which emerge as a natural evolutionary source of pathogen defense organizations in higher eukaryotes (Nürnberger and Brunner 2002). Several times, the pathogens reduce the MTI/PTI immune response by a system and supporting infections by the discharge of effector molecules in plant cells resulting in effector-triggered susceptibility (ETS) (Figure 3.2). The effector molecule initiates ETI, the R-gene-based vertical immunity, which

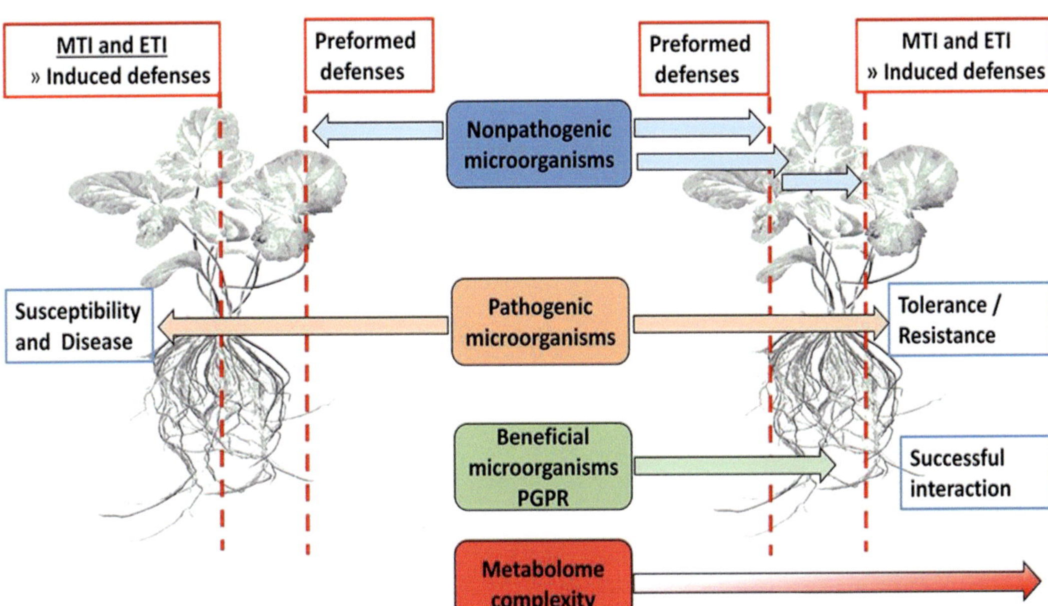

FIGURE 3.2 Overview of physical barriers and innate immunity defenses (Mhlongo et al. 2018).

induces hypersensitive response (HR) – a form of cell death (Hogenhout et al. 2009; Muthamilarasan and Prasad 2013). The direct and indirect interactions between R and Avr gene products are precisely documented (Keen 1990; Jia et al. 2000; Dodds et al. 2006); however, the effector molecules have specific targets within the host, and this assumption of effector mechanism is defined via the guard model (Dangl and Jones 2001). The guard model proposed that in the presence of a functional R-gene, the pathogen interaction study was stronger; however, in the absence of a functional R-gene, the study did not provide strong observation. This is evolutionarily a volatile situation that can be highly defined by the decoy model, which mimics the effector targets to trap the pathogen into a recognition event (van der Hoorn and Kamoun 2008). The SAR gets activated at the infection site and stops the growth of infection in the healthy tissues through the usage of the activation and expression of pathogenesis-related (PR) proteins (Durrant and Dong 2004; Park et al. 2007). The advances in genomic apparatus such as dual RNA-sequence of plants and pathogens and the position of non-coding RNAs (ncRNAs) in monitoring the plant defense responses have strengthened the knowledge of simultaneous relative data evaluation of plants and pathogens as well as their defense responses (Westermann et al. 2012; 2016). The NGS technique such as dual RNA-sequence for the concurrent application of host and pathogen transcriptomes during their cooperation is probe independent and can be conveniently used for any plant-pathogen communication research as a proper concept along with disclosure of differentially expressed genes in addition to determining the modifications in transcriptional regulatory events (Westermann et al. 2012; Das et al. 2015; Enguita et al. 2016).

Likewise, non-coding RNAs such as microRNAs, phasiRNAs, and prolonged intergenic non-coding RNAs (lincRNAs) are extremely significant for the protection of plant responses to special pathogens and provide innate immunity mechanism like PAMP- and effector-prompted defense responses to plants. The investigation of these non-coding RNAs demonstrated that how the epigenetic consequences alter the plant genes implicated in defense against pathogens (Enguita et al. 2016; Lenandias-Briere et al. 2016). Besides the MAMPs, precipitated immunity, and ETI, various investigations proposed the characteristics of transgenerational immune memory in plants and this memory has been studied for the environmental stresses, including pathogens (Slaughter et al. 2012). Mainly, it provides an efficient adaptation to plants against a particular stress, and in later generations, it is introduced as an acquired immune power (Muthamilarasan and Prasad 2013). When *Arabidopsis* is confronted with a virulent pressure of *Pseudomonas syringae*, the next era plant will produce a direct and enhanced accumulation of salicylic acid (SA) signaling pathway transcripts with advanced disease resistance (Slaughter et al. 2012). Such kind of immunity in plants additionally proposed the capacity of plants to inherit the resistance to the next generation. The plant-pathogen interactions and plant immunity are always considered as the essential components of studies, and suggest a few interesting statistics like version of specific immune mechanisms of plants closer to pathogenic strains, R-protein-mediated action, small interfering RNA (siRNA) silencing, post-transcriptional gene silencing (PTGS) involving cell RNAs and transgenerational immune memory (Muthamilarasan and Prasad 2013). In spite of many significant innovations within the subject of plant immune system, few mechanisms are yet to be identified like identification of many Avr genes in plant-pathogen interactions, plant root immune mechanisms, molecular mechanisms of colonization pathogens in plants, regulation of cellular interest and gene expression, and signaling systems concerned with plant immune system. Therefore, in addition to this, improvement in post-genomic technology will surface the way to highly recognize the plant immunity and plant interactions with microbes.

3.5 Plant-Microbe Interactions at Metabolic Level (Genome-Scale Modeling)

Currently, there is sufficient information on plant-pathogen interactions at the molecular stage, along with the numerous signaling pathways which otherwise coordinate completely in defense responses of plants. Various studies have been conducted in the course of the genomic and post-genomic era that there may be almost sufficient research studies at the molecular stage because of the high adaptive complexity and interactions of cellular components to massive numbers of internal and external conditions (Collakova et al. 2012). Various current and complicated procedures are developing to investigate and examine the plant metabolites. Genome-scale modeling is practiced mathematically to model the metabolism, and is mostly *in silico* metabolic flux version, derived from the available genomic data. Genome-scale models are becoming a challenging task to use the phenotype at a particular point of host-pathogen interactions (Collakova et al. 2012). The studies of genomics, transcriptomics, proteomics, and metabolomics mostly recommend tools for the metabolic community modeling of plants and pathosystems. Some of the selected plant-microbe pairs broadly studied and represented as suitable pathosystems are represented in Table 3.1. The genome-wide investigation of the interaction between plants and pathogens has been feasible in recent years due to the development of genome sequencing and annotations (Kanehisa et al. 2002; Shendure et al. 2004). The whole genome collection functions in the genome-extensive annotations of all the proteins, enzymes, and much correlated metabolic reactions (Duan et al. 2013). The incorporation of the metabolic network of plant and accompanied pathogens has been verified as an essential step for the test of effective conclusions of metabolic networks (Duan et al. 2013).

3.5.1 Unravel the Secrets of Plant-Microbe Interactions by Secretomics

Secretomics characterizes and examines the proteins secreted by a cell, tissue or an organism, and has previously emerged as a discipline, but its activity is evolving quickly. The adaptability of oomycetes, fungi, and bacteria permits to interact with plants in different ways depending on their development as a biotroph, hemibiotroph, necrotroph, or saprotroph. When interacting with a living organism, a microbe penetrates its plant host and controls metabolism, either harmfully if it a pathogen or beneficially if it is a symbiont. Deciphering secretomes became a crucial biological question when an increasing body of evidence indicated that secreted proteins were the main effectors initiating interactions, whether of pathogenic or symbiotic nature, between microbes and their plant hosts.

3.5.2 Rhizobia–Plant Interaction and Proteins

Proteomics is a unique approach to understanding a part of a complex network involved in a plant-microbial reaction that identifies a group of proteins under precise conditions. Rhizobia injection has been shown to increase protein levels in stem hair of soybeans, roots, and bacterioids (Mathesius 2009). In extension, extensive knowledge was gained by the regulation of signal transduction involved in bacterial invasion, nodule organogenesis, and long-distance signaling to regulate nodule number (Oka-Kira and Kawaguchi 2006; Popp and Ott 2011). But various studies examined proteins secreted within the rhizosphere or the ones which might be related to the bacterial outer membrane, and these experiments

TABLE 3.1

Overview of Selected Plant–Microbe Pairs Extensively Studied and Represented as Most Suitable Patho system (Imam et al. 2016)

S. No.	Pathogen	Plant	Pathogen Type	Reference
1.	*Pseudomoas syringae*	*Arabidopsis thaliana*	Bacterium/ hemi-biotrophic	Glazebrook 2005
2.	*Hyaloperonspora parasitica*	*Arabidopsis thaliana*	Oomycetes	Rentel et al. 2008
3.	*Phytophthora infestans*	*Arabidopsis thaliana*	Oomycetes	Huitema et al. 2004
4.	*Xanthomonas oryzae* pv. *Oryzae*	*Oryza sativa*	Bacterium/biotrophic	Salzberg et al. 2008
5.	*Ustilago maydis*	*Zea mays*	Fungus/biotrophic	Mueller et al. 2008
6.	*Melampsora larici-populina*	*Populus trichocarpa*	Fungus/biotrophic	Hacquard et al. 2012
7.	*Sclerotinia sclerotiorum*	*Glycine max*	Fungus/biotrophic	Amselem et al. 2011
8.	*Magnaporthe oryzae*	*Oryza sativa*	Fungus/biotrophic	Yoshida et al. 2009

resulted in the growth of plant in liquid media, protein concentration through precipitation or lyophilization, desalting, two-dimensional gel electrophorosis, and protein identification through mass spectrometry (Jayaraman et al. 2012). Applying exceptional methods, various proteins secreted to the rhizosphere involved in plant-microbe interaction have been identified (Fauvart and Michiels 2008; Tseng et al. 2009; Downie 2010). Rhizobial cells secrete adhesins, including rhicadhesin, that are essential for the attachment with the root hairs (Smit et al. 1992), hydrolytic proteins, viz. cellulases that infect the noncrystalline cellulose in the cellular wall of root hair permitting penetration of rhizobial (Robledo et al. 2008), and glycanases that cut emerging EPS (extracellular polysaccharides) generated through rhizobia and are helpful for the formation of biofilm (Russo et al. 2006). Different extracellular glycanases which are helpful in the nodulation and EPS modification were characterized and recognized in rhizobia such as PlyA and PlyB of *Rhizobium leguminosarum*. The secreted nodulation-signalling protein (NodO) has been purified from the cultures of *R. leguminosarum* (Sutton et al. 1994). NodO is a calcium-binding protein that describes cation-selective channels in membranes and may supplement Nod factor feature by promoting cations motion across the substructure hair membrane (Downie 2010). *Medicago sativa* inoculation with *Sinorhizobium meliloti* resulted in an increase in the secretion of plant hydrolases, peroxidase precursors, pathogenesis-related proteins, bacterial superoxide, lectins, dismutase, glycine betaine-binding ABC transporter, and a putative outer membrane lipoprotein transmembrane (De la Peña et al. 2008).

3.5.3 The Act of Root Exudates (Early Signaling Events)

The roots of plants import water and vitamins from the soil solution; however, additionally, they release low- and high-molecular weight compounds in rhizosphere. Root exudates contain a variety of root-secreted molecules that behave as complicated chemical combinations that mediate connections taking place within the rhizosphere and maintain soil microbial associations (Okumoto and Pilot 2011). Their chemical composition is affected by natural factors such as plant genotype and the multipartite interactions taking place within the rhizosphere between distinct factors. Carbon-based compounds are the principle component of these complex combinations; however ions, oxygen, and inorganic acids are additionally crucial additives with applicable roles in the course of rhizosphere interactions (Badri and Vivanco 2009). Exuded molecules consist of low-molecular weight compounds, which include sugars and phenolics, and high-molecular weight compounds that include polysaccharides and proteins, which are usually the largest fraction of the total exudates mass (Cai et al. 2012). Although there are various chemicals that are secreted by the roots, rhizobacteria further add more novel metabolites to the pool of molecules that intercede rhizosphere interactions (Badri et al. 2009).

3.6 Conclusion and Future Aspects

The new information on proteins involved in two-component signal transduction systems that allow sensing and responding to different stimuli, transcriptional regulators, and plant-derived peptides is far from completing the picture of the microbe–plant interaction. In this review, just a few contemporary and applicable earlier statistics linked to the molecules concerned with plant-microbe interplay are applied to demonstrate an incomplete panorama. Substances released by vegetation and microorganisms offer a combination of molecules (carbohydrates, flavonoids, amino acids, phytohormones, and proteins) that represent the terms of a chemical conversation between plants and microorganisms within the rhizosphere. Experts are interpreting the information material and importance of the cells' signaling and responses. New developments in analytical abilities and molecular and biochemical strategies have supplied new equipment for comparing the herbal functions of this material and the mechanisms under which it is controlled. In short, the process of plant-microbe interaction includes a massive variety of molecules that stretches our idea. Each year an ultra-modern signaling molecule is discovered and general information becomes more complex. Particularly, the genes take part in plant metabolic pathways and have an impact on the root exudation composition up to some specific signals within the rhizosphere. Thus, continued development in the functionality to accumulate the exudates will be critical for the evaluation of the molecules vegetation used for the cross-talk with soil microorganisms, and additionally, for the process that decrypts those signals. Further research is required to improve the catabolic genes expression within the rhizosphere of the plant-microbe mixtures and will describe the use of this approach. The use of endophytes in the biological degradation of pollutants is a problem that has not been fully explored. Progress in this area will help to describe the assimilation and delivery of the hazardous chemical by the plants. But, it can pose a serious threat, if the compound is transported to the shoot where it becomes accessible to animals. The outcome of contaminants in phytoremediation procedures should be studied in detail to avoid undesirable effects. Another area of interest is exploring molecular interactions between plants and microorganisms to achieve high results in the elimination of contaminants. The new "omics" practices will even permit the tracking or choice of catabolic genes to enhance remediation techniques (Kiely et al. 2006). The development of metagenomic assessment reveals new degradative capabilities that are worth introducing into strains with different interesting traits. The signals that plant and microbes exchanged during communication will be elucidated, and the molecular basis of an appropriate interaction between positive plant genotypes and specific bacteria will also be dissected. The information obtained from these studies provides insight into successful rhizoremediation strategies. Finally, more research needs to be done on the use of recombinant microorganisms with protective measures against indigenous microbial groups, especially to address highly toxic chemicals, including dioxins and PCBs (polychlorinated biphenyls).

REFERENCES

Afroz, A., Ali, A. G. M., Mir, A. and Komatsu, S. 2011. Application of proteomics to investigate stress-induced proteins for improvement in crop protection. *Plant Cell Rep* 30:745–763.

Agrawal, G. K., Jwa, N. S., Lebrun, M. H., Job, D. and Rakwal, R. 2013. Plant secretome: unlocking secrets of the secreted proteins. *Proteomics* 10:799–827.

Akhon, M. A. Y. and Machray, G. C. 2009. Biotech crops: technologies, achievements and prospects. *Euphytica* 166:47–59.

Allwood, J. W., Ellis, D. I. and Goodacre, R. 2008. Metabolomic technologies and their application to the study of plants and plant-host interactions. *Physiol Plant* 132:117–135.

Amselem, J., Cuomo, C. A., vanKan, J. A. L., Viaud, M., Benito, E. P., Couloux, A., Coutinho, P. M., de Vries, R. P., Dyer, P. S., Fillinger, S. and Fournier, E. 2011. Genomic analysis of the necrotrophic fungal pathogens *Sclerotinia sclerotiorum* and *Botrytis cinerea*. *PLoS Genet* 7:e1002230.

Andrew, F. B., Jinrong, W. and Mark, D. F. 2002. Probing plant pathogen interaction and downstream defense signaling using DNA microarray. *Funct Integr Genomic* 2:259–273.

Badri, D. and Vivanco, J. 2009. Regulation and function of root exudates. *Plant Cell Environ* 32:666–681.

Badri, D., Weir, T., van der Lelie, D. and Vivanco, J. 2009. Rhizosphere chemical dialogues: plant–microbe interactions. *Curr Opin Biotechnol* 20:642–650.

Baggerman, G., Vierstraete, E., De Loof, A. and Schoofs, L. 2005. Gel based versus gel-free proteomics: a review. *Comb Chem High Throughput Screen* 8:669–677.

Beck, M., Heard, W., Mbengue, M. and Robatzek, S. 2012. The ins and outs of pattern recognition receptors at the cell surface. *Curr Opin Plant Biol* 15:367–374.

Bhadauria, V., Banniza, S., Wang, L. X., Wei, Y. D. and Peng, Y. L. 2010. Proteomic studies of phytopathogenic fungi, oomycetes and their interactions with hosts. *Eur J Plant Pathol* 126:81–95.

Bonfante, P. and Genre, A. 2010. Mechanisms underlying beneficial plant-fungus interactions in mycorrhizal symbiosis. *Nat Commun* 1:48.

Brencic, A., Eberhard, A. and Winans, S. C. 2004. Signal quenching, detoxification and mineralization of Vir gene-inducing phenolics by the VirH2 protein of *Agrobacterium tumefaciens*. *Mol Microbiol* 51:1103–1115.

Budiharjo, A., Chowdhury, S. P., Dietel, K., Beator, B., Dolgova, O., Fan, B., Bleiss, W., Ziegler, J., Schmid, M., Hartmann, A. and Borriss, R. 2014. Transposon mutagenesis of the plant-associated *Bacillus amyloliquefaciens* ssp. *plantarum* FZB42 revealed that the nfrA and RBAM17410 genes are involved in plant-microbe-interactions. *PLoS One* 9(5): e98267.

Bulgarelli, D., Schlaeppi, K., Spaepen,S., Ver Loren van Themaat, E. and Schulze-Lefert, P. 2013. Structure and functions of the bacterial microbiota of plants. *Annu Rev Plant Biol* 64:807–838.

Cai, Z., Kastell, A., Knorr, D. and Smetanska, I. 2012. Exudation: an expanding technique for continuous production and release of secondary metabolites from plant cell suspension and hairy root cultures. *Plant Cell Rep* 31(3):461–477.

Chisholm, S. T., Coaker, G., Day, B. and Staskawicz, B. J. 2006. Host-microbe interactions: shaping the evolution of the plant immune response. *Cell* 124:803–814.

Cho, H. and Winans, S. C. 2005. VirA and VirG activate the Ti plasmid repABC operon, elevating plasmid copy number in response to wound-released chemical signals. *Proc Natl Acad Sci USA* 102:14843–14848.

Collakova, E., Yen, J. Y. and Senger, R. S. 2012. Are we ready for genome-scale modeling in plants. *Plant Sci* 19:53–70.

Dangl, J. L. and Jones, J. D. G. 2001. Plant pathogens and integrated defence responses to infection. *Nature* 411:826–833.

Das, A., Chai, J. C., Kim, S. H., Park, K. S., Lee, Y. S., Jung, K. H. and Chai, Y. G. 2015. Dual RNA sequencing reveals the expression of unique transcriptomic signatures in lipopolysaccharide-induced BV-2 microglial cells. *PLoS One* 10:e0121117.

De la Peña, C., Lei, Z., Watson, B. S., Sumner, L. W. and Vivanco, J. M. 2008. Root-microbe communication through protein secretion. *J Biol Chem* 283:25247–25255.

Dodds, P. N. and Rathjen, J. P. 2010. Plant immunity: towards an integrated view of plant-pathogen interactions. *Nat Rev Genet* 11:539–548.

Dodds, P. N., Lawrence, G. J., Catanzariti, A. M., Teh, T., Wang, C. I., Ayliffe, M. A., Kobe, B. and Ellis, J. G. 2006. Direct protein interaction underlies gene-for-gene specificity and coevolution of the flax resistance genes and flax rust avirulence genes. *Proc Natl Acad Sci USA* 103:8888–8893.

Downie, J. 2010. The roles of extracellular proteins, polysaccharides and signals in the interactions of rhizobia with legume roots. *FEMS Microbiol Rev* 34:150–170.

Duan, G., Christian, N., Schwachtje, J., Walther, D. and Ebenhöh, O. 2013. The metabolic interplay between plants and phytopathogens. *Metabolites* 3:1–23.

Durrant, W. E. and Dong, X. 2004. Systemic acquired resistance. *Annu Rev Phytopathol* 42:185–209.

Enguita, F. J., Costa, M. C., Fusco-Almeida, A. M., Mendes-Giannini, M. J. and Lúcia Leitão, A. 2016. Transcriptomic crosstalk between fungal invasive pathogens and their host cells: opportunities and challenges for next-generation sequencing methods. *J Fungi* 2(7):10.

Farrar, K., Bryant, D. and Cope-Selby, N. 2014. Understanding and engineering beneficial plant–microbe interactions: plant growth promotion in energy crops. *Plant Biotechnol J* 12:1193–1206.

Fauvart, M. and Michiels, J. 2008. Rhizobial secreted proteins as determinants of host specificity in the rhizobium-legume symbiosis. *FEMS Microbiol Lett* 285:1–9.

Fox, R. T. V., Manners, J. G. and Myers, A. 1971. Ultrastructure of entry and spread of *Erwinia carotovora* var. *atroseptica* into potato tubers. *Potato Res* 14:61–73.

Galagan, J. E., Henn, M. R., Ma, L. J., Cuomo, C. A. and Birren, B. 2005. Genomics of the fungal kingdom: insights into eukaryotic biology. *Genome Res* 15:1620–1631.

Getz, S., Fulbright, D. W. and Stephens, C. T. 1983. Scanning electron microscopy of infection sites and lesion development on tomato fruit infected with *Pseudomonas syringae* pv. tomato. *Phytopathol* 73:39–43.

Girard, V., Dieryckx, C., Job, C. and Job, D. 2013. Secretomes: the fungal strike force. *Proteomics* 13:597–608.

Glazebrook, J. 2005. Contrasting mechanisms of defense against biotrophic and necrotrophic pathogens. *Annu Rev Phytopathol* 43:205–207.

Gonzalez-Fernandez, R., Prats, P. and Jorrin-Novo, J. V. 2010. Proteomics of plant pathogenic fungi. *J Biomed Biotechnol* 932527.

Gust, A. A., Brunner, F. and Nürnberger, T. 2010. Biotechnological concepts for improving plant innate immunity. *Curr Opin Biotechnol* 21:204–210.

Hacquard, S., Joly, D. L., Lin, Y. C., Tisserant, E., Feau, N., Delaruelle, C., Kohler, V., Tanguay, A., Petre, P. B. and Frey, P. 2012. A comprehensive analysis of genes encoding small secreted proteins identifies candidate effectors in *Melampsora laricispopulina* (poplar leafrust). *Mol Plant Microbe Interact* 25:279–293.

Haggag, W. M., Abouziena, H. F., Abd-El-Kreem, F. and El Habbasha, S. 2015. Agriculture biotechnology for management of multiple biotic and abiotic environmental stress in crops. *Chem Pharm Res* 7:882–889.

Hammond-Kosack, K. E. and Jones, J. D. G. 1997. Plant disease resistance genes. *Annu Rev Plant Physiol Plant Mol Biol* 48:575–607.

Hammond-Kosack, K. E. and Parker, J. E. 2003. Deciphering plant-pathogen communication: fresh perspectives for molecular resistance breeding. *Curr Opin Biotechnol* 14:177–193.

He, F., Nair, G. R., Soto, C. S., Chang, Y., Hsu, L., Ronzone, E., DeGrado, W. F. and Binns, A. N.2009. Molecular basis of ChvE function in sugar binding, sugar utilization and virulence in *Agrobacterium tumefaciens*. *J Bacteriol* 191:5802–5813.

Hogenhout, S. A., Van der Hoorn, R. A., Terauchi, R. and Kamoun, S. 2009. Emerging concepts in effector biology of plant-associated organisms. *MPMI* 22:115–122.

Hu, X., Zhao, J., Degrado, W. F. and Binns, A. N. 2013. *Agrobacterium tumefaciens* recognizes its host environment using ChvE to bind diverse plant sugars as virulence signals. *Proc Natl Acad Sci USA* 110:678–683.

Huitema, E., Bos, J. I., Tian, M., Win, J., Waugh, M. E. and Kamoun, S. 2004. Linking sequence to phenotype in *Phytophthora*-plant interactions. *Trends Microbiol* 12:193–200.

Imam, J., Alam, S., Mandal, N. P., Shukla, P., Sharma, T. R. and Variar, M. 2015a. Molecular identification and virulence analysis of Avr genes in rice blast pathogen, *Magnaporthe oryzae* from Eastern India. *Euphy* 206:21–31.

Imam, J., Alam, S., Mandal, N. P., Variar, M. and Shukla, P. 2014. Molecular screening for identification of blast resistance genes in North East and Eastern Indian rice germplasm (*Oryza sativa* L.) with PCR based markers. *Euphy* 196:199–211.

Imam, J., Alam, S., Mandal, N. P., Variar, M. and Shukla, P. 2015b. Advances in molecular mechanism toward understanding plant–microbe interaction: a study of *M. oryzae* vs rice, In: Shukla P. (ed), *Frontier Discoveries and Innovations in Interdisciplinary Microbiology* Springer, New Delhi, pp:79–96.

Imam, J., Singh, P. K. and Shukla, P. 2016. Plant microbe interactions in post genomic era: perspectives and applications. *Front Microbiol* 7:1488.

Jayaraman, D., Forshey, K. L., Grimsrud, P. A. and Ane, J. M. 2012. Leveraging proteomics to understand plant-microbe interactions. *Front Plant Sci* 3:1–6.

Jia, Y., McAdams, S. A., Bryan, G. T., Hershey, H. P. and Valent, B. 2000. Direct interaction of resistance gene and avirulence gene products confers rice blast resistance. *EMBO J* 19:4004–4014.

Kachroo, A. and Robin, G. P. 2013. Systemic signaling during plant defense. *Curr Opin Plant Biol* 16:527–533.

Kanehisa, M., Goto, S., Kawashima, S. and Nakaya, A. 2002. The KEGG databases at genome net. *Nucleic Acids Res* 30:42–46.

Kav, N. N. V., Srivastava, S., Yajima, W. and Sharma, N. 2007. Application of proteomics to investigate plant–microbe interactions. *Curr Proteom* 4:28–43.

Keen, N. T. 1990. Gene-for-gene complementarity in plant-pathogen interactions. *Annu Rev Genet* 24:447–473.

Kiely, P. D., Haynes, J. M., Higgins, C. H., Franks, A., Mark, G. L., Morrissey, J. P. and O'Gara, F. 2006. Exploiting new systems-based strategies to elucidate plant-bacterial interactions in the rhizosphere. *Microb Ecol* 51:257–266.

Kloepper, J. W. and Ryu, C. M. 2006. Bacterial endophytes as elicitors of induced systemic resistance. *Soil Biol* 9:33–52.

Knief, C. 2014. Analysis of plant microbe interactions in the era of next generation sequencing technologies. *Front Plant Sci* 5:216.

Lee, C. W., Efetova, M., Engelmann, J. C., Kramell, R., Wasternack, C., Ludwig-Muller, J., Hedrich, R. and Deeken, R. 2009. *Agrobacterium tumefaciens* promotes tumor induction by modulating pathogen defense in *Arabidopsis thaliana*. *Plant Cell* 21:2948–2962.

Lenandias-Briere, C., Moreau, J., Hartmann, C. and Crespi, M. 2016. Noncoding RNAs, emerging regulators in root endosymbiosis. *MPMI* 29:170–180.

Li, J., Krichevsky, A., Vaidya, M., Tzfira, T. and Citovsky, V. 2005. Uncoupling of the functions of the *Arabidopsis* VIP1 protein in transient and stable plant genetic transformation by *Agrobacterium*. *Proc Natl Acad Sci USA* 102:5733–5738.

Libault, M., Farmer, A., Brechenmacher, L., Drnevich, J., Langley, R. J., Bilgin, D. D., Radwan, O., Neece, D. J., Clough, S. J., May, G. D. and Stacey, G. 2010. Complete transcriptome of the soybean root hair cell, a single-cell model, and its alteration in response to *Bradyrhizobium japonicum* infection. *Plant Physiol* 152:541–552.

Lu, R., Martin-Hernandez, A. M., Peart, J. R., Malcuit, I. and Baulcombe, D. C. 2003. Virus-induced gene silencing in plants. *Methods* 30:296–303.

Madden, L. V. and Wheelis, M. 2003. The threat of plant pathogens as weapons against U.S. Crops. *Annu Rev Phytopathol* 41:155–176.

Mathesius, U. 2009. Comparative proteomic studies of root-microbe interactions. *J Proteomics* 72:353–366.

Mendgen, K., Hahn, M. and Deising, H. 1996. Morphogenesis and mechanisms of penetration by plant pathogenic fungi. *Annu Rev Phytopathol* 34:364–386.

Mhlongo, M. I., Piater, L. A., Madala, N. E., Labuschagne, N. and Dubery, I. A. 2018. The chemistry of plant–microbe interactions in the rhizosphere and the potential for metabolomics to reveal signaling related to defense priming and induce dsystemic resistance. *Front Plant Sci* 9:112.

Mueller, O., Kahmann, R., Aguilar, G., Trejo-Aguilar, B., Wu, A. and deVries, R. P. 2008. The secretome of the maize pathogen *Ustilago maydis*. *Fungal Genet Biol* 45(Suppl. 1):S63–S70.

Muthamilarasan, M. and Prasad, M. 2013. Plant innate immunity: an updated insight into defense mechanism. *J Bio sci* 38:433–449.

Newton, A. C., Fitt, B. D. L., Atkins, S. D., Walters, D. R. and Daniell, T. J. 2010. Pathogenesis, parasitism and mutualism in the trophic space of microbe-plant interactions. *Trends Microbiol* 18:365–373.

Nürnberger, T. and Brunner, F. 2002. Innate immunity in plants and animals: emerging parallels between the recognition of general elicitors and pathogen-associated molecular patterns. *Curr Opin Plant Boil* 5:318–324.

Nurnberger, T., Brunner, F., Kemmerling, B. and Piater, L. 2004. Innate immunity in plants and animals: striking similarities and obvious differences. *Immuno Rev* 198:249–266.

Oka-Kira, E. and Kawaguchi, M. 2006. Long-distance signaling to control root nodule number. *Curr Opin Plant Biol* 9:496–502.

Okumoto, S. and Pilot, G. 2011. Amino acid export in plants: a missing link in nitrogen cycling. *Mol Plant* 4:453–463.

Oldroyd, E. D. G., Murray, J. D., Poole, P. S. and Downie, J. A. 2011. The rules of engagement in the legume-rhizobial symbiosis. *Annu Rev Genet* 45:119–144.

Padmanabhan, C., Zhang, X. and Jin, H. 2009. Host small RNAs are big contributors to plant innate immunity. *Curr Opin Plant Biol* 12:465–472.

Park, S. W., Kaimoyo, E., Kumar, D., Mosher, S. and Klessig, D. F. 2007. Methyl salicylate is a critical mobile signal for plant systemic acquired resistance. *Science* 318:113–116.

Pitzschke, A. 2013. *Agrobacterium tumefaciens* and plant defense-transformation success hangs by a thread. *Front Plant Sci* 4:519.

Popp, C. and Ott, T. 2011. Regulation of signal transduction and bacterial infection during root nodule symbiosis. *Curr Opin Plant Biol* 14:458–467.

Quirino, B. F., Candido, E. S., Campos, P. F., Franco, O. L. and Kruger, R. H. 2010. Proteomic approaches to study plant-pathogen interactions. *Phytochemistry* 71:351–362.

Ramalingam, A., Kudapa, H., Pazhamala, L. T., Weckworth, W. and Varshney, R. K. 2015. Proteomics and metabolomics: two emerging areas for legume improvement. *Front Plant Sci* 6:1116.

Ramey, B. E., Koutsoudis, M., von Bodman, S. B. and Fuqua, C. 2004. Biofilm formation in plant–microbe associations. *Curr Opin Microbiol* 7:602–609.

Reid, A. 2011. Microbes helping to improve crop productivity. *Microbe* 6:435–439.

Rentel, M. C., Leonelli, L., Dahlbeck, D., Zhao, B. and Staskawicz, B. J. 2008. Recognition of the *Hyaloperonospora parasitica* effector ATR13 triggers resistance against oomycete, bacterial and viral pathogens. *Proc Natl Acad Sci USA* 105:1091–1096.

Rey, T. and Schornack, S. 2013. Interactions of beneficial and detrimental root-colonizing filamentous microbes with plant hosts. *Genome Biol* 14:121.

Riely, B. K., Ané, J. M., Penmetsa, R. V. and Cook, D. R. 2004. Genetic and genomic analysis in model legumes brings nod-factor signaling to center stage. *Curr Opin Plant Biol* 7:408–413.

Riely, B. K., Mun, J. H. and Ané, J. M. 2006. Unravelling the molecular basis for symbiotic signal transduction in legumes. *Mol Plant Pathol* 7:197–207.

Robledo, M., Jimenez-Zurdo, J. I., Velazquez, E., Trujillo, M. E., Zurdo-Piñeiro, J. L., Ramírez-Bahena, M. H., Ramos, B., Díaz-Mínguez, J. M., Dazzo, F., Martínez-Molina, E. and Mateos, P. F. 2008. Rhizobium cellulase $CelC_2$ is essential for primary symbiotic infection of legume host roots. *Proc Natl Acad Sci USA* 105:7064–7069.

Russo, D. M., Williams, A., Edwards, A., Posadas, D. M., Finnie, C., Dankert, M., Downie, J. A. and Zorreguieta, A. 2006. Proteins exported via de PrsD-PrsE type I secretion system and the acidic exopolysaccharide are involved in biofilm formation by *Rhizobium leguminosarum*. *J Bacteriol* 188:4474–4486.

Sahu, P. P., Puranik, S., Khan, M. and Prasad, M. 2012. Recent advances in tomato functional genomics: utilization of VIGS. *Protoplasma* 249:1017–1027.

Salzberg, S. L., Sommer, D. D., Schatz, M. C., Phillippy, A. M., Rabinowicz, P. D. and Tsuge, S., Furutani, A., Ochiai, H., Delcher, A. L., Kelley, D. and Madupu, R. 2008. Genome sequence and rapid evolution of the rice pathogen *Xanthomonas oryzae* pv. Oryzae PXO99A. *BMC Genomics* 9:204.

Schenk, P. M., McGrath, K. C. and Lorito, M. 2008. Plant–microbe and plant-insect interactions meet common grounds. *New Phytol* 179:251–255.

Scherp, P., Ku, G., Coleman, L. and Kheterpal, I. 2011. Gel-based and gel-free proteomic technologies. *Methods Mol Biol* 702:163–190.

Shendure, J., Mitra, R. D., Varma, C. and Church, G. M. 2004. Advanced sequencing technologies: methods and goals. *Nat Rev Genet* 5:335–344.

Slaughter, A., Daniel, X., Flors, V., Luna, E., Hohn, B. and Mauch-Mani, B. 2012. Descendants of primed plants exhibit enhanced resistance to biotic stress. *Plant Physiol* 158:835–843.

Smit, G., Swant, S., Lugtenberg, B. J. J. and Kijne, J. W. 1992. Molecular mechanisms of attachment of rhizobium bacteria to plant roots. *Mol Microbiol* 6:2897–2903.

Smith, S. E. and Smith, F. A. 2011. Roles of arbuscular mycorrhizas in plant nutrition and growth: new paradigms from cellular to ecosystem scales. *Annu Rev Plant Biol* 62:227–250.

Strange, R. N. and Scott, P. R. 2005. Plant disease: a threat to global food security. *Annu Rev Phytopathol* 43:83–116.

Subramoni, S., Nathoo, N., Klimov, E. and Yuan, Z. 2014. *Agrobacterium tumefaciens* responses to plant-derived signaling molecules. *Front Plant Sci* 5:322.

Sutton, J. M., Lea, E. J. and Downie, J. A. 1994. The nodulation-signaling protein NodO from *Rhizobium leguminosarum* biovarviciae forms ion channels in membranes. *Proc Natl Acad Sci USA* 91:9990–9994.

Tan, K. C., Ipcho, S. V., Trengove, R. D., Oliver, R. P. and Solomon, P. S. 2009. Assessing the impact of transcriptomics, proteomics and metabolomics on fungal phytopathology. *Mol Plant Pathol* 10:703–715.

Tseng, T.-T., Tyler, B. M. and Setubal, J. C. 2009. Protein secretion systems in bacterial-host associations, and their description in the gene ontology. *BMC Microbiol* 9:S2.

van derHoorn, R. A. L. and Kamoun, S. 2008. From guard to decoy: a novel model for perception of plant pathogen effectors. *Plant Cell* 20:2009–2017.

Vorholt, J. A. 2012. Microbial life in the phyllosphere. *Nat Rev Microbiol* 10:828–840.

Wang, Z., Gerstein, M. and Snyder, M. 2009. RNA-seq: a revolutionary tool for transcriptomics. *Nat Rev Genet* 10(1): 57–63.

Westermann, A. J., Forstner, K. U., Amman, F., Barquist, L., Chao, Y., Schulte, L. N., Müller, L., Reinhardt, R., Stadler, P. F. and Voge, J. 2016. Dual RNA-seq unveils noncoding RNA functions in host-pathogen interactions. *Nature* 529:496–501.

Westermann, A. J., Gorski, S. A. and Vogel, J. 2012. Dual RNA-seq of pathogen and host. *Nat Rev Microbiol* 10:618–630.

Winans, S. C. 1992. Two-way chemical signaling in Agrobacterium–plant interactions. *Microbiol Rev* 56:12–31.

Wirthmueller, L., Maqbool, A. and Banfield, M. J. 2013. On the front line: structural insights into plant–pathogen interactions. *Microbiol Nat Rev* 11:761–776.

Wu, C. F., Lin, J. S., Shaw, G. C. and Lai, E. M. 2012. Acid-induced type VI secretion system is regulated by ExoR-ChvG/ChvI signaling cascade in *Agrobacterium tumefaciens*. *PLoS Pathog* 8:e1002938.

Yoshida, K., Saitoh, H., Fujisawa, S., Kanzaki, H., Matsumura, H., Yoshida, K., Tosa, Y., Chuma, I., Takano, Y., Win, J. and Kamoun, S.. 2009. Association genetics reveals three novel a virulence genes from the rice blast fungal pathogen *Magnaporthe oryzae*. *Plant Cell* 2:1573–1591.

Yuan, Z. C. and Williams, M. 2012. A really useful pathogen, *Agrobacterium tumefaciens*. *Plant Cell* 24(10):112.

Yuan, Z. C., Haudecoeur, E., Faure, D., Kerr, K. F. and Nester, E. W.2008a. Comparative transcriptome analysis of *Agrobacterium tumefaciens* in response to plant signal salicylic acid, indole-3-acetic acid and gamma-amino butyric acid reveals signalling cross-talk and Agrobacterium–plant co-evolution. *Cell Microbiol* 10:2339–2354.

Zhao, F., Chen, L., Perl, A., Chen, S. and Ma, H. 2011. Proteomic changes in grape embryogenic callus in response to *Agrobacterium tumefaciens*-mediated transformation. *Plant Sci* 181:485–449.

4

Phosphate Solubilizing Bacteria: A Neglected Bioresource for Ameliorating Biotic Stress in Plants

Rajinder Kaur, Sukhminderjit Kaur, and Navneet Kaur
Chandigarh University, Punjab, India

CONTENTS

4.1 Introduction

Biotic stress is referred to as departure from optimal physiological condition in plants as instigated by the damage caused by living organisms either microscopic (bacteria, virus, and fungi) or macroscopic (parasites, nematodes, insects, and weeds) (Newton et al. 2011). Biotic stress is a major impediment for commercial crops as it suppresses the expression of genes responsible for high yield, and also results in metabolic changes resulting in overall poor quality of the crop. In India, insects account for about 25% decrease in crops, nematodes up to 20%, and vertebrate pest from 6% to8%. Primarily, pesticides such as monocrotophos, aldrin, dichloro-diphenyl-trichloro-ethane (DDT), dieldrin, β-benzene hexachloride (BHC), 2,4-dichlorophenoxyacetic acid (2,4-D), chlordane, and endrinare (Delaplane 2000); weedicides such as atrazine, simazine, alachlor, metolachlor, and trifluralin; insecticides such as diazinon, parathion methyl; and fungicides such as captan, triadimefron, metaxyl, and glyodin are extensively used against biological stressors, but they possess a great threat to biodiversity. Due to high persistence in soil (Ferencz and Balog 2010), water contamination (Anasco et al. 2010), toxicity to beneficial soil microorganisms (Pal et al. 2005), effect on non-target soil community (Evans et al. 2010), and decline in beneficial non-target pollinators (Brittain et al. 2010), there is a need to look for alternatives to mitigate the biotic stress in a sustainable manner (Timmusk et al. 2017). Significance of plant growth promoting bacteria (PGPB) has been well-recognized in terms of both nutritional benefits and non-nutritional benefits (Numan et al. 2018). The term PGPB was first described by Kloepper and Schroth (1978) as the living agents enabling plant growth and plant protection from biotic and abiotic stress. The use of bioinoculants has been suggested as en route to plant stress management as they produce antibiotics, siderophores, iron chelating agents, detoxifying agents, and many lytic enzymes (Alori et al. 2017). Various rhizospheric thriving bacteria such as *Pseudomonas* spp. *Bacillus* spp., *Burkholderia phytofirmans* PsJN also trigger induced systematic resistance (ISR) in plants in response to broad spectrum pathogenic agents, including bacteria, viruses, fungi, nematodes, and parasites (Borriss 2020). Among PGPB, phosphate solubilizing bacteria (PSB) form an important group of bacteria with mineral phosphate solubilizing and organic phosphorus mineralization activity (Kaur and Kaur 2018). These bacteria are present in the soil and help in transformation of insoluble phosphorus into soluble form which can be easily utilized by plants. A large number of PSB have been isolated from the rhizosphere of several crops, and these constitute about 20% to 40% of the cultivable population of soil microorganisms. The important PSB belong to *Burkholderiaceae*, *Pseudomonadaceae*, *Enterobacteriaceae*, *Streptococcaceae*, *Flavobacteriaceae*, *Bacillaceae*, *Paenibacillaceae*, and *Micrococcaceae* families (Kaur and Kaur 2020). This chapter aims to understand the potential of PSB in mitigating biotic stress in order to achieve a bioformulation that accounts for both nutritional and non-nutritional benefits.

4.2 PSB in Mitigation of Biotic Stress

Rhizospheric bacteria are known to have the capacity to tolerate the environmental changes and adapt themselves to the new environment. Also, microbial interactions with plant are known to enhance the plant resilience to biotic stress. The bacterial diversity improves the soil quality and increases the soil fertility by enhancing availability of soil nutrients and affecting the soil agglomeration, which help in the growth of plant (Kirk et al. 2004). The PSB are widely studied for their plant growth promoting properties under *in vitro* conditions such as indole acetic acid, exopolysaccharides, siderophore and hydrogen cyanide (HCN) production, extracellular enzyme activity, increased plant growth, plant biomass, and availability of nutrients. In the past few years, PSB have gripped the attention of researchers due to their antagonistic potential against various phytopathogens implicated through a number of mechanisms (Figure 4.1). The roles of several PSB belonging to the genera *Bacillus*, *Pseudomonas*, *Lycobacter*, *Burkholderia*, *Skermanella*, *Paenibacillus*, *Enterobacter*, and *Citrobacter* have been documented previously to mitigate the biotic stress in plants. Recently, Cao et al. (2018) have demonstrated the role of *Bacillus velezensis* against phytopathogens such as *Ralstonia solanacearum* and *R. oxysporum* by the secretion of surfactin, iturin, and fengycin. Biocontrol potentials of *B. acidiceler* and *B. megaterium* on the *Zea mays* and *Triticum aestivum* have also been reported (Akinrinlola 2018). The rhizobacteria-induced biotic stress resilience in plants is a broad spectrum mechanism, which functions by the secretions of different metabolites by the bacteria. The biocontrol action can be exhibited by the production of antibiotics such as 2,4-diacetylphloroglucinol (DAPG), phenazine, pyoluteorin, tropolone, oxytetracycline, oomycin A, amphisin, pyrrolnitrin, kasugamycin, tensin, and streptomycin. Some of these antibiotics arrest the protein synthesis mechanism in the pathogen by binding to the ribosomes. Antigonistic activity of these microorganisms can also be exhibited by several cyclic lipopeptidic compounds, including surfactin, zwittermicin A, kanosamine, iturin, fengycin, xanthobaccin, and oligomycin (Meena and Kanwar 2015). Apart from the production of secondary metabolites such as phenolic compounds, flavonoids, phenylalanine ammonia lyase, and siderophores, bacteria also elicit induced systemic resistance (ISR) in plants. HCN production by phosphate solubilizing fluorescent *Pseudomonas fluorescens* has shown biocontrol activity against phytopathogenic fungi affecting various crops (Naik et al. 2008). Increase in phenolics and plant defense enzymes such as polyphenol oxidase, peroxidase, superoxide dismutase, β-1,3-glucanase, chitinase; phenol, callose, and lignin deposition has been known to increase in response to these biocontrol agents, which enhances plants antagonistic response toward phytopathogens. Exogenous application of bacteria also induces a cellular defense response in plants (Wu et al. 2019). These methods can be broadly classified as direct and indirect methods of stress tolerance (Figure 4.2).

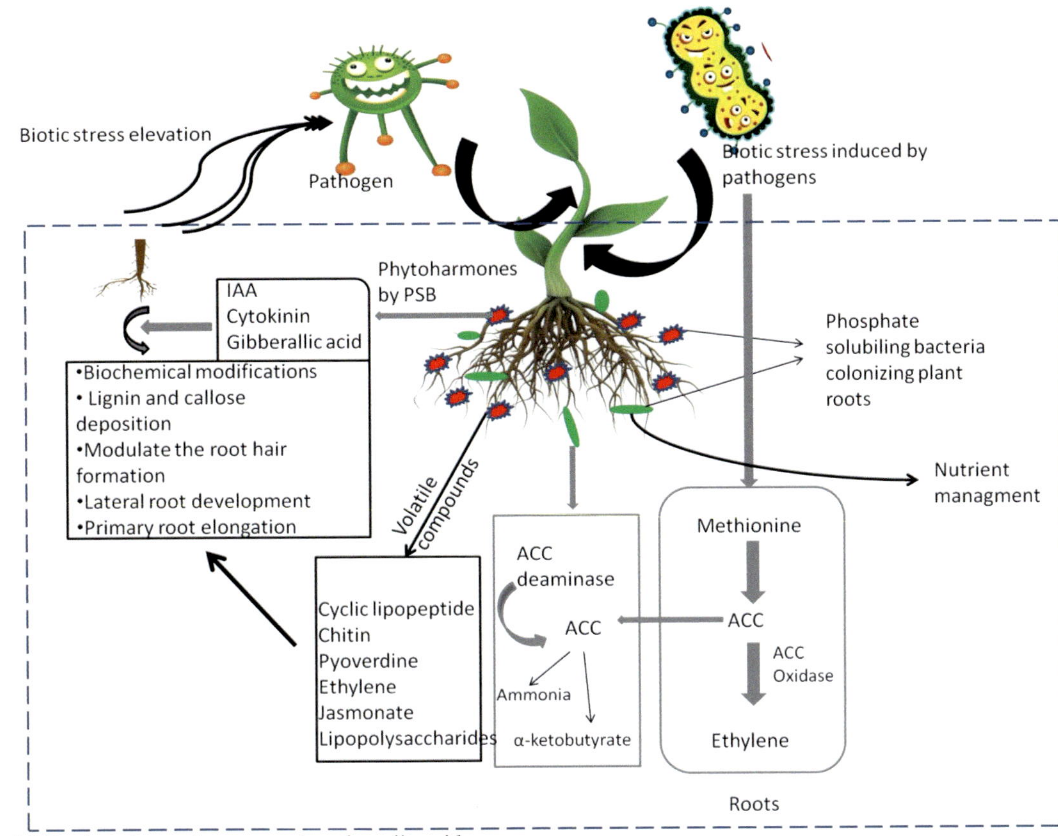

ACC – 1-aminocyclopropane-1-carboxylic acid
PSB- Phosphate solubilizing Bacteria
IAA- Indole acetic acid

FIGURE 4.1 Mechanism for biotic stress elevation in plants by PSB.

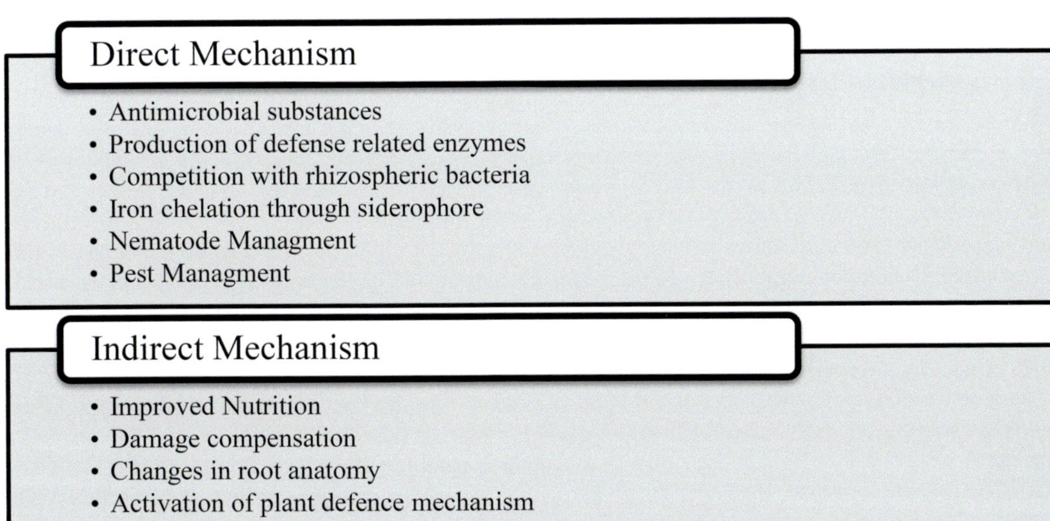

FIGURE 4.2 Direct and indirect methods of biotic stress tolerance in plants through PSB.

4.3 Direct Mechanism

4.3.1 Antimicrobial Substances

The antagonistic activity of PSB is due to the release of biologically active compounds called antibiotics that have antipythopathogenic action. Antimicrobial compounds produced by PSB against different phytopathogens are given in Table 4.1. These antimicrobial agents can be volatile or nonvolatile components with antibacterial, antiviral, antitumor, antihelminthic, and antioxidant properties. Aldehydes, HCN, alcohols, and ketones are volatile antibiotics, while lipopeptides, hydrophilic ionic compounds, cyclic nitrogenous compounds, cyclic lipopeptides, and amino polyols are nonvolatile antibiotics (Nielsen and Sorensen 2003). Recently, Moreno et al. (2019) have reported the secretion of three broad spectrum nonvolatile antimicrobials: Streptothricin F, D, and E from phosphate solubilizing *Streptomyces* A20 strain. Streptothricin is effective in controlling rice phytopathogens like *Burkholderia glumae* and *P. fuscovaginae* and has shown antifungal activity against *Fusarium oxysporum*, *Phomopsis* sp., *Gaeumannomyces* sp., *Colletotrichum* sp. 26B, *Rhizoctonia solani*, and *Ulocladium* sp. (Moreno et al. 2019). Another PSB genera *Pseudomonas* has been reported to produce several volatile antimicrobial metabolites from different groups such as DAPG, acetamidophenol, viscosinamide, and HCN (Paulin et al. 2017). DAPG is considered as an important group in *Pseudomonas* biocontrol role against broad spectrum bacteria, fungi, and nematodes. HCN has various antifungal properties and helps in black root rot prevention in Tobacco. The production of DAPG and HCN in *P. brassicacearum* LBUM300 has been associated with repression of bacterial canker symptoms in tomato plant (Paulin et al. 2017). Other volatile antimicrobial compounds, S-methyl thioacetate (MTA), dimethyl disulfide, and dimethyl trisulfide produced by *Pseudomonas* are strongly antifungal and antinomycetes showing effective potential against plant pathogens *R. solani*, *F. culmorum*, *Verticillium dahliae*, and *Pythium ultimum* (Ossowicki et al. 2017).

4.3.2 Production of Defense Enzymes

PSB produce various hydrolytic enzymes such as chitinase, chitosanases, lipases, protease, pectinase, β-1,3-glucanase that disrupt the glycosidic bonds in the cell wall fungal pathogens. These hydrolytic enzymes also prevent the germination of fungal spores and aid in protein degradation (Li et al. 2018). The production of hydrolytic enzymes as the measure of biocontrol has a significant value as it prevents spore germination inhibiting fungal pathogens to infect fruit wounds and lenticels.

4.3.3 Competition with Rhizosphere Bacteria

PSB are able to sustain the tremendous microbial diversity through the phenomenon of competitive exclusion. Competitive exclusion involves multiple mechanisms to sustain in the rhizosphere. PSB produce antimicrobial toxins and regulate their functioning by quorum sensing. Also, resource competition is a vital strategy of PSB to sustain in the rhizospheric soil. Antimicrobial production and rapid growth rate are the processes to maximize nutrient uptake by PSB at the expense of other microbes. PSB compete with other soil microbes for carbon and nitrogen sources. PSB can grow well utilizing a variety of carbon sources. The sequestration of phosphorous and iron by the secretion of organic acids and siderophores, respectively, enables their survival over other species (Oehmen et al. 2007). Another competitive mechanism of PSB is the siderophore production that is associated with cooperative behavior. Production of high-affinity siderophores by PSB lowers the ability of low-affinity siderophore producing bacterial strain to grow in low-iron conditions (Joshi et al. 2006). Another important aspect about PSB is their ability to take and hold the stable position at favorable sites in the environment, and after their entry into favorable rhizospheric root locations, they colonize new niches and displace the existing colonizers. They require secretion of receptors, signal molecules like peptides, butyrolactones, butanolides, furans, and adhesions that enable their access to plant roots (Hibbing et al. 2010).

TABLE 4.1

Antimicrobial Compounds Produced by PSB against Different Pythopathogens

PSB	Antimicrobial Compound	Phytopathogen	Crop	Reference
Streptomyces A20	Streptothricin F Streptothricin D Streptothricin E	**Bacterial pathogens** *Burkholderia glumae, B. gladioli, Chromobacterium violaceum, Enterobacter aerogenes, Escherichia coli, Klebsiella pneumonie, Pseudomonas putida, P. aeruginosa, P. fuscovaginae, Staphylococcus aureus* **Fungal pathogens** *Fusarium* sp. *Gaeumannomyces* sp *Phomopsis* sp. *Ulocladium* sp. *Rhizoctonia solani, Colletotrichum* sp. *Colletotrichum* sp.	Rice	Moreno et al. 2019
Paenibacillus jamilae HS-26	N, N-diethyl-1, 4-phenylenediamine, 2, 3-butanediol, acetoin, undecanone, N, N-diethyl-1, 4-phenylenediamine	*Bipolaris sorokiniana and R. solani, Fusarium oxysporum*	Wheat	Wang et al. 2019
Pseudomonas donghuensis P482	dimethyl sulfide S-methyl thioacetate methyl thiocyanate dimethyl trisulfide 1-undecan HCN	*Pseudomonas syringae, R. solani, Verticulum dahliae JR, Pythium ultimum P17*	NA	Ossowicki et al. 2017
Pseudomonas chlororaphis GBPI 507 (MCC2693)	phenazine-1-carboxylic acid, HCN	*Alternaria altenata, Phytophthora* sp., *F. solani, F. oxysporum*	NA	Jain and Pandey 2016
Pseudomonas spp.	Phenazines, hydrogen cyanide, diacetylphloroglucinol (DAPG)	*Streptomyces scabies, Fusarium graminearum, Seiridium cardinal, Pythium myriotylum*	Wheat, tomato, ginger, chickpea, bean, potato	Arseneault et al. 2016
Pseudomonas fluorescens	2,4-diacetypholoroglucinol (DAPG)	*Plasmopara viticola, Aphanomyces cochlioides*	Grapevine (*Vitisvinifera* cv. Müller-Thurgau	Islam and Tidemann 2011
Pseudomonas stutzeri and *Bacillus cepacia*	Allelochemicals	*R. solani*	Potato	Tariq et al. 2010
Pseudomonas fluorescens	HCN, phloroglucinols	*Thielaviopsis basicola*	Tobacco	Ramette et al. 2006
Pseudomonas aeruginosa PUPa3	Phenazine-1-carboxamide	*Macrophomina phaseolina, M. grisea, R. solani, Sitophilus oryzae, Botrytis cinera, Colletotrichum capsici, C. gleosporoides, C. falcatum CFL, Phaeosphaerella theae PTS*	Rice, ground nut, Tobcco, chilli, sugarcane, Tea	Kumar et al. 2005

4.3.4 Iron Chelation through Siderophores

Siderophores are low-molecular weight (600–1500 Da) water-soluble peptidic iron chelators. They are recognized by a corresponding outer membrane protein side chains and functional groups that behave as ligand for iron sequestration. Siderophores can solubilize iron aggregates (insoluble hydroxides and oxyhydroxide), generating quickly diffusing iron-siderophore chelates potentially and significantly increasing iron in the environment. Many strains of PSB produce siderophores as a method of biocontrol and some of them are given in Table 4.2. *Azotobacter vinelandil*, a potential phosphate solubilizing bacterium, produces three siderophores: the tricatecholate, the dicatecholate azotochelin, and the monocatecholate aminochellin. These siderophores have higher affinity to quench the iron, due to which it becomes unavailable for fungal/bacterial pathogens to utilize it for their growth, which prevents it from fungal/bacterial pathogens to iron. Thus, low levels of iron restrain the phytopathogens to proliferate in the root zone of the plant (Dowling et al. 1996). In contrary, plants can use this iron siderophore complex for their nutrition. PSB like *Klebsiella michiganensis* W14, *Serratia nematodiphila* DSM21420, *Acinetobacter soli* CIP110264, *Citrobacter freundii*, *Pseudomonas hunanensis*, *P. oryzihabitans*, *Bacillus megaterium*, *A. johnsonii* CIP 64.6, and *P. otitidis* MCC10330 have been reported to produce siderophores in response to oxidative stress (Sharma et al. 2013). The catecholate, azotochelin, and protochelin are highly produced for chemical protection from oxidative damage. A small amount of

TABLE 4.2

Siderophore Producing Strains of PSB

PSB	Siderophore Produced	Target Pathogen	Crop	Reference
Streptomyces sp.	NA	*Aspergillus fumigates* *Gaeumannomyces graminis* *Fusarum moniliforme* *Penicillium notatum*	Olive plant	Dede et al. 2020
Bacillus sp. GZDF3	Catecholets	*Candida albicans*	*Pinellia ternate*	Sheng et al. 2020
Streptomyces sp.	NA	*Burkholderia glumae*, *Fusarium*, *Pythium*	Rice	Moreno et al. 2019
Bacillus niabensis *Bacillus subtilis* *Bacillus mojayensis*	Catecholets and salicylate	*Ralstonia solanacearum*	Banana	Kesaulya et al. 2018
Azotobacter and *Bacillus*	NA	*Rhizoctonia solani*	Maize	Bjelić et al. 2018
Brevibacillus agristrain TTD *Aneurinibacillus aneurinilyticus strain TTD21* *Sporosarcina koreensis strain BT22* *Bacillus megaterium*	NA	*Nigrospora sphaerica*, *Pestalotiopsis theae*, *Curvulariaeragostidis*, *Glomerella cingulata*, *R. solani*, *and Fusarium oxysporum*	Tea	Dutta and Thakur 2017
*Pseudomonas fluorescens Mst*8.2	Hydroxamate	*R. solani*	Wheat	Gull and Hafeez 2012
Brevibacterium antiquum	NA	*Macrophomonia phaseolina*	Sorghum	Gopalakrishnan et al. 2011
Pseudomonas stutzeri *Bacillus cepacia*	NA	*R. solani*	Potato	Tariq et al. 2010
Pseudomonas sp.	Piochelin	*Pythium*	Tomato	Buysens et al. 1996
Pseudomonas sp.	Pyoyerdine	*Fusarium*, *Gaeumannomyces*, *Pythium*	Tomato Pea Peanuts Maize	Duijff et al. 1993

protochelin is sufficient to prevent iron catalyzed oxidative damage under the low Fe^{3+} growth conditions. The PSB, including *Sinorhizobium meliloti*, *Rhizobium leguminosarum*, and *R. tropici*, produce siderophores, contribute to antagonistic effect against pathogenic bacteria, and have insecticidal activity (Gopalakrishnan et al. 2015). The siderophores are produced and excreted outside the microbial cell to form complex with iron, and this complex is transported to the insect. It helps to destroy pathogens that create biotic stress and eco-friendly alternative to synthetic chemical insecticides. Application of siderophore-based insect antagonists against insect pests can be employed as one of the best alternatives and eco-friendly strategies to control insect pests or provide tolerance to biotic stress and prevent plant diseases (Sonawane et al. 2018).

4.3.5 Nematode Management

Biocontrol agents such as bacteria that produce lytic enzymes are recommended as the alternative approach for synthetic nematicides. Chitosanase and protease are helpful to alter the structure of egg and inhibit nematode egg hatching. Application of chitinase is associated with increased mortality of *Meloidogyne incognita*. Lytic enzymes, proteases, and chitinases degrade the cuticle of nematodes (Chen et al. 2015). PSB such as *Bacillus subtilis* and *B. megaterium* are recommended for the management of root-knot nematode, *M. incognita* due to the secretion of secondary metabolites with nematocidal activity, including lytic enzymes cyclic lipopeptides, surfactin, and iturin (El-Hadad et al. 2011). Secretion of chitinase and glucanase in *B. cereus* has demonstrated its antagonistic activity against *M. javanica*. These enzymes inhibit egg hatching and thereby decrease the percentage of juvenile survival rate (Basyony and Abo-Zaid 2018). Applications of *B. subtilis* and *Pseudomonas aeruginosa* in the tomato fields have been successfully able to control *M. incognita* eggs and suppress root-knot nematodes (Abo-Elyousr et al. 2010).

4.3.5.1 Insect/Pest Management

Many plant growth promoting rhizobacteria (PGPR) are known to be involved in increasing the plant resistance toward insect by secreting phytohormones and increasing nutrient solubilization and iron chelation mechanisms. Rhizobacteria-mediated increased uptake of macro and micronutrients enhances the production of defense enzymes in host plant against pathogens (Pineda et al. 2010). PGPR also induce the accumulation of soluble and cell-bound phenolic compounds and antioxidative substances like phytoalexins in the host plant. Pathogenesis-related genes encoding chitinase, glucanase, phenylalanine ammonia-lyase and peroxidase, lignification, and ethylene production are activated by these bacteria that provide physical and chemical barriers toward insects in plants (Meena et al. 2000; Rajendran et al. 2007). PSB stains of *Bacillus* and *Pseudomonas* suppress the insect attack by siderophores production, antimicrobial metabolites, ISR, and induction of lignification in the host. Applications of Pseudomonas-like bacterial strain P29, *P. putida* strain PS9, *B. amyloliquefaciens*, *B. subtilis*, and *B. pumilus* have reported significant decrease in the population of green peach aphid, cotton aphids, blue green aphid, spotted cucumber beetle, whitefly on tomato, cotton, and cucumber (Kavitha 2010).

4.4 Indirect Mechanism

4.4.1 Improved Nutrition

It has been well-documented that the optimal concentration of trace metal ions is crucial for the plant growth, development, and defense against pathogens. Cross-talk between plant response to biotic stress and nutrition is an area of upcoming interest. Sodium (Na), aluminum (Al), zinc (Zn), selenium (Se), and silicon (Si) are considered vital for plant growth, while cobalt (Co) is crucial for sustaining the microbial partnerships. These metal ions also have functional relevance against biotic stress through different mechanisms, including metal fortification, trade off or metal therapy. The exogenous application of these metal ions can generate defense response in plants by producing various secondary metabolites such as thionins, defensins, phytoalexins, and glucosinolates. Reactive oxygen species (ROS) produced in response to metal ions switch on antioxidant defense system. The Al ions trigger shikimate pathway essential for the production of stress signaling salicylic acid, phenolics such as flavonoid, and essential amino acids which increase plants' resistance to biotic stress (Kaur et al. 2016). Exogenous application of Al retards spore germination, mycelium, and vegetative growth in *Phytophthora infestans* and *Thielaviopris basicola* infecting potato (Andrivon 1995). The accumulation of Se in *Brassica juncea* L. protects the plant from *Fusarium*, *Alternaria brassicicola* infections and attack of caterpillers, snails, and aphids (Hanson et al. 2003;2004). Co-inoculation of PSB such as *Puffinus auricularis* and *Bacillus aryabhattai* enhances the nutrient uptake (N, P, K) besides improving the plant metabolism (Wu et al. 2019). Inoculation of soil with PGPR strains (including PSB) activates the nutrients that are already present in the soil and thus decreasing the use of fertilizer by 25%–50% (Wang et al. 2020).

4.4.2 Damage Compensation

The attack of pathogen on the plant stimulates the production of ethylene in response, which further magnifies the stress condition. The simplest way to combat this damage is to knock down the release of ethylene. This can be done by suppressing the release of 1-aminocyclopropane-1-carboxylic acid (ACC), immediate precursor of ethylene by the action of bacteria that express genes for the ACC deaminase enzyme. ACC deaminase reduces the damage caused by the ethylene by degrading ACC to ammonia and α-ketobutyrate. This strategy has been used on different crops, including soybean, potato, tomato, beans, and carrot that were infected by *Sclerotium rolfsii*, *F. oxysporum*, *P. ultimum*, and *R. solani* (Kang et al. 2019). It has also been reported that PGPR with both indole-acetic acid (IAA) and ACC deaminase production are more effective in protecting the plant against biotic stress damage. Production of both ensures the cell growth and proliferation in stress conditions. ACC deaminase, by degrading ACC, confirms the transcription of auxin factors, which would be suppressed by the high levels of ethylene. A number of PSB strains have been reported to produce ACC deaminase and IAA that can be used to develop strategies to mitigate the biotic stress (Zabihi et al. 2011; Ehsan et al. 2014; Kang et al. 2019; Pandey and Gupta 2019).

4.4.3 Changes in Root Anatomy

The root morphology significantly enhances root length and surface area of roots. Production of phytoharmones, lytic enzymes, and secondary metabolites by PGPR modulates the root hair formation, lateral root development, and primary root elongation. This enhances the nutrient supply to the plant during the conditions of stress. Apart from anatomical changes, PGPR also induce biochemical modifications in the root structure. PSB like *Bacillus* sp. are able to induce lignin deposition in root epidermal cells as a mechanism of physical defense in pearl millets, melon, and pea (Garcia-Gutierrez et al. 2013). In some plants, lignin accumulation is accompanied by callose deposition, when colonized by *Bacillus pumilus* INR-7 strain and resistance against fungal pathogens is increased by this mechanism. Similar cell wall modifications have been observed in bean and pea when inoculated by *Pseudomonas* sp. (Vacheron et al. 2013).

4.4.4 Activation of Plant Defense Mechanism

Pathogenic attack triggers the defense, i.e., systemic acquired resistance (SAR) system in plants. Phenotypically a similar non-specific mechanism, ISR has been induced in plants by PGPR. PGPR like PSB release signal molecules such as cyclic lipopeptides, chitin, pyoverdine, ethylene, jasmonate, and lipopolysaccharides, which help in the induction of ISR in plants. It doesn't involve direct killing of pathogen, but provides enhanced defensive capability to plant through the physical and chemical protection. These elicitors, are broadly termed as microbe-associated molecular patterns (MAMPs – cyclic lipopeptides, chitin, flagellin, pyoverdine, and lipopolysaccharides), volatile organic compounds (VOCs – secondary metabolites such as alcohols, ketones, esters, sesquiterpenes, alkanes, and terpenoids) or siderophores (Tyagi et al. 2018). PSB including *Pseudomonas simiae*, *P. fluorescens*, *Paenibacillus polymyxa*, *Bacillus velezensis*, *B. subtilis*, and *Serratia* are well-known to induce ISR in plants (Suryadi et al. 2019).

4.5 Conclusion

It has been well-established that PSB strains have multidimensional role in plant nutrition, growth, proliferation, and immunity toward phytopathogens. These bacteria can be used as biofertilizer and biocontrol agent to address the problems like nutrient deficiency and pathogen attack. However, it has been observed that PSB biocontrol agents are crop-specific and have narrow spectrum of action. Further studies should be carried out to develop a broad spectrum PSB biocontrol agent that can be as efficient as chemical pesticides. Work can also be modulated toward developing consortia of PSB that can have a broad spectrum action and high efficiency under varied environmental conditions. In a nutshell, it can be concluded that PSB as the biocontrol agent can be a useful resource in future approaches toward sustainable agriculture.

REFERENCES

Abo-Elyousr, K. A., Khan, Z. and Abedel-Moneim, M. F. 2010. Evaluation of plant extracts and *Pseudomonas* for control of root-knot nematode. Meloidogyne incognita on tomato. *Nematropica* 40(2):289–299.

Akinrinlola, R. J. 2018. Evaluation of *Bacillus* strains for plant growth-promotion potentials on corn (*Zea mays*), wheat (*Triticumae stivum*), and soybean (*Glycine max*). Master's Thesis, University of Nebraska, United States.

Alori, E. T., Dare, M. O. and Babalola, O. O. 2017. Microbial inoculants for soil quality and plant health. In: *Sustainable Agriculture Reviews*. Springer, Cham, pp:281–307.

Anasco, N., Uno, S., Koyama, J., Matsuoka, T. and Kuwahara, N. 2010. Assessment of pesticide residues in freshwater areas affected by rice paddy effluents in Southern Japan. *Environ Monit Assess* 160:371–383.

Andrivon, D.1995. Inhibition by aluminum of mycelial growth and of sporangial production and germination in *Phytophthora infestans*. *Eur J Plant Pathol* 101(5):527–533.

Arseneault, T., Goyer, C. and Filion, M. 2016. Biocontrol of potato common scab is associated with high *Pseudomonas fluorescens* LBUM223 populations and phenazine-1-carboxylic acid biosynthetic transcript accumulation in the potato geocaulosphere. *Phytopathology* 106(9):963–970.

Basyony, A. G. and Abo-Zaid, G. A. 2018. Biocontrol of the root-knot nematode, *Meloidogyne incognita*, using an eco-friendly formulation from *Bacillus subtilis*, lab. and greenhouse studies. *Egypt J Biol Pest Control* 28:87–99.

Bjelić, D., Marinkovic, J., Tintor, B. and Mrkovacki, N. 2018. Antifungal and plant growth promoting activities of indigenous rhizobacteria isolated from maize (*Zea mays* L.) rhizosphere. *Commun Soil Sci Anal* 49(1):88–98.

Borriss, R. (ed). 2020. Phytostimulation and biocontrol by the plant-associated bacillus amyloliquefaciens FZB42: an update. In: *Phyto-Microbiome in Stress Regulation*. Springer, Singapore, pp:1–20.

Brittain, C. A., Vighi, M., Bommarco, R., Settele, J. and Potts, S. G. 2010. Impacts of a pesticide on pollinator species richness at different spatial scales. *Basic Appl Ecol* 11:106–115.

Buysens, S., Heungens, K., Poppe, J. and Hofte, M. 1996. Involvement of pyochelin and pyoverdin in suppression of Pythium-induced damping-off of tomato by *Pseudomonas aeruginosa* 7NSK2. *Appl Environ Microbiol* 62(3): 865–871.

Cao, Y., Pi, H., Chandrangsu, P., Li, Y., Wang, Y., Zhou, H. and Cai, Y. 2018. Antagonism of two plant-growth promoting *Bacillus velezensis* isolates against *Ralstonia solanacearum* and *Fusarium oxysporum*. *Sci Rep* 8(1):1–14.

Chen, L., Jiang, H., Cheng, Q., Chen, J., Wu, G., Kumar, A., Sun, M. and Liu, Z. 2015. Enhanced nematicidal potential of the chitinase pachi from *Pseudomonas aeruginosa* in association with Cry21Aa. *Sci Rep* 5:14395–14406.

Dede, A., Güven, K. and Şahin, N. 2020. Isolation, plant growth-promoting traits, antagonistic effects on clinical and plant pathogenic organisms and identification of actinomycetes from olive rhizosphere. *Microb Pathog* 143:104–134.

Delaplane, K. S. 2000. *Pesticide Usage in the United States: History, Benefits, Risks, and Trends. Cooperative Extension Service*. The University of Georgia, Athens, Georgia, pp: 1–6.

Dowling, D. N., Sexton, R. and Fenton, A. 1996. Iron regulation in plant-associated *Pseudomonas fluorescens* M114: implications for biological control. In: Nakazawa T., Furukawa K., Haas D. and Silver S. (eds), *Molecular Biology of Pseudomonads*. American Society for Microbiology Press, Washington, pp:502–511.

Duijff, B. J., Meijer, J. W., Bakker, P. A. and Schippers, B. 1993. Siderophore-mediated competition for iron and induced resistance in the suppression of *Fusarium* wilt of carnation by fluorescent *Pseudomonas* spp. Neth. *Plant Pathol* 99(5–6):277–289.

Dutta, J. and Thakur, D. 2017. Evaluation of multifarious plant growth promoting traits, antagonistic potential and phylogenetic affiliation of rhizobacteria associated with commercial tea plants grown in Darjeeling, India. *PLoS One* 12(8):1–24.

Ehsan, S., Saleem, I. and Zafar, H. 2014. Effect of phosphate solubilizing bacteria along with ACC-deaminase on wheat growth under axenic conditions. *Asian J Agric Biol* 2(3):183–190.

El-Hadad, M. E., Mustafa, M. I., Selim, S. M., El-Tayeb, T. S., Mahgoob, A. E. A. and Aziz, N. H. A. 2011. The nematicidal effect of some bacterial biofertilizers on *Meloidogyne incognita* in sandy soil. *Brazil J Microbiol* 42(1):105–113.

Evans, S. C., Shaw, E. M. and Rypstra, A. L. 2010. Exposure to a glyphosate-based herbicide affects agrobiont predatory arthropod behavior and long-term survival. *Ecotoxicology* 19:1249–1257.

Ferencz, L. and Balog, A. 2010. A pesticide survey in soil, water and foodstuffs from central Romania. *Carpath J Earth Environ* 5(1):111–118.

Garcia-Gutierrez, L., Zeriouh, H., Romero, D., Cubero, J., de Vicente, A. and García, A. 2013. The antagonistic strain *Bacillus subtilis* UMAF 6639 also confers protection to melon plants against cucurbit powdery mildew by activation of jasmonate-and salicylic acid-dependent defence responses. *Micro Biotechnol* 6(3):264–274.

Gopalakrishnan, S., Humayun, P., Kiran, B. K., Kannan, I. G. K., Vidya, M. S., Deepthi, K. and Rupela, O. 2011. Evaluation of bacteria isolated from rice rhizosphere for biological control of charcoal rot of sorghum caused by *Macrophomina phaseolina* (Tassi) Goid. *World J Microbiol Biotechnol* 27(6):1313–1321.

Gopalakrishnan, S., Sathya, A., Vijayabharathi, R., Varshney, R. K., Gowda, C. L. and Krishnamurthy, L. 2015. Plant growth promoting rhizobia: challenges and opportunities. *3 Biotech* 5(4):355–377.

Gull, M. and Hafeez, F. Y. 2012. Characterization of siderophore producing bacterial strain *Pseudomonas fluorescens* Mst 8.2 as plant growth promoting and biocontrol agent in wheat. *Afr J Microbiol Res* 6(33):6308–6318.

Hanson, B., Garifullina, G. F., Lindblom, S. D., Wangeline, A., Ackley, A., Kramer, K. and Pilon Smits, E. A. 2003. Selenium accumulation protects *Brassica juncea* from invertebrate herbivory and fungal infection. *New Phytologist* 159(2):461–469.

Hanson, B., Lindblom, S. D., Loeffler, M. L. and Pilon Smits, E. A. 2004. Selenium protects plants from phloem feeding aphids due to both deterrence and toxicity. *New Phytologist* 162(3):655–662.

Hibbing, M. E., Fuqua, C., Parsek, M. R. and Peterson, S. B. 2010. Bacterial competition: surviving and thriving in the microbial jungle. *Nat Rev Microbiol* 8(1):15–25.

Islam, M. T. and Von Tiedemann, A. 2011. 2, 4-diacetylphloroglucinol suppresses zoosporogenesis and impairs motility of Peronosporomycete zoospores. *World J Microb Biotechnol* 27(9):2071–2079.

Jain, R. and Pandey, A. 2016. A phenazine-1-carboxylic acid producing polyextremophilic *Pseudomonas chlororaphis* (MCC2693) strain, isolated from mountain ecosystem, possesses biocontrol and plant growth promotion abilities. *Microbiol Res* 190:63–71.

Joshi, F., Archana, G.and Desai, A. 2006. Siderophore cross-utilization amongst rhizospheric bacteria and the role of their differential affinities for Fe^{3+} on growth stimulation under iron-limited conditions. *Curr Microbiol* 53(2): 141–147.

Kang, S. M., Shahzad, R., Bilal, S., Khan, A. L., Park, Y. G., Lee, K. E. and Lee, I. J. 2019. Indole-3-acetic-acid and ACC deaminase producing *Leclercia adecarboxylata* MO1 improves *Solanum lycopersicum L.* growth and salinity stress tolerance by endogenous secondary metabolites regulation. *BMC Microbiol* 19(1):80–84.

Kaur, R. and Kaur, S. 2018. Biological alternates to synthetic fertilizers: efficiency and future scopes. *Indian J Agric Sci* 52(6):587–595.

Kaur, R. and Kaur, S. 2020. Variation in the phosphate solubilizing bacteria from virgin and the agricultural soils of Punjab. *Curr Microbiol* 77(9), 2118–2127.

Kaur, S., Kaur, N., Siddique, K. H. and Nayyar, H. 2016. Beneficial elements for agricultural crops and their functional relevance in defence against stresses. *Arch Agron Soil Sci* 62(7):905–920.

Kavitha, S. J. 2010. Management of major sucking pests in cowpea vigna unguiculata (L.) walp. with entomopathogens and plant defense inducing rhizobacteria, Doctoral dissertation, College of Agriculture, Vellayani.

Kesaulya, H., Hasinu, J. V.and Tuhumury, G. N. 2018. Potential of *Bacillus* spp produces siderophores in suppressing the wilt disease of banana plants. *IOP Conf Ser Earth Environ Sci* 102(1):12–16.

Kirk, J. L., Beaudette, L. A., Hart, M., Moutoglis, P., Klironomos, J. N., Lee, H. and Trevors, J. T. 2004. Methods of studying soil microbial diversity. *J Microbial Methods* 58(2):169–188.

Kloepper, J. W. and Schroth, M. N. 1978. Plant growth promoting rhizobacteria on radishes. In: Proceedings of the Forth International Conference on Plant Pathogenic Bacteria, *Angers France,* pp:879–882.

Kumar, R. S., Ayyadurai, N., Pandiaraja, P., Reddy, A. V., Venkateswarlu, Y., Prakash, O. and Sakthivel, N. 2005. Characterization of antifungal metabolite produced by a new strain *Pseudomonas aeruginosa* PUPa3 that exhibits broad spectrum antifungal activity and biofertilizing traits. *J Appl Microbiol* 98(1):145–154.

Li, J., Yuan, X., Dong, Z., Mugabe, W. and Shao, T. 2018. The effects of fibrolytic enzymes, cellulolytic fungi and bacteria on the fermentation characteristics, structural carbohydrates degradation, and enzymatic conversion yields of Pennisetum sinese silage. *Bioresour Technol* 264:123–130.

Meena, B., Radhajeyalakshmi, R., Marimuthu, T., Vidhyasekaran, P., Doraiswamy, S. and Velazhahan, R. 2000. Induction of pathogenesis-related proteins, phenolics and phenylalanine ammonia-lyase in groundnut by *Pseudomonas fluorescens. J Plant Dis Protect* 107 (5): 514–527.

Meena, K. R. and Kanwar, S. S. 2015. Lipopeptides as the antifungal and antibacterial agents: applications in food safety and therapeutics. *Biomed Res Int* 1–9.

Moreno, S. Z. R., Vinchira-Villarraga, D. M., Vergara-Morales, D. I., Castellanos, L., Ramos, F. A., Guarnaccia, C. and Moreno-Sarmiento, N. 2019. Plant-growth promotion and biocontrol properties of three *Streptomyces* spp. isolates to control bacterial rice pathogens. *Front Microbiol* 10:290–297.

Naik, P. R., Raman, G., Narayanan, K. B. and Sakthivel, N. 2008. Assessment of genetic and functional diversity of phosphate solubilizing fluorescent *Pseudomonads* isolated from rhizospheric soil. *BMC Microbiol* 8(1):230–244.

Newton, A. C., Johnson, S. N. and Gregory, P. J. 2011. Implications of climate change for diseases, crop yields and food security. *Euphytica* 179(1): 3–18.

Nielsen, T. H. and Sorensen, J. 2003. Production of cyclic lipopeptides by *Pseudomonas fluorescens* strains in bulk soil and in the sugar beet rhizosphere. *Appl Environ Microbiol* 69(2):861–868.

Numan, M., Bashir, S., Khan, Y., Mumtaz, R., Shinwari, Z. K., Khan, A. L. and Ahmed, A. H. 2018. Plant growth promoting bacteria as an alternative strategy for salt tolerance in plants: a review. *Microbiol Res* 209:21–32.

Oehmen, A., Lemos, P. C., Carvalho, G., Yuan, Z., Keller, J., Blackall, L. L. and Reis, M. A. 2007. Advances in enhanced biological phosphorus removal: from micro to macro scale. *Water Res* 41(11):2271–300.

Ossowicki, A., Jafra, S. and Garbeva, P. 2017. The antimicrobial volatile power of the rhizospheric isolate *Pseudomonas donghuensis* P482. *PLoS One* 12(3):e0174362.

Pal, R., Chakrabarti, K., Chakraborty, A. and Chowdhury, A. 2005. Degradation of pencycuron in soil: effect of application rate and soil conditions. *Pest Manag Sci* 61:1220–1223.

Pandey, S. and Gupta, S. 2019. ACC deaminase producing bacteria with multifarious plant growth promoting traits alleviates salinity stress in French bean (*Phaseolus vulgaris*) plants. *Front Microbiol* 10:1506–1510.

Paulin, M. M., Novinscak, A., Lanteigne, C., Gadkar, V. J. and Filion, M. 2017. Interaction between DAPG and HCN producing *Pseudomonas brassicacearum* LBUM300 and *Clavibacter michiganensis* subsp. michiganensis in the rhizosphere of tomato. *Appl Environ Microbiol* 83(13):e00073–17.

Pineda, A., Zheng, S. J., Van Loon, J. J., Pieterse, C. M. and Dicke, M. 2010. Helping plants to deal with insects: the role of beneficial soil-borne microbes. *Trends Plant Sci* 15(9):507–514.

Rajendran, L., Samiyappan, R., Raguchander, T. and Saravanakumar, D. 2007. Endophytic bacteria mediate plant resistance against cotton bollworm. *J Plant Inter* 2(1):1–10.

Ramette, A., Moënne-Loccoz, Y. and Defago, G. 2006. Genetic diversity and biocontrol potential of fluorescent *Pseudomonads* producing phloroglucinols and hydrogen cyanide from Swiss soils naturally suppressive or conducive to *Thielaviopsis basicola*-mediated black root rot of tobacco. *FEMS Microbial Ecol* 55(3):369–381.

Sharma, S. B., Sayyed, R. Z., Trivedi, M. H. and Gobi, T. A. 2013. Phosphate solubilizing microbes: sustainable approach for managing phosphorus deficiency in agricultural soils. *Springer Plus* 2(1):587–593.

Sheng, M. M., Jia, H. K., Zhang, G. Y., Zeng, L. N., Zhang, T. T., Long, Y. H. and Liu, H. M. 2020. Siderophore production by rhizosphere biological control bacteria *Brevibacillus brevis* GZDF3 of *Pinellia ternata* and its antifungal effects on *Candida albicans*. *J Microbiol Biotechnol* 30(5):689–699.

Sonawane, M. S., Chaudhary, R. D., Shouche, Y. S. and Sayyed, R. Z. 2018. Insect gut bacteria: a novel source for siderophore production. *Proc Natl Acad Sci India Sect B Biol Sci* 88(2):567–572.

Suryadi, Y., Susilowati, D. N. and Fauziah, F. 2019. Management of plant diseases by PGPR-mediated induced resistance with special reference to tea and rice crops. In: Sayyed R. (ed), *Plant Growth Promoting Rhizobacteria for Sustainable Stress Management. Microorganisms for Sustainability*, Springer, Singapore, vol. 13.

Tariq, M., Yasmin, S. and Hafeez, F. Y. 2010. Biological control of potato black scurf by rhizosphere associated bacteria. *Braz J Microbiol* 41(2):439–451.

Timmusk, S., Behers, L., Muthoni, J., Muraya, A. and Aronsson, A. C. 2017. Perspectives and challenges of microbial application for crop improvement. *Front Plant Sci* 8:49–52.

Tyagi, S., Mulla, S. I., Lee, K. J., Chae, J. C. and Shukla, P. 2018. VOCs-mediated hormonal signaling and crosstalk with plant growth promoting microbes. *Crit Rev Biotechnol* 38:1277–1296.

Vacheron, J., Desbrosses, G., Bouffaud, M. L., Touraine, B., Moënne-Loccoz, Y., Muller, D. and Prigent-Combaret, C. 2013. Plant growth-promoting rhizobacteria and root system functioning. *Front Plant Sci* 4:356–359.

Wang, J., Li, R., Zhang, H., Wei, G. and Li, Z. 2020. Beneficial bacteria activate nutrients and promote wheat growth under conditions of reduced fertilizer application. *BMC Microbiol* 20(1):38–50.

Wang, X., Li, Q., Sui, J., Zhang, J., Liu, Z., Du, J. and Liu, X. 2019. Isolation and characterization of antagonistic bacteria *Paenibacillus jamilae* HS-26 and their effects on plant growth. *Biomed Res Int* 19:1–13.

Wu, F., Li, J., Chen, Y., Zhang, L., Zhang, Y., Wang, S. and Liang, J. 2019. Effects of phosphatesolubilizingbacteria on the growth, photosynthesis, and nutrient uptake of *Camellia oleiferaabel*. *Forests* 10(4):348–358.

Zabihi, H. R., Savaghebi, G. R., Khavazi, K., Ganjali, A. and Miransari, M. 2011. Pseudomonas bacteria and phosphorous fertilization, affecting wheat (*Triticum aestivum L.*) yield and P uptake under greenhouse and field conditions. *Acta Physiologiae Plantarum* 33(1):145–152.

5

Biochar: A Strategy for Rejuvenating Soil Health and Carbon Sequestration

Mahima Begum[1], Milon Jyoti Konwar[1], and Gayatri Goswami Kandali[2]
[1]*Department of Agronomy, Assam Agricultural University, Jorhat, India*
[2]*Department of Soil Science, Assam Agricultural University, Jorhat, India*

CONTENTS

5.1 Introduction

"Biochar" is a solid carbonaceous product developed through pyrolysis from organic biomass. It is known because of its numerous beneficial effects on the physical (Major et al. 2010), chemical, and biological properties of soil (Pietikainen et al. 2000). For example, neutralization of acidic soil (Jayasanka et al. 2020), increased porosity due to the reduction in soil compaction (Laird et al. 2010), reduction in density of heavy clay soil (Obia et al. 2018), increased the available nutrient content in soils like loamy sand soil (Chan et al. 2008; Timilsina et al. 2017), reduction in strength as well penetration resistance (PR) of soil (Chan et al. 2008; Busscher et al. 2010), increased hydraulic conductivity in heavy textured soil (Brockhoff et al. 2010), and increased water retention of sandy soil (Arthur and Ahmed 2017). Due to the numerous beneficial effects of biochar on soil health, it has been considered as an important tool for rejuvenating the soil health, particularly the degraded and contaminated soils (Van Zwieten et al. 2010). With the global climate change, the soil sustainability through soil health management is the alarming issue to feed the global population. In this context, biochar is gaining attention as an important tool for rejuvenation of the soil health. But the properties of biochar differ depending on the feedstock and pyrolysis conditions and accordingly, their amelioration effect on soil also varies (Koji et al. 2017). Many researchers also reported various results on soil properties due to type of biochar and pyrolysis temperature (Table 5.1). The biochar produced from rice straw was found to be better than biochar derived from maize stover, groundnut shell, coconut shell, and coir waste in terms of pH and electrical conductivity (Jayasanka et al. 2020). In calcareous soil, significant improvement in soil quality due to the application of biochar derived from sugarcane bagasse was reported by Chen et al. (2010) and Kameyama et al. (2012). In severely degraded soil, biochar @ 30 t ha^{-1} considerably increased the cocoyam production, and also improved the soil properties by reducing soil loss (Aruna et al. 2020).

5.2 Improvement in Soil Physical Properties

The biochar is a very porous structure with high specific area (Lehmann and Joseph 2009; Van Zwieten et al. 2010) that improves the soil structure by altering the soil porosity, bulk density, pore size distribution (Major et al. 2010), and ultimately increases the water retention capacity (Herath et al. 2013). The effect of biochar on the physical properties of soil highly varies depending on the feedstock used for pyrolysis and its process characteristic (Chan et al. 2007; Mukherjee and Lal 2013), type and amount of biochar, as well as type of soil used (Aslam et al. 2014). Recently, it was found that particle size of biochar also influenced the physical properties of soil (Duarte et al. 2019), e.g. the biochar derived from poultry manure was not suitable for improving the physical properties of soil as compared to compost or raw poultry manure (Area et al. 2017). Similarly, Lehmann (2007) stated that biochar produced at 400°C is not appropriate for soil amendment because of low surface area.

5.2.1 Effect of Biochar on Bulk Density of Soil

The reduction in soil bulk density due to biochar amendment was reported by many researchers (Jones et al. 2011; Case et al. 2012; Mukherjee et al. 2014; Burrell et al. 2016). However, the extent of reduction in bulk density due to biochar highly depends on the type and amount of biochar used, including the soil type. The bulk density of soil greatly reduces with high rate of biochar due to increase in soil aeration, water retention, and porosity of soil (Leonard 2013). The combined effect of biochar @ 10 t ha^{-1} and the recommended dose of fertilizers did not show any positive effect on soil bulk density as compared to other organic amendments such as vermicompost @ 5 t ha^{-1}, farm yard manure (FYM) @ 10 t ha^{-1}, and tank silt @ 50 t ha^{-1}. The non-significant effect on soil physical properties might be due to sub-optimal or lower dose of biochar indicating that the rate of biochar is one of the main factors which determine the soil properties (Laxman Rao et al. 2017). Even the higher dose of biochar up to 15 t ha^{-1} did not influence the soil physical properties in Ultisol soil although it reduced the soil bulk density (1.38 g cm^{-3}) and mean weight diameter (2.79 mm) as compared to soil without biochar treatment (Njoku et al. 2015). In sandy loam Alfisol, biochar @ 30 t ha^{-1} reduced the soil bulk density up to 46.3% than the soil without biochar treatment (Aruna et al. 2020). In loamy sand soil, treating the soil with 20 Mg ha^{-1} of biochar significantly reduced the soil bulk density (1.235 g cm^{-3}) as compared to control (1.415 g cm^{-3}) (Timilsina et al. 2017). The reduced bulk density due to biochar application was the consequence of modification in the soil aggregate size (Tejada and Gonzalez 2007). The biochar derived from sewage and sludge significantly reduced the bulk density in loamy sand soil, but there was no difference in bulk density recorded between two higher rates of 40 t ha^{-1} and 60 t ha^{-1}, which clearly indicated that the best result could be obtained with the use of optimum dose only (You et al. 2019). As stated earlier, the effect of biochar is mainly determined by both the types of biochar and soil where it has been applied. The biochar derived from wood was found to have more pronounced effect than the straw biochar in reducing the soil bulk density (Quin et al. 2014; Burrell et al. 2016). It was also stated that the amount of reduction was more in course-textured soil than the heavy-textured soil. The straw biochar due to its low density had more pronounced effect on improving the soil aggregate stability (Liu et al. 2014) and availability of water in coarse-textured soil than the wood biochar (Burrell et al. 2016). The low density of biochar was the reason for the reduction in bulk density of soil (Ulyett et al. 2014), and biochar had greater effect on lowering the bulk density of soil in Alfisol than Andisol (Herath et al. 2013). The bulk density between biochar and sandy soil was higher than that between the biochar and clay soil, which results in more reduction in bulk density of sandy soil as compared to clay soil (Blanco-Canqui 2017). Soil degradation is a common phenomenon under intensified cropping and mechanization due to soil compaction that restricts the water storage and air circulation making the soil unproductive, and the soil compaction was found to be increased with increase in the soil bulk density (Gomez et al. 2002) resulting in a decline in soil porosity, diffusion, and other microbial actions (Hamza and Anderson 2005). But the application of biochar reduced the tensile strength of soil cores, which have the capability of reducing the soil compaction resulting in the reduction in soil bulk density (Chan et al. 2008). However, due to biochar application, the soil compaction reduced significantly by developing soil PR (Figure 5.1). The reduction in PR due to biochar is more in upper 0–40 cm soil depth (34.7%) as compared to upper 0–20 cm soil (28.34%) with the same rate of application (Hamza et al. 2019). The southeastern coastal soils generally have low carbon, poor water-holding capacities, and high PR, and addition of biochar to such soils significantly resulted in improved characters (Busscher et al. 2010).

In artificially degraded soil, the addition of biochar reduced the soil bulk density up to 13% as compared to control (Mukherjee et al. 2014). According to the study, the reduction in soil bulk density was mainly due to increased production of soil organic matter. The biochar amendment at the rate of 10 t ha^{-1} improved the soil quality of degraded soil in Southern Guinea Savanna Zone, Nigeria by increasing the soil porosity as well as by reducing the soil bulk density as a result of alteration to the size of soil particles that mainly involves increasing the silt and clay particles and reducing the sand particles (Ndor et al. 2015). In clay soil, the bulk density decreased linearly with the increase in the dose of biochar (Barnes et al. 2014). In highly weathered soil, the amendment of biochar reduced the bulk density and improved the porosity and hydraulic properties of soil (Jien and Wang 2013).

5.2.2 Effect of Biochar on Soil Porosity

Generally, the porosity and soil aggregates define the quality of soil structure, and the soil porosity includes both macropores and micropores that can be modified by biochar amendment. However, scientists have reported various results in case of increase in soil porosity. Omondi et al. (2016) reported 8.4% increase in soil porosity, while Quin et al. (2014) reported 5% increase in soil porosity due to biochar amendment, but the extent of modification greatly depends on the soil type and the kind of biochar used (Herath et al. 2013). The soil porosity is directly proportional to biochar application and increases linearly with the increase in the rate of biochar dosage (Leonard 2013). In tropical sandy loam Alfisol, addition of 30 t ha^{-1} biochar was found to be the best for increasing the maximum soil porosity (46.5%) as compared to the soil without biochar (Aruna et al. 2020). However, the enhancement of porosity percentage is mostly confined to the upper soil surface only (Ajayi et al. 2016), and the biochar, because of its very porous structure and high-specific area (Lehmann and Joseph 2009), can manipulate the structure as well as porosity of soil by influencing the soil bulk density as well as pore size distribution (Major et al. 2010). The biochar amendment reduces the soil compaction, thereby increasing the soil porosity (Laird et al. 2010). In degraded slope land soil after application with biochar, the soil porosity and aggregate size increased as a result of increase in macropores (Zeng et al. 2014), whereas in sandy loam soil, proportion of macropores increased up to 35% (Lei and Zhang 2013). In addition to this, in clay soil also, the macroporosity greatly decreased with the decrease in the rate of biochar (Obia et al. 2018). The reduction in bulk density and increase in porosity of soil are mainly due to porous structure of biochar (Nyambo et al. 2018).

5.2.3 Soil Aggregate Stability

Among the physical properties of the soil, soil aggregation is directly related to soil texture. Therefore, the effect of biochar on soil aggregation varies depending on the soil texture (Nimmo 1997). Like all other physical properties of the soil, the pyrolytic condition also plays a role in the formation of aggregate stability. The biochar derived from corn stover at 550°C more significantly increased the soil aggregate stability in Alfisol than Andisol, whereas the biochar derived at 350°C had more pronounced effects on soil aggregate stability in Andisol as compared to Alfisol (Herath et al. 2013). However, the response of biochar to soil aggregates also varies depending on the soil type and its characteristics (Hartley et al. 2016). Besides this, the formation of soil aggregate is greatly influenced by various factors such as soil organic matter, microorganisms, and minerals (Verheijen et al. 2009). In the coarse-textured planosol, the addition of straw biochar resulted in 98% increase in soil aggregate stability (Burrell et al. 2016). In silt loam soil also, biochar amendment extensively increased the soil aggregate sizes resulting in the improvement in soil health (Liu et al. 2012). In clay soil, the amendment with biochar produced from sunflower residues significantly increased the soil aggregate stability as compared to control, but the stability decreased with high dose of application (Abdulwahhab and Seker 2019), and the reason might be the reduction in soil microbial activities as a result of unfavorable conditions created by excessive dosage of biochar (Ferreiro et al. 2012). The improvement in soil aggregate stability is mainly due to augmentation of soil microbial activities, which helps in the formation of soil aggregates (Downie et al. 2009). The presence of high amount of carbon in biochar enhanced the soil aggregate stability (Alghamdi 2018). Besides this, the biochar acts as a binding agent that helps in binding of soil micro-aggregates to form new macro-aggregates (Njoku et al. 2015). The formation and stabilization of the soil macro-aggregates in the sandy loam soil due to biochar amendment was reported by Ouyang et al. (2013). In sandy loam soil, the amount of macro-aggregates formed was higher than the silty clay soil due to 2% (on dry weight basis) application of biochar derived from dairy manure (Ouyang et al. 2013). In dry land areas, irrigation management plays a vital role in determining the agricultural productivity because of the cost associated with each irrigation. However, the application of biochar derived from forest biomass waste improved the soil porosity, water retention capacity, and soil structural index (SI) of desert sandy/dry soils, which clearly indicates the improvement in soil aggregate stability and reduction in the irrigation needs in those areas (Baiamontea et al. 2019).

5.2.4 Hydrological Properties

Soil hydraulic properties are considered as an important factor for determining the soil health for crop growth and development (Cullotta et al. 2016). They mainly include water retention capacity and hydraulic conductivity of soil, and the enhancement of moisture content and infiltration rate due to biochar in sandy loam Alfisol was reported by Aruna et al. (2020). In fallow soil of sub-humid region of Kenya, continuous application of biochar for ten years improved the water-holding capacity of soil in comparison to control where no biochar was added (Katterer et al. 2019). The water retention capacity of sandy soil greatly increased due to biochar (Chen et al. 2018) and the improvement in hydraulic properties in such soil was also reported by many researchers (Barnes et al. 2014; Liu et al. 2016; Arthur and Ahmed 2017). The hydraulic conductivity in sandy soil is normally high, but the water and nutrient retention capacity is very poor, which ultimately hinders the crop growth and development. However, this problem could be alleviated by applying biochar as it reduced the saturated hydraulic conductivity and enhanced the water and nutrient retention capacity of the soil (Arthur and Ahmed 2017). In sandy soil, the wide-coarse pores greatly reduced, while in silt substrate, they greatly increased along with decline in narrow-coarse pores and slight increase in medium-coarse pores due to biochar amendment (Ajayi and Horn 2016) resulting in the improvement in water retention capacity of both the soils. The rearrangement of soil particles due to biochar application increased the soil aggregate stability as well as proportion macropores that facilitate the permeability of the soil (An et al. 2008). In degraded slope land soil, the available water content significantly increased due to biochar amendment. This happened mainly because of increased proportion of soil micropores (Zeng et al. 2014). The alteration to the pore-size distribution of soil due to biochar resulted in the increase in water permeability (Asai et al. 2009). The improvement in soil water movement is mostly feasible in coarse-textured soils or soils having a large number of macropores (Glaser et al. 2002) as the macropores help in the movement of water in the soil (Obia et al. 2018). The biochar derived from pine and oak also increased the available water content in sandy soil, whereas it was decreased in clay soil, and remain unaffected in loamy soil suggesting that the biochar has positive effect on coarse-textured soil in case of soil hydraulic property (Tryon 1948). In silty clay soil, the saturated water content on an average increased by 2.2% as compared to 7.4% in sandy soil, which implied that the biochar was more effective in sandy loam soil than the silty clay soil (Ouyang et al. 2013). Besides this, even in heavy clay soil, the application of 3% bagasse biochar registered 39% increase in the available soil moisture content (Chen et al. 2010). In heavy soil, poor drainage and aeration are the major problems which restrict the crop growth and development. In heavy clay soil, the porosity and moisture content at field capacity and wilting point were unaffected due to biochar treatment (Obia et al. 2018). However, with higher rate of application, the wood-derived biochar increased the saturated hydraulic conductivity in clay soil (Barnes et al. 2014). In silt loam soil, the hydraulic conductivity also increased due to the application of biochar @16 Mg ha^{-1}(Asai et al. 2009). The enhancement of saturated hydraulic conductivity due to biochar was reported by Major et al. (2010). The biochar increased the soil water-holding capacity at lower tension in Alfisol because of better drainage facility in poorly drained soil (Herath et al. 2013). However, in bare soil, the water content of soil remains unaffected due to biochar which might be due to well-structured tested soil (Ventura et al. 2013). The improvement in soil particles, more specifically the medium and fine particles, helps in water retention, which indicates that the biochar can be considered as a tool for moisture stress management under the changing climatic conditions. The boosting of water content of soil due to biochar is mainly because of the presence of large structure pore volume which is highly porous in nature (Joseph et al. 2013), and the porous structure created by tubular pores resemble a semi-permeable membrane which helps in the retention of water in sandy soil (Carvalho et al. 2014). Besides this, the particle size of biochar is the key factor that determines the physical and hydraulic properties of soil (Duarte et al. 2019), and with the increase in the particle size of biochar from 0.5 mm to 2 mm, water content and water retention capacity of sandy soil are greatly reduced (Głab et al. 2016). Increasing the size of biochar particles reduces the hydraulic conductivity of the sandy soils especially under dry climatic conditions but might remained unchanged, increase or decrease under wet conditions (Chen et al. 2018). The high specific surface area of the small biochar particle enhanced the water retention capacity of the soil (Blanco-Canqui 2017) because of increased volume of macropores and mesopores (Duarte et al. 2019); however, the specific surface area of biochar is inversely proportional to the particle size.

5.3 Improvement in Soil Chemical Properties

In many parts of the world, the abiotic stresses such as drought, high salinity, and extreme temperatures are the most vital factors that hinder the proper growth and development of crop (Thalmann and Santelia 2017). Besides them, soil contaminated with heavy metals due to various anthropogenic activities affects the plant growth and development, and ultimately, reduces the crop yield (Moon et al. 2013). The soil strength is directly related to nutrient availability and plants require soil nutrients like nitrogen (N), phosphorus (P), and potassium (K) for their growth, but the levels of the nutrient may decrease over time after crop harvesting. In many parts of India, the soils are deficient in both macronutrients and micronutrients (Pathak 2010). Thus, to fulfill the gap between the nutrient demand and supply, huge quantities of chemical fertilizers are added to the soil which deteriorate the environment causing infinite problems. Besides this, the prolong use of both chemical fertilizers and pesticides reduces the nutrient composition of the crops and degrades the soil fertility due to accumulation of some heavy metals (Yargholi and Azarneshan 2014). Now-a-days, biochar is considered as the most significant alternative for remediation of heavy metal contaminated soils to decrease the bioavailability of heavy metals to plants (Lahori et al. 2017). Biochar is a carbon-rich organic material and a by-product produced from biomass by pyrolysis under high-temperature and low-oxygen conditions. It is obtained from the carbonization of organic wastes which can not only influence the sequestration of soil carbon but also modify its physicochemical and biological properties (Garcia et al. 2016; Zhang et al. 2017). In case of chemical properties, the addition of biochar significantly increased the pH (Jatav et al. 2017) and cation exchange capacity (CEC) of soil (Nigussie et al. 2012). Ajema (2018) reported the improvement in soil quality due to biochar amendment because of increase in soil pH, ability to retain moisture, and CEC. It has shown a greater ability to adsorb cations per unit carbon as compared to other soil organic matters because of its greater surface area, negative surface charge, and charge density (Liang et al. 2006), thereby offering the possibility of improving yields (Lehmann et al. 2006). The excessive accumulation of lead deteriorates both quality and fertility of soil; therefore, the lead metal (Pb) remediation is a challenging task all over the world. In this context, Dume et al. (2016) reported that the addition of biochar to the military camp soil not only improved the physio-chemical properties of soil but also stabilized the Pb content of soil. Earlier, Verheijen et al. (2009) reported that the biochar application affected the toxicity, transport, and fate of various heavy metals in the soil due to improved soil absorption capacity. It has been also reported that biochar reduces soil acidity by increasing the pH and helps the soil to retain nutrients and fertilizers (Lehmann 2007). The application of biochar improves soil fertility through two mechanisms: (1) adding of nutrients to the soil (such as K, to a limited extent P, and many micronutrients) and (2) retaining nutrients from other sources, including nutrients from the soil itself. Biochar increases the availability of C, N, Ca, Mg, K, and P to plants because it adsorbs and slowly releases fertilizers (DeLuca et al. 2015). Due to its high surface area, CEC and stability, biochar helps in improving water adsorption and nutrient retention in the soil (Ding et al. 2017). Its surface can exhibit hydrophilic, hydrophobic, acidic, and basic properties; hence, biochar can adsorb solutes from soil solutions, affecting both micro- and macroelement retention (Uchimiya et al. 2011; Yao et al. 2012). Many authors have reported that one of the notable effects of biochar addition to the soil is an increase in organic carbon (Utomo 2010; Sukaronto et al. 2011).

5.3.1 How Will Biochar Improve the Chemical Properties of Soil?

Biochar improve the chemical properties of soil because of its ability to adsorb soluble organic matter and inorganic nutrients (Thies and Rillig 2009). Lehmann and Rondon (2006) reported that biochar can adsorb both NH_4^+ and NH_3^- from the soil solution. Biochar is very efficient at adsorbing dissolved solute nutrients such as ammonium (Lehmann et al. 2002), nitrate (Mizuta et al. 2004), phosphate (Beaton et al. 1960), and other ionic solutes (Radovic et al. 2001). Jia et al. (2015) reported that biochar, because of its high surface area and porosity, enables the absorption of water as well as NH_4^+ and NH_3^- ions, which enhances the metabolic and physiological activities of microorganisms. The addition of biochar to Alfisol significantly improved the soil quality by increasing the soil pH, organic carbon, and exchangeable Na, K, and Ca and decreasing the amount of exchangeable Al (Chan et al. 2007). The presence of cation exchange sites on the surface of biochar is responsible for increasing the quantity of K, Ca, Mg, and CEC in biochar amended soil (Sohi et al. 2010; Jones et al. 2012). Moreover, the high amount of cations in biochar amended soil might be due to the presence of carboxyl group which is indicated by high oxygen and carbon ratios on the surface of the biochar after microbial degradation (Liang et al. 2006; Preston and Schmidt 2006).

5.3.2 Improvement in Nutrient Status of Soil

In acidic soil, to overcome the negative effect of low pH on soil nutrient availability and crop growth, generally lime is broadly used (Zhang et al. 2017), but due to certain drawbacks, the lime has been replaced by biochar as it could also increase the soil pH by increasing buffering capacity of soil (Yuan and Xu 2011), and ultimately, the crop growth (Zhang et al. 2016). Like the soil acidity, the salt accumulation is also considered as a major threat that hinders crop productivity in most of the arable croplands worldwide (Mahmoodabadi et al. 2013). All over the world, a considerable portion of agricultural land is affected by salt, which reduces the nutrient availability to the crop, and ultimately, affects the crop growth (Fageria et al. 2011). Various factors, including low-organic matter input from plant biomass (Lakhdar et al. 2009) and huge loss of organic matter mainly from sodic soil (Wong et al. 2010), are responsible for low-nutrient status of salt-affected soils. But addition of high rate of charcoal greatly increased the nutrients uptake by plants and also reduced the leaching of nutrients such as N and Ca (Lehmann et al. 2003). Therefore, the addition of biochar to salt-affected soils not only improves the soil fertility status but also enhances the nutrient uptake by plants grown in such problematic soil (Chan et al. 2008; Joseph et al. 2010). Lashari et al. (2013) also reported a remarkable increase in soil nutrient content, especially in salt-affected soils, due to the biochar amended. The availability of nutrient like P is highly dependent on the pH of soil and organic matter content, and it decreases with the increase in the soil (pH 7) (Naidu et al. 1991). The high pH in salt-affected soils and low organic matter decrease the availability of P, which restricts the proper plant growth and development in such soil (Elgharably 2008). The addition of biochar derived from poultry manure compost followed by pyroligneous solution in salt-affected Entisol improved the soil quality as well as increased the crop yield by decreasing the soil salinity, pH, and bulk density while increasing the availability of P and soil organic carbon (Lashari et al. 2013). The increase in the population of some beneficial microbes like phosphate-solubilizing bacteria helps in higher availability and release of P to the crop (Liu et al. 2017). In salt-affected soil, the excessive amount of Na in soil solution reduces the K availability to the plants (Cakmak 2005). Therefore, it is always necessary to maintain an optimum amount of soil K in such salt-affected soils (Wakeel 2013) for proper growth and development. The accumulation of salt is one of the critical factors responsible for soil degradation, which increased the uptake of trace element and decreased the uptake of essential elements by crop. But the addition of biochar to such degraded soils reduced the bioavailability of trace element and increased the essential elements available to plants (Abbas et al. 2017). However, the type of biochar and properties of soil can play an important role in altering K availability (Taghavimehr 2015).

5.3.3 Moderation in Soil pH, Electrical Conductivity (EC), Sodium Adsorption Ratio (SAR), and Exchangeable Sodium Percentage (ESP)

The crop growth and nutrient availability are much dependent on the soil pH. Biochar is known for its capacity to increase the soil pH (Laird et al. 2010), especially in soil having low pH. Liu et al. (2017) found that the application of rice husk biochar in acid red soil, organic brown soil, and saline soil significantly modified the soil pH in all these three types of soil. The addition of biochar decreased the soil pH of saline and sodic soil whose initial pH was high, and reverse happened in case of soil having low pH (Abrishamkesh et al. 2015; Liu et al. 2017). Besides this, many researchers reported a significant decrease in soil pH in salt-affected soils with the application of biochar (Chaganti and Crohn 2015; Khalifa and Yousef 2015). The mechanisms responsible for pH reduction are elusive because of high pH of saline-sodic and sodic soils and are primarily associated with high ESP) (Na salts) (Shaygan et al. 2017). It was also stated that decrease in soil ESP due to addition of biochar might have reduced the soil pH in such soil (Lashari et al. 2015). However, the modification in soil pH also depends on the pH of biochar. The addition of acidic biochar having low pH reported a remarkable decrease in pH of salt-affected soil as compared to alkaline biochar having high pH (Amini 2015). Therefore, the difference between the pH values of biochar and soil might be responsible for alteration to soil pH (Liu and Zhang 2012). The acidity or basicity of char sample was mainly due to its different surface polar-group contents (Chun et al. 2004). Serkalem (2015) also reported a significant reduction in pH of saline soil after biochar application because of augmentation in acidic functional groups released during oxidation of biochar. Besides this, the decomposition of organic matter present in soil and biochar may also contribute toward reduction in soil pH (Tang and Rengel 2003). Another reason for the decrease in pH may be the relative paucity of the hydroxyl (OH^-) ions. Even though biochar contains carbonates and soluble base cations (van Zwieten et al. 2010; Yuan and Xu 2011), sparingly soluble carbonates may form in some soils, restricting the hydrolyzation of carbonates and decreasing the content of hydroxyl in the soil. Thus, the soil pH decreases after the addition of biochar. The accumulation of excessive salt in saline soil is the major thread for crop production and that could be ameliorated by adopting proper cropping as well as leaching method (Qadir et al. 2000). Chhabra (2004) stated that adopting proper strategy for leaching of soluble salts and application amendments for reduction in ESP are the most suitable options in case of saline-sodic and sodic soils. Studies have also demonstrated that biochar can also improve leaching of soluble salts to decrease soil EC (Lashari et al. 2015). In a recent study, Yue et al. (2016) reported that the addition of biochar derived from sunflower straw reduced more amounts of sodium ions (Na^+) and bicarbonate ions (HCO_3^-) in saline soil as compared to biochar derived from cow manure and rice straw. Chaganti and Crohn (2015) also reported that irrespective of source biochar, biosolids compost, and green waste compost considerably improved all the physio-chemical properties of saline-sodic soil when leached with reclaimed water due to improvement in soil hydraulic conductivity that facilitates leaching of salts from such soil.

5.3.4 Biochar and Soil Degradation

In agricultural systems, conservation of soil organic carbon has been recognized as a strategy to reduce the increasing soil degradation. The addition of different source and dose of organic matter significantly increased the soil aggregate stability and nutrient availability, irrespective of their source and rate of application (Karami et al. 2012). Similarly, like other organic sources, biochar plays a significant role in rejuvenating the degraded soil, and in acidic soil, the biochar amendment increased the amount of dissolved organic matter by increasing the soil pH (Smebye et al. 2016). It was also stated that biochar amendment in degraded soil has the potentiality to improve the soil fertility and crop productivity by improving the soil organic carbon (Tang and Rengel 2003). So, the use of biochar to improve the chemical properties of soil is the need of the hour in order to attain sustainability in agricultural sector.

5.4 Improvement in Biological Properties of Soil

Soil health depends on the varied population and different kinds of soil microbes. Continuous cultivation practices by modern agricultural methods lead to deleterious effect on soil microbial population and enzymes, which play a crucial role in maintaining the properties and health of soil. To mitigate this effect, ecologically safe, and user-friendly feasible alternative is biochar.

5.4.1 Soil Microflora

The presence of pores in between the biochar particles due to their large surface area makes a congenial environment for the shelter of microorganisms (Shoi et al. 2009). This helps in adsorbing and increasing amount of soil organic matter resulting in good soil microbial population, which helps in reducing the loss of soil N, increasing the amount of nutrients accessible for the plants (Winsley 2007), and favoring colonization in roots by mycorrhizal fungi (Warnock et al. 2007). It was also reported that biochar increased arbuscular mycorrhizal fungi (AMF)-mediated host plant resistance against *Fusarium* (Matsubara et al. 2002). Many studies reported that there is a change in soil microbial population due to biochar amendment. Steinbeiss et al. (2009) reported that irrespective of forest and arable soil type, biochar derived from yeast favored the growth of fungi compared to biochar derived from glucose which has no significant effect on soil microorganisms. Application of biochar increased the biological nitrogen fixation in *Phaseolus vulgaris* (Rondon et al. 2007), and leaching of NH_4 could be efficiently checked by the application of biochar (Lehmann et al. 2003). Mycorrhizal growth increased significantly in clover bioassay plants by the use of biochar which probably created a congenial environment for root colonization (Solaiman et al. 2010). Application of rice straw biochar revealed that the probacteria, acidobacteria, zygomycota, and basidomycota increased with the increase in proportion of biochar amount and promoted fungal community diversity in tobacco-planting soils (Gao et al. 2017). Biochar improves soil acidity due to the change in pH from acidic to neutral (Zhang et al. 2019), which brings about a congenial environment for bacterial group of soil organisms (Rousk et al. 2010) as acidic environment is inhibitory for bacterial growth. A remarkable improvement in the growth of carrot and legumes in an acidic soil by the use of biochar was also reported by Rondon et al. (2007). Biochar acts as a store house of soil nutrients in acidic soil due to favorable soil pH and active functioning of bacteria (Jeffery et al. 2011). The role of bacteria in biochar amended acidic soil is manifold. Some reported that bacteria belonging to the family *Oxalobacteraceae* are involved in soil carbon sulfur cycles and nitrogen fixation (Favet et al. 2013; Jia et al. 2018). Along with chemical fertilizer, biochar can increase the concentration of bacteria belonging to the family *Geobacteraceae*, *Comamonadaceae*, and *Chitinophagaceae* which have great roles in degeneration of plant residues and nutrient cycling (Zhang et al. 2017). A notable relationship exists among pH, soil organic matter, acid phosphatase enzyme, and bacterial population in acidic soil (Zhang et al. 2019), which is contributed by the high carbon content of biochar and provides food for soil microbes (Cookson et al. 2005; Zhang et al. 2019). Application of biochar in combination with fertilizer has more pronounced effect on population of bacteria in acidic condition as compared to alkaline soil. Biochar along with fertilizer increased the population of some beneficial bacteria of the family

Oxalobacteraceae through cycling of nutrients (Zhang et al. 2019). Studies over the last few decades showed that addition of biochar can increase soil microbial population and their activities (Li et al. 2018). Increase in soil microbial activity by the addition of biochar ultimately helps in improving soil quality and crop productivity (Palansooriya et al. 2019). Tropical soils are often poor and acidic in large part due to heavy rainfall that leaches the nutrients and organic materials out of the soil. In general, tropical soils have low organic matter, CEC, and nutrient content which can be improved by the addition of biochar (Ferreiro et al. 2015). Besides improving physical and chemical properties, biochar also enhances the biological properties of the soil. Increase in enzymatic activity of tropical soils by addition of biochar has been described by Ferreiro et al. (2015). They also reported higher microbial biomass with earthworm than without earthworm which was explained by the fact that the earthworms have a stronger effect on soils amended with biochar. Soil respiration, which is an indicator of the level of microbial activity, has been found to increase with biochar application. With the increase in earthworm population, arbuscular mycorrhiza (AM), actinomycetes, and fungal to bacterial biomass increased, which implied that both biochar and earthworm can alter soil microbial community in inherently poor acidic tropical soils. A study of interaction between earthworms and biochar was carried out by Van Zwieten et al. (2010) and they reported that the earthworms ingest not only soils but also biochar which can alter the distribution of the soil profile. Biochar has proved to be an efficient soil modification method that increases the microbial activity in alkaline soils which are poor in organic carbon and available N and P (Devika et al. 2018). There was significant increase in bacterial and fungal population in treatments with biochar over control at all the stages of growth of sweet corn (Devika et al. 2018). Increase in bacterial population might have been due to the increase in volatile material in biochar which was also confirmed by the findings of Ouyang et al. (2014). The increase in fungal population was due to the adaptability of fungi to altered environmental condition and fungal capacity to degrade biochar and sporulate inside (Watzinger et al. 2014). However, activity of actinomycetes could not show satisfactory result, as the resistant nature of carbon in the biochar failed to enhance actinomycetes population (Devika et al. 2018).

5.4.2 Soil Enzymes

Soil enzymes play a vital role in many important processes in soils such as mineralization of soil organic matter (SOM) and recycling of nutrients (Makoi and Ndakidemi 2008). Biochar has large specific surface area, which increases the adsorptive capacity of biochar to adsorb labile substrates, which in turn increase the soil enzyme activity (Lammirato et al. 2011). A wide range of soil enzymes has been described in studies relating to biochar application, including enzymes such as oxidases and hydrolases. These enzymes help in decomposition of various substances present in soil organic matter (Sinsabaugh 2010) and their performances are essentially influenced by addition of biochar (Song et al. 2018). The presence of certain labile substances in biochar increases the microbial and enzymatic activity and makes biochar a good amendment to reclaim and restore infertile degraded soil and is more effective in grassland as compared to fallow land (Cybulak et al. 2019). Several studies have also contradictory views regarding biochar application on soil enzyme activities such as Pukalchik et al. (2018) reported a remarkable increase in soil enzymes on application of biochar; however, Zheng et al. (2016) and Benavente et al. (2018) were of the opinion that biochar decreased soil enzyme activity and suggested that the type of substrate, pyrolysis condition, and rate of biochar application greatly affect microbial and enzymatic activities (Singh et al. 2010; Gul et al. 2015). Biochar produced at low temperature (< 350°C) increased dehydrogenase activity; however, Moeskops et al. (2010) reported that pyrolysis of biochar at low temperature activates the production of dehydrogenase enzyme and increases the metabolic activity of the microorganisms. Segun (2019) studied the effect of different rates of rice husk biochar (0, 3, 6, and 12 t ha^{-1}) application on soil enzymes, viz. catalase, urease, alkaline phosphatase, and invertase, on rain-fed upland rice and found that all of the enzymes and rhizospheric secretions which take part in N, P, and C cycle increased significantly ($P < 0.05$) when biochar was applied at the rate of 12 t ha^{-1}. In biochar treated plot, the urease activity increased over control with a reduction in acid phosphatase activity which might be due to higher pH of the biochar (Devika et al. 2018). However, biochar increased acid and alkaline phosphatase activity which may be due to enhanced root activity in the crop releasing extracellular enzyme (Devika et al. 2018). Use of biochar in soil of arid alkaline region poor in organic carbon and nutrients has increased microbial activity, biomass carbon, urease, and dehydrogenase activity (Muhammad et al. 2016). Biochar amended soil increases soil respiration and labile substrate which helps in absorbing the salts from the soil, builds up available nutrients, improves the physical condition of the soil, and makes a favorable environment for the microbial community to overcome the negative effect of salinity (Singh et al. 2010).

5.4.3 Soil Pathogens and Parasites

Biochar has also been reported to be very effective against certain pathogens and parasites. The synergistic effect of biochar and mycorrhizal fungi helps to protect the fungal hyphae from predators present in soil such as nematodes, collembolan, mites, other protozoans by colonizing inside the biochar particles (Saito 1990). Biochar made of ground hard wood ash decreased the root lesion in asparagus caused by *Fusarium oxysporum*, *F. asparagi*, and *F. proliferatum* (Elmer and Joseph 2011). Biochar has the ability to control air-borne pathogens like *Botrytis cinerea* and different species of powdery mildew and soil-borne pathogens like *Rhizoctonia solani*, and species of *Fusarium* and *Phytophthora* (Elmer et al. 2010; Bonanomi et al. 2015). Biochar application at the rate of 0.32%, 1.60%, and 3.20% (w/w) effectively controlled *Fusarium* root rot disease, and also increased the biomass of asparagus plant (Elmer et al. 2010). Thies and Rillig (2009) found drastic reduction in *Fusarium* root rot disease in asparagus when biochar was inoculated with mycorrhizal fungi.

5.5 Biochar in Carbon Sequestration

Carbon sequestration refers to long-term storage of CO_2 or other forms of carbon to either mitigate or defer global warming and avoid dangerous climate change. Accumulation of greenhouse gases (GHGs) by burning of fossil fuels is reduced as a result of carbon sequestration (Sedjo and Sohngen 2012). The intensified cropping system to meet the food demand of rising world population leads to the removal of huge amount of organic carbon resulting in soil degradation. In order to overcome the soil health-related issue, the application of carbonized materials or biochars has long-term stability (Glaser et al. 2002).

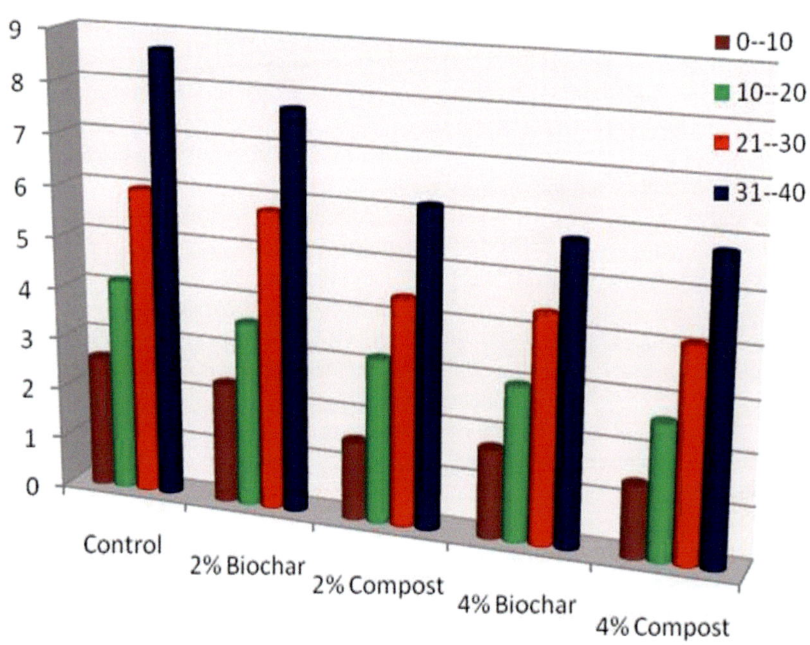

FIGURE 5.1 Effects of organic amendment on penetration resistance of soil (MPa) in 0–40 cm soil depth (Hamza et al. 2019).

TABLE 5.1

Effect of Type of Biochar and Pyrolysis Temperature on Soil Health (Gai et al. 2014)

Type of Biochar	Pyrolysis Temperature	Effect on Soil	Reference
Biochar derived from poultry litter	450°C	Enhancement in soil pH, available P, reduction in soil strength	Chan et al. 2008
Biochar derived from oak (*Quercus* spp.) and hickory (*Carya* spp.)	450–500°C	Reduction in N_2O in fine loamy Clarion soil	Rogovska et al. 2011
Biochar derived from swine manure and willow wood	350°C and 700°C	Increased the microbial biomass carbon in sandy soil	Ameloot et al. 2013
Biochar derived from wheat-straw, corn-straw, and peanut-shell	400°C and 500°C	Improved the cation exchange capacity of soil	Gai et al. 2014
Biochar derived from rice hull	400°C	Reduced the bulk density in mudstone slopeland soil	Zeng et al. 2014
Biochar derived from apple tree branches	500°C	Increased soil pH, Cu, Mn, and Fe	Zhao et al. 2017
Biochar derived from wheat (*Triticum aestivum* L.) straw	450°C and 600°C	Increased the water retention capacity of soil at field capacity or under dry conditions in sandy soil	Chen et al. 2018

Besides soil amendment, biochar is also known because of its role in long-term carbon sequestration (Dover 2007; Lal 2009). Biochar is a high carbon-rich material which has the capacity to sequester carbon and counter-balance the effect of change in climate and retain the fertility of the soil. The recalcitrant nature and high carbon content of biochar makes it a suitable substrate for carbon sequestration (Gupta et al. 2020). The high carbon content of biochar helps to break down and sequester carbon in soil for many years (Lehmann 2007) as the half-life of carbon in charcoal is more than 1000 years (Glaser et al. 2002). Soils of agro-ecosystem have a global potential to sequester about 2.1 billion tons carbon per year (Lal 2010). Yadav et al. (2017) reported that soils with relatively lower organic carbon have the capability to sequester more organic carbon than soils with higher organic carbon. Though there are a number of means for carbon sequestration in agro-ecosystem (Lal 2010), biochar produced by pyrolysis at a low temperature and under anaerobic condition (Zimmerman 2010) is considered as a suitable modern approach for carbon sequestration. Lehmann et al. (2006) reported that carbon sequestration depends on the type of feedstock used and approximately 50%–80% of carbon is sequestered from feedstock biomass. The carbon content of forest soil was increased from 17% to 41% by the addition of biochar, and so biochar can be used for climate change mitigation in forests by sequestering carbon in the soil (Jessica et al. 2019). According to Lal (2010), two points need consideration for carbon sequestration under wider agro-climatic situations. The first and the most important point is that farmers should be convinced to grow crops that show high efficiency in photosynthesis as the plants have to be charred to produce biochar. The photosynthetically efficient plant biomass has a high capacity to emit its carbon as it decomposes very fast. Second consideration is how long carbon can reside in the soil as biochar to sequester carbon and emit GHG from the soil system to reduce the bad effect of climate change (Yadav et al. 2017). Higher the stability of the charred biochar product, greater is the potential for carbon sequestration. According to Luo et al. (2011), production of biochar by high pyrolysis temperature can increase the stability of biochar. To obtain maximum possible benefit of carbon sequestration through addition of biochar, stability of biochar and yield of the crop should be optimized (Yadav et al. 2017). Besides this, the biochar produced from various organic materials such as wood and grasses, paper sludge, manure, or green wastes reduced the GHG emissions mainly N_2O

and CH_4 (Yanai et al. 2007). It was reported that in paddy fields, the addition of biochar significantly decreased the CH_4 emissions (Liang et al. 2014). The addition of biochar has the capability to sequester up to 12% of GHG emission in a system which is economically and ecologically sustainable (Woolf et al. 2010). Therefore, biochar is recognized as a climate change mitigation technology at a significant global scale (Molina et al. 2009).

5.6 Biochar Application and Crop Yield

The rate of biochar application depends mainly on the nature and characteristics of the source of biochar material, especially the acidity or alkalinity and ash content of the material. Biochar applied at the rate of 5–50 t ha^{-1} along with recommended dose of nutrients had a good effect on the yield of crops. However, Rondon et al. (2007) found a negative effect on application of biochar at the rate of 165 t ha^{-1} without proper nutrient management. Biochar derived from peanut hull and pine chip when applied at the rate of 11–22 t ha^{-1} reduced corn yields even beyond control (Gaskin et al. 2010). So, both the source of biochar and the crop to be applied should be considered before fixing the dose of biochar. Biochar applied at the rate of 90 kg ha^{-1} increased nitrogen fixation in *P. vulgaris* from 50% to 72% along with a significant increase in bean yield and production of biomass (Rondon et al. 2007). Biochar applied at the rate of 20 t ha^{-1} as a single application to a Colombian Savanna soil increased maize yield by 28–140% (Major et al. 2010). Application of biochar at 3–6 t ha^{-1} increased rice grain yield by 46% and straw yield by 47% (Segun 2019). Biochar along with phosphate solubilizing fungi increased yield of *Vigna radiata* and *Glycine max* (Saxena et al. 2017).

5.7 Conclusion

The application of biochar is considered as one of the potential strategies for rejuvenating the soil health by improving the physico-chemical and biological properties of the soil besides enhancing carbon sequestration for mitigating climate change. Thus, the use of biochar has the potentiality to rejuvenate soil health of degraded soil by making it productive. Moreover, it also reduces the nutrients leaching, enhances soil quality in highly weathered tropical soils, and helps in rehabilitation of degraded land.

REFERENCES

Abbas, T., Rizwan, M., Ali, S., Adrees, M., Zia-ur-Rehman, M., Qayyu, M. F., Ok, Y. S. and Murtaza, G. 2017. Effect of biochar on alleviation of cadmium toxicity in wheat (*Triticum aestivum* L.) grown on Cd-contaminated saline soil. *Environ Sci Pollut Res* 25(26):25668–25680.

Abdulwahhab, Q. R. and Seker, C. 2019. Effect of biochar applications on soil aggregation status, Conference paper, International Soil Congress. 17–19 June.

Abrishamkesh, S., Gorji, M. and Asadi, H. 2015. Effects of rice husk biochar application on the properties of alkaline soil and lentil growth. *Plant Soil Environ* 61:475–482.

Ajayi, A. E. and Horn, R. 2016. Modification of chemical and hydrophysical properties of two texturally differentiated soils due to varying magnitudes of added biochar. *Soil Till Res* 164:34–44.

Ajayi, A. E., Holthusen, D. and Horn, R. 2016. Changes in microstructural behaviour and hydraulic functions of biochar amended soils. *Soil Till Res* 155:166–175.

Ajema, L. 2018. Effects of biochar application on beneficial soil organism review. *In J Res Stud Sci Eng Technol* 5:9–18.

Alghamdi, A. G. 2018. Biochar as a potential soil additive for improving soil physical properties – a review. *Arab J Geosci* 11(24):766.

Ameloot, N., DeNeve, S., Jegajeevagan, K., Yildiz, G., Buchan, D., Funkuin, Y. N. and Sleutel, S. 2013. Short-term CO_2 and N_2O emissions and microbial properties of biochar amended sandy loam soils. *Soil Biol Biochem* 57:401–410.

Amini, S. 2015. Carbon dynamics in salt-affected soils (doctoral dissertation). Griffith School of Environment. Australia.

An, S. S., Huang, Y. M., Zheng, F. L. and Yang, J. G. 2008. Aggregate characteristics during natural revegetation on the loess plateau. *Pedosphere* 18:809–816.

Area, K. S., Adelana, A. O., Fademi, O. O. and Ainab, O. 2017. Improving physical properties of degraded soil: potential of poultry manure and biochar. *Agric Nat Res* 51:454–462.

Arthur, E. and Ahmed, F. 2017. Rice straw biochar affects water retention and air movement in a sand-textured tropical soil. *Arch Agron Soil Sci* 63:2035–2047.

Aruna, O. A., Taiwo, M. A., Adeniyi, O., Wutem, S. E., Timothy, A., Titilayo, T. A. and Jerry, F. A. 2020. Effect of biochar on soil properties, soil loss, and cocoyam yield on a tropical sandy loam alfisol. *Hindawi Sci World J* Article ID:9391630" 1–9.

Asai, H., Samson, B. K., Stephan, H. M., Songyikhangsuthor, K., Homma, K., Kiyon, Y., Inoue, Y., Shiraiwa, T. and Horie, T. 2009. Biochar amendment techniques for upland rice production in Northern Laos 1. Soil physical properties, leaf SPAD and grain yield. *Field Crops Res* 111:81–84.

Aslam, Z., Muhammad, K. and Muhammad, A. 2014. Impact of biochar on soil physical properties. *J Agric Sci* 4(5):280–284.

Baiamontea, G., Crescimannoa, G., Parrinob, F. and Pasquale, C. D. 2019. Effect of biochar on the physical and structural properties of a desert sandy soil. *Catena* 175:294–303.

Barnes, R. T., Gallagher, M. E., Masiello, C. A., Liu, Z. and Dugan, B. 2014. Biochar-induced changes in soil hydraulic conductivity and dissolved nutrient fluxes constrained by laboratory experiments. *PLoS One* 9(9):e108340.

Beaton, J. D., Peterson, H. B. and Bauer, N. 1960. Some aspect of phosphate adsorption by charcoal. *Soil Sci Soc Am J* 24:340–346.

Benavente, I., Gascó, G., Plaza, C., Paz-Ferreiro, J. and Méndez, A. 2018. Choice of pyrolysis parameters for urban wastes affects soil enzymes and plant germination in a Mediterranean soil. *Sci Total Environ* 634:1308–1314.

Blanco-Canqui, H. 2017. Biochar and soil physical properties. *Soil Sci Soc Am J* 81:687–711.

Bonanomi, G., Ippolito, F. and Scala, F. 2015. Biochar as a new soil amendment for controlling plant diseases. *J Plant Pathol* 97(2):223–234.

Brockhoff, S. R., Christian, N. E., Killorn, R. J., Horton, R. and Davis, D. D. 2010. Physical and mineral-nutrition properties of sand-base turf grass root zones amended with biochar. *Agron J* 102:1627–1631.

Burrell, L. D., Zehetner, F., Rampazzo, N., Wimmer, B. and Soja, G. 2016. Long-term effects of biochar on soil physical properties. *Geoderma* 282:96–102.

Busscher, W. J., Novak, J. M., Evans, D. E., Watts, D. W., Niandou, M. A. S. and Ahmendna, M. 2010. Influence of pecan biochar on physical properties of a Norfolk loamy sand. *Soil Sci* 175:10–14.

Cakmak, I. 2005. The role of potassium in alleviating detrimental effects of abiotic stresses in plants. *J Plant Nutr Sci* 168:521–530.

Carvalho, M. A., Holanda Nunes, M., Madari, B. E., Bastiaans, L., Van Oort, P. A. J., Heinemann, B., Silva, M., Petter, A. and Meink, F. A. 2014. Biochar increases plant available water in a sandy soil under an aerobic rice cropping system. *Solid Earth Discuss* 6:887–917.

Case, S. D. C., McNamara, N. P., Reay, D. S. and Whitaker, J. 2012. The effect of biochar addition on N_2O and CO_2 emissions from a sandy loam soil – the role of soil aeration. *Soil Biol Biochem* 51:125–134.

Chhabra, R. 2004. Classification of salt affected soils. *Arid Land Res Manag* 19(1):61–79.

Chaganti, V. N. and Crohn, D. M. 2015. Evaluating the relative contribution of physiochemical and biological factors in ameliorating a saline–sodic soil amended with composts and biochar and leached with reclaimed water. *Geoderma* 259:45–55.

Chan, K. Y., VanZwieten, L., Meszaros, I., Downie, A. and Joseph, S. 2007. Agronomic values of green waste biochar as a soil amendment. *Aust J Soil Res* 45:629–634.

Chan, K., Van Zwieten, L., Meszaros, I., Downie, A. and Joseph, S. 2008. Using poultry litter biochars as soil amendments. *Aust J Soil Res* 46(5):437–444.

Chen, Y., Shinogi, Y and Taira, M. 2010. Influence of biochar use on sugarcane growth, soil parameters, and groundwater quality. *Aust J Soil Res* 48(7):526–530

Chen, C., Wang, R., Shang, J., Liu, K., Irshad, M. K., Hu, K. and Arthur, E. 2018. Effect of biochar application on hydraulic properties of sandy soil under dry and wet conditions. *Vadose Zone J* 17(1):1–8.

Chun, Y., Sheng, G., Chiou, G. T. and Xing, B. 2004. Compositions and sportive properties of crop residue-derived chars. *Environ Sci Technol* 38:4649–4655.

Cookson, W. R., Abaye, D. A., Marschner, P., Murphy, D. V., Stockdale, E. A. and Goulding, K. W. T. 2005. The contribution of soil organic matter fractions to carbon and nitrogen mineralization and microbial community size and structure. *Soil Biol Biochem* 37:1726–1737.

Cullotta, S. V., Bagarello, G., Baiamontea, G., Gugliuzzab, M., Iovinoc, D. S. and MelaVecac, L. 2016. Comparing different methods to determine soil physical quality in a Mediterranean forest and pasture land. *Soil Sci Soc Am J* 80:1038–1056.

Cybulak, M., Sokołowska, Z. and Boguta, P. 2019. Impact of biochar on physicochemical properties of haplic luvisol soil under different land use: a plot experiment. *Agronomy* 9(531):1–16.

DeLuca, T. H., Gundale, M. J., MacKenzie, M. D. and Jones, D. L. 2015. Biochar effects on soil nutrient transformation. In: Lehmann J. and Joseph S. (eds), *Biochar for Environmental Management: Science and Technology*, 2nd ed. Routledge, London, pp:421–454.

Devika, O., Prasad, S., Ratna, P., Rani, P., Prasuna, P. and Lakshmi, P. R. 2018. Refinement of soil biological properties with biochar addition in sweetcorn. *Pharm Innov J* 7(10):708–711.

Ding, Y., Liu, Y., Liu, S., Huang, X., Li, Z., Tan, X., Zeng, G. and Zhou, L. 2017. Potential benefits of biochar in agricultural soils: a review. *Pedosphere* 27:645–661.

Dover, M. 2007. Anyone for char? Dark earth holds carbon storage hope. *Inwood Mag* 77:33–34.

Downie, A., Crosky, A. and Munroe, P. 2009. Physical properties of biochar. In: Lehmann L. and Joseph S. (eds), *Biochar for Environmental Management Science and Technology*. Earthscan Publishers, London, pp:13–32.

Duarte, S. J., Glaser, B. and Cerri, C. E. P. 2019. Effect of biochar particle size on physical, hydrological and chemical properties of loamy and sandy tropical soils. *Agronomy* 9(165):1–15.

Dume, B., Mosissa, T. and Nebiyu, A. 2016. Effect of biochar on soil properties and lead (Pb) availability in a military camp in South West Ethiopia. *Afr J Environ Sci Technol* 10(3):77–85.

Elgharably, A. G. 2008. Nutrient availability and wheat growth as affected by plant residues and inorganic fertilizers in saline soils (doctoral dissertation). The University of Adelaide, South Australia, Australia.

Elmer, W., White, J. C. and Pignatello, J. J. 2010. Impact of biochar addition to soil on the bioavailability of chemicals important in agriculture. Report. University of Connecticut, New Haven.

Elmer, W. H. and Joseph, J. P. 2011. Effect of biochar amendments on mycorrhizal associations and Fusarium crown and root rot of asparagus in replant soils. *Plant Dis* 95(8):960–966.

Fageria, N. K., Gheyi, H. R. and Moreira, A. 2011. Nutrient bioavailability in salt affected soils. *J Plant Nutr* 34:945–962.

Favet, J., Lapanje, A., Giongo, A., Kennedy, S., Aung, Y. Y., Cattaneo, A., Richardson, D. A., Brown, T. C., Kort, R., Brumsack, H. J., Schnetger, B., Chappell, A., Kroijenga, J., Beck, A., Schwibbert, K., Mohamed, A., Kirchner, T., Quadros, P. D., Triplett, E., Broughton, E. W. and Gorbushina, A. A. 2013. Microbial hitchhikers on intercontinental dust: catching a lift in Chad. *Inter Soc Micro Eco J* 7:50–867.

Ferreiro, P. J., Gascó, G., Gutiérrez, B. and Méndez, A. 2012. Soil biochemical activities and the geometric mean of enzyme activities after application of sewage sludge and sewage sludge biochar to soil. *Biol Fert Soils* 48(5):511–517.

Ferreiro, J., Liang, C., Shenglei, F., Mendez, A. and Gasco, G. 2015. The effect of biochar and its interaction with the earthworm *Pontoscolex corethrurus* on soil microbial community structure in tropical soils. *PLoS One* 10(4):e0124891.

Garcia, A. C., Souza, L. G. A., Pereira, M. G., Castro, R. N., Garcia-Mina, J. M., Zonta, E., Lisboa, J. G. F. and Berbara, L. L. R. 2016. Structure-property-function relationship in humic substances to explain the biological activity in plants. *Sci Rep* 6:20798.

Gaskin, J. W., Speir, R. A., Harris, K., Das, K. C., Lee, R. D. and Morris, L. A. 2010. Effect of peanut hull and pine chip biochar on soil nutrients, corn nutrient status, and yield. *Agron J* 102:623–633.

Gai, X., Wang, H., Liu, J., Zhai, L. and Liu, S. 2014. Effects of feedstock and pyrolysis and temperature on biochar adsorption of ammonium and nitrate. *PLoS One* 9(12):e11388.

Gao, L., Wang, R., Shen, G., Zhang, J., Meng, G. and Zhang, J. 2017. Effects of biochar on nutrients and the microbial community structure of tobacco-planting soils. *J Soil Sci Plant Nutr* 17(4):884–896.

Głab, T., Palmowska, J., Zaleski, T. and Gondek, K. 2016. Effect of biochar application on soil hydrological properties and physical quality of sandy soil. *Geoderma* 281:11–20.

Glaser, B., Lehmann, J. and Zech, W. 2002. Ameliorating physical and chemical properties of highly weathered soils in the tropics with charcoal – a review. *Biol Fert Soils* 35:219–230.

Gomez, A., Powers, R. F., Singer, M. J. and Horwath, W. R. 2002. Soil compaction effects on growth of young ponderosa pine following litter removal in California's Sierra Nevada. *Soil Sci Soc Am J* 66:1334–1343.

Gul, S., Whalen, J. K., Thomas, B. W., Sachdeva, V. and Deng, H. 2015. Physico-chemical properties and microbial responses in biochar-amended soils: mechanisms and future directions. *Agric Ecosyst Environ* 206:46–59.

Gupta, K. D., Gupta, C. K., Dubey, R., Fagodiya, R. K., Sharma, G., Keerthika, A., Mohamed, M. B., Dev, R. and Shukla, A. K. 2020. Role of biochar in carbon sequestration and greenhouse gas mitigation. In: Singh J. and Singh C. (eds), *Biochar Applications in Agriculture and Environment Management*. Springer, Cham, pp:141–166.

Hamza, M. A. and Anderson., W. K. 2005. Soil compaction in cropping systems: a review of the nature, causes and possible solutions. *Soil Till Res* 82:121–145.

Hamza, N., Coved., S., Gumuş, I., Manirakiza, N. and Mucevher, O. 2019. Effects of biochar and compost applications on penetration resistance and physical quality of a sandy clay loam soil. *Commun Soil Sci Plant Anal* 51(1):38–44.

Hartley, W., Riby, P. and Waterson, J. 2016. Effects of three different biochars on aggregate stability, organic carbon mobility and micronutrient bioavailability. *J Environ Manag* 181:770–778.

Herath, H. M. S. K., Arbestain, M. C. and Hedley, M. 2013. Effect of biochar on soil physical properties in two contrasting soils, an alfisol and an andisol. *Geoderma* 209:188–197.

Jatav, H. S., Jayant, H., Kumar, S., Kumar, V., Chattopadhya, A., Dhawal, S. K. and Singh, Y. V. 2017. Role of biochar: in agriculture sector its implication and perspective. *Int J Chem Stud* 5:14–18.

Jayasanka, D. J., Jayasinghe, G. Y. and Maheepala, S. A. D. S. S. 2020. Cinnamon wood biochar characteristics and application rates influence on crop growth and properties of problematic soil in Southern Sri Lanka. *Int J Innov Technol Explor Eng* 9(3S):30–33.

Jeffery, S., Verheijen, F. G., Der Velde, M. V. and Bastos, A. C. 2011. A quantitative review of the effects of biochar application to soils on crop productivity using meta-analysis. *Agric Econ Environ* 144(1):175–187.

Jessica, L. S., Dumroese, D. S. and Coleman, M. D. 2019. Soil greenhouse gas, carbon content, and tree growth response to biochar amendment in western United States forests. *Glob Change Biol Bioenergy* 11(5):660–671.

Jia, X., Yuan, W. and Ju, X. 2015. Short report: effects of biochar addition on manure composting and associated N_2O emissions. *J Sust Bioenergy Syst* 5(2):56–61.

Jia, T., Wang, R. H. and Chai, B. F. 2018. Various phyllosphere and soil bacterial communities of natural grasses and the impact factors in a copper tailings dam. *Curr Microbiol* 76:7–14.

Jien, S. H. and Wang, C. S. 2013. Effects of biochar on soil properties and erosion potential in a highly weathered soil. *Catena* 110:225–233.

Jones, D. L., Murphy, D. V., Khalid, M., Ahmad, W., Edwards-Jones, G. and DeLuca, T. H. 2011. Short-term biochar-induced increase in soil CO_2 release is both biotically and abiotically mediated. *Soil Biol Biochem* 43:1723–1731.

Jones, D. L., Rousk, J. G., Jones, E., DeLuca, T. H. and Murphy, D. V. 2012. Biochar-mediated changes in soil quality and plant growth in a three year field trial. *Soil Biol Biochem* 45:113–124.

Joseph, S. D., Camps, M., Lin, Y., Munroe, P., Chia, C. H., Hook, J., Van Zwieten, S., Cowie, A., Singh, B. P. and Lehmann, J. 2010. An investigation into the reactions of biochar in soil. *Soil Res* 48:501–515.

Joseph, S., Graber, E. R., Chia, C., Munroe, P., Donne, S., Thomas, T., Nielsen, S., Marjo, C., Rutlidge, H., Pan, G. X., Li, L., Taylor, P., Rawal, A. and Hook, J. 2013. Shifting paradigms: development of high-efficiency biochar fertilizers based on nano-structures and soluble components. *Carb Manag* 4:323–343.

Kameyama, K., Miyamoto, T., Shiono, T. and Shinogi, Y. 2012. Influence of sugarcane bagasse-derived biochar application on nitrate leaching in calcaric dark red soil. *J Environ Qual* 41(4):1131–1137.

Karami, A., Homaee, M., Afzalinia, S., Ruhipour, H. and Basirat, S. 2012. Organic resource management: impacts on soil aggregate stability and other soil physicochemical properties. *Agric Ecosyst Environ* 148:22–28.

Katterer, T., Roobroeck, D., Andrén, O., Kimutai, G., Karltun, E., Kirchmann, H., Nyberg, G., Vanlauwe, B. and Roing, N. K. 2019. Biochar addition persistently increased soil fertility and yields in maize soybean rotations over 10 years in sub-humid regions of Kenya. *Field Crop Res* 235:18–26.

Khalifa, N. and Yousef, L. F. 2015. A short report on changes of quality indicators for a sandy textured soil after treatment with biochar produced from fronds of date palm. *Energy Procedia* 74:960–965.

Koji, K., Yukiyoshi, I. W. A. T. A. and Teruhito, M. 2017. Biochar amendment of soils according to their physicochemical properties. *Jpn Agric Res Q* 51(2):117–127.

Lahori, A. H., Zhanyu, G., Zhang, Z., Li, R., Mahar, A., Awasthi, M., Feng, S., Ali, T., Farhana, K., Ping, W. and Shuncheng, J. 2017. Use of biochar as an amendment for remediation of heavy metal contaminated soils: prospects and challenges. *Pedosphere* 27(6):991–1014.

Laird, D. A., Fleming, P., Davis, D. D., Horton, R., Wang, B. and Karlen, D. L. 2010. Impact of biochar amendments on the quality of a typical Midwestern agricultural soil. *Geoderma* 158:436–442.

Lakhdar, A., Rabhi, M., Ghnaya, T., Montemurro, F., Jedidi, N. and Abdelly, C. 2009. Effectiveness of compost use in salt-affected soil. *J Hazard Mater* 171:29–37.

Lal, R. 2009. Challenges and opportunities in soil organic matter research. *Euro J Soil Sci* 60(2):158–169.

Lal, R. 2010. Managing soils and ecosystems for mitigating anthropogenic carbon emissions and advancing global food security. *Biol Sci* 60 (9):708–721.

Lammirato, C., Miltner, A. and Kaestner, M. 2011. Effects of wood char and activated carbon on the hydrolysis of cellobiose by b-glucosidase from *Aspergillus niger. Soil Biol Biochem* 43(9):1936–1942.

Lashari, M. S., Liu, Y., Li, L., Pan, W., Fu, J., Pan, G., Zheng, J., Zheng, J., Zhang, X. and Yu, X. 2013. Effects of amendment of biochar-manure compost in conjunction with pyroligneous solution on soil quality and wheat yield of a salt-stressed cropland from Central China Great Plain. *Field Crop Res* 144:113–118.

Lashari, M. S., Ye, Y., Ji, H., Li, L., Kibue, G. W., Lu, H., Zheng, J. and Pan, G. 2015. Biochar–manure compost in conjunction with pyroligneous solution alleviated salt stress and improved leaf bioactivity of maize in a saline soil from central China: a 2-year field experiment. *J Sci Food Agric* 95(6):1321–1327.

Laxman Rao, P., Jayasree, G., Pratibha, G. and Ram Prakash, T. 2017. Effect of soil amendments on physical properties of soil in maize (*Zea mays* L.). *Int J Curr Microb Appl Sci* 6(6):2082–2091.

Lehmann, J. 2007. Bio-energy in the black. *Front Ecol Environ* 5(7):381–387.

Lehmann, J., DaSilva, J. P., Rondon, M., Da Siva, C. M., Greenwood, J., Steiner, C. and Glaser, B. 2002. Slash-and char: a feasible alternative for soil fertility management in the Central Amazon? In: Proceedings of the 17th World Congress of Soil Science, Bangkok, Thailand, 449:1–12

Lehmann, J., Pereira da Silva, J., Steiner, C., Nehls, T., Zech, W. and Glaser, B. 2003. Nutrient availability and leaching in an archaeological Anthrosol and a Ferralsol of the Central Amazon basin: fertilizer, manure and charcoal amendments. *Plant Soil* 249:343–357.

Lehmann, J. and Rondon, M. 2006. Biochar soil management on highly weathered soils in the humid tropics. In: Uphoff N., Ball A. S., Fernandes E., Herren H., Husson O., Mark L., Palm C., Jules P., Pedro S., Sanginga N. and Thies J. (eds), *Biological Approaches to Sustainable Soil Systems.* CRC Press, Boca Raton, FL, pp:517–530.

Lehmann, J., Gaunt, J. and Rondon, M. 2006. Bio-char sequestration in terrestrial ecosystems – a review. *Mitig Adapt Strat Glob Chang* 11(2):403–427.

Lehmann, J. and Joseph, S. 2009. Biochar for environmental management: science and technology. *Earthscan* 25(1):15801–15811.

Lei, O. and Zhang, R. 2013. Effects of biochars derived from different feedstocks and pyrolysis temperatures on soil physical and hydraulic properties. *J Soils Sedim* 13(9):1561–1572.

Leonard, G. 2013. Effect of biochar application rate on soil physical and hydraulic properties of a sandy loam. *Arch Agron Soil Sci* 60(4):57–470.

Li, Y., Chang, S. X., Yang, Y., Fu, S., Jiang, P., Luo, Y., Yang, M., Chen, Z. and Hu, S. 2018. Biochar reduces soil heterotrophic respiration in a subtropical plantation through increasing soil organic carbon recalcitrancy and decreasing carbon-degrading microbial activity. *Soil Biol Biochem* 122:173–185.

Liang, B., Lehmann, J., Solomon, D., Kinyangi, J., Grossman, J. and O'Neill, B. 2006. Black carbon increases cation exchange capacity in soils. *Soil Sci Soc Am J* 70(5):1719–1730.

Liang, F., Li, G. T., Lin, Q. M. and Zhao, X. R. 2014. Crop yield and soil properties in the first 3 years after biochar application to a calcareous soil. *J Integr Agric* 13(3):525–532.

Liu, X. H., Han, F. P. and Zhang, X. C. 2012. Effects of biochar on soil aggregates in the Loess Plateau: results from incubation experiments. *Int J Agric Biol* 14(6):975–979.

Liu, X. H. and Zhang, X. C. 2012. Effect of biochar on ph of alkaline soils in the Loess Plateau: results from incubation experiments. *Int J Agric Biol* 14(5):745–750.

Liu, Z., Chen, X., Jing, Y., Li, Q., Zhang, J. and Huang, Q. 2014. Effects of biochar amendment on rapeseed and sweet potato yields and water stable aggregate in upland red soil. *Catena* 123:45–51.

Liu, Z., Dugan, B., Masiello, C. A., Barnes, R. T., Gallagher, M. E. and Gonnermann, H. 2016. Impacts of biochar concentration and particle size on hydraulic conductivity and DOC leaching of biochar–sand mixtures. *J Hydrol* 533:461–472.

Liu, S., Meng, J., Jiang, L., Yang, X., Lan, Y., Cheng, X. and Chen, W. 2017. Rice husk biochar impacts soil phosphorous availability, phosphatase activities and bacterial community characteristics in three different soil types. *Appl Soil Ecol* 116:12–22.

Luo, Y., Durenkamp, M., De Nobili, M., Lin, Q. and Brookes, P. C. 2011. Short term soil priming effects and the mineralisation of biochar following its incorporation to soils of different pH. *Soil Biol Biochem* 43 (11):2304–2314.

Mahmoodabadi, M., Yazdanpanah, N., Sinobas, L., Pazira, L. and Neshat, A. 2013. Reclamation of calcareous saline sodic soil with different amendments (1); redistribution of soluble cations within the soil profile. *Agric Water Manag* 120(C):30–38.

Major, J., Rondon, M., Molina, D., Riha, S. J. and Lehmann, J. 2010. Maize yield and nutrition during four years after biochar application to a Colombian savanna oxisol. *Plant Soil* 333(1):117–128.

Makoi, J. H. and Ndakidemi, P. A. 2008. Selected soil enzymes: examples of their potential roles in the ecosystem. *Afric J Biotechnol* 7(3):181–191.

Matsubara, Y., Hasegawa, N. and Fukui, H. 2002. Incidence of Fusarium root rot in asparagus seedlings infected with arbuscular mycorrhizal fungus as affected by several soil amendments. *J Jpn Soc Hort Sci* 71:370–374.

Mizuta, K., Matsumoto, T., Hatate, Y., Nishihara, K. and Nakanishi, T. 2004. Removal of nitrate-nitrogen from drinking water using bamboo powder charcoal. *Biores Technol* 95(3):255–257.

Moeskops, B., Buchan, D., Sleutel, S., Herawaty, L., Husen, E., Saraswati, R., Setyorini, D. and De Neve, S. 2010. Soil microbial communities and activities under intensive organic and conventional vegetable farming in West Java, Indonesia. *Appl Soil Ecol* 45(2):112–120.

Molina, M., Zaelke, D., Sarma, K. M., Andersen, S. O., Ramanathan, V. and Kaniaru, D. 2009. Reducing abrupt climate change risk using the Montreal Protocol and other regulatory actions to complement cuts in CO_2 emissions. *Proc Natl Acad Sci USA* 106:20616–20621.

Moon, D. H., Park, J. W., Chang, Y. Y., Ok, Y. S., Lee, S. S. and Ahmad, M. 2013. Immobilization of lead in contaminated firing range soil using biochar. *Environ Sci Pollut Res* 20(12):8464–8671.

Muhammad, A., Rifat, H., Hussain, Q., Ahmed, M., Muhammad, I. and Crowley, D. E. 2016. Effect of biochar amendment on soil microbial biomass, abundance and enzyme activity in the mash bean field. *J Biodiv Environ Sci* 8(6):1–13.

Mukherjee, A. and Lal, R. 2013. Biochar impacts on soil physical properties and greenhouse gas emissions. *Agronomy* 3(2):313–339.

Mukherjee, A., Lal, R. and Zimmerman, A. R. 2014. Effects of biochar and other amendments on the physical properties and greenhouse gas emissions of an artificially degraded soil. *Sci Total Environ* 487:26–36.

Naidu, R., Syers, J. K., Tillman, R. W. and Kirkman, J. H. 1991. Assessment of plant available phosphate using several soil-testing procedures. *Fert Res* 30:47–53.

Ndor, E., Amana, S. M. and Asadu, C. L. A. 2015. Effect of biochar on soil properties and organic carbon sink in degraded soil of Southern Guinea Savanna Zone, Nigeria. *Int J Plant Soil Sci* 4(3):252–258.

Nigussie, A., Kissi, E., Misganaw, M. and Ambaw, G. 2012. Effect of biochar application on soil properties and nutrient uptake of lettuces (*Lactuca sativa*) grown in chromium polluted soils. *Am Eur J Agric Environ Sci* 12(3):369–376.

Nimmo, J. R. 1997. Modeling structural influences on soil water retention. *Soil Sci Soc Am J* 61:712–719.

Njoku, C., Mbah, C. N., Igboji, P. O., Nwite, J. N., Chibuike, C. C. and Uguru, B. N. 2015. Effect of biochar on selected soil physical properties and maize yield in an ultisol in Abakaliki Southeastern Nigeria. *Glob Adv Res J Agric Sci* 4(12):864–870.

Nyambo, P., Taeni, T., Chiduza, C. and Araya, T. 2018. Effects of maize residue biochar amendments on soil properties and soil loss on acidic Hutton soil. *Agronomy* 8(11):256.

Obia, A., Mulder, J., Hale, S. E., Nurida, N. L. and Cornelissen, G. 2018. The potential of biochar in improving drainage, aeration and maize yields in heavy clay soils. *PLoS One* 13(5):e0196794.

Omondi, M. O., Xia, X., Nahayo, A., Liu, X., Korai, P. K. and Pan, G. 2016. Quantification of biochar effects on soil hydrological properties using meta-analysis of literature data. *Geoderma* 274:28–34.

Ouyang, F., Wang, L., Tang, J., Yu, L. and Zhang, R. 2013. Effects of biochar amendment on soil aggregates and hydraulic properties. *J Soil Sci Plant Nutr* 13(4):991–1002.

Ouyang, L., Tang, Q., Yu, L. and Zhang, R. 2014. Effects of amendment of different biochars on soil enzyme activities related to carbon mineralization. *Soil Res* 52(7):706–716.

Palansooriya, K. N., Ok, Y. S., Awad, Y. M., Lee, S. S., Sung, J. K., Koutsospyros, A. and Moon, D. H. 2019. Impacts of biochar application on upland agriculture: a review. *J Environ Manag* 234:52–64.

Pathak, H. 2010. Trend of fertility status of Indian soils. *Curr Adv Agric Sci* 2(1):10–12.

Pietikainen, J., Kiikkila, O. and Fritze, H. 2000. Charcoal as a habitat for microbes and its effect on the microbial community of the underlying humus. *Oikos* 89:231–242.

Preston, C. M. and Schmidt, M. W. I. 2006. Black (pyrogenic) carbon: a synthesis of current knowledge and uncertainties with special consideration of boreal regions. *Biogeosciences* 3(4):397–420.

Pukalchik, M., Mercl, F., Terekhova, V. and Tlustos, P. 2018. Biochar, wood ash and humic substances mitigating trace elements stress in contaminated sandy loam soil: evidence from an integrative approach. *Chemosphere* 203:228–238.

Qadir, M., Ghafoor, A. and Murtaza, G. 2000. Amelioration strategies for saline soils: a review. *Land Degrad Dev* 11(6):501–521.

Quin, P. R., Cowie, A. L., Flavel, R. J., Keen, B. P., Macdonald, L. M., Morris, S. G., Singh, B. P., Young, I. M. and Van Zwieten, L. 2014. Oil malleebiochar improves soil structural properties – a study with X-ray micro-CT. *Agric Ecosyst Environ* 191:142–149.

Radovic, L. R., Moreno, C. C. and Rivera, U. J. 2001. Carbon materials as adsorbents in aqueous solutions. In: *Chemistry and Physics of Carbon*. Ed. Marcel Dekker, Ljubisa R. Radovic New York, vol. 27, pp:227–405.

Rogovska, N., Laird, D., Cruse, R., Fleming, P., Parkin, T. and Meek, D. 2011. Impact of biochar on manure carbon stabilization and greenhouse gas emissions. *Soil Sci Soc Am J 75*(3):871–879.

Rondon, M., Lehmann, J., Ramírez, J. and Hurtado, M. 2007. Biological nitrogen fixation by common beans (*Phaseolus vulgaris* L.) increases with bio-char additions. *Biol Fert Soils* 43(6):699–708.

Rousk, J., Baath, E., Brookes, P. C., Lauber, C. L., Lozupone, C. A., Caporaso, J. G., Knight, R. and Fierer, N. 2010. Soil bacterial and fungal communities across a pH gradient in an arable soil. *ISME J* 4(10):1340–1351.

Saito, M. 1990. Charcoal as a micro habitat for VA mycorrhizal fungi, and its practical application. *Agric Ecosyst Environ* 29:341–344.

Saxena, J., Rawat, J. and Kumar, R. 2017. Conversion of biomass waste into biochar and the effect on mung bean crop production. *Clean-Soil Air Water* 45(7):1501020.

Sedjo, R. and Sohngen, B. 2012. Carbon sequestration in forest and soils. *Annu Rev Res Econ* 4:127–144.

Segun, O. O. 2019. Effect of biochar amendment on soil enzymatic activities, carboxylate secretions and upland rice performance in a sandy clay loam Alfisol of Southwest Nigeria. *Sci Afr* 4(e00107):1–13.

Serkalem, W. M. 2015. Effect of *Prosopis juliflora* biochar amendment on some soil properties: the case of Salic Fluvisols from Melkawerer Research Station, Ethiopia. Research thesis. Addis Ababa University, Ethiopia.

Shaygan, M., Reading, L. P. and Baumgartl, T. 2017. Effect of physical amendments on salt leaching characteristics for reclamation. *Geoderma* 292(C): 96–110.

Shoi, S., Loez-Capel, E., Krull, E. and Bol, R. 2009. Biochar roles in soil and climate change. A review of research needs. *CSIRO Land Water Science* Report 05/09, pp. 1–64.

Singh, B., Singh, B. P. and Cowie, A. L. 2010. Characterization and evaluation of biochars for their application as a soil amendment. *Soil Res* 48(7):516–525.

Sinsabaugh, R. L. 2010. Phenol oxidase, peroxidase and organic matter dynamics of soil. *Soil Biol Biochem* 42(3):391–404.

Smebye, A., Alling, V., Vogt, R. D., Gadmar, T. C., Mulder, J., Cornelissen, G. and Hale, S. E. 2016. Biochar amendment to soil changes dissolved organic matter content and composition. *Chemosphere* 142:100–105.

Sohi, S. P., Krull, E. E., Capel, L. and Bol, R. 2010. A review of biochar and its use and function in soil. *Adv Agron* 105:47–82.

Solaiman, Z. M., Blackwell, P., Abbott, L. K. and Storer, P. 2010. Direct and residual effect of biochar application on mycorrhizal root colonisation, growth and nutrition of wheat. *Aust J Soil Res* 48:546–554.

Song, D., Tang, J., Xi, X., Zhang, S., Liang, G., Zhou, W. and Wang, X. 2018. Responses of soil nutrients and microbial activities to additions of maize straw biochar and chemical fertilization in a calcareous soil. *Eur J Soil Biol* 84:1–10.

Steinbeiss, S., Gleixner, G. and Antonietti, M. 2009. Effect of biochar amendment on soil carbon balance and soil microbial activity. *Soil Biol Biochem* 41:1301–1310.

Sukaronto, W. H., Utomo, W. H., Kusuma, Z. and Nugroho, W. H. 2011. Soil fertility status, nutrient uptake and maize (*Zea mays* L.) yield following biochar and cattle manure application on sandy soils of Lombok, Indonesia. *J Trop Agric* 49(1–2): 47–52.

Taghavimehr, J. 2015. Effect of biochar on soil microbial communities, nutrient availability, and greenhouse gases in short rotation coppice systems of Central Alberta (doctoral dissertation). University of Alberta, Alberta, Canada.

Tang, C. and Rengel, Z. 2003. Role of plant cation/anion uptake ratio in soil acidification. In: Rengel Z. (ed), *Handbook of Soil Acidity*. Marcel Dekker, New York, pp:57–81.

Tejada, M. and Gonzalez, J. L. 2007. Influence of organic amendments on soil structure and soil loss under simulated rain. *Soil Tillage Res* 93(1):197–205.

Thalmann, M. and Santelia, D. 2017. Starch as a determinant of plant fitness under abiotic stress. *New Phytol* 214(3):943–951.

Thies, J. E. and Rillig, M. 2009. Characteristics of biochar: biological properties. In: Lehmann M. and Joseph S. (eds), *Biochar for Environmental Management Science and Technology*. Earthscan, London, pp:85–105.

Timilsina, S., Khanal, B. R., Shah, S. C., Shrivastav, C. P. and Khanal, A. 2017. Effects of biochar applicaton on soil properties and production of radish (*Raphanus sativus* L.) on loamy sand soil. *J Agric Forest Univ* 1:103–111.

Tryon, E. H. 1948. Effect of charcoal on certain physical, chemical, and biological properties of forest soils. *Ecol Monogr* 18:81–115.

Uchimiya, M., Chang, S. C. and Klasson, K. T. 2011. Screening biochars for heavy metal retention in soil: role of oxygen functional groups. *J Hazard Mater* 190:432–441.

Ulyett, J., Sakrabani, R., Kibblewhite, M. and Hann, M. 2014. Impact of biochar addition on water retention, nitrification and carbon dioxide evolution from two sandy loam soils. *Eur J Soil Sci* 65(1):96–104.

Utomo, W. H. 2010. Rice husk biochar for rice based cropping system in acid soil. The characteristics of rice husk biochar and its influence on the properties of acid sulphate soils and rice growth in West Kalimantan. Indonesia. *J Agric Sci* 2(1):39–47.

Van Zwieten, L., Morris, S., Chan, K. Y., Dowine, A., Rust, J., Joseph, S. and Cowie, A. 2010. Effects of biochar from slow pyrolysis of paper mill waste on agronomic performance and soil fertility. *Plant Soil* 327(1–2):235–246.

Ventura, F., Salvatorelli, F., Piana, S., Pieri, L. and Pisa, P. L. 2013. The effects of biochar on the physical properties of bare soil. *Earth Environ Sci Trans R Soc Edinb* 103(1):5–11.

Verheijen, F., Jeffery, S., Bastos, A. C., Vander Velde, M. and Diafas, I. 2009. *Biochar Application to Soils—A Critical Scientific Review of Effects on Soil Properties, Processes and Functions*. Luxemburg: EUR 24099 EN, Office for the Official Publications of the European Communities, Luxembourg, pp:1–149.

Wakeel, A. 2013. Potassium–sodium interactions in soil and plant under saline-sodic conditions. *J Plant Nutr Soil Sci* 176(3):344–354.

Warnock, D. D., Lehmann, J., Kuyper, T. W. and Rillig, M. C. 2007. Mycorrhizal responses to biochar in soil concepts and mechanisms. *Plant Soil* 300:9–20.

Watzinger, A., Feichtmair, S., Kitzler, B., Zehetner, F., Kloss, S. and Wimmer, B. 2014. Soil microbial communities responded to biochar application in temperate soils and slowly metabolized C13 labelled biochar as revealed by C13 PLFA analyses: results from a short term incubation and pot experiment. *Eur J Soil Sci* 65(1):40–51.

Winsley, P. 2007. Biochar and bio energy production for climate change mitigation. *New Zeal Sci Rev* 64:5–10.

Wong, V. N. L., Greene, R. S. B., Dalal, R. C. and Murphy, B. W. 2010. Soil carbon dynamics in saline and sodic soils: a review. *Soil Use Manag* 26:2–11.

Woolf, D., Amonette, J. E., Street-Perrott, F. A., Lehmann, J. and Joseph, S. 2010. Sustainable biochar to mitigate global climate change. *Nat Commun* 1(56):1–9.

Yanai, Y. K., Toyota, K. and Okazaki, M. M. 2007. Effects of charcoal addition on N_2O emissions from soil resulting from rewetting air-dried soil in short-term laboratory experiments. *Soil Sci Plant Nutr* 53(2):181–188.

Yadav, R. K., Yadav, M. R., Kumar, R., Parihar, C. M., Yadav, N., Bajiya, R., Ram, H., Meena, R. K., Yadav, D. K. and Yadav, B. 2017. Role of biochar in mitigation of climate change through carbon sequestration. *Int J Curr Micro Appl Sci* 6(4):859–866.

Yao, Y., Gao, B., Zhang, M., Inyang, M. and Zimmerman, A. R. 2012. Effect of biochar amendment on sorption and leaching of nitrate, ammonium and phosphate in a sandy soil. *Chemosphere* 89(11):1467–1471.

Yargholi, B. and Azarneshan, S. 2014. Long term effects of pesticides and chemical fertilizers usage on some soil properties and accumulation of heavy metals in the soil (case study of Moghan plain's (Iran) irrigation and drainage network). *Int J Agri Crit Sci* 7:518–523.

You, J., Sun, L., Liu, X., Hu, X. and Xu, Q. 2019. Effects of sewage sludge biochar on soil characteristics and crop yield in loamy sand soil. *Pol J Environ Stud* 28(4): 2973–2980.

Yuan, J. H. and Xu, R. K. 2011. The amelioration effects of low temperature biochar generated from nine crop residues on an acidic Ultisol. *Soil Use Manag* 27(1):110–115.

Yue, Y., Guo, W. N., Lin, Q. M., Li, G. T. and Zhao, X. R. 2016. Improving salt leaching in a simulated saline soil column by three biochars derived from rice straw (*Oryza sativa* L.), sunflower straw (*Helianthus annuus*), and cow manure. *J Soil Water Conserv* 71(6):467–475.

Zeng, Y. H., Shih-Hao, J., Hsin, W. C. and Ruei, C. L. 2014. Impacts of biochar on physical properties and erosion potential of a mudstone slope land soil. Hindawi Publishing Corporation. *Sci World J* 602197:1–10.

Zhang, H., Voroney, R. Price, G. 2015. Effects of temperature and processing conditions on biochar chemical properties and their influence on soil C and N transformations. Soil Biol Biochem. 83: 19–28.

Zhang, R., Zhang, Y., Song, L., Song, X., Hanninen, H. and Wu, J. 2017. Biochar enhances nut quality of Torreyagrandis and soil fertility under simulated nitrogen deposition. *For Ecol Manag* 391:321–329.

Zhang, M., Muhammad, R., Zhang, L., Xia, H., Cong, M. and Jiang, C. 2019. Investigating the effect of biochar and fertilizer on the composition and function of bacteria in red soil. *Appl Soil Ecol* 139:107–116.

Zhao, S., Ta, N. and Wang, X. 2017. Effect of temperature on the structural and physicochemical properties of biochar with apple tree branches as feedstock material. *Energy* 10(9):1293.

Zheng, J., Chen, J., Pan, G., Liu, X., Zhang, X., Li, L., Bian, R., Cheng, K. and Jinwei, Z. 2016. Biochar decreased microbial metabolic quotient and shifted community composition four years after a single incorporation in a slightly acid rice paddy from southwest China. *Sci Total Environ* 571:206–217.

Zimmerman, A. R. 2010. Abiotic and microbial oxidation of laboratory-produced black carbon biochar. *Environ Sci Technol* 44(4): 1295–1301.

6

Role of Metabolomics in Deciphering the Plant-Microbe Interaction

Younas Rasheed Tantray[1], Naseema Wani[2], Nazir Ahmed Malik[3], Mohammad Saleem Wani[2], and Ishrat Jan[4]
[1] Government Degree College, Tangdhar, Jammu and Kashmir, India
[2]Punjabi University Patiala, Punjab, India
[3]Dolphin PG College of Science & Agriculture, Punjab, India
[4]Kashmir University, Jammu and Kashmir, India

CONTENTS

6.1 Introduction

The field of metabolomics is regarded as abroad, non-biased study of complex metabolite mixtures which allow the identification as well as quantification of each single metabolite. Metabolomics is regarded as an evolving field among omics sciences that is mainly concerned with the high-throughput snapshot of a metabolome (Shafi and Zahoor 2020). This field has come out as a functional genomics methodology, paving the way for understanding the complicated molecular interactions occurring inside the living systems (Hall et al. 2002). Metabolites are basically the final products of various processes occurring within a cell. The level of metabolites within the living system can be considered as the final reaction of living entity to various genomic or ecological changes (Fiehn 2001). The metabolites are highly vibrant and display a vast variety of structures raising the huge task for analytical procedures in their measurement (Stitt and Fernie 2003). Therefore, the field of metabolomics signifies the reasonable advancement from an extensive study of proteins and RNA at the system level (Bino et al. 2004), but in contrast to the products of transcription and translation, the classy techniques like mass spectrometry (MS) and nuclear magnetic resonance (NMR) spectroscopy are needed to decipher the vast molecular identity of metabolites

(Lei et al. 2011). Metabolomics allows uniform evaluation of a cellular condition within the framework of an instant climate, considering gene expression, enzyme production and control, and improvements in metabolic reactions (Mendes et al. 1992; 1996; Griffin and Shockcor 2004). In comparison to genomics or proteomics, metabolomic studies revealed variations in morphology and role of a specific tissue or organism. Hence, the omic disciplines are being utilized to perform studies in a balancing way to define a phenotype. Changes in gene expression and proteins will definitely carry improvements in metabolite profile within a cell, tissue, or organism (Griffin and Shockcor 2004; Spratlin et al. 2009). Additionally, metabolomics offers similar benefits like proteomics and transcriptomics as they all are having capacity to examine biofluids and are relatively inexpensive, quick, and automatic techniques. Toxicology, pharmacology, and inborn metabolic errors were the first studies in which metabolomics was applied (Spratlin et al. 2009). Currently, it is of great importance to use metabolomics to describe different pathological conditions in human diseases such as tumor, diabetes, self-immune, and coronary disease. In addition, the metabolomics area can offer useful resources in a broad range of applications such as plant biology, microbial biotechnology, food processing, pharmacology, enzyme detection, and plant biotechnology. The current plant research includes widespread usage of metabolomics as a result of numerous human health benefits from several plant-derived products such as food, pharmaceutical products, and industrial products, as well as its role in crop improvement and nutritional assessments. Moreover, the large chemical diversity of plants in relation to other living organisms also leads to the significance of metabolomics in plant sciences (Akiyama et al. 2008). The new technical developments in NMR spectroscopy and MS, the two most commonly known techniques used in metabolite analysis, have enhanced the impact and spectral resolution of analytical tests on metabolomic samples and increased efforts to achieve comprehensive biochemical estimation. Despite the fact that not all the systems used in metabolomics are fully recognized, it is anyway conceivable to sum up the most renowned ones (Fiehn 2001; Harrigan and Goodacre 2003; Goodacre 2007), including target analysis of metabolites, profiling of metabolites, metabolomics, metabolite flux analysis, and metabolic fingerprinting. Metabolomics was initially evolved from metabolic profiling and gas chromatography–mass spectrometry (GC–MS) techniques were used in the mid-1970s to study metabolites of hormones, toxins, and neutral and acidic urinary drugs (Horning and Horning 1970; Devaux et al. 1971); however, metabolite profiling was not used until the mid-1990s to identify plant life forms (Roessner et al. 2000;2001). Plant metabolites are associated with numerous resistance and stress reactions and furthermore, improve shading, taste, smell, and aroma of flowers and fruits (Bino et al. 2004).

As we described earlier, an organism's biochemical phenotype is the definitive product of communications between the genotype and the environment; however, it is further regulated by the intracellular physiological variances that form a part of homeostasis (Weckwerth 2003). In this way, it is imperative to know the metabolome dynamics, examine fluctuations in metabolism, and unravel the location of each metabolite following various stimuli (Fiehn and Weckwerth 2003). The metabolomic test is designed to identify variations in biological and metabolic processes that can contribute to a cell, tissue, or living being's physiological and developmental phenotype (Weckwerth 2003; Bino et al. 2004). The combination of metabolomics with transcription profiles will provide intimations for recognizing the elements of the unidentified genes, whether these are derived from model or non-model plants (Saito et al. 2008). Plants produce in excess of two lakh metabolites; several of them perform explicit roles in enabling adaptation to particular niches in the environment (Oksman-Caldentey and Inze 2004). Hence, the key challenges faced when describing the plant metabolome indicate that the metabolome in existence is profoundly unstable compared to the proteomics or transcriptomics due to the tremendous diversity of chemical molecules. There is also a wide variety of metabolite concentrations, which can vary in magnitude over nine orders (pM to mM). Such enormous variations in the existence and centralization of analytes to be analyzed make all the analytical techniques used in metabolomic methodologies challenging (Dunn et al. 2005). In plant biotic communications, metabolites play an integral role, and thus, can be viewed as a phenotypic marker of a living being that reflects inherited changes, histone modifications, and transcriptomic and proteomic components. The description and annotation of the whole cell metabolome, along with a thoughtless interpretation of the functions metabolites perform in biotic interactions, guarantee a portion of open inquiries, while at the same time representing a noteworthy test due to the high structural and molecular decent metabolite variety. Three different factors that fetter metabolomic finding are: (1) wide variety of physiological concentrations of metabolites, (2) the mind-boggling and swift complexity of metabolic regulation, with rapid enzymatic responses that modify the representation in virtually no time interconnect with each other, producing elaborate systems, and (3) the significant level of chemical interference impurities, which are considerably less important in nucleic acid or protein analysis. Neither any existing metabolite extraction techniques nor structured stages may convey a full metabolome summary. One reason is that the high diversity of plant-generated metabolites administers numerous wonders in plant biotic interactions, from herbivores that favor specific plants while disregarding other (Pearse et al. 2013), defending molecules which prevent explicit herbivores while attracting others (Shroff et al. 2015). The halfway metabolome image we have nowadays leaves multiple inquiries unresponded. In general, metabolites are divided into two categories: primary (needed for cell viability) and secondary (essential for the sustainability of the living beings in the ecosystem). Fluctuations in secondary metabolism are apparent mediators when considering plant biotic interactions from the start – the plant produces exacerbation that prevent microbes and herbivores (Sugimoto et al. 2014), attract the foes of herbivores (Cusumano et al. 2015), build mechanical boundaries against microbe intrusion (Bourdenx et al. 2011), legitimately kill invading microbes (Prince et al. 2014), and convey message to organs, distant plant parts, and neighboring plants about a disease (Erb et al. 2015), attract useful symbionts (D'Alessandro et al. 2014), and can serve these activities associatively (Clay et al. 2009). Incidentally, microorganisms (Geng et al. 2012) and herbivores themselves develop and utilize metabolites of secondary origin to weaken the immune system of plants. Coronatine, an analog of jasmonic acid (JA) of bacteria, gives a well-examined model (Geng et al. 2012); it mediates in the sensitive harmony between the two fundamental defense pathways in the plant, the JA and the salicylic acid (SA) pathways, by "claiming" to be JA, thus smothering the SA pathway that is destructive to this bacterium (Geng et al. 2012). Then again, the plant's secondary metabolome is also prey to parasite modifications, and this is frequently interfered with effector proteins that interfere with biosynthetic pathways of plant metabolites also (Djamei et al. 2011). Furthermore, the compositions of the primary metabolites present in the plant and the microbes that usually contaminate them have co-developed, and the microorganisms are somewhat capable of adjusting to the rapid changes in the plant's primary metabolite concentrations, e.g. during fruit maturation (Bi et al. 2015). Table 6.1 depicts the role of secondary metabolites produced from microorganisms involved in various plant-microbe interactions (Olanrewaju et al. 2017).

6.2 Techniques in the Plant Metabolome Determination

The basic techniques used to estimate plant metabolites are chromatography, including GC and high-performance chromatography, together with various spectroscopy techniques such as infrared (IR), near infrared (NIR), and MS. The utilization of chromatographic approaches with MS can significantly enhance the deepness of metabolome analysis (Bedair and Sumner 2008). GC–MS is considered as the most supreme and extensively used analytical method for determination of plant metabolites. The method is used for analyzing a wide variety of volatile and non-volatile metabolites with high thermal stability, both qualitatively and quantitatively. Mass spectrophotometers identify the eluted compounds after separation, and high analytical reproducibility and reduced cost make this technique more efficient and acceptable in comparison to rest of the other methods such as liquid chromatography–mass spectroscopy(LC–MS) or liquid chromatography–nuclear magnetic resonance (LC–NMR) spectroscopy (Fiehn et al. 2000; Bedair and Sumner 2008). Additionally, researchers may also use direct injection MS analysis for plant phenotyping, including Fourier transformed mass spectroscopy, which delivers accurate detection limits and precision mass measurement to support analysis of metabolites (Aharoni et al. 2002; Bedair and Sumner 2008). The use of LC–MS in the study of plant metabolites has increased in the past few years. Compared with GC–MS, the LC–MS method is more appropriate for labile metabolites and for those which were hard to derive. The use of LC–MS in metabolomics includes the characterization of metabolites involved in stress (Kim et al. 2007; Sanchez et al. 2008), the identification of flavonoid glycoconjugates in roots of *Medicago truncatula* (Staszkow et al. 2011), description and identification of metabolome database of tomato (MoTo DB) (Moco et al. 2006), recognition and accumulation

TABLE 6.1
Role of secondary Metabolites Produced from Microorganisms Involved in Plant-Microbe Communication (Olanrewaju et al. 2017)

S. No.	Secondary Metabolites	Microorganisms	Function
1	Auxins	*Azospirillum*, *Bradyrhizobium*, *Comamonas*, *Enterobacter*, *Pseudomonas*, and *Rhizobium*	Stimulation of root growth and development, advancement of plant through augmenting the length and width of root hairs; growth in mechanisms of water and mineral nutrients uptake from a large bulk of soil
2	Gibberellins	*Achromobacter xylosoxidans*, *Agrobacterium*, *Arthrobacter*, *Acinetobacter calcoaceticus*, *Azotobacter*, *Azospirillum*, *Bacillus*, *Burkholderia*, *Clostridium*, *Flavobacterium*, *Gluconobacter diazotrophicus*, *Herbaspirillum seropedicae*, *Micrococcus*, *Pseudomonas*, *Rhizobia*, and *Xanthomonas*	Early elongation of cells within the subapical meristem that results in stem elongation and boost nitrogen uptake by the plant
3	Cytokinins	*Bacillus subtilis*, *Pseudomonas*, and *Sinorhizobium meliloti*	Upgrading of the cell division, root development, and root hair formation; enhancement of drought stress
4	Abscisic acid	*Achromobacter xylosoxidans*, *Azospirillum brasilense*, *Bacillus pumilus*, *Brevibacterium halotolerans*, *Bacillus subtilis*, *B. licheniformis*, *B. pumilus*, *Lysinibacillus fusiformis*, and *Pseudomonas putida*	Closure of stomata in water-stressed conditions, water uptake regulation and transport in plant, found in arid and semiarid climates
5	1-aminocyclopropane-1-carboxylic acid (ACC) deaminase	*Acinetobacter*, *Achromobacter*, *Agrobacterium*, *Alcaligenes*, *Azospirillum*, *Bacillus*, *Burkholderia*, *Enterobacter*, *Pseudomonas*, *Ralstonia*, *Rhizobium*, and *Serratia*	Pathogen decline; protecting plant from stress conditions; nodulation and mycorrhizal colonization is accelerated

of oxylipins in the genus *Arabidopsis* (Thiocone et al. 2008), detection of flavonoid and isoflavonoid compounds in *Medicago truncatula* (Farag et al. 2007), and untargeted large-scale plant metabolome research in particular. Capillary electrophoresis (CE) is another method which is being used for quantitative study of plant metabolites along with mass spectroscopy (CE–MS) (Soga 2007). CE–MS has been used in cell cultures of plants to classify the amino acid profile (Williams et al. 2007), to determine the metabolite profile of orange leaves infected with bacteria (Cevallos-Cevallos et al. 2009), and to examine the metabolite status in *Illicium anisatum* (Urakami et al. 2010). In conjunction with MS, CE possesses high resolution and high reliability (Soga 2007) to a wide variety of sample types than the other techniques such as Raman spectroscopy, NMR spectroscopy, Fourier transformed-infrared spectroscopy (FT-IR), and NIR resonance, which are continually evolving fast, leading to a non-destructive and high-throughput study. For the plant sciences FT-IR was also identified as a metabolic fingerprinting technique (Kaderbhai et al. 2003; Baranska et al. 2004). Raman spectroscopy and microscopy has proved to be one of the successful techniques, enabling the quantification and identification of plant chemicals and their delivery directly from plant tissues (Schulz et al. 2003; 2005; Baranska et al. 2005). The NMR technique can detect and measure a large variety of micro molar organic compounds (Schripsema 2010; Kim et al. 2011); recently, it was used to evaluate metabolites involved in tolerance mechanism of host plant (Leiss et al. 2011), and also to identify and depict the metabolites produced in *Zea mays* after exposure to salt stress (Gavaghan et al. 2011). This technique is used for many purposes such as food safety investigations and the standardization and regulation of various phytochemical preparations

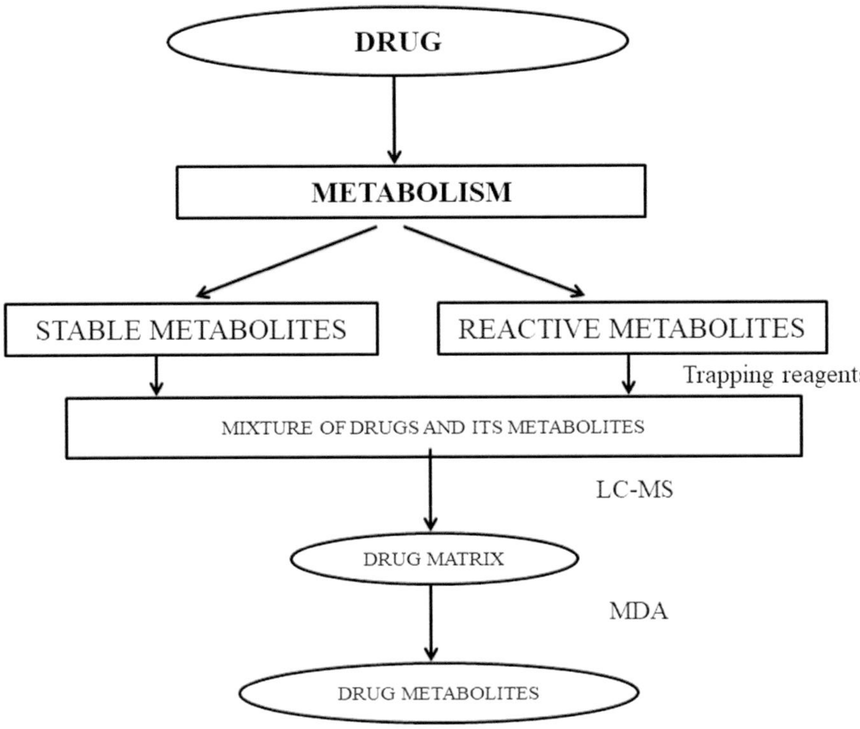

FIGURE 6.1 Overall strategy for studying drug metabolism using a metabolomic approach. The mixture of a drug and its metabolites is analyzed by LC–MS and the data are acquired and processed to generate a data matrix that is further analyzed by multivariate data analysis (MDA) to screen drug metabolites. For reactive metabolites, trapping reagents are needed because of their unstable properties (Feng et al. 2012).

(Ward et al. 2007). Statistical research can be carried out on the spectra of unknown compounds in non-target analyses and in target analyses; the collection of spectral data related to various compounds are used as metabolic profiles for specific compounds in further research (Mochida and Shinozaki 2010). All the above listed technologies used for metabolomic determination involve data processing investigation, e.g. some incorrectness like mass drift and chromatogram shift. The overall strategy for studying drug metabolism using a metabolomic approach is shown in Figure 6.1.

6.3 Metabolomics in Plant Pathogenic Study

Metabolomics is a quite new effort in plant pathology research. Although the significance of individual metabolites has been recognized long before by researchers in plant diseases, it is merely just that plant pathologists have started to pursue universal metabolomic investigations. The old approaches for studying plant infection rely on morphological studies involving assessments of development of symptoms amongst vulnerable and resilient varieties and different molecular statistics of vegetal protection like cell wall by accumulation of callose at the site of infection (Chen and Kim 2009; Luna et al. 2011) synthesis of antimicrobial and signaling agents in the form of reactive oxygen species (ROS) (Torres et al. 2006), and the secretion of other antimicrobial compounds (González-Lamothe et al. 2009). The role of plant hormones in plant disease interactions has been well-established, with certain hormone signs associated with an attack by a biotrophic or necrotrophic pathogen (Denancé et al. 2013; Zhang and Sonnewald 2017). All of these are plant metabolism products; however, these responses raise the question of how they adjust in a broader plan of metabolite plant physiology.

6.3.1 Omics Technology in Study of Plant Pathogens

Until metabolomics started, the growth of proteomics, genomics, and transcriptomics significantly assisted our knowledge of plant infections as well as the successful acquisition of pathogenic nutrients and the elimination of plant immunity (Castro-Moretti et al. 2020). Genetic studies that analyze the structure of plants and pathogenic genes can be used to observe how organisms adapt to disease stress (Bergelson et al. 2001; Moller and Stukenbrock 2017). Transcriptomic studies provide information about how pathogenic host genes manage disease status or how they reprogram for a successful immune response. Transcriptomic study has shown that the main concern of Chinese corn growers is that the sugarcane mosaic virus (SCMV) can significantly control the photosynthetic genes having chlorotic lesion morphology. In addition, the workers estimated translational responses through ribosome profiling, apparently, two transcriptionally downregulated enzymes-4-coumarate coenzyme a ligase (4LC) and phenylalanine ammonia lyase (PAL) associated with phenylpropanoid biosynthesis were upregulated on the translational stage *Pantoea stewartii*, believed to be a substantial pathogen of *Z. mays* in the USA, likewise decreases the expression of genes associated with photosynthesis but additionally causes the expression of several phenylpropanoid metabolic enzymes including-PAL and 4LC (Xu et al. 2019). Similar studies highlight the complicated way of plant reactions to microbes and emphasize the requirement of direct observation of metabolites for optimal features of the pathosystem. Proteomic studies can provide vital facts about interactors, parasite-host targets, and elicitors of infection due to infectious agents (Rampitsch and Bykova 2012). The experts reviewed the effect of such an elicitor –the proteinaceous Tox A released by *Pyrenophora tritici-repentis* –the fungal pathogen of wheat proteome, and found a reduction in the photosystem II complex level that might raise ROS level and begin the progress of symptoms (Manning et al. 2009).

6.4 Progress in Plant Pathology Using Metabolomic Methods

After the identification of plant genes responsible for resistance and virulence targets of plant pathogens, there is a need to further investigate the phenotypes linked with these communications. Metabolomics can offer an overview about the metabolism of plant throughout its developmental process in response to several living and non-living stimuli, including environmental or nutrient challenges (Méchin et al. 2014; Tsogtbaatar et al. 2015; Leary et al. 2016; Lowe-Power et al. 2017; Wu et al. 2018). Metabolomic reports tend to be classified based on the completed analysis, and in case of targeted metabolomics, a list of finite compounds is collected for evaluation, while non-target evaluations are scanned for several undefined features (Ribbenstedt et al. 2018). However, the range of compounds that can be detected using trial parameters, such as extract preparation and equipment, is limited (Liu and Locasale 2017). For a certain biological sample, and based on the organic nature of the extract solvent, various metabolites are taken. Water has been shown to effectively absorb a variety of metabolites of primary origin from the leaves of plants, including phosphorylated compounds, organic acids, amino acids, and sugars (Cocuron et al. 2014). Specific metabolites can be removed more powerfully by using methanol from plant material (Cocuron et al. 2019). Along with solvent selection, the instrumentation used to identify and/or calculate compounds of interest is a significant factor in characterizing the metabolite profile of the relationship between the plant and the pathogen. LC–tandem MS is mostly used for generating small quantitative sets of information on low processed samples, while GC–MS is usually used to obtain broad qualitative sets of information from samples that are volatile in nature (Leary et al. 2016; Lowe-Power et al. 2017; Wu et al. 2018). Metabolomics is specially helpful in the field of natural products to find out the novel compounds related with the activity of plant extracts employed for human health and disease treatment (Mumtaz et al. 2017).

6.4.1 Metabolomic Study in Host-Pathogen Attack

Effectors are the substances naturally produced by pathogens throughout the infection that interact with host activities in order to impede the protection or facilitate water and nutrient accessibility (Pritchard and Birch 2011). Effectors are likely to be proteins, but additionally they contain a number of non-protein metabolites (Collemare et al. 2019). *Pseudomonas syringae* which secretes polyketide-phytotoxin coronatine (COR) creates metabolic instabilities in contaminated plants by forming a strong coordination with JA receptor (Nomura et al. 2005; Geng et al. 2012). In addition, COR and other effectors ease the entry of pathogens into host tissues and cells (Toruño et al. 2016), and metabolomics may detect the metabolic dysfunction brought on by the effectors and their producers. For example, a number of effectors may be involved in affecting the stability between SA and JA-signaling in plants; COR and the effectors of microbial proteins target various forms of JA signals (Macho et al. 2010; Jiang et al. 2013; Yang et al. 2017; Nobori et al. 2018). Extensive work on maize and corn smut causative agent (*Ustilago maydis*) has indicated that effectors from this pathogen inhibit SA biosynthesis (Djamei et al. 2011; Djamei and Kahmann 2012). Toxins are also used by pathogens to test out invasion and colonization; although molecular genetics has been shown to be essential for determining the activity of toxins in pathogenesis, metabolomics plays a vital role in the characterization and identification of toxins (Kimura et al. 2001). For example, Rubrofusarin and many contaminants along with terpenes and various secondary metabolites were identified in Fusarium species. *Fusarium* species cause various diseases in cereal crops, and produce an extensive variety of toxins (Placinta et al. 1999; Zapparata et al. 2017). As a consequence of mycotoxin-related health issues, to diagnose lethal molecules in drinks whose by-products may be transmitted by plant-pathogenic fungi, LC–MS/MS protocols have been proposed (Al-Taher et al. 2013; Habler et al. 2017).

6.4.2 Metabolomics in the Induced Resistance Route

Metabolomics has helped in unwinding many aspects of priming or incited resistance. A plant should be prepared for microbe or herbivore attacks when past encounters with a specific raider have made the plant increasingly impervious to the forthcoming dangers. The priming is called systemic acquired resistance (SAR) if the elicitor is a microorganism, and as induced systemic resistance (ISR) if the intruder is a symbiotic fungus or plant growth promoting rhizobacteria (PGPR). Infection at one site induces resistance at the other far off site indicating that priming also works at a distance. Azelaic acid which is a subtle SAR-conferring signal molecule was detected by means of metabolomic profiling. In a similar manner, by using helpful ISR-inducing bacteria and growth promoting factors, possible volatile ISR metabolites around the plants have been detected by extracting them from the air and subjecting them to metabolomic profiling.

6.4.3 Metabolomics in Beneficial Biotic Interactions

There have been several studies on important biotic interactions in plants, especially mycorrhiza, plant growth promoting rhizobia (PGPR), and many forms of filamentous fungi, which include different *Trichodermic* strains. Numerous theoretical viewpoints such as nutrient intake, receptor recognition, and effect or transfer are all shared between mortal infection and advantageous relations. Many relations are described in the ways when a plant is primed against pathogens by a symbiont (first partner) associated with the plant (second partner). Metabolomics has played a key role in elucidating the connection and the differentiation between the disease and the symbiosis. The communications between *Arabidopsis*, PGPR, and various microbes were investigated in a substantial effort that acts as an effective illustration of the integration of different types of omic data (Van de Mortel et al. 2012). Many studies have described protection and signaling molecules in plants using metabolomics profiling and microarray, viz. SA, camalexin, and glucosinolates that facilitate the PGPR-stimulated priming of the plant against many pathogens. As metabolomics has shown that the relationship among plants and symbionts can be quite specific, e.g. it has been shown that two strains of PGPR isolated from two explicit rice cultivars (*Oryza sativa*) encourage rice yields, especially in the cultivar from which they were isolated, and this was also evident at the metabolic level (Chamam et al. 2013). A parallel protocol for arbuscular mycorrhiza (AM) was suggested much more extensively in which five distinct plant species were inhabited with a single common AM and the metabolomic profiling indicated remarkably divergent and species-specific metabolic changes regardless of a retained "core metabolome" (Schweiger et al. 2014). In an increasingly direct method, induced metabolites in AM-inhabited plants were checked for the metabolites and the

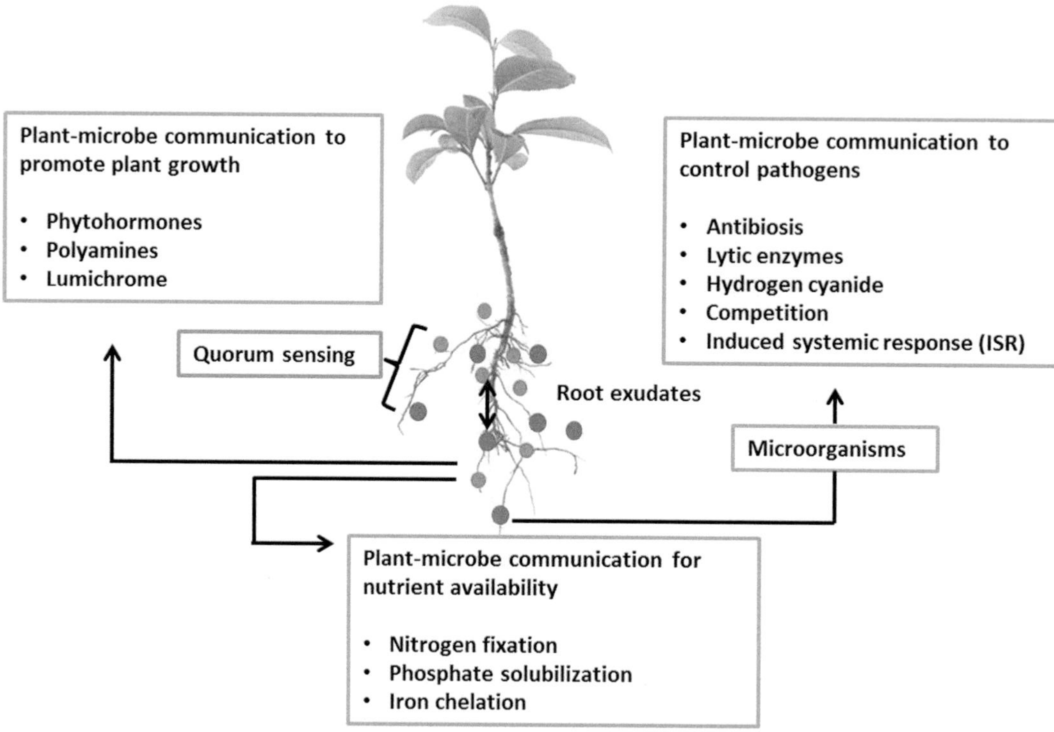

FIGURE 6.2 Plant-microbe interactions analyzed for improvement in agriculture (Verma et al. 2018).

detected metabolites were then added *in vitro* to the AM fungus, and several of them were responsible to cause colonization-related traits in the fungus (Laparre et al. 2014). The nature of three-way communication was further described in another study showing that the metabolic content of a diseased plant and that of the same plant associated with beneficial to *Trichoderma* fungi do not simply add up when the plant is challenged with both, exacerbating the dynamic complexity of the relationship (Brotman et al. 2012). Volatiles formed from root microbes in maize (*Z. mays*) were evaluated and they were known to affect maize resistance to pathogens and herbivores, and in tritrophic interactions, they were enticing natural foes of the herbivores. In the case of amarginal type of symbiotic cooperation including plants, the metabolomic status of epiphytic, commensal microorganisms living on plant parts above the ground was described using a metabolomic method of the environment (Ryffel et al. 2015). The complicated mechanisms by which plants "pick" the microorganisms with which their partners are progressively being revealed (Haney et al. 2015; Lebeis et al. 2015) give a fruitful ground to the application of metabolomics. The plant-microbe interaction is shown in Figure 6.2.

6.4.4 Metabolomics in Plant Defense

Plant protection metabolites are comprehensive and complex, as would be anticipated for an immobile organism that is subjected to a multitude of stresses, including microbial threats. In order to understand how metabolites operate in plant-pathogen relationships, it is important to identify molecules based on their structure, biosynthesis, restriction, or capability. Here, we depict a couple of classes associated with plant defense system utilizing metabolomic methods.

6.4.4.1 Phytoalexins and Phytoanticipins

We know that unique and complex metabolites are being produced by plants from a comparatively limited pool of building blocks. The two major forms of protection molecules resulting from metabolites of secondary origin are phytoalexins and phytoanticipins. Phytoalexins are the substances transmitted by the host as an immediate reaction to perception of the microorganism, but phytoanticipins are formed ahead of time of an attack and are just changed over to their poisonous structures post pathogen recognition (Pedras et al. 2008). Camalexin and indole-glucosinolates are biosynthesized from tryptophan and are the examples of such resistance compound groups (Schlaeppi and Mauch. 2010). Such molecules show synergistic resistance against *Phytophthora brassica* in *Arabidopsis*. Buxdorf et al. (2013) demonstrated that the *Brassicaceae*-specific pathogen *Alternaria brassicicola* is increasingly tolerant of certain glucosinolate hydrolysis products produced by *Arabidopsis thaliana* compared to the multi-host microbe *Botrytis cinerea* during metabolomics research into necrotrophic fungal microbes (Buxdorf et al. 2013). In response to the biotrophic fungal microbe *Plasmodiophora brassicae* which causes clubroot, it was reported in another metabolomics-based investigation that canola, a *Brassica* species, uses glucosinolates inside the root tissue, and besides the identification of realized defense compounds, this metabolomics method also revealed antifungal metabolites that had not previously been identified in plants and could be named as new phytoalexins.

6.4.4.2 Volatile Organic Compounds (VOCs)

Due to the dynamic nature of VOCs, they are considered as ideal targets for research into metabolomics and an important feature of plant protection molecules. Being vaporous in nature, the volatile compounds are not limited to the small space of pathogens attack or colonization sites like some other defense molecules; instead, they are secreted for local and long-term effects on the natural environment. The latest study has shown that plants releasing volatile compounds have beneficial effects on neighboring herbivory-prone plants, e.g. maize (*Z. mays*) plants exposed to molasses grass (*Melinis minutiflora*) for 21 days showed reduced egg laying by stem borers (*Chilo partellus*) in comparison to maize plants grown in the absence of molasses grass. Remarkably, this observation suggested that VOCs from exposed grass plants contained significant quantities of molecules such as (R)-linalool, a compound believed to have an effect on insect pest activity (Huff and Pitts 2019; Tolosa et al. 2019). VOCs not only protect the plant from herbivores but also help in attracting useful microorganisms (López-Gresa et al. 2018).

6.5 Improvement and Trends in Metabolomics

Since the beginning of the field of metabolomics, a number of novel developments have been identified and these are described in the framework of plant biotic interaction studies. Studies of metabolomics in plants under stress condition have advanced over the years from frill-free listings of metabolites such as flavonoid screening with subsequent application of a biotic inducer in *Lupinu salbus* (Bednarek et al. 2001) and metabolomic fingerprinting of *Catharanthus roseus* contaminated with phytoplasma (Choi et al. 2004) to studies which position metabolites in a mechanistic context, e.g. explanation of antifungal protection mechanisms in maize using metabolomic profiling of secondary metabolites (Balmer et al. 2013), and the analysis of tobacco metabolite (*Nicotiana tabacum*) after the use of ergosterol, a fungal molecule that induces defense responses in plants(Tugizimana et al. 2014). A shift from single- to multi-omics approach was the need of the hour to explore diverse metabolites with immense properties. To sum up, the metabolomics field has provided countless scientific as well as conceptual steps over the last decade and allowed the discovery of spatial, temporal, cellular, and ecological dimensions and the incorporation of data from neighboring omics fields in deep studies using different platforms.

6.6 Challenges and Future Perspectives

The key challenge while studying plant-microbe interaction is to differentiate the metabolites released by the host and the pathogen. However, Pang et al. (2018) have shown that labeling could overcome this problem in the stomatal cells of *Arabidopsis* after inoculation with *P. syringae*. The marking of bacterial metabolites with heavy isotopes helped to distinguish them from the metabolites of the host, using a selective metabolite detection method. This technique helped to distinguish the organic and amino acids of bacteria, which would otherwise have been confused as products of plants. Later, it was predicted that *P. syringae* reprogrammed primary metabolism and cell signaling of host to modulate stomatal movement (Pang et al. 2018). This was not the first effort to isolate metabolites in this pathosystem; the FT-IS was used by Allwood et al. (2010) to acquire different metabolic fingerprints of *Arabidopsis* and *P. syringae*. Although this technique led to a separate analysis of metabolic profiles of plants and pathogens, it did not really distinguish each compound; instead, it observed differences in level. The detection of unknown compounds is another main problem faced by metabolomics, which is important particularly in the field of plant pathology. The definite metabolite repositories for pathogens or plants like those utilized by humans and yeast are not yet available for studies (Jewison et al. 2012; Wishart et al. 2018). In addition to the complex relationship among host species and their invaders, the essence of secondary metabolism, which serves as the main participant in plant protection and microbiota infection, certainly does not make annotation of metabolites easier. Recognition of specific metabolites, therefore, usually requires labor-intensive metabolite extraction, filtration, and NMR, or predictive synthesis and validation by matching LC–MS mass transitions. Due to this purpose, several organizations linked to such work have their own handmade collections, which often contain metabolites very unique to their research interests and are not necessarily open to the public (Johnson and Lange 2015). The use of metabolomic profiles for detection and identification of pathogenic compounds produced throughout infection/colonization has recently progressed and can offer more proof on the microbial side of the pathogens relationship with their hosts. In order to understand the host-pathogen relationship, the incorporation of proteome, transcriptome, and metabolome is a crucial factor. However, it is not an easy task to incorporate all the omics datasets though the genotype integration with phenotype is not as easy as linking genomic data with proteomics, as there is no clear correlation between a gene or a group of genes and a specific metabolite (Pinu et al. 2019). Many systems are designed to integrate omics data, but are not easy to use (Misra et al. 2019) because they need specialized bioinformatics skills as well. The Metabo Analyst software still has a pleasant configuration, which could be used online or with R software and now started to incorporate omics data in its database and pathway analysis platforms (Chong et al. 2018). Advances in the integration of multi-omics data are required to assist studies of plant-microbe interaction. Incorporating genomic and proteomic data into a metabolomic approach will require coordination, knowledge allocation, and details. All this will allow scientists to build databanks and platforms that integrate the fundamental evidence for recognition of host-pathogen communications and to develop new methodologies for improving plant resistance to biotic stress.

6.7 Conclusion

The numerous metabolites recognized so far only denote the tip of the iceberg. Though researchers have studied many classes of secondary metabolites extensively, still other groups remain ignored. In addition, plant metabolite annotation deficiency is a major bottleneck; several steps have been taken over the past few years, including standardizing and reporting new methods for metabolomic studies. The technology is persistently improving, and this progress will not only benefit active researchers but also increase the accessibility of metabolomics and assist in its widespread application. The latest methods make the work much simpler by testing a large number of samples in much less time; however, these methods are costly and laborious. Traditional methods are also capable of monitoring single rhizosphere signaling network interaction threads at a time, but do not provide complete preview of these interactions. Therefore, more work is required to understand the impact of root-released metabolites on microorganisms and microbial community signals in order to get different plant species for sustainable agriculture engineering in the future.

REFERENCES

Aharoni, A., De Vos, C. H. R. and Verhoeven, H. A. 2002. Nontargeted metabolome analysis by use of Fourier transform ion cyclotron mass spectrometry. *OMICS* 6:217–234.

Akiyama, K., Chikayama, E. and Yuasa, H. 2008. PRIMe: a website that assembles tools for metabolomics and transcriptomics. *In Silico Biol* 8:339–345.

Allwood, J., Clarke, A., Goodacre, R. and Mur, L. A. J. 2010. Dual metabolomics: a novel approach to understanding plant-pathogen interactions. *Phytochemistry* 71:590–597.

Al-Taher, F., Banaszewski, K., Jackson, L., Zweigenbaum, J., Ryu, D. and Cappozzo, J. 2013. Rapid method for the determination of multiple mycotoxins in wines and beers by LC-MS/MS using a stable isotope dilution assay. *J Agric Food Chem* 61:2378–2384.

Balmer, D., Villacres de Papajewski, D., Planchamp, C., Glauser, G. and Mauch-Mani, B. 2013. Induced resistance in maize is based on organ-specific defence responses. *Plant J* 74:213–225.

Baranska, M., Schulz, H., Rosch, P., Strehle, M. A. and Popp, J. 2004. Identification of secondary metabolites in medicinal and spice plants by NIR-FT-Raman microspectroscopic mapping. *Analyst* 129:926–930.

Baranska, M., Schulz, H., Kruger, H. and Quilitzsch, R. 2005. Chemo-taxonomy of aromatic plants of the genus *Origanum* via vibrational spectroscopy. *Anal Bioanal Chem* 381:1241–1247.

Bedair, M. and Sumner, L. W. 2008. Current and emerging mass spectrometry technologies for metabolomics. *Trends Anal Chem* 27:238–250.

Bednarek, P., Frańskib, R., Kerhoasc, L., Einhornc, J., Wojtaszeka, P. and Stobiecki, M. 2001. Profiling changes in metabolism of isoflavonoids and their conjugates in *Lupinu salbus* treated with biotic elicitor. *Phytochemistry* 56:77–85.

Bergelson, J., Kreitman, M., Stahl, E. A. and Tian, D. 2001. Evolutionary dynamics of plant R-genes. *Science* 292:2281–2285.

Bi, F., Barad, S., Ment, D., Luria, N., Dubey, A., Casado, V., Glam, N., Mínguez, J. D., Espeso, E. A., Fluhr, R. and Prusky, D. 2015. Carbon regulation of environmental pH by secreted small molecules that modulate pathogenicity in phytopathogenic fungi. *Mol Plant Pathol* 17(8):1178–1195.

Bino, R. J., Hall, R. D. and Fiehn, O. 2004. Potential of metabolomics as a functional genomics tool. *Trends Plant Sci* 9:418–425.

Bourdenx, B., Bernard, A., Domergue, F., Pascal, S., Léger, A., Roby, D., Pervent, M., Vile, D., Haslam, R. P., Napier, J. A. and Lessire, R. 2011. Over-expression of *Arabidopsis* ECERIFERUM1 promotes wax very-long-chain alkane biosynthesis and influences plant response to biotic and abiotic stresses. *Plant Physiol* 156:29–45.

Brotman, Y., Lisec, J., Méret, M., Chet, I., Willmitzer, L. and Viterbo, A. 2012. Transcript and metabolite analysis of the Trichoderma-induced systemic resistance response to *Pseudomonas syringae* in *Arabidopsis thaliana*. *Microbiology* 158:139–146.

Buxdorf, K., Yaffe, H., Barda, O. and Levy, M. 2013. The effects of glucosinolates and their breakdown products on necrotrophic fungi. *PLoS One* 8:e70771.

Castro-Moretti, F. R., Gentzel, I. N., Mackey, D. and Alonso, A. P. 2020. Metabolomics as an emerging tool for the study of plant–pathogen interactions. *Metabolites* 10:52.

Cevallos-Cevallos, J. M., Rouseff, R. and Reyes-DeCorcuera, J. I. 2009. Untargeted metabolite analysis of healthy and Huanglongbing-infected orange leaves by CE-DAD. *Electrophoresis* 30:1240–1247.

Chamam, A., Sanguina, H., Bellvert, F., Meiffren, G., Comte, G., Wisniewski-Dyé, F., Bertrand, C. and Prigent-Combaret, C. 2013. Plant secondary metabolite profiling evidences strain-dependent effect in the *Azospirillum–Oryza sativa* association. *Phytochemistry* 87:65–77.

Chen, X. Y. Y. and Kim, J. Y. Y. 2009. Callose synthesis in higher plants. *Plant Signal Behav* 4:489–492.

Choi, Y. H., Tapias, E. C., Kim, H. K., Lefeber, A. W., Erkelens, C., Verhoeven, J. T. J., Brzin, J., Zel, J. and Verpoorte, R. 2004. Metabolic discrimination of *Catharanthus roseus* leaves infected by phytoplasma using 1H-NMR spectroscopy and multivariate data analysis. *Plant Physiol* 135:2398–2410.

Chong, J., Soufan, O., Li, C., Caraus, I., Li, S., Bourque, G., Wishart, D. S. and Xia, J. 2018. MetaboAnalyst 4.0: towards more transparent and integrative metabolomics analysis. *Nucleic Acids Res* 46:486–494.

Clay, N. K., Adio, A. M., Denoux, C., Jander, G. and Ausubel, F. M. 2009. Glucosinolate metabolites required for an *Arabidopsis* innate immune response. *Science* 323:95–101.

Cocuron, J. C., Anderson, B., Boyd, A. and Alonso, A. P. 2014. Targeted metabolomics of *Physaria fendleri*, an industrial crop producing hydroxy fatty acids. *Plant Cell Physiol* 55:620–633.

Cocuron, J. C., Casas, M. I., Yang, F., Grotewold, E. and Alonso, A. P. 2019. Beyond the wall: high-throughput quantification of plant soluble and cell-wall bound phenolics by liquid chromatography tandem mass spectrometry. *J Chromatogr A* 1589:93–104.

Collemare, J., O'Connell, R. and Lebrun, M. 2019. Nonproteinaceous effectors: the *terra incognita* of plant–fungal interactions. *New Phytol* 223:590–596.

Cusumano, A., Weldegergis, B. T., Colazza, S., Dicke, M. and Fatouros, N. E. 2015. Attraction of egg-killing parasitoids toward induced plant volatiles in a multi-herbivore context. *Oecologia* 179:163–174.

D'Alessandro, M., Erb, M., Ton, J., Brandenburg, A., Karlen, D., Zopfi, J. and Turlings, T. C. J. 2014. Volatiles produced by soil-borne endophytic bacteria increase plant pathogen resistance and affect tritrophic interactions. *Plant Cell Environ* 37:813–826.

Denancé, N., Sánchez-Vallet, A., Goffner, D. and Molina, A. 2013. Disease resistance or growth: the role of plant hormones in balancing immune responses and fitness costs. *Front Plant Sci* 4:1–12.

Devaux, P. G., Homing, M. G. and Homing, B. C. 1971. Benyzloxime derivative of steroids, a new metabolic profile procedure for human urinary steroids. *Anal Lett* 4:151.

Djamei, A., Schipper, K., Rabe, F., Ghosh, A., Vincon, V., Kahnt, J., Osorio, S., Tohge, T., Fernie, A. R., Feussner, I. and Feussner, K. 2011. Metabolic priming by a secreted fungal effector. *Nature* 478:395–398.

Djamei, A. and Kahmann, R. 2012. Ustilago maydis: dissecting the molecular interface between pathogen and plant. *PLoS Pathog* 8:11–14.

Dunn, W. B., Bailey, N. J. C. and Johnson, H. E. 2005. Measuring the metabolome: current analytical technologies. *Analyst* 130:606–625.

Erb, M., Veyrat, N., Robert, C. A. M., Xu, H., Frey, M., Ton, J. and Turlingse, T. C. J. 2015. Indole is an essential herbivore-induced volatile priming signal in maize. *Nat Commun* 6:6273.

Farag, M. A., Huhman, D. V., Lei, Z. and Sumner, L. W. 2007. Metabolic profiling and systematic identification of flavonoids and isoflavonoids in roots and cell suspension cultures of *Medicago truncatula* using HPLC-UV-ESI-MS and GC-MS. *Phytochemistry* 68:342–354.

Feng, L., Frank, J. G. and Xiaochao, M. 2012. LC–MS-based metabolomics in profiling of drug metabolism and bioactivation. *Acta Pharm Sin B* 2:116–123.

Fiehn, O. 2001. Combining genomics, metabolome analysis, and biochemical modelling to understand metabolic networks. *Comp Funct Genom* 2:155–168.

Fiehn, O. and Weckwerth, W. 2003. Deciphering metabolic networks. *Eur J Biochem* 270:579–588.

Fiehn, O., Kopka, J., Trethewey, R. N. and Willmitzer, L. 2000. Identification of uncommon plant metabolites based on calculation of elemental compositions using gas chromatography and quadrupole mass spectrometry. *Anal Chem* 72:3573–3580.

Gavaghan, C. L., Li, J. V. and Hadfield, S. T. 2011. Application of NMR-based metabolomics to the investigation of salt stress in maize (*Zea mays*). *Phytochem Anal* 22:214–224.

Geng, X., Cheng, J., Gangadharan, A. and Mackey, D. 2012. The coronatine toxin of *Pseudomonas syringaeis* a multifunctional suppressor of Arabidopsis defense. *Plant Cell* 24:4763–4774.

González-Lamothe, R., Mitchell, G., Gattuso, M., Diarra, M., Malouin, F. and Bouarab, K. 2009. Plant antimicrobial agents and their effects on plant and human pathogens. *Int J Mol Sci* 10:3400–3419.

Goodacre, R. 2007. Metabolomics of a superorganism. *J Nutr* 137:259–266.

Griffin, J. L. and Shockcor, J. P. 2004. Metabolic profiles of cancer cells. *Nat Rev Cancer* 4:551–561.

Habler, K., Gotthardt, M., Schüler, J. and Rychlik, M. 2017. Multi-mycotoxin stable isotope dilution LC–MS/MS method for *Fusarium* toxins in beer. *Food Chem* 218:447–454.

Hall, R., Beale, M., Fiehn, O., Hardy, N., Sumner, L. and Bino, R. 2002. Plant metabolomics: the missing link in functional genomics strategies. *Plant Cell* 14:1437–1440.

Haney, C. H., Samuel, B. S., Bush, J. and Ausubel, F. M. 2015. Associations with rhizosphere bacteria can confer an adaptive advantage to plants. *Nat Plants* 1:15051.

Harrigan, G. G. and Goodacre, R. 2003. *Metabolic Profiling: Its Role in Biomarker Discovery and Gene Function Analysis*. Kluwer Academic Publishers, Springer US.

Horning, B. C. and Horning, M. G. 1970. *Metabolic Profiles: Chromatographic Methods for Isolation and Characterization of a Variety of Metabolites in Man*. Yearbook Medical Publishers, Chicago.

Huff, R. M. and Pitts, R. J. 2019. An odorant receptor from *Anopheles gambiae* that demonstrates enantioselectivity to the plant volatile, linalool. *PLoS One* 14:e0225637.

Jewison, T., Knox, C., Neveu, V., Djoumbou, Y., Guo, A. C., Lee, J., Liu, P., Mandal, R., Krishnamurthy, R., Sinelnikov, I. and Wilson, M. 2012. YMDB: the yeast metabolome database. *Nucleic Acids Res* 40:815–820.

Jiang, S., Yao, J., Ma, K. W., Zhou, H., Song, J., He, S. Y. and Ma, W. 2013. Bacterial effector activates jasmonate signaling by directly targeting JAZ transcriptional repressors. *PLoS Pathog* 9:e1003715.

Johnson, S. R. and Lange, B. M. 2015. Open-access metabolomics databases for natural product research: present capabilities and future potential. *Front Bioeng Biotechnol* 3:1–10.

Kaderbhai, N. N., Broadhurst, D. I., Ellis, D. I., Goodacre, R. and Kell, D. B. 2003. Functional genomics via metabolic footprinting: monitoring metabolite secretion by *Escherichia coli* tryptophan metabolism mutants using FT-IR and direct injection electrospray mass spectrometry. *Comp Funct Genom* 4:376–391.

Kim, H. K., Choi, Y. H. and Verpoorte, R. 2011. NMR-based plant metabolomics: where do we stand, where do we go? *Trends Biotechnol* 29:267–275.

Kim, J. K., Bamba, T., Harada, K., Fukusaki, E. and Kobayashi, A. 2007. Time-course metabolic profiling in *Arabidopsis thaliana* cell cultures after salt stress treatment. *J Exp Bot* 58:415–424.

Kimura, M., Anzai, H. and Yamaguchi, I. 2001. Microbial toxins in plant-pathogen interactions: biosynthesis, resistance mechanisms, and significance. *J Gen Appl Microbiol* 47:149–160.

Laparre, J., Malbreil, M., Letissecde,F., Portais, J. C., Roux, C., Bécard, G. and Puech-Pagès, V. 2014. Combining metabolomics and gene expression analysis reveals that propionyl- and butyryl-carnitines are involved in late stages of *arbuscular mycorrhizal* symbiosis. *Mol Plant* 7:554–566.

Leary, B. M., Neale, H. C., Geilfus, C. M., Jackson, R. W., Arnold, D. L. and Preston, G. M. 2016. Early changes in apoplast composition associated with defence and disease in interactions between Phaseolus vulgaris and the halo blight pathogen *Pseudomonas syringae* Pv. phaseolicola. *Plant Cell Environ* 39:2172–2184.

Lebeis, S. L., Paredes, S. H., Lundberg, D. S., Breakfield, N., Gehring, J., McDonald, M., Malfatti, S., Del Rio, T. G., Jones, C. D., Tringe, S. G. and Dangl, J. L. 2015. Salicylic acid modulates colonization of the root microbiome by specific bacterial taxa. *Science* 349:860–864.

Lei, Z., Huhman, D. V. and Sumner, L. W. 2011. Mass spectrometry strategies in metabolomics. *J Biol Chem* 29:25435–25442.

Leiss, K. A., Choi, Y. H., Verpoorte, R. and Klinkhamer, P. G. L. 2011. An overview of NMR-based metabolomics to identify secondary plant compounds involved in host plant resistance. *Phytochem Rev* 10:205–216.

Liu, X. and Locasale, J. W. 2017. Metabolomics: a primer. *Trends Biochem Sci* 42:274–284.

López-Gresa, M. P., Payá, C., Ozáez, M., Rodrigo, I., Conejero, V., Klee, H., Bellés, J. M. and Lisón, P. 2018. A new role for green leaf volatile esters in tomato stomatal defense against *Pseudomonas syringe* pv. tomato. *Front Plant Sci* 9:1855.

Lowe-Power, T. M., Hendrich, C. G., von Roepenack-Lahaye, E., Li, B., Wu, D., Mitra, R., Dalsing, B. L., Ricca, P., Naidoo, J., Cook, D. and Jancewicz, A. 2017. Metabolomics of tomato xylem sap during bacterial wilt reveals *Ralstonia solanacearum* produces abundant putrescine, a metabolite that accelerates wilt disease. *Environ Microbiol* 20:1330–1349.

Luna, E., Pastor, V., Robert, J., Flors, V., Mauch-Mani, B. and Ton, J. 2011. Callose deposition: a multifaceted plant defense response. *Mol Plant-Microbe Interact* 24:183–193.

Macho, A. P., Guevara, C. M., Tornero, P., Ruiz-Albert, J. and Beuzón, C. R. 2010. The *Pseudomonas syringae* effector protein HopZ1a suppresses effector-triggered immunity. *New Phytol* 187:1018–1033.

Manning, V. A., Chu, A. L., Steeves, J. E., Wolpert, T. J. and Ciuffetti, L. M. 2009. A host-selective toxin of *Pyrenophora tritici-repentis*, PtrToxA, induces photosystem changes and reactive oxygen species accumulation in sensitive wheat. *Mol Plant-Microbe Interact* 22:665–676.

Méchin, V., Laluc, A., Legée, F., Cézard, L., Denoue, D., Barrière, Y. and Lapierre, C. 2014. Impact of the brown-midrib bm 5 mutation on maize lignins. *J Agric Food Chem* 62:5102–5107.

Mendes, P., Kell, D. B. and Westerhoff, H. V. 1992. Channelling can decrease pool size. *Eur J Biochem* 204:257–266.

Mendes, P., Kell, D. B. and Westerhoff, H. V. 1996. Why and when channeling can decrease pool size at constant net flux in a simple dynamic channel. *Biochimicaet Biophysica Acta* 1289:175–186.

Misra, B. B., Langefeld, C., Olivier, M. and Cox, L. A. 2019. Integrated omics: tools, advances and future approaches. *J Mol Endocrinol* 62:21–45.

Mochida, K. and Shinozaki, K. 2010. Genomics and bioinformatics resources for crop improvement. *Plant Cell Physiol* 51:497–523.

Moco, S., Bino, R. J. and Vorst, O. 2006. A liquid chromatography-mass spectrometry-based metabolome database for tomato. *Plant Physiol* 141:1205–1218.

Moller, M. and Stukenbrock, E. H. 2017. Evolution and genome architecture in fungal plant pathogens. *Nat Rev Microbiol* 15:756–771.

Mumtaz, M. W., Hamid, A. A., Akhtar, M. T., Anwar, F., Rashid, U. and Al-Zuaidy, M. H. 2017. An overview of recent developments in metabolomics and proteomics–phytotherapic research perspectives. *Front Life Sci* 10:1–37.

Nobori, T., Mine, A. and Tsuda, K. 2018. Molecular networks in plant–pathogen holobiont. *FEBS Lett* 592:1937–1953.

Nomura, K., Melotto, M. and He, S. Y. 2005. Suppression of host defense in compatible plant-*Pseudomonas syringae* interactions. *Curr Opin Plant Biol* 8:361–368.

Oksman-Caldentey, K. M. and Inze, D. 2004. Plant cell factories in the post-genomic era: new ways to produce designer secondary metabolites. *Trends Plant Sci* 9:433–440.

Olanrewaju, O. S., Glick, B. R. and Babalola, O. O. 2017. Mechanisms of action of plant growth promoting bacteria. *World J Microb Biot* 33:197.

Pang, Q., Zhang, T., Wang, Y., Kong, W., Guan, Q., Yan, X. and Chen, S. 2018. Metabolomics of early stage plant cell–microbe interaction using stable isotope labeling. *Front Plant Sci* 9:760.

Pearse, I. S., Gee, W. S. and Beck, J. J. 2013. Headspace volatiles from 52 oak species advertise induction, species identity, and evolution, but not defense. *J Chem Ecol* 39:90–100.

Pedras, M. S. C., Zheng, Q. A. and Strelkov, S. 2008. Metabolic changes in roots of the oilseed canola infected with the biotroph *Plasmodiophora brassicae*: phytoalexins and phytoanticipins. *J Agric Food Chem* 56:9949–9961.

Pinu, F. R., Beale, D. J., Paten, A. M., Kouremenos, K., Swarup, S., Schirra, H. J. and Wishart, D. 2019. Systems biology and multi-omics integration: viewpoints from the metabolomics research community. *Metabolites* 9:76–84.

Placinta, C. M., D'Mello, J. P. F. and MacDonald, A. M. C. 1999. A review of worldwide contamination of cereal grains and animal feed with *Fusarium mycotoxins*. *Anim Feed Sci Technol* 78:21–37.

Prince, D. C., Drurey, C., Zipfel, C. and Hogenhout,S. A. 2014. The leucine-rich repeat receptor-like kinase brassinosteroid insensitive1-associated kinase1 and the cytochrome P450 phytoalexin deficient3 contribute to innate immunity to aphids in *Arabidopsis*. *Plant Physiol* 164:2207–2219.

Pritchard, L. and Birch, P. A. 2011. Systems biology perspective on plant–microbe interactions: biochemical and structural targets of pathogen effectors. *Plant Sci* 180:584–603.

Rampitsch, C. and Bykova, N. V. 2012. Proteomics and plant disease: advances in combating a major threat to the global food supply. *Proteomics* 12:673–690.

Ribbenstedt, A., Ziarrusta, H. and Benskin, J. P. 2018. Development, characterization and comparisons of targeted and non-targeted metabolomics methods. *PLoS One* 13:e0207082.

Roessner, U., Wagner, C., Kopka, J., Trethewey, R. N. and Willmitzer, L. 2000. Simultaneous analysis of metabolites in potato tuber by gas chromatography-mass spectrometry. *Plant J* 23:131–142.

Roessner, B. U., Luedemann, A., Brust, D., Fiehn, O., Linke, T., Willmitzer, L. and Fernie, A. R. 2001. Metabolic profiling allows comprehensive phenotyping of genetically or environmentally modified plant systems. *Plant Cell* 13:11–29.

Ryffel, F., Helfrich, E. J. N., Kiefer, P., Peyriga, L., Portais, J. C., Piel, J. and Vorholt, J. A. 2015. Metabolic footprint of epiphytic bacteria on *Arabidopsis thaliana* leaves. *ISME J* 10:632–643.

Saito, K., Hirai, M. Y. and Yonekura-Sakakibara, K. 2008. Decoding genes with coexpression networks and metabolomics-majority report by precogs. *Trends Plant Sci* 13:36–43.

Sanchez, D. H., Siahpoosh, M. R., Roessner, U., Udvardi, M. and Kopka, J. 2008. Plant metabolomics reveals conserved and divergent metabolic responses to salinity. *Physiol Plant* 132:209–219.

Schlaeppi, K. and Mauch, F. 2010. Indolic secondary metabolites protect Arabidopsis from the oomycete pathogen *Phytophthora brassicae*. *Plant Signal Behav* 5:1099–1101.

Schripsema, J. 2010. Application of NMR in plant metabolomics: techniques, problems and prospects. *Phytochem Anal* 21:14–21.

Schulz, H., Schrader, B., Quilitzsch, R., Pfeffer, S. and Kruger, H. 2003. Rapid classification of basil chemotypes by various vibrational spectroscopy methods. *J Agric Food Chem* 51:2475–2481.

Schulz, H., Baranska, M., Quilitzsch, R., Schutze, W. and Losing, G. 2005. Characterization of peppercorn, pepper oil, and pepper oleoresin by vibrational spectroscopy methods. *J Agric Food Chem* 53:3358–3363.

Schweiger, R., Baier, M. C., Persicke, M. and Müller, C. 2014. High specificity in plant leaf metabolic responses to arbuscular mycorrhiza. *Nat Commun* 5:1–11.

Shafi, A. and Zahoor, I. 2020. Metabolomics of medicinal and aromatic plants: goldmines of secondary metabolites for herbal medicine research. In: Aftab T. and Hakeem K. H. (eds), *Medicinal and Aromatic Plants*. Academic Press, Elsevier, pp:261–287.

Shroff, R., Schramm, K., Jeschke, V., Nemes, P., Vertes, A., Gershenzon, J. and Svatoš, A. 2015. Quantification of plant surface metabolites by matrix-assisted laser desorption–ionization mass spectrometry imaging: glucosinolates on *Arabidopsis thaliana* leaves. *Plant J* 81:961–972.

Soga, T. 2007. Capillary electrophoresis-mass spectrometry for metabolomics. *Methods Mol Biol* 358:129–137.

Spratlin, J. L., Serkova, N. J. and Eckhardt, S. G. 2009. Clinical applications of metabolomics in oncology: a review. *Clin Cancer Res* 15:431–440.

Staszkow, A., Swarcewicz, B., Banasiak, J., Muth, D., Jasiński, M. and Stobiecki, M. 2011. LC/MS profiling of flavonoid glycol conjugates isolated from hairy roots, suspension root cell cultures and seedling roots of *Medicago truncatula*. *Metabolomics* 7:604–613.

Stitt, M. and Fernie, A. R. 2003. From measurements of metabolites to metabolomics: an "on the fly" perspective illustrated by recent studies of carbon nitrogen interactions. *Curr Opin Biotechnol* 14:136–144.

Sugimoto, K., Matsui, K., Iijima, Y., Akakabe, Y., Muramoto, S., Ozawa, R., Uefune, M., Sasaki, R., Alamgir, K. M., Akitake, S. and Nobuke, T. 2014. Intake and transformation to a glycoside of (Z)-3-hexenol from infested neighbors reveals a mode of plant odor reception and defense. *Proc Natl Acad Sci USA* 111:7144–7149.

Thiocone, A., Farmer, E. E. and Wolfender, J. L. 2008. Screening for wound-induced oxylipinsin *Arabidopsis thaliana* by differential HPLC-APCI/MS profiling of crude leaf extracts and subsequent characterization by capillary-scale NMR. *Phytochem Anal* 19:198–205.

Tolosa, T. A., Tamiru, A., Midega, C. A. O., Van Den Berg, J., Birkett, M. A., Woodcock, C. M., Bruce, T. J., Kelemu, S., Pickett, J. A. and Khan, Z. R. 2019. Molasses grass induces direct and indirect defense responses in neighbouring maize plants. *J Chem Ecol* 45:982–992.

Torres, M. A., Jones, J. D. G. and Dangl, J. L. 2006. Reactive oxygen species signaling in response to pathogens. *Plant Physiol* 141:373–378.

Toruño, T. Y., Stergiopoulos, I. and Coaker, G. 2016. Plant-pathogen effectors: cellular probes interfering with plant defenses in spatial and temporal manners. *Annu Rev Phytopathol* 54:419–441.

Tsogtbaatar, E., Cocuron, J. C., Sonera, M. C. and Alonso, A. P. 2015. Metabolite fingerprinting of pennycress (*Thlaspiarvense* L.) embryos to assess active pathways during oil synthesis. *J Exp Bot* 66:4267–4277.

Tugizimana, F., Steenkamp, P. A., Piater, L. A. and Dubery, I. A. 2014. Multi-platform metabolomic analyses of ergosterol-induced dynamic changes in *Nicotiana tabacum* cells. *PLoS One* 9:e87846.

Urakami, K., Zangiacomi, V., Yamaguchi, K. and Kusuhara, M. 2010. Quantitative metabolome profiling of *Illicium anisatum* by capillary electrophoresis time-of-flight mass spectrometry. *Biomed Res* 31:161–163.

Van de Mortel, J. E., de Vos, R. C. H., Dekkers, E., Pineda, A., Guillod, L., Bouwmeester, K., van Loon, J. J. A., Dicke, M. and Raaijmakers, J. M. 2012. Metabolic and transcriptomic changes induced in Arabidopsis by the rhizobacterium *Pseudomonas fluorescens* SS101. *Plant Physiol* 160:2173–2188.

Verma, A., Kumar, S., Hemansi Kumar, G., Saini, J. K., Agrawal, R., Satlewal, A. and Ansari, M. W. 2018. Rhizosphere metabolite profiling: an opportunity to understand plant-microbe interactions for crop improvement. *Crop Improv Microb Biotechnol* 17:343–361.

Ward, J. L., Baker, J. M. and Beale, M. H. 2007. Recent applications of NMR spectroscopy in plant metabolomics. *FEBS J* 274:1126–1131.

Weckwerth, W. 2003. Metabolomics in systems biology. *Annu Rev Plant Biol* 54:669–689.

Williams, B. J., Cameron, C. J., Workman, R., Broeckling, C. D., Sumner, L. W. and Smith, J. T. 2007. Amino acid profiling in plant cell cultures: an inter-laboratory comparison of CE/MS and GC/MS. *Electrophoresis* 28:1371–1379.

Wishart, D. S., Feunang, Y. D., Marcu, A., Guo, A. C., Liang, K., Vázquez-Fresno, R., Sajed, T., Johnson, D., Li, C., Karu, N. and Sayeeda, Z. 2018. HMDB 4.0: the human metabolome database for 2018. *Nucleic Acids Res* 46:608–617.

Wu, S., Tohge, T., Cuadros-Inostroza, Á, Tong, H., Tenenboim, H., Kooke, R., Méret, M., Keurentjes, J. B., Nikoloski, Z., Fernie, A. R. and Willmitzer, L. 2018. Mapping the *Arabidopsis* metabolic landscape by untargeted metabolomics at different environmental conditions. *Mol Plant* 11:118–134.

Xu, T., Lei, L., Shi, J., Wang, X., Chen, J., Xue, M., Sun, S., Zhan, B., Xia, Z., Jiang, N. and Zhou, T. 2019. Characterization of maize translational responses to sugarcane mosaic virus infection. *Virus Res* 259:97–107.

Yang, L., Teixeira, P. J. P. L., Biswas, S., Finkel, O. M., He, Y., Salas-Gonzalez, I., English, M. E., Epple, P., Mieczkowski, P. and Dangl, J. L. 2017. *Pseudomonas syringae* type III effector HopBB1 promotes host transcriptional repressor degradation to regulate phytohormone responses and virulence. *Cell Host Microbe* 21:156–168.

Zapparata, A., Da Lio, D., Somma, S., Muñoz, I. V., Malfatti, L., Vannacci, G., Moretti, A., Baroncelli, R. and Sarrocco, S. 2017. Genome sequence of *Fusarium graminearum* ITEM 124 (ATCC 56091), a mycotoxigenic plant pathogen. *Genome Announc* 5:e01209–17.

Zhang, H. and Sonnewald, U. 2017. Differences and commonalities of plant responses to single and combined stresses. *Plant J* 90:839–855.

7

Arsenic Phytotoxicity via the Membrane Transporters and Its Mechanism of Action and Detoxification in Plant Cells

Amna, Sadia Qamar, and M. Irfan Qureshi
Jamia Millia Islamia, New Delhi, India

CONTENTS

7.1 Introduction

Arsenic (As) is an element existing naturally and distributed widely within the crust of the earth. Arsenic, a class 1 carcinogen (International Agency for Research on Cancer (IARC) 2004), is widely distributed in the environment. Inorganic compounds of arsenic are formed in the environment by the combination of arsenic with chlorine, oxygen, and sulfur. In plants and animals, arsenic combines with hydrogen and carbon leading to the formation of organic compounds of arsenic. Inorganic compounds of arsenic are majorly involved in wood preservation. Copper chromated arsenate (CCA) is utilized in making "pressure-treated" lumber. In the United States, CCA has not been used anymore for residential purposes though it is still utilized for industrial applications (Barton 2014). Organic compounds of arsenic are utilized as pesticides, mainly on orchards and cotton fields. Arsenic is found to be highly harmful to the plants. Phytochelatin (PC) and other compounds of thiol undergo complexation in order to detoxify arsenic present in the plants. In some areas of Southeast Asia, agricultural soils are contaminated with arsenic as a result of mining activities, irrigation with groundwater contaminated with arsenic and industrial activities that result in the increased risk of arsenic accumulation in crop plants (Meharg and Rahman 2003; Williams et al. 2009). High arsenic concentrations in crop plants can pose a significant risk to human health through dietary exposure (Meharg et al. 2009; Zhao et al. 2010). Therefore, it is necessary to understand how plants cope up with arsenic toxicity. Arsenite [As (III)] and arsenate [As (V)] are the major arsenic species present in the soil and ever-increasing human activities contribute to the heavy metal input to the agricultural soils, which results in the contamination of the soil, water, and eventually affects the health of plants. Due to human and natural activities occurring continuously within the environment, the heavy metals ubiquitously exist within the environment via several pathways and ultimately heavy metals come in direct contact with plants, animals, and humans (Wilson and Pyatt 2007). In plants, several outwardly expressed responses to the stress are caused by these toxic heavy metal ions possessing easily detectable symptoms like damaged root, chlorosis of leaves, inhibition of the plant growth, or even complete yield loss. Arsenic enters the human food chain, contaminates the fodder and crops, and becomes a part of the farming systems by various means, which involve usage of arsenic-based pesticides, mining activities, irrigation with arsenic-laded groundwater, naturally occurring geochemical activities (Smedley and Kinniburgh 2002), and by carrying out fertilization using municipal solid wastes (Meharg et al. 2009). Water contaminated by arsenic and utilized for the irrigation of crops might be considered hazardous for the quality of crops as well as for the environment. A report indicated that a crop with high concentration (20 ppm) of arsenic showed a loss of 20% in terms of crop production (Davis et al. 1998). For irrigation, the prolonged utilization of arsenic contaminated water leads to higher concentration of arsenic in the soil of agricultural lands (Meharg and Rahman 2003; Roychowdhury et al. 2005; Dahal et al. 2008). As a result of growing plants continuously on the arsenic contaminated soils, arsenic gets accumulated in the fruits, vegetables, and grains of crops such as wheat and rice (Roychowdhury et al. 2002; Meharg 2004; Norra et al. 2005). This chapter deals with the mechanism involved in the uptake, distribution, and detoxification of arsenic in the plants. It also includes the strategies which can be used to reduce the uptake and accumulation of arsenic in plants.

7.2 Environmental Exposure and Effect of Arsenic

On the basis of the conditions of the environment, the inorganic arsenic species, i.e., arsenite and arsenate are inter-convertible. The chemical analog of inorganic phosphate (Pi) is As (V), which enters the cells via the phosphate transporters. In contrast to As (V), As (III) is present predominantly as an undissociated neutral molecule under normal environmental conditions and it enters the plant roots via the aquaporin channels (Bienert et al. 2008). Natural activities involving geothermal waters, erosion, and weathering of rocks and minerals as well as volcanic eruptions serve as the main source of arsenic contamination in water and soil (Garelick et al. 2008; Khalid et al. 2017b; Shakoor et al. 2016). Human activities such as the flow-off of the industries, indiscriminate utilization of fertilizers and pesticides, chemicals used in agricultural processes, and wood preservatives such as copper chromate arsenate majorly contributed to the arsenic contamination in the water and soil environments (Garelick et al. 2008; Niazi et al. 2015; Shakoor et al. 2015; Figure 7.1). Soils are further contaminated by the incorporation of oxides of arsenic as a result of fallout from the cotton waste and coal burning and by smelting of ores, and higher levels of arsenic contamination results in phytotoxicity. Compounds of varying solubilities are formed by the interaction of arsenic with the soil and its solutions (Chukhlantsev 1956; Woolson et al. 1971). Aluminum, magnesium, calcium, and iron have the ability to interact with arsenic within the soil along with the hydrous aluminum and iron oxides. With time, the concentration of arsenic within the soil solution decreases, although the organic matter of the soil, its pH, and the cations have also been found to influence the concentration of the soil solution. Nonetheless, the phytotoxic effect of arsenic within the soil decreases with time (Kardos et al. 1941; Woolson et al. 1973a). The application of arsenic to the plants and soil occurs for several purposes such as leaf desiccation in cotton plants or utilization as a weed killer on the vegetation carried out by the application of arsenic acid ($As_2O_3.H_2O$). After the emergence of the plant, the organic arsenic compounds such as hydroxyl dimethylarsine oxide and monosodium methane arsonate (MSMA) are used as selective or general herbicides, respectively, irrespective of the form in which the arsenic is applied; it is eventually oxidized and metabolized to arsenate (Von Endt et al. 1968; Woolson and Kearney 1973b).

7.2.1 Fate of Arsenic and the Environment

Arsenic is present naturally in minerals and soil. It may contaminate water, land, and air from the dust carried by wind and from the draining water via the process of leaching. Within the environment, arsenic is only capable of changing its form since it is indestructible (Salnikow and Zhitkovich 2008). The dust

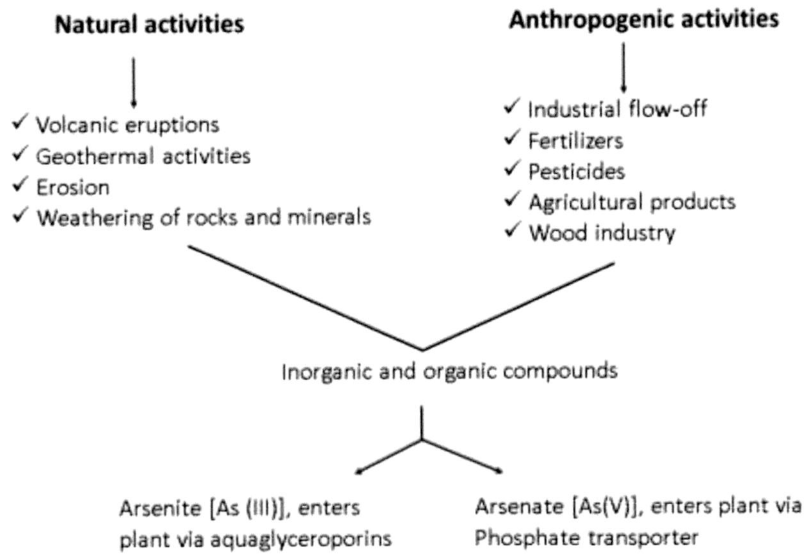

FIGURE 7.1 Distribution of arsenic in the environment.

particles of arsenic in the air are eliminated by snow and rain since several compounds of arsenic are dissolvable in the water. Arsenic-contaminated water eventually settles as sediment or accumulates in the soil. Arsenic can also accumulate within shellfish or other fish majorly as arsenobetaine, an organic form of arsenic which is harmful to a lot lesser extent (Zhang et al. 2019).

7.2.2 Effect on Health

Entry of high levels of inorganic arsenic in the human body via the breathing system can result in irritated lungs or a sore throat (Wang et al. 2020). Breathing in a very high amount of arsenic can lead to death. When exposed to low level of arsenic, it can lead to nausea and vomiting, abnormality in the heartbeat, and damage to blood vessels (Lee et al. 2003; Ratnaike 2003; Freeman 2007). Consuming low concentration of the inorganic form of arsenic for a prolonged duration of time might lead to darkening, swelling, or reddening of skin, and also results in the formation of little corns, warts on the torso, soles, and the palms (Saha et al. 1999). The effects of the organic arsenic compounds on the human health are mostly unknown and the studies carried out on animals prove that inorganic arsenic compounds are more toxic than some simple organic compounds (Majidi and Shemirani 2011). Damaged kidneys and diarrhea are a result of consuming dimethyl and methyl arsenic compounds (Khairul et al. 2015). Many studies showed that breathing of inorganic arsenic may lead to an increased risk of liver, skin, lung, and bladder cancer (Hong et al. 2014). Inorganic form of arsenic has been declared as a carcinogen in humans by the IARC, the Environmental Protection Agency (EPA), and the Department of Health and Human Services (DHHS). A few evidences suggest that prolonged arsenic exposure in children resulted in lesser IQ scores, while young adults showed elevated mortality rates who were exposed to arsenic during early childhood (Nahar et al. 2013; Steinmaus et al. 2014). Also, a few evidences suggest that exposure of pregnant women to arsenic can be harmful to them and their unborn babies although these studies are not definitive (Navasumrit et al. 2019). Arsenic has the ability to cross the placenta and was found in fetal tissues (Punshon et al. 2015). Low level of arsenic has also been found in breast milk (Carignan et al. 2015).The EPA has recommended an optimum range of arsenic, which the industries can discharge in the environment. It has also cancelled or restricted the usage of arsenic compounds in the preparation of pesticides. For drinking water, its limit is restricted to 0.01 ppm by the EPA. At a working place, the permissible exposure limit (PEL) of the air has been set by the Occupational Safety and Health Administration (OSHA) as 10 ug/m^3 for a shift of 8 hours working day and 40 hours a week

(Agency for Toxic Substances and Disease Registry 2007). Several reports have been undertaken that highlight the plants grown in heavy metal-contaminated sites might often take up high amount of metals than the maximum permissible limits (MPLs) resulting in the high risk of diseases in humans (Ramirez-Andreotta et al. 2013; Khalid et al. 2017a). Millions of people across the world, particularly within the Southeast Asia, have been found to be at high risk of health diseases through the consumption of food crops contaminated with arsenic (Rehman et al. 2016; Ma et al. 2017).

7.3 Absorption and Distribution of Arsenic in the Plants

7.3.1 Uptake of Arsenic from Soil into Root Cells

As the nonessential elements possess no biological functions, they hitch-hike into the cell via the transporters involved in transporting essential or beneficial elements (Clemens 2006). A survey was conducted on a variety of plant species indicating that the major forms of methylated arsenic such as monomethylarsonic acid (MMA) and dimethylarsinic acid (DMA), and also As (V) are found in the soil and can be taken up by the plants; out of these, As (V) has, on an average, the greatest efficiency of entering into the roots of the plants (Raab et al. 2007). It has been described that the uptake of As (V) occurs via the transporters involved in the transport of phosphate, the chemical analog of arsenic, e.g. studies were done on several plants, including *Holcus lanatus*, and revealed that they exhibited natural hyper-tolerance to arsenic. This work partly explained the trait of arsenic uptake through the phosphate transporter by suppressing the uptake of high-affinity phosphate (Meharg and Macnair 1992). Successive studies were done on *Arabidopsis thaliana* and showed that when mutation was done in the phosphate transporters, the plant was found to possess low sensitivity to As (V) (Shin et al. 2004; Zhao et al. 2010). Moreover, As (III) possesses higher bioavailability than As (V), due to which in rice plants, uptake of arsenic from soil to the plant is almost ten times more compared to other cereals (Zhao et al. 2010). In the rice roots, As (III) enters with the aid of nodulin 26-like intrinsic protein (NIP) aquaglyceroporin Lsi1 (OsNIP2;1) (Ma et al. 2008) and is taken up passively via the concentration gradient. The uptake of silicic acid is basically done through the transporter, Lsi1 (Ma et al. 2006) and the higher amount of silicon (Si) accumulation in rice results in the attainment of immunity against abiotic and biotic stresses (Ma and Yamaji 2015). A recent finding by Li et al. (2009) revealed that methylated arsenic is also taken up by the Lsi1 transporter. It was revealed that the methylated arsenic (MMA and DMA) uptake was notably higher in the roots of wild type plants as compared to that of Lsi1 mutated plants. Also, in the oocyte of *Xenopus laevis*, the expression of Lsi1 conferred the uptake activity of MMA and DMA (Li et al. 2009). This process of Lsi1-mediated As (III) efflux has been observed in many plant species (Zhao et al. 2009) and revealed that in majority of plants, the most efficient translocation is of the methylated species of arsenic among the other arsenic species (Raab et al. 2007). Retention of the majority of As (III) within the root cells is due to the sequestration of As (III)-PC complexes within the vacuoles. After researching for longer period of time, Song et al. (2010) discovered ABC transporters which were found to be associated with the compartmentalization of metal/metalloid-PC complexes in the vacuoles (Figure 7.2).

7.3.2 Long-Distance Transportation of Heavy Metals

Toxic element transportation from one organ to another occurs through long-distance transportation through phloem and xylem. Apart from this, phloem is involved principally in mediating the movement from leaves and redistribution from the shoots to the roots (Mendoza-C′ozatl et al. 2011; Uraguchi and Fujiwara 2013). However, a few feeding experiments have been conducted on rice with flag leaves and have revealed that inorganic arsenic is immobile as compared to organic arsenic (Carey et al. 2011). Presumably, its uptake occurs into the companion cells of the phloem; however, the transporters involved in this uptake are yet to be identified (Khan et al. 2014) except for rice. In rice, the transportation of inorganic arsenic into the phloem occurs via the intervascular transport to node I, which serves as a crucial route involved in the transportation of the toxic substances toward the grains of the crop. Several

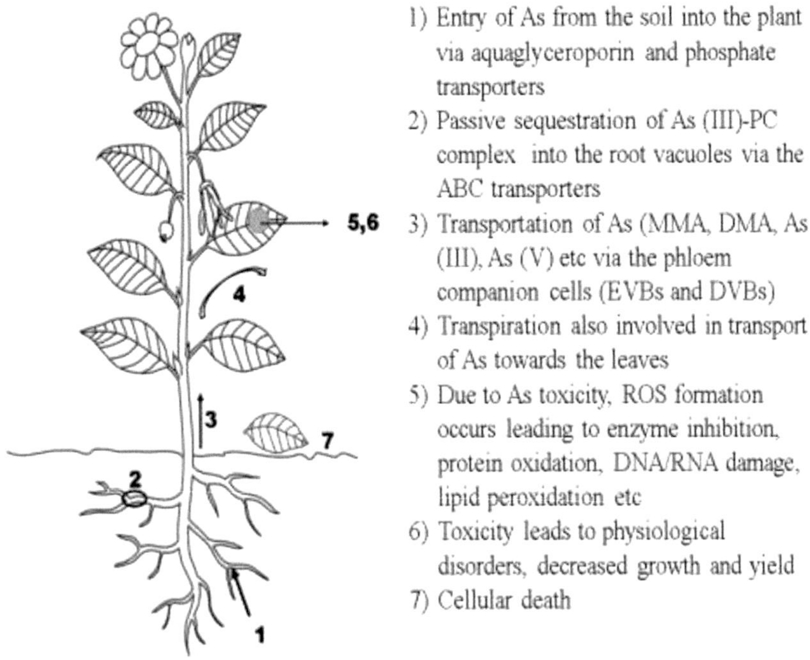

FIGURE 7.2 Arsenic toxicity from soil to the plant.

nodes, particularly node I, is involved in controlling the allocation of substances either to the grains or to the leaves. This is important as the process of transpiration is involved in the transport of the minerals, thus firmly favoring the uptake of substances to the leaves possessing higher rate of transpiration. The nodes found in rice contain two prime vascular bundles: Diffused vascular bundles (DVBs) and enlarged vascular bundles (EVBs). EVBs are connected to the leaf and originate from the lower nodes, while DVBs that surround the EVBs are linked to the panicles or the upper portion of the nodes. Hence, the movement of harmful substances to the grains needs an intervascular uptake at the nodes to the DVBs from EVBs, which by presumption necessitates the involvement of several transporters, and it might be the reason that arsenic concentration was found to be more in the nodes in comparison to the internodes (Fujimaki et al. 2010; Moore et al. 2014). Moore et al. (2014) observed arsenic in the various vascular bundles of the phloem, primarily within its companion cells in the rice nodes, and the transport of arsenic to the grains is restricted by the suggested role of the node filtering in relation to the trapping of As(III)-PC complexes. Recently, Song et al. (2014) revealed the role of OsABCC1 in this process and it was found in the phloem companion cells of the DVBs within the nodes.

7.3.3 Heavy Metals and Storage in the Seeds

Amidst the toxic heavy metals, the available data are confined mainly to arsenic. In the rice grains, the analysis of arsenic based on the speciation found methylated and inorganic species of arsenic and the proportions differed significantly in various grain parts, pointing to differences in the loading pathways and/or mobility for the organic and the inorganic species of arsenic in the grain. A study based on X-ray absorption near edge structure (XANES) revealed that a major number of complexes of As-thiol are derived from inorganic species of arsenic, indicating that PCs are involved in the deposition of arsenic within the crop grains (Lombi et al. 2009). Another study based on higher-resolution nanoscale secondary ion mass spectrometry (nanoSIMS) led to the detection of arsenic association with the fraction of protein and was found to be in accordance with the complexation of the thiols, while the localization of phosphorus varied distinctly, as per the dominant character of As(III) mobility within plants (Moore et al. 2010).

7.4 Mechanism of Action

As (V) competes with Pi for phosphorylation reactions and can replace Pi from biomolecules (Finnegan and Chen 2012). By contrast, As(III) binds to dithiol reactive compounds and enzymes containing closely spaced cysteine residues, disrupting their structures and catalytic functions (Chen et al. 2010; Finnegan and Chen 2012; Shen et al. 2013). In plant cells, As (III) detoxification occurs essentially via the complex formation of PCs and several other compounds of thiol (Clemens et al. 1999; Ha et al. 1999; Liu et al. 2010b) and several As (III)-PC complexes have been detected in plant tissues (Raab et al. 2005; Liu et al. 2010a). Complexes of As (III)-PC have been sequestered inside the vacuoles through the tonoplast transporters, ABCC1 and ABCC2 (Song et al. 2010). In arsenic-non hyperaccumulators, translocation of arsenic to the shoots from the roots is generally limited (Zhao et al. 2009). Despite limited translocation, arsenic after reaching the shoot tissues could disrupt a range of physiological and biochemical processes and ultimately resulted in decreased chlorophyll content, suppression of the photochemistry in the photosystem II (PSII), and increased oxidative stress (Mishra et al. 2014; Iriel et al. 2015; Mishra et al. 2016; Srivastava et al. 2017; Patel et al. 2018). A common symptom of arsenic toxicity is leaf chlorosis (Jain and Gadre 1997; Srivastava et al. 2017) suggesting that the primary target of toxicity might be the chloroplasts; however, little is known about how the chloroplasts cope up with arsenic toxicity. Several evidences exist which showed that arsenic interferes negatively with various physiological and biochemical processes within the plant leading to reduced plant yield and growth (Flora 2011; Marmiroli et al. 2011; Khalid et al. 2017b). Within the plant cell, heavy metals such as arsenic resulted in the formation of high amount of reactive oxygen species (ROS) that ultimately leads to oxidative stress, and might cause cellular death through several oxidative activities such as enzyme inhibition, oxidation of the proteins, lipid peroxidation, and damage to the RNA and DNA (Flora 2011; Marmiroli et al. 2011). Arsenic, in general, accumulates majorly within the roots and a small portion was found in the above ground parts of the plant that leads to the inhibition of plant growth (dry and fresh accumulation of the biomass) (Stoeva et al. 2003), physiological disorders (Wells and Gilmor 1997), decreased crop productivity (Stepanok 1998) as well as reduced photosynthesis rate (Miteva and Merakchiyska 2002). Arsenic damages the membranes of the chloroplasts and disorganizes the structure of the membrane, and the effect of high concentrations of arsenic damaging the structure of the chloroplasts implies the functional changes of the integral photosynthetic process. Numerous physiological processes are susceptible to arsenic toxicity and plants exposed to it have damaged cellular membranes leading to the leakage of electrolytes, which is followed by an elevated level of malondialdehyde, a by-product of the lipid peroxidation suggesting the occurrence of oxidative stress (Singh et al. 2006). Toxic effect of arsenate occurs because of suppression of high affinity phosphate due to the uptake of arsenic in the system (Meharg and Macnair 1992). Once it reaches the cellular cytoplasm, its competition begins with the phosphate, e.g. the phosphate in a molecule of ATP is replaced with arsenic to form an unstable complex of ADP-As, thereby causing a disturbance in the cellular flow of energy (Meharg 1994). In plants, the biochemical responses toward the stress caused by arsenic are not fully studied (Hartley-Whitaker et al. 2001). Although arsenic was not a redox metal, various studies have been carried out that revealed when plants are exposed to inorganic compounds of arsenic, formation of reactive oxygen species (ROS) took place. This oxidative process was linked with the change in the valency of arsenic, a process readily occurring within the plants (Sharma 2012; Mishra et al. 2019). ROS has the potential to directly cause harm to the cellular proteins, nucleic acids, amino acids, and also leads to the peroxidation of membrane lipids (Dat et al. 2000). It was observed that As (V) induced lipid peroxidation in the arsenic hyperaccumulator, *Pteris vittata* (Singh et al. 2006, Srivastava et al. 2005). It was also indicated that a general feature of the plant in response to arsenic involves the generation of ROS and also the extent of the redox disturbances within the plant cells might turn out to be an essential determining factor of the toxicity caused by the ROS formation. In order to combat these effects, plants generate several antioxidant mechanisms (non-enzymatic and enzymatic) for the sequestration of the ROS. Glutathione S-transferase (GST), catalase (CAT), superoxide dismutase (SOD), etc. are enzymatic forms of antioxidants, while ascorbate, proline, and glutathione are the non-enzymatic forms (Handa et al. 2018; Goharrizi et al. 2020; Loutfy et al. 2020). The toxicity of arsenic is majorly and importantly attributed to the capability of arsenate to replace phosphate in various reactions in which root and shoot length together with biomass drastically reduced, especially at higher arsenic doses, i.e., 400 μM and 600 μM (Srivastava and Sharma 2013). According to Lidon and Hendriques (1991), heavy metal toxicity resulted in inhibited chlorophyll synthesis, and also chlorophyll a and chlorophyll b, i.e., total chlorophyll along with the carotenoids showed a decline in arsenic treated plants as compared to control plants and thus hampering the photosynthesis processes. Similar studies showed that increased concentrations of arsenic lead to the variations in the shape of chloroplast. It resulted in the bending of the concave membrane and partial damage of the chloroplast. Also, it was associated with the differed flow and accumulation of assimilates that lead to the decreased contents of chlorophyll within the leaves of rice plant (Miteva and Merakchiyska 2002). However, contrary to this, a few reports on plants have also showed that lower arsenic concentrations in soil stimulate synthesis of pigments in plants (Miteva 2002).

7.4.1 Metabolism and Detoxification

Between arsenite [As (III)] and arsenate [As (V)], the former is more toxic in nature than the latter, while both these forms of arsenic are highly toxic as compared to its organic forms (Vromman et al. 2018). Reduction of arsenate to arsenite in the presence of arsenate reductase (AR) is considered as the initial step in arsenic detoxifying mechanism within the plants. Apart from the reduction of arsenate to arsenite, another reduction process was noticed with respect to the methylation within the plant bodies. Many hydroponic-based studies were performed, wherein the plants were majorly incorporated with inorganic form of arsenic to observe a little concentration of Dimethylarsinic acid and Monomethylarsonic acid (DMA and MMA) in the xylem sap and plant tissues, indicating the occurrence of arsenic biomethylation within the plants (Li et al. 2009; Zhao et al. 2010). As per Challenger (1945), arsenic methylation in fungi brings about the catalyzation of successive reduction of As (V) along with the incorporation of a methyl group to As (III). This reaction occurs repeatedly causing the formation of volatile trimethylarsine (TMA). Based on a biochemical analysis performed *in vitro*, Hayakawa et al. (2005) suggested another pathway involving methylation of arsenic, which occurs alternatively apart from that of the Challenger pathway. According to Hayakawa et al. (2005), substrate for the enzyme A3MT includes GSH complex (ATG) and As (III), which are involved in the catalyzation of the methyl group into DMA and MMA, the final products of the reaction. Ajees et al. (2012) supported the above pathway in the algae, *Cyanidioschyzon merolae* on the basis of the structure of the enzyme ArsM (Arsenite (As [III]) S-adenosylmethionine methyltransferase). Also, in the recent study carried out on *Prochlorococcus*, Saunders and Rocap (2016) described a strategy for the remediation of arsenic managed by the enzyme ArsM, which catalyzed the conversion of inorganic arsenic to a less harmful organic species and enabled a better and extensive detoxification method. Uptake of arsenic in plants occurs as As (V) where its reduction to As (III) is carried out via the reductase enzyme, although in plants, toxicity was observed because of the change of As (V) to As (III). Cultivation of plant species on sites contaminated with arsenic can be adapted by suppressing the phosphate transporters and thereby limiting the uptake of As (V) (Meharg and Hartley-Whitaker 2002). Sequestration of the vacuoles and formation of complexes with PCs are other mechanisms adopted by the plants to detoxify arsenic. In addition to this, in Arabidopsis plants, As (V) tolerance was enhanced by the upregulation of the expression of PC synthase (AtPCS1) and/or γ-glutamylcysteine synthetase (γ-ECS gene) (Li et al. 2004, 2006 Gasic and Korban 2007). The transport of the complex, As III-PCs, within the vacuoles plays a very crucial role in the detoxification metabolism of arsenic (Song et al. 2014) and relies on the transporter proteins in charge for the movement of As III–PCs into the vacuoles.

7.5 Future Prospects of Arsenic-Plant Studies

There is an increase in the concerns regarding the heavy metal intoxication over the past years, particularly the harmful effects of toxicity caused by arsenic to the human health. Arsenic contamination is a grave issue faced across the world and there is necessity for bringing up urgent solutions for addressing this issue in order to ensure that vulnerable irrigated land can carry on with the cultivation. New and advanced methods are necessary for tackling the harmful effects of stress caused by heavy metals on crop plants, along with the agronomic practices of management through which arsenic toxicity could be reduced via water-related strategies (Polizzotto et al. 2013) and appropriate fertilization techniques (Ma et al. 2008). The two crop improvement strategies that can be used include plant breeding technique for developing tolerant lines through the genetic engineering, and phytoremediation technique which utilizes green plants for achieving such objectives (Jin et al. 2019; Sychta et al. 2019; Liu et al. 2020; Muthusaravanan et al. 2020).

7.5.1 Phytoremediation

Phytoremediation involves many subsequent processes utilizing plants for the purpose of remediating environmental toxicity, which includes surface water, groundwater, and soil (Blaylock and Wang 2000; Ensley 2000; McGrath et al. 2002; Butcher 2009). The most common phytoremediation method utilized against arsenic toxicity is phytoextraction technique, which ensures the availability of toxic substance of our interest to the plants. Phytoextraction of arsenic occurs in a unique way since arsenic normally exists as an anion within the environment, and therefore, is handled in a different manner than the other elements, which exist usually as cations. Another commonly applied phytoremediation technique is called phytostablization or in-place activation, in which green plants are used for reducing the exposure to humans through arsenic localization (Berti and Cunningham 2000). Soils contaminated with heavy metals are aimed at reducing the heavy metal's solubility and the methods recommended to immobilize the arsenic include organic materials (e.g., manure, compost), hydrous iron oxides, as well as inorganic clay substances through the mechanism of sorption that plays an important role in the inactivation of arsenic (Wuana and Okieimen 2011; Chibuike and Obiora 2014). Many authors have examined the hyperaccumulators used for the phytoremediation of arsenic (Huili et al. 2019; Patel et al. 2020). Rice been a staple crop is consumed daily by around half of the globe's population, specifically by the people of Asian countries. Human exposure to arsenic is enhanced by consuming not only rice but also the food based on the grain. Various microorganisms involving microalgae, bacteria, and fungi are referred to as plant growth promoting microorganisms (PGPMs), and found in association with the plant's rhizospheric area. These microorganisms are involved in improving the growth of plants via different mechanisms such as changing the biochemical properties of plants, increasing the nutrients, enhancing the quality of soil, production of hormones, siderophore, and various other mechanisms (Upadhyay et al. 2018). One more benefit provided by the PGPMs is the varied arsenic speciation within the soil by successively changing the arsenic bioavailability to the plants and the methylation process, and changing the speciation of arsenic plays a role in affecting the its uptake and transport in plants (Upadhyay et al. 2018).

7.5.2 Strategies for Reducing Accumulation of Toxic Elements

For decreasing the accumulation of toxic elements in the crops, many agronomic mitigation strategies are employed such as silicon fertilization used for saturating the silicon transport pathway, which plays a role in mediating the accumulation of As (III), and proper management of irrigation regimes for controlling the arsenic bioavailability (Zhao et al. 2015). In the past, the methods mentioned above are based on the molecular understanding that has become popular as it promises a way for achieving higher food safety. In the breeding programs, the loss in the function of the alleles of metal transporters was easier to employ comparatively. Such alleles are derived from mutated plants, ideally with the genetic background of the frequently grown good plant cultivars. Allele can be isolated by an impartial mutant screening such as the one carried out in case of OsNramp5 (Ishikawa et al. 2012) or either through the specific gene targeting such as in the targeting induced local lesions in genomes (TILLING) populations. OsNramp5 is a transporter in the cell membrane of plants involved in the transport of iron and manganese. With the help of screening at molecular level of the natural variation of heavy metal accumulation, more interesting alleles concerned with lower accumulation of heavy metals are thought to be found. Earlier as mentioned about OsNramp5, a disadvantage related to allele's functional loss involves possible disturbance in the homeostasis of important minerals as represented through OsHMA2. Hence, it might be favorable in altering the specificity of the membrane transporter proteins associated with the movement of the heavy metals (Khan et al. 2014). The relevant variants could either be incorporated into the crops via genome editing techniques or by searching the variants in the existing populations. Pottier et al. (2015) were successful in isolating forms of AtNramp4, which possessed change in the transport of zinc (Zn) and cadmium (Cd) with an unaltered transportation of iron (Fe) in the wild-type plant and was immensely reliable, considering the important role of Nramps played in the cadmium accumulation. Hoch et al. (2012) identified the amino acid residues for a zinc transporter in mammals, belonging to the family of cation diffusion facilitator (CDF); however, in plants, it belongs to the metal tolerance protein (MTP) family and is found to be an important factor in cadmium discrimination although each transporter is not responsive to selective engineering. For example, in case of aquaglyceroporins, it was rather demanding in finding its different forms which can differentiate arsenite (Mitani-Ueno et al. 2011). Changing the transporter protein's expression pattern via the methods of transgene introduction could be a success, although there is no alteration in its selective behavior. For the cultivars of rice with higher expression of OsHMA3 that normally possess lesser expression pattern, there was an immense decrease in the grain cadmium without causing any effect on iron and zinc nutrition, and also there was no change in the yield of rice after the cultivation of crops on cadmium-contaminated soil (Ueno et al. 2010). This result with no change can be due to the compensatory upregulation of the four genes in the over-expressed lines involved in zinc transport (Sasaki et al. 2014). There has been an enormous progress in the past decades in discriminating the molecular pathways involved in the accumulation of arsenic and in the identification of few molecular substances participating in the process, especially in rice.

Likewise, there has been studies that led to the identification of several loci related to the arsenic accumulation in the rice grain, although there was not any co-localization of the genes associated with the arsenic accumulation (Norton et al. 2014). Hence, several factors affecting the accumulation of toxic element lack understanding and are still unidentified in various other crop plants. Few factors upon which detailed work can be done in the future are:

1. A complete understanding of the pathway associated with the accumulation of the toxic element requiring knowledge of the metal transporter and ligand functioning possessing resolution via the complex cellular associations in directing the metal/metalloid flow across the rice nodes.
2. The regulation of the methods used in the transportation of metal in reaction to change of the nutrient flow. Also, the outcomes of such regulations involving accumulation of toxic elements needs to be explained.
3. The ligands and the transporters involved in the transport of harmful substances through the apoplast to the seeds are still not known.
4. The mechanisms of uptake of the toxic elements into the tubers and leaves of crops remain unidentified.

Generally, there is an alteration in the growth and yield of the plant even if the arsenic concentration did not pass the standard limit set by the Food and Drug Administration (FDA) for the plant's edible parts. A few reports exhibited arsenic to be accumulated within the root crops, such as radishes, with more than 2 mg of accumulation per kilogram (Smith et al. 2008, 2009). Arsenic has garnered immense popularity due to its biogeochemical cycle in the surrounding. Awareness regarding the chemical and physical nature of arsenic along with its interaction with the biological systems will help in the remediation process via novel goal-oriented analysis within animals as well as plants. The arsenic bioavailability in the soil along with speciation of arsenic are distinctly altered by the surrounding environment that imparts knowledge on the amount of accumulation of arsenic within plants. Food chain is affected due to the accumulation of the toxic elements within the plants and notably cellular advances have been observed in the past decades involving speciation of arsenic, its mobility, and the detoxifying mechanism within the plants. In aerobic environment, the inorganic phosphate transporters are responsible for the entry of As (V)in the root system of the plants, although in anaerobic environment, methylated arsenic species and As (III) are majorly transported via the aquaporins. Further investigations are needed to be done in the present knowledge gaps, specifically regarding mechanism related to the sequestration of arsenic in the vacuoles, its mobility and transport in the phloem and xylem, and its accumulation within the grains. Study of cellular and molecular mechanisms for metabolism and transport of arsenic may be helpful in providing new chances for developing crop cultivars possessing enhanced resistance and accumulation of arsenic to a lesser extent in the grains or the edible portions of the crops, leading to sustainably produced and safely utilized products of the crop plants. Sustainable approaches need to be developed in the future for remediation of the arsenic-contaminated sites, and also for the safe agricultural production in such sites. PGPMs and arsenic resistance provide a great hope in this regard, which indicates a strong future for the PGPMs in enabling safer cultivation of rice with lower accumulation of arsenic within the grains. Simultaneously, PGPMs can lead to enhanced accumulation of arsenic in phytoremediator plants such as *P. vittata*. Therefore, it can be speculated that the future belongs to PGPMs-driven regulation of arsenic concentrations and improving the toxicity of arsenic within the plants. Although detailed researches shall be needed for the advancement of this technology via standardization of PGPM-based strategy suitable for various environments, identifying possible mix involving PGPMs for a specific class of organisms and hunting the mix of PGPMs from different clusters such as algae-fungi, fungi-bacteria, algae-bacteria-fungi, bacteria-AMF, etc. also lead to better understanding of the molecular mechanisms involved in the PGPMs. However, the evolved approaches need to be economically successful and the participation of the farmers on a large scale would lead to the success of sustainable agriculture programs.

REFERENCES

Ajees, A. A., Marapakala, K., Packianathan, C., Sankaran, B. and Rosen, B. P. 2012. Structure of an As(III) S-adenosylmethionine methyltransferase: insights into the mechanism of arsenic biotransformation. *Biochemistry* 51:5476–5485.

ATSDR. 2007. Toxicological profile for arsenic, agency for toxic substances and disease registry.

Barton, C. 2014. CCA-treated wood. In: Philip W. (ed.), *Encyclopedia of Toxicology*. Academic press, USA, pp:751–752.

Berti, W. R. and Cunningham, S. D. 2000. Phytostabilization of metals. In: Raskin I. and Ensley B. D. (eds), *Phytoremediation of Toxic Metals: Using Plants to Clean Up the Environment*. John Wiley & Sons, New York, pp:71–88.

Bienert, G. P., Thorsen, M., Schüssler, M. D., Nilsson, H. R., Wagner, A., Tamás, M. J. and Jahn, T. P. 2008. A subgroup of plant aquaporins facilitate the bi-directional diffusion of $As(OH)_3$ and $Sb(OH)_3$ across membranes. *BMC Biol* 6:26.

Blaylock, M. J. and Wang, J. W. 2000. Phytoextraction of metals. In: Raskin I., Ensley B. D. (eds), *Phytoremediation of Toxic Metals: Using Plants to Clean Up the Environment*. John Wiley & Sons, New York, pp:53–70.

Butcher, D. J. 2009. Phytoremediation of lead in soil: recent applications and future prospects. *Appl Spectros Rev* 44:123–139.

Carey, A., Norton, G. J., Deacon, C., Scheckel, K. G., Lombi, E., Punshon, T., Guerinot, M. L., Lanzirotti, A., Newville, M., Choi, Y., Price, A. H. and Meharg, A. A. 2011. Phloem transport of arsenic species from flag leaf to grain during grain filling. *New Phytol* 192:87–98.

Carignan, C.C., Cottingham, K.L., Jackson, B.P., Farzan, S.F., Gandolfi, A.J., Punshon, T., Folt, C.L., Karagas, M.R. 2015. Estimated exposure to arsenic in breastfed and formula-fed infants in a United States cohort. *Environ Health Perspect* 123(5):500–506.

Chen, W., Chi, Y., Taylor, N. L., Lambers, H. and Finnegan, P. M. 2010. Disruption of ptLPD1 or ptLPD2, genes that encode isoforms of the plastidial lipoamide dehydrogenase, confers arsenate hypersensitivity in Arabidopsis. *Plant Physiol* 153:1385–1397.

Chibuike, G. U. and Obiora, S. 2014. Heavy metal polluted soils: effect on plants and bioremediation methods. *Appl Environ Soil Sci* 1–12.

Chukhlantsev, V. G. 1956. Solubility products of a series of arsenates. *Zh Anal Khim* 11:529–535.

Clemens, S. 2006. Toxic metal accumulation, responses to exposure and mechanisms of tolerance in plants. *Biochimie* 88(11):1707–1719.

Clemens, S., Kim, E. J., Neumann, D. and Schroeder, J. I. 1999. Tolerance to toxic metals by a gene family of phytochelatin synthases from plants and yeast. *EMBO J* 18(12):3325–3333.

Dahal, B. M., Fuerhacker, M., Mentler, A., Karki, K. B., Shrestha, R. R. and Blum, W. E. H. 2008. Arsenic contamination of soils and agricultural plants through irrigation water in Nepal. *Environ Pollut* 155:157–163.

Dat, J., Vandenabeele, S., Vranová, E., Montagu, M. V., Inzé, D. and Breusegem, F. V. 2000. Dual action of the active oxygen species during plant stress responses. *Cell Mol Life Sci* 57:779–795.

Davis, J. A., Coston, J. A., Kent, D. B. and Fuller, C. C. 1998. Application of the surface complexation concept to complex mineral assemblages. *Environ Sci Technol* 32(19):2820–2828.

Ensley, B. D. 2000. *Rationale for Use of Phytoremediation*. In: Raskin I. and Ensley B. D. (eds), Phytoremediation of toxic metals - Using plants to clean up the environment. New York: John Wiley & Sons, 2000. p.3–11.

Finnegan, P. M. and Chen, W. 2012. Arsenic toxicity: the effects on plant metabolism. *Front Physiol* 3:182.

Flora, S. J. 2011. Arsenic-induced oxidative stress and its reversibility. *Free Radic Biol Med* 51(2):257–281.

Freeman, K. 2007. Arsenic upsets heartbeat. Possible early warning for cardiovascular risk. *Environ Health Perspect* 115(5):A262.

Fujimaki, S., Suzui, N., Ishioka, N. S., Kawachi, N., Ito, S., Chino, M. and Nakamura, S. I. 2010. Tracing cadmium from culture to spikelet: non-invasive imaging and quantitative characterization of absorption, transport, and accumulation of cadmium in an intact rice plant. *Plant Physiol* 152:1796–1806.

Garelick, H., Jones, H., Dybowska, A. and Valsami-Jones, E. 2008. Arsenic pollution sources. *Rev Environ Contam Toxicol* 197:17–60.

Gasic, K. and Korban, S. S. 2007. Transgenic Indian mustard (*Brassica juncea*) plants expressing an Arabidopsis phytochelatin synthase (AtPCS1) exhibit enhanced As and Cd tolerance. *Plant Mol Biol* 64:361–369.

Goharrizi, J. K., Moosavi, S. S. and Amirmahani, F. et al. 2020. Assessment of changes in growth traits, oxidative stress parameters, and enzymatic and non-enzymatic antioxidant defense mechanisms in *Lepidium draba* plant under osmotic stress induced by polyethylene glycol. *Protoplasma* 257:459–473.

Ha, S. B., Smith, A. P., Howden, R., Dietrich, W. M., Bugg, S., O'Connell, M. J., Goldsbrough, P. B. and Cobbett, C. S. 1999. Phytochelatin synthase genes from Arabidopsis and the yeast *Schizosaccharomyces pombe*. *Plant Cell* 11(6):1153–1163.

Handa, N., Kohli, S., Sharma, A., Thukral, A., Bhardwaj, R., AbdAllah, E., Alqaawi, A. and Ahmad, P. 2018. Selenium modulates dynamics of antioxidative defence expression, photosynthetic attributes and secondary metabolites to mitigate chromium toxicity in *Brassica juncea* L. Plants. *Environ Exp Bot* 161:180–192.

Hartley-Whitaker, J., Ainsworth, G. and Meharg, A. A. 2001. Copper- and arsenate induced oxidative stress in Holcuslanatus L. clones with differential sensitivity. *Plant Cell Environ* 24:713–722.

Hayakawa, T., Kobayashi, Y., Cui, X. and Hirano, S. 2005. A new metabolic pathway of arsenite: arsenic-glutathione complexes are substrates for human arsenic methyltransferase Cyt19. *Arch Toxicol* 79:183–191.

Hoch, E., Lin, W., Chai, J., Hershfinkel, M., Fu, D. and Sekler, I. 2012. Histidine pairing at the metal transport site of mammalian ZnT transporters controls Zn^{2+} over Cd^{2+} selectivity. *Proc Natl Acad Sci USA* 109:7202–7207.

Hong, Y. S., Song, K. H. and Chung, J. Y. 2014. Health effects of chronic arsenic exposure. *J Prev Med Public Health* 47:245–252.

Huili, Y., Gao, Y., Wu, L., Wang, L., Zhang, T., Changhua, D., Xu, W., Feng, L., Ma, M. and He, Z. 2019. Potential use of the *Pteris vittata* arsenic hyperaccumulation-regulation network for phytoremediation. *J Hazard Mater* 368:386–396.

International Agency for Cancer Research (IARC). 2004. Some drinking-water disinfectants and contaminants, including arsenic. In: IARC Monographs on the Evaluation of Carcinogenic Risks to Humans, vol. 84, International Agency for Cancer Research (IARC), Lyon, France.

Iriel, A., Dundas, G., Cirelli, A. F. and Lagorio, M. G. 2015. Effect of arsenic on reflectance spectra and chlorophyll fluorescence of aquatic plants. *Chemosphere* 119:697–703.

Ishikawa, S., Ishimaru, Y., Igura, M., Kuramata, M., Abe, T., Senoura, T., Hase, Y., Arao, T., Nishizawa, N. K. and Nakanishi, H. 2012. Ion-beam irradiation, gene identification, and marker-assisted breeding in the development of low-cadmium rice. *Proc Natl Acad Sci USA* 109:19166–19171.

Jain, M. and Gadre, R. 1997. Effect of As on chlorophyll and protein contents and enzymic activities in greening maize tissues. *Water Air Soil Pollut* 93:109–115.

Jin, Z., Deng, S., Wen, Y., Jin, Y., Pan, L., Zhang, Y., Black, L., Jones, K., Zhang, H. and Zhang, D. 2019. Application of Simplicillium chinense for Cd and Pb biosorption and enhancing heavy metal phytoremediation of soils. *Sci Total Environ* 697:134148.

Kardos, L. T., Van decaveye, S. C. and Benson, N. 1941. Causes and remedies of the unproductiveness of certain soils following the removal of mature (fruit) trees. *Wash Agr Exp Sta Bull* 410:142.

Khairul, I., Wang, Q., Jiang, Y., Wang, C. and Naranmandura, H. 2015. Metabolism, toxicity and anticancer activities of arsenic compounds. *Oncotarget* 8(14):23905–23926.

Khalid, S., Shahid, M., Dumat, C., Niazi, N. K., Bibi, I., Bakhat, F. S., Abbas, G., Murtaza, B. and Javeed, H. M. R. 2017a. Influence of groundwater and wastewater irrigation on lead accumulation in soil and vegetables: implications for health risk assessment and phytoremediation. *Int J Phytorem* 19(11):1037–1046.

Khalid, S., Shahid, M., Niazi, N. K., Rafiq, M., Bakhat, H. F., Imran, M., Abbas, T., Bibi, I. and Dumat, C. 2017b. Enhancing clean-up of environmental pollutants: non biological approaches. In: Anjum et al. (eds), *Arsenic Behaviour in Soil-Plant System: Biogeochemical Reactions and Chemical Speciation Influences.* Springer, Cham, Switzerland, pp:97–140.

Khan, M. A., Castro-Guerrero, N. A. and Mendoza-C´ozatl, D. 2014. Moving toward a precise nutrition: preferential loading of seeds with essential nutrients over non-essential toxic elements. *Front Plant Sci* 5:51.

Lee, M. Y., Jung, B. I., Chung, S. M., Bae, O. N., Lee, J. Y., Park, J. D., Yang, J. S., Lee, H. and Chung, J. H. 2003. Arsenic-induced dysfunction in relaxation of blood vessels. *Environ Health Perspect* 111:513–517.

Li, R., Ago, Y., Liu, W., Mitani, N., Feldmann, J., McGrath, S. P., Ma, J. F. and Zhao, F. J. 2009. The rice aquaporin Lsi1 mediates uptake of methylated arsenic species. *Plant Physiol* 150:2071–2080.

Li, Y. J., Dankher, O. P., Carreira, L., Smith, A. P. and Meagher, R. B. 2006. The shoot-specific expression of gamma glutamyl cysteine synthetase directs the long-distance transport of thiol-peptides to roots conferring tolerance to mercury and arsenic. *Plant Physiol* 141:288–298.

Li, Y. J., Dhankher, O. P., Carreira, L., Lee, D., Chen, A., Schroeder, J. I., Balish, R. S. and Meagher, R. B. 2004. Overexpression of phytochelatin synthase in Arabidopsis leads to enhanced arsenic tolerance and cadmium hypersensitivity. *Plant Cell Physiol* 45(12):1787–1797.

Lidon, F. C. and Hendriques, F. S. 1991. Effects of copper on protein and biomass yields of rice plants. *Acta Bot Hung* 37:373–382.

Liu, W. J., Wood, B. A., Raab, A., McGrath, S. P., Zhao, F. J. and Feldmann, J. 2010a. Complexation of arsenite with phytochelatins reduces arsenite efflux and translocation from roots to shoots in Arabidopsis. *Plant Physiol* 152:2211–2221.

Liu, X., Yu, F. and Rodermel, S. 2010b. An Arabidopsis pentatricopeptide repeat protein, suppressor of variegation7, is required for FtsH-mediated chloroplast biogenesis. *Plant Physiol* 154:1588–1601.

Liu, Y., Kang, T.,Cheng, J., Yi, Y., Han, J., Cheng, H., Li, Q., Tang, N. and Liang, M. . 2020. Heterologous expression of the metallothionein *PpMT2* gene from *Physcomitrella patens* confers enhanced tolerance to heavy metal stress on transgenic Arabidopsis plants. *Plant Growth Regul* 90:63–72.

Lombi, E., Scheckel, K. G., Pallon, J., Carey, A. M., Zhu, Y. G. and Meharg, A. A. 2009. Speciation and distribution of arsenic and localization of nutrients in rice grains. *New Phytol* 184:193–201.

Loutfy, N., Azooz, M. and Alhamd, M. 2020. Exogenously-applied salicylic acid and ascorbic acid modulate some physiological traits and antioxidative defense system in *Zea mays* L. seedlings under drought stress. *Egyptian J Bot* 60(1):313–324.

Ma, J. F., Tamai, K., Yamaji, N., Mitani, N., Konishi, S., Katsuhara, M., Ishiguro, M., Murata, Y. and Yano, M. 2006. A silicon transporter in rice. *Nature* 440:688–691.

Ma, J. F. and Yamaji, N. 2015. A cooperative system of silicon transport in plants. *Trends Plant Sci* 20:435–442.

Ma, J. F., Yamaji, N., Mitani, N., Xu, X. Y., Su, Y. H., McGrath, S. P. and Zhao, F. J. 2008. Transporters of arsenite in rice and their role in arsenic accumulation in rice grain. *Proc Natl Acad Sci USA* 105(29):9931–9935

Ma, L., Yang, Z., Kong, Q. and Wang, L. 2017. Extraction and determination of arsenic species in leafy vegetables: method development and application. *Food Chem* 217:524–530.

Majidi, B. and Shemirani, F. 2011. In situ solvent formation microextraction in the presence of ionic liquid for preconcentration and speciation of arsenic in saline samples and total arsenic in biological samples by electrothermal atomic absorption spectrometry. *Biol Trace Elem Res* 143:579–590.

Marmiroli, M., Pietrini, F., Maestri, E., Zacchini, M., Marmiroli, N. and Massacci, A. 2011. Growth, physiological and molecular traits in Salicaceae trees investigated for phytoremediation of heavy metals and organics. *Tree Physiol* 31(12):1319–1334

McGrath, S. P., Zhao, F. J. and Lombi, E. 2002. Phytoremediation of metals, metalloids, and radionucleotides. *Adv Agron* 75:1–56.

Meharg, A. A. and Macnair, M. 1992. Suppression of the high affinity phosphate uptake system: a mechanism of arsenic tolerance in *Holcuslanatus* L. *J Exp Bot* 43(4):519–524.

Meharg, A. A. 1994. Integrated tolerance mechanisms-constitutive and adaptive plant- responses to elevated metal concentrations in the environment. *Plant Cell Environ* 17(9):989–993.

Meharg, A. A. 2004. Arsenic in rice–understanding a new disaster for South-East Asia. *Trends Plant Sci* 9(9):415–417.

Meharg, A. A. and Hartley-Whitaker, J. 2002. Arsenic uptake and metabolism in arsenic resistant and non-resistant plant species. *New Phytol* 154:29–43.

Meharg, A. A. and Rahman, M. M. 2003. Arsenic contamination of Bangladesh paddy field soils: implications for rice contribution to arsenic consumption. *Environ Sci Technol* 37(2):229–234.

Meharg, A. A., Williams, P. N., Adomako, E., Lawgali, Y. Y., Deacon, C., Villada, A., Cambell, R. C. J., Sun, G., Zhu, Y. G., Feldmann, J., Raab, A., Zhao, F. J., Islam, R., Hossain, S. and Yanai, J. 2009. Geographical variation in total and inorganic arsenic content of polished (white) rice. *Environ Sci Technol* 43(5):1612–1617.

Mendoza-C´ozatl, D. G., Jobe, T. O., Hauser, F. and Schroeder, J. I. 2011. Long-distance transport, vacuolar sequestration, tolerance, and transcriptional responses induced by cadmium and arsenic. *Curr Opin Plant Biol* 14:554–562.

Mishra, S., Alfeld, M., Sobotka, R., Andresen, E., Falkenberg, G. and Kupper, H. 2016. Analysis of sublethal arsenic toxicity to *Ceratophyllum demersum*: subcellular distribution of arsenic and inhibition of chlorophyll biosynthesis. *J Exp Bot* 67:4639–4646.

Mishra, S., Stark, H. J. and Kupper, H. 2014. A different sequence of events than previously reported leads to arsenic-induced damage in *Ceratophyllum demersum* L. *Metallomics* 6:444–454.

Mishra, S., Dwivedi, S., Mallick, S. and Tripathi, R. D. 2019. Redox homeostasis in plants under arsenic stress. In: Panda S. and Yamamoto Y. (eds), *Redox Homeostasis in Plants. Signaling and Communication in Plants*. Springer, Cham.

Mitani-Ueno, N., Yamaji, N., Zhao, F. J. and Ma, J. F. 2011. The aromatic/arginine selectivity filter of NIP aquaporins plays a critical role in substrate selectivity for silicon, boron, and arsenic. *J Exp Bot* 62:4391–98.

Miteva, E. 2002. Accumulation and effect of arsenic on tomatoes. *Commun Soil Sci Plant Anal* 33:1917–1926.

Miteva, E. and Merakchiyska, M. 2002. Response of chloroplasts and photosynthetic mechanism of bean plants to excess arsenic in soil. *Bulg J Agric Sci* 8:151–156.

Moore, K. L., Chen, Y., Van De Meene, A. M. L., Hughes, L., Liu, W., Geraki, T., Mosselmans, F., McGrath, S. P., Grovenor, C. and Zhao, F. J. 2014. Combined nano SIMS and synchrotron X-ray fluorescence reveal distinct cellular and subcellular distribution patterns of trace elements in rice tissues. *New Phytol* 201:104–115.

Moore, K. L., Schr¨oder, M., Lombi, E., Zhao, F. J., McGrath, S. P., Hawkesford, M. J., Shewry, P. R. and Grovenor, C. R. M. 2010. Nano SIMS analysis of arsenic and selenium in cereal grain. *New Phytol* 185:434–445.

Nahar, M., Inaoka, T. and Fujimura, M. 2013. A consecutive study on arsenic exposure and intelligence quotient (IQ) of children in Bangladesh. *Environ Health Prev Med* 19:194–199.

Navasumrit, P., Chaisatra, K. and Promvijit, J., Parnlob, V., Waraprasit, S., Chompoobut, C., Binh, T. T., Hai, D. N., Bao, N. D., Hai, N. K., Kim, K. W., Samson, L. D., Graziano, J. H., Mahidol, C., Ruchirawat, M. et al. 2019. Exposure to arsenic in utero is associated with various types of DNA damage and micronuclei in newborns: a birth cohort study. *Environ Health* 18:51.

Niazi, N. K., Singh, B. and Minasny, B. 2015. Mid-infrared spectroscopy and partial least-squares regression to estimate soil arsenic at a highly variable arsenic-contaminated site. *Int J Environ Sci Technol* 12:1965–1974.

Norra, S., Aggrawala, P., Wagner, F., Berner, Z., Chandrasekharam, D. and Stüben, D. 2005. Impact of irrigation with As rich groundwater on soil and crops: a geochemical case study in Malda District, West Bengal. *App Geochem* 20:1890–1906.

Norton, G. J., Douglas, A., Lahner, B., Yakubova, E., Guerinot, M. L., Pinson, S. R. M., Eizenga, L., McGrath, G. C., Zhao, S. P., Islam, F. J., Islam, M. R., Duan, S., Zhu, G., Salt, Y., Meharg, D. E. and Price, A. H., 2014. Genome wide association mapping of grain arsenic, copper, molybdenum and zinc in rice (*Oryza sativa* L.) grown at four international field sites. *PLoS One* 9(2):e89685.

Patel, M., Kumari, A. and Parida, A. K. 2020. arsenic tolerance mechanisms in plants and potential role of arsenic hyperaccumulating plants for phytoremediation of arsenic-contaminated soil. In: Hasanuzzaman M. (ed), *Plant Ecophysiology and Adaptation under Climate Change: Mechanisms and Perspectives II*. Springer, Singapore.

Patel, A., Tiwari, S. and Prasad, S. M. 2018. Toxicity assessment of arsenate and arsenite on growth, chlorophyll a fluorescence and antioxidant machinery in *Nostoc muscorum*. *Ecotoxicol Environ Saf* 157:369–379.

Polizzotto, M. L., Lineberger, E. M., Matteson, A. R., Neumann, R. B., Badruzzaman, A. B. M. and Ali, M. A. 2013. Arsenic transport in irrigation water across rice-field soils in Bangladesh. *Environ Pollut* 179:210–217.

Pottier, M., Oomen, R., Picco, C., Giraudat, J., Scholz-Starke, J., Richaud, P., Carpaneto, A. and Thomine, S. 2015. Identification of mutations allowing natural resistance associated macrophage proteins (NRAMP) to discriminate against cadmium. *Plant J* 83:625–637.

Punshon, T., Davis, M., Marsit, C., Theiler, S., Baker, E., Jackson, B., Conway, D. and Karagas, M. 2015. Placental arsenic concentrations in relation to both maternal and infant biomarkers of exposure in a US cohort. *J Expo Sci Environ Epidemiol* 25:599–603.

Raab, A., Schat, H., Meharg, A. A. and Feldmann, J. 2005. Uptake, translocation and transformation of arsenate and arsenite in sunflower (*Helianthus annuus*): formation of arsenic-phytochelatin complexes during exposure to high arsenic concentrations. *New Phytol* 168:551–558.

Raab, A., Williams, P. N., Meharg, A. and Feldmann, J. 2007. Uptake and translocation of inorganic and methylated arsenic species by plants. *Environ Chem* 4:197–203.

Ramirez-Andreotta, M. D., Brusseau, M. L., Beamer, P. and Maier, R. M. 2013. Home gardening near a mining site in an arsenic-endemic region of Arizona: assessing arsenic exposure dose and risk via ingestion of home garden vegetables, soils, and water. *Sci Total Environ* 454:373–382.

Ratnaike, R. 2003. Acute and chronic arsenic toxicity. *Postgrad Med J* 79:391–396.

Rehman, Z. U., Khan, S., Qin, K., Brusseau, M. L., Shah, M. T. and Din, I. 2016. Quantification of inorganic arsenic exposure and cancer risk via consumption of vegetables in southern selected districts of Pakistan. *Sci Total Environ* 550:321–329.

Roychowdhury, T., Tokunaga, H., Uchino, T. and Ando, M. 2005. Effect of arsenic contaminated irrigation water on agricultural land soil and plants in West Bengal, India. *Chemosphere* 58:799–810.

Roychowdhury, T., Uchino, T., Tokunaga, H. and Ando, M. 2002. Survey of arsenic in food composites from an arsenic-affected area of West Bengal, India. *Food Chem Toxicol* 40:1611–1621.

Saha, J. C., Dikshit, A. K., Bandyopadhyay, M. and Saha, K. C. 1999. Review of arsenic poisoning and its effects on human health: critical reviews in environmental science and technology. *Crit Rev Environ Sci Technol* 29(3):281–313.

Salnikow, K. and Zhitkovich, A. 2008. Genetic and epigenetic mechanisms in metal carcinogenesis and cocarcinogenesis: nickel, arsenic, and chromium. *Chem Res Toxicol* 21:28–44.

Sasaki, A., Yamaji, N. and Ma, J. F. 2014. Overexpression of OsHMA3 enhances Cd tolerance and expression of Zn transporter genes in rice. *J Exp Bot* 65:6013–6021.

Saunders, J. K. and Rocap, G. 2016. Genomic potential for arsenic efflux and methylation varies among global *Prochlorococcus* populations. *ISME J* 10:197–209.

Shakoor, M. B., Niazi, N. K., Bibi, I., Murtaza, G., Kunhikrishnan, A., Seshadri, B., Shahid, M., Ali, S., Bolan, N. S. and Ok, Y. S. 2016. Remediation of arsenic contaminated water using agricultural wastes as biosorbents. *Crit Rev Environ Sci Technol* 46:467–499.

Shakoor, M. B., Niazi, N. K., Bibi, I., Rahman, M. M., Naidu, R., Dong, Z., Shahid, M. and Arshad, M. 2015. Unraveling health risk and speciation of arsenic from groundwater in rural areas of Punjab, Pakistan. *Int J Environ Res Public Health* 12:12371–12390.

Sharma, I. 2012. Arsenic induced oxidative stress in plants. *Biologia* 67:447–453.

Shen, S., Li, X. F., Cullen, W. R., Weinfeld, M. and Le, X. C. 2013. Arsenic binding to proteins. *Chem Rev* 113:7769–7792.

Shin, H., Shin, H. S., Dewbre, G. R. and Harrison, M. J. 2004. Phosphate transport in Arabidopsis: Pht1;1 and Pht1;4 play a major role in phosphate acquisition from both low- and high-phosphate environments. *Plant J* 39:629–642.

Singh, N., Ma, L. Q., Srivastava, M. and Rathinasabapathi, B. 2006. Metabolic adaptations to arsenic-induced oxidative stress in *Pteris vittata L* and *Pteris ensiformis* L. *Plant Sci* 170:274–282.

Sivaramakrishnan, M., Sivarajasekar, S., Vivek, J., Vasudha, P. S., Thanaraj, P., Dhakal, N. and Naushad, Mu. 2020. Research updates on heavy metal phytoremediation: enhancements, efficient post-harvesting strategies and economic opportunities. In: Naushad M. and Lichtfouse E. (eds), *Green Materials for Wastewater Treatment. Environmental Chemistry for a Sustainable World*, vol 38. Springer, Cham.

Smedley, P. L. and Kinniburgh, D. G. 2002. A review of the source, behaviour and distribution of arsenic in natural waters. *Appl Geochem* 17:517–568.

Smith, P. G., Koch, I. and Reimer, K. J. 2008. Uptake, transport and transformation of arsenate in radishes (*Raphanus sativus*). *Sci Total Environ* 390:188–197.

Smith, E., Juhasz, A. L. and Weber, J. 2009. Arsenic uptake and speciation in vegetables grown under greenhouse conditions. *Environ Geochem Health* 31:125–132.

Song, W., Park, J., Mendoza-Cozatl, D. G., Suter-Grotemeyer, M., Shim, D., Hörtensteiner, S., Geisler, M., Weder, B., Rea, P. A., Rentsch, D., Schroeder, J. I., Lee, Y. and Martinoia, E. 2010. Arsenic tolerance in Arabidopsisis mediated by two ABCC-type phytochelatin transporters. *Proc Natl Acad Sci USA* 107:21187–21192.

Song, W. Y., Mendoza-Cózatl, D. G., Lee, Y., Schroeder, J. I., Ahn, S. N., Lee, H. S., Wicker, T. and Martinoia, E. 2014. Phytochelatin–metal(loid) transport into vacuoles shows different substrate preferences in barley and Arabidopsis. *Plant Cell Environ* 37:1192–1201.

Srivastava, M., Ma, L. Q., Singh, N. and Singh, S. 2005. Antioxidant responses of hyper-accumulator and sensitive fern species to arsenic. *J Exp Bot* 56(415):1335–1342

Srivastava, S. and Sharma, Y. 2013. Altered growth, photosynthetic machinery and induced oxidative stress in spinach in response to arsenic stress. *J Plant Physiol Pathol* 1:4.

Srivastava, S., Sinha, P. and Sharma, Y. K. 2017. Status of photosynthetic pigments, lipid peroxidation and anti-oxidative enzymes in *Vigna mungo* in presence of arsenic. *J Plant Nutr* 40:298–306.

Steinmaus, C., Ferreccio, C., Acevedo, J., Yuan, Y., Liaw, J., Duran, V., Cuevas, S., Garcia, J., Meza, R., Valdés, R., Valdes, G., Benitez, H., VanderLinde, V., Villagra, V., Cantor, K., Moore, L., Pérez, S., Steinmaus, S. and Smith, A. 2014. Increased lung and bladder cancer incidence in adults after in utero and early-life arsenic exposure. *Cancer Epidemiol Biomark Prev* 23(8):1529–1538.

Stepanok, V. 1998. The effect of arsenic on the yield and elemental composition of agricultural crops. *Agrokhimiya* 12:57–63.

Stoeva, N., Berova, M. and Zlatev, Z. 2003. Effect of arsenic on some physiological indexes in maize (*Zea mays*). *J Environ Prot Ecol* 4:796–801.

Sychta, K., Słomka, A., Sliwinska, E., Migdałek, G. and Kuta, E. 2019. From cells highly tolerant to Zn and Pb to fully fertile plants – selection of tolerant lines with in vitro culture. *Plant Physiol Biochem* 146:231–237.

Ueno, D., Yamaji, N., Kono, I., Huang, C. F., Ando, T., Yano, M. and Ma, J. F. 2010. Gene limiting cadmium accumulation in rice. *Proc Natl Acad Sci USA* 107:16500–16505.

Upadhyay, M., Yadav, P., Shukla, A. and Srivastava, S. 2018. Utilizing the potential of microorganisms for managing arsenic contamination: a feasible and sustainable approach. *Front Environ Sci* 6:24.

Uraguchi, S. and Fujiwara, T. 2013. Rice breaks ground for cadmium-free cereals. *Curr Opin Plant Biol* 16:328–334.

Von Endt, D. W., Kearney, P. C. and Kaufman, D. D. 1968. Degradation of monosodium methanearsonic acid by soil microorganisms. *J Agr Food Chem* 16:17–20.

Vromman, D., Martínez, J., Kumar, M., Slejkovec, Z. and Lutts, S. 2018. Comparative effects of arsenite (As(III)) and arsenate (As(V)) on whole plants and cell lines of the arsenic-resistant halophyte plant species *Atriplex atacamensis*. *Environ Sci Pollut Res* 25(34):34473–34486.

Wang, W., Wang, Q., Zou, Z., Zheng, F. and Zhang, A. 2020. Human arsenic exposure and lung function impairment in coal-burning areas in Guizhou, China. *Ecotoxicol Environ Saf* 190:110174.

Wells, B. and Gilmor, J. 1997. Sterility in rice cultivars as influenced by MSMA rate and water management. *Agron J* 69:451–454.

Williams, P. N., Lei, M., Sun, G. X., Huang, Q., Lu, Y., Deacon, C., Meharg, A. A. and Zhu, Y. G. 2009. Occurrence and partitioning of cadmium, arsenic and lead in mine impacted paddy rice: Hunan, China. *Environ Sci Technol* 43:637–642.

Wilson, B. and Pyatt, F. B. 2007. Heavy metal dispersion, persistance, and bioaccumulation around an ancient copper mine situated in Anglesey, UK. *Ecotox Environ Safe* 66:224–231.

Woolson, E. A., Axley, J. H. and Kearney, P. C. 1971. Correlation between available soil arsenic, estimated by six methods, and response to corn (*Zea mays* L.). *Soil Sci Soc Amer Proc* 35:101–105.

Woolson, E. A., Axley, J. H. and Kearney, P. C. 1973a. The chemistry and phytotoxicity of arsenic in soils. II. Effect of time and phosphorus. *Soil Sci Soc Am Proc* 37:254–259.

Woolson, E. A. and Kearney, P. C. 1973b. Persistence and reactions of [14C]-cacodylic acid in soils. *Environ Sci Technol* 7:47–50.

Wuana, R. and Okieimen, F. 2011. Heavy metals in contaminated soils: a review of sources, chemistry, risks and best available strategies for remediation, 1–20.

Zhang, W., Guo, Z. and Wu, Y., Qiao, Y. and Zhang, Li. 2019. Arsenic bioaccumulation and biotransformation in clams (*Asaphis violascens*) exposed to inorganic arsenic: effects of species and concentrations. *Bull Environ Contam Toxicol* 103:114–119.

Zhao, F. J., Ma, J. F., Meharg, A. A. and McGrath, S. P. 2009. Arsenic uptake and metabolism in plants. *New Phytol* 181:777–794.

Zhao, F. J., Ma, Y., Zhu, Y. G., Tang, Z. and McGrath, S. P. 2015. Soil contamination in China: current status and mitigation strategies. *Environ Sci Technol* 49:750–759.

Zhao, F. J., McGrath, S. P. and Meharg, A. A. 2010. Arsenic as a food chain contaminant: mechanisms of plant uptake and metabolism and mitigation strategies. *Annu Rev Plant Biol* 61:535–559.

8

Phytoremediation: A Unified Remedial System to Tackle Pesticide Pollution

Sajid Muhammad[1], Faisal Islam[1], Pingchuan Deng[1], Mukhtar Ahmad[2], Md. Mosfeq-Ul-Hasan[1,3], Ali Muhammad[4], Mehmood Jan[1], Muhammad Afzal[1], Asad Ullah Khan[1], Muhammad Zeeshan[1], Syed Muhammad Azam[5], and Syed Mehar Ali Shah[6]

[1] *Zhejiang University, Hangzhou, China*

[2] *University of the Punjab, Lahore, Pakistan*

[3] *Hajee Mohammad Danesh Science and Technology University, Dinajpur, Bangladesh*

[4] *Huazhong Agricultural University, Wuhan, China*

[5] *Fujian Agriculture and Forestry University, Fuzhou, China*

[6] *University of Agriculture, Peshawar, Pakistan*

CONTENTS

8.1 Introduction

Green revolution in the 20th century enabled a sharp increase in the yield of cereals to address the increasing caloric needs of human population. Since then, pest management mainly depends on the demanding use of synthetic pesticides. Together with this, advances in plant varieties, crop nutrition, irrigation system, and mechanization have helped to enhance major crops yields (Zhang et al. 2015; Durán-Lara et al. 2020). This evolution in agro-systems has accelerated agricultural production over the past 70 years, favoring selective sweeps whereby new pathogen races outnumbered displacing the existing strains. Recent sweeps include potato late blight (*Phytophthora infestans*), yellow rust in wheat and barley (*Puccinia striiformis*), and head blight in wheat and barley (*Fusarium graminearum*) (Khan et al. 2017). On the contrary, over the next few decades, food production will have to explode significantly in order to meet the needs of rising human population. There will be no "universal remedy" to the impending food production challenge. This should happen without damaging other public assets – environmental or social that farming brings.

In preference, some advancements must be done to meet farmer's requirements according to local circumstances.

- Improvement in the pest, weeds, and invertebrates and vertebrates management can increase food availability as these species cause about 40% loss in the world's crop production.
- Other restraints such as poor soil, extreme weather, and less technicalities of farmers also challenge the global food security.
- In some areas, utilization of pesticides is too high, which raises concerns over chemical residues in food, thereby decreasing the production alternatively.
- Socio-cultural factors such as labor and land use can also affect the overall food production.
- The use of the latest technologies and environment-friendly cultural practices also aid in food production.
- Introduction of pest resilient cultivar can also increase food production and decrease pesticide load.

The application of pesticides has played a crucial role for farmers in boosting up the agricultural production (Zhang et al. 2015). Some of these farm chemicals are involved in the improvement in agricultural production by controlling soil-borne pathogens. In paddies, nearly 15% of the total plant protection, synthetic chemicals are used for plant production (Muhammad et al. 2019).

This excessive and persistent use of pesticides has contaminated rural areas, agricultural soils, and natural water reservoirs (Pascal-Lorber and Laurent 2011). Some of these pesticides once used were banned because of having venomous effects on human health. However, to minimize the ill-effects of these xenobiotics in contaminated areas, many conventional cleaning methods were used such as burning, land filling, composting, and other chemical adaptations (Wehtje et al. 2000). Being *ex situ*, it is very costly to excavate and transport these contaminants. Also, these methods are vicious damaging soil quality and texture causing an overall risk to ecosystem. Subsequently, there has been an increase in the *in situ* remediation technologies over the past few decades because they are environment-friendly, cost-effective, low-maintenance, sociable, and natural (Chaudhry et al. 2005). Phytoremediation, being an integral part of bioremediation, seems to be a "universal remedy" for the removal of contaminants. In some cases, phytoremediation has proved to be effective irrespective of plant-associated microorganisms. In either case, it relies on the innate capabilities of plants and soil microorganisms to absorb contagions like they do with other beneficial nutrients to accumulate, digest, or even co-metabolize (McGuinness and Dowling 2009). However, the effective use of bioremediation technologies is very complicated because of the diversification of pollutants existing in soils, which makes it difficult to decide between the bioremediation technologies. This chapter describes the remedial solutions for diverse pesticides and advancements in phytotechnologies with or without the assistance of microorganisms. Important plant species used for phytoremediation on pilot scale are also discussed in detail along with future challenges and perspectives.

8.2 Pesticides Application

Pesticides are the chemicals used to control unwanted materials, and excessive use of pesticides and fertilizers lead to their accumulation in environmental commodities, polluting food products and raising concerns over the safety of biodiversity. However, their low biodegradability has classified these chemical substances as persistent toxic elements. Collectively, the term "Pesticides" is used for all those chemicals that respond to a certain group of organisms, e.g. fungicides (against fungi), herbicides (against plants), and insecticides (against insects) are the three primary classes of pesticides (Lushchak et al. 2018). While discussing these primary classes individually, there is still a variability among these compounds and their chemical properties. Further, insecticides are of four main groups: organochlorines, organophosphates, pyrethroids, and carbamate esters. Among insecticides, organochlorines are commonly used and have created an alarming situation for the entire ecosystem. Very common compounds in this group are hexacyclochlorohexanes: a, b, g, d, t (HCHs), dichlorodiphenyltrichloroethane (DDT), dichlorodiphenyldichloroethylene (DDE), dichlorodiphenyldichloroethane (DDD), and chlordane (Pascal-Lorber and Laurent 2011). Herbicides, usually known as weed killers, are substances that are used to control unwanted plants. Common herbicide contaminations come with atrazine (2-chloro-4-(aminoethyl)-6-(aminoisopropyl)-s-1,3,5-triazine), which is a photosynthesis inhibiting compound (Pascal-Lorber and Laurent 2011) used against broad-leaved weeds and annual grasses on commercial sites and along rail tracks that causes widespread contamination of surface and underground water (Eevers et al. 2017). Conversely, fungicides are chemical compounds or biological organisms used to kill parasitic fungi. But, these chemicals can cause serious damage to agriculture resulting in critical yield losses. Important members of fungicide are hexachlorobenzene (HCB) and pentachlorophenol (PCP) (Leemans et al. 2019). HCB is a water-repellent organic compound, which is identified for its bioavailability and can be detected in air, soil, fish, birds, and human milk. Besides that, PCP can be used as fungicide, herbicide, and insecticide raising considerable concerns over its adverse effects on the ecosystem. It causes contamination in soil, food, surface waters, sediments, drinking water, rainwater, aquatic organisms, as well as in human milk and urine (Eevers et al. 2017). It is well-known that the use of farm chemicals is an extensive approach against insect pests without harming crop physiology and metabolic processes; however, they have severe adverse effects on environment and ecosystem, as depicted in Figure 8.1

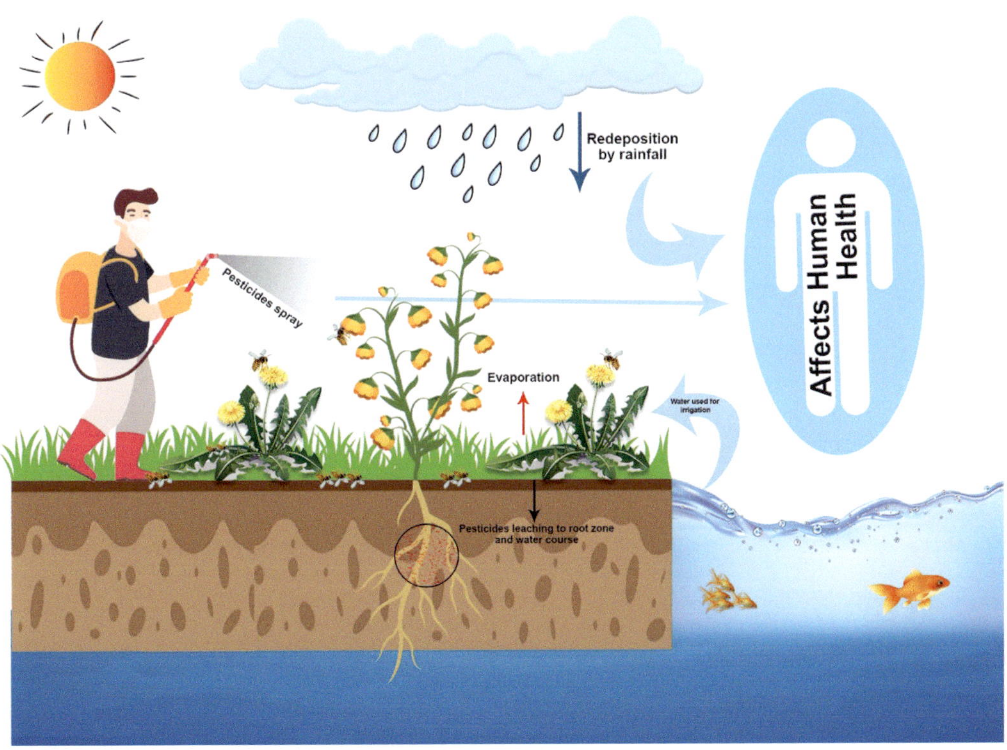

FIGURE 8.1 An overview of pesticide recycling in the atmosphere impacting human health and biosphere. Pesticides can kill insects and pests adding agricultural production, but they also kill beneficial microorganisms. They leach down to root zone and water courses, absorbed by plants roots, accumulated in plant tissues, evaporated to environment, and redeposited to soil in the form of rain.

(Muhammad et al. 2019). Reasoning from the fact, intensive rely on chemical control is a major driver of biodiversity loss and decline in functioning of related ecosystems. Scientific community has made big progresses in solving problems related to pesticides, biodiversity, and ecosystem functioning relevant to the management of ecosystems. For this purpose, different biophysical and biochemical processes were used, including adsorption, transport, hyper-accumulation, transformation, and mineralization, but all these technologies have serious constraints such as they may be expensive, laborious, and dangerous to soil properties causing disturbance of soil native microflora (Salamanca et al. 2015); only phytoremediation is "as fit as a fiddle" to control environmental pollution.

8.2.1 Pesticides Remediation

Remedial approaches for pesticides contamination can be physical, chemical, biological, or a combination of some or all of these methodologies (Pascal-Lorber and Laurent 2011). They may be used as conventional methods for the removal of pesticides from soils, exhumed and transported to particular landfills on site (McGuinness and Dowling 2009). Being effective, these strategies have considerable limitations which make them laborious, costly, and detrimental to ecosystem. Therefore, the world needs possible alternatives to these conventional methods that would be more promising in terms of reliability and sustainability of the biome. Phytoremediation, a part of bioremediation, is an innovative and appropriate approach that can tackle elevating pesticide pollution in environments through proper management and use of selectable plant species among which some are given in Table 8.1 (Hussain et al. 2009).

TABLE 8.1

Plant Species with Pesticide Remediation Potential

Plant Species	Pesticide	Reference
Cucurbita pepo	DDE	(Eevers et al. 2018)
Glycine max	Azoxystrobin	(Romeh 2015)
Lotus tenuis	HCHs	(Balázs et al. 2018)
Artemisia vulgaris	HCHs	(Balázs et al. 2018)
	DDD, DDE, DDT	(Mitton et al. 2014)
Brassica campestris	Endosulfan	(Mukherjee and Kumar 2012)
Helianthus annus	Azoxystrobin	(Romeh 2015)
	DDD, DDE, DDT	(Mitton et al. 2014)
Medicago sativa	DDD, DDE, DDT	(Mitton et al. 2014)
	Napropamide	(Cui and Yang 2011)
Oryza sativa	DDT	(Chu et al. 2006)
	PCBs	(Chu et al. 2006)
	Atrazine	(Huang et al. 2016)
	Isoproturon	(Huang et al. 2016)
Avena sativa	HCHs	(Pereira et al. 2006)
Cajanus cajan	DDE	(White et al. 2005)
Allium fistulosum	Phoxim	(Wang et al. 2011)
Vigna radiate	Aldicarb	(Sun et al. 2004)
Daucus carota	DDD, DDE, DDT	(Moklyachuk et al. 2012)
	HCHs	(Bogdevich and Cadocinicov 2010)
	Phoxim	(Wang et al. 2011)
Acorus calamus	Atrazine	(Wang et al. 2012)
Arachis hypogaea	DDE	(White et al. 2005)
Sorghum bicolor	DDD, DDE, DDT, HCHs	(Bogdevich and Cadocinicov 2010)
Cytisus striatus	HCHs	(Becerra-Castro et al. 2013)
Achillea millefolium	DDT	(Moklyachuk et al. 2012)
Brassica juncea	DDE	(White et al. 2005)

(Continued)

TABLE 8.1 (*Continued*)
Plant Species with Pesticide Remediation Potential

Plant Species	Pesticide	Reference
Cucumis sativus	DDE	(Gent et al. 2007)
	Endosulfan sulfate	(Somtrakoon et al. 2014)
	HCHs	(Bogdevich and Cadocinicov 2010; Moklyachuk et al. 2010)
Hordeum vulgare	DDD, DDE, DDT, HCHs	(Moklyachuk et al. 2010)
Eichhornia crassipes	Aldrin, DDT, dieldrin, endosulfan, malathion, methyl parathion	(Mercado-Borrayo et al. 2015)
Iris pseudacorus	Atrazine	(Wang et al. 2012)
	Chlorpyrifos	(Wang et al. 2013)
Holcus lanatus	HCHs	(Kidd et al. 2008)
Zea mays	HCHs	(Alvarez et al. 2015)
	Endosulfan sulfate	(Somtrakoon et al. 2014)
	DDD, DDE, DDT	(Bogdevich and Cadocinicov 2010)
Sedum alfredii	DDD, DDE, DDT	(Zhu et al. 2012)
Jathropa curcas	HCHs	(Abhilash et al. 2013)
Trifolium incarnatum	DDE	(White et al. 2005)
Vigna sinensis	Endosulfan sulfate	(Somtrakoon et al. 2014)
Elodea canadensis	Copper sulfate,	(Morton et al. 2014)
Vigna unguiculata	Carbofuran	(Huan et al. 2016)
Typha latifolia	Hexachlorobenzene	(Zhou et al. 2013)
Spirodela polyrhiza	Fenpropathrin	(Xu et al. 2015)
Pisum sativum	Carbendazim	(Shahid et al. 2018)
	Hexaconazole	(Shahid et al. 2018)
	2,4-dichlorophenoxyacetic acid	(Germaine et al. 2006)
Lolium multiflorum	Chlorpyrifos	(Ahmad et al. 2012; Jabeen et al. 2016)
	Terbuthylazine	(Mimmo et al. 2015)
Orychophragmus violaceus	DDD, DDE, DDT, HCHs	(Sun et al. 2015)
Lythrum salicaria	Atrazine	(Wang et al. 2012)
Erigeron canadensis	DDD, DDE, DDT	(Moklyachuk et al. 2012)
Nicotiana tabacum	HCHs	(Singh et al. 2011)
Sesamum indicum	HCHs	(Abhilash and Singh 2010)
Solanum lycopersicum	DDD, DDE, DDT	(Mitton et al. 2014)
Ricinus communis	DDT	(Rissato et al. 2015)

8.3 Phytoremediation

Phytoremediation is an emerging and "all in one remedy" that uses green plants and associated microorganisms to eliminate environmental contaminants from soil, water, and air through extraction, degradation, immobilization, or transformation of contaminants to less toxic form. The word phytoremediation, which consists of *phyto* from Ancient Greek means "plant," and *remedium* from Latin means "restoring balance," refers to phytotechnologies (da Conceição Gomes et al. 2016). Phytoremediation in general terms in collaboration with other technologies such as mycoremediation (fungi-based remediation), bioleaching (metals extraction using living organisms), composting (organic matter decomposition), bioaugmentation (addition of bacteria to speed up degradation of contaminants), biostimulation (involves alterations to environment to boost up the existing bacteria capable of bioremediation) is collectively called "bioremediation" that involves the removal of contaminants from all kinds of environments (Srichandan et al. 2019; Akhtar and Mannan 2020;

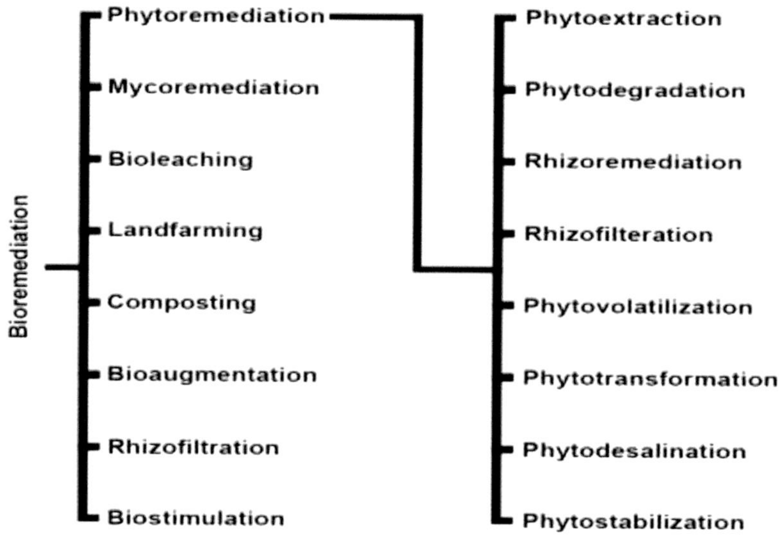

FIGURE 8.2 An illustration of different examples of bioremediation and phytoremediation.

Cai et al. 2020; Michalska et al. 2020). Bioremediation and its important examples are presented in Figure 8.2. Phytoremediation approach is the need of the hour as it is cost-effective (Yasin et al. 2015), environment-friendly (Parise 2016), suitable for a wide range and concentration (Olivares et al. 2013) of multiple pollutants (Tripathi et al. 2015), capable for combination treatment methods (Mojiri et al. 2016), *in situ* treatment method (Cheng et al. 2015), and easy large-scale applicability (Willscher et al. 2013), nutrients, organic amendments (Park et al. 2011) with effective success in current environmental pollution. Phytoremediation as a new remedy for pesticides pollution is socially expected to overcome all hurdles with higher efficiency and lower environmental impact as well as economic feasibility. Consequently, phytoremediation research has a wide range of applications as it has multidisciplinary approach that involves plant biology, ecology, genetic engineering (GE), environmental engineering, soil chemistry, and soil microbiology (Figure 8.3). Generally, organic pollutants are the principal targets of phytoremediation. Moreover, *in situ* trials have established proof of using green plants to bioaccumulate environmental wastes. The application of this green technology has wide ranges from static water to polluted soils and other contaminated environments concentrated with pesticides, heavy metals, solvents, explosives, and crude oil along with their derivatives. However, due to physiological differences, not all plants can accumulate pollutants; even plants from the same species have varying capabilities to pesticides accumulation (Lone et al. 2008; Dosnon-Olette et al. 2010; Moklyachuk et al. 2010; Nurzhanova et al. 2010; Moore and Locke 2012; Mukherjee and Kumar 2012; Wang et al. 2012; Mitton et al. 2014; Eevers et al. 2017). Conversely, many plants have proven to be successful in hyperaccumulation of contaminants at toxic waste sites contaminated with heavy metals such as lead, uranium, and arsenic and are listed in Table 8.2 (Shi et al. 2011). Apart from advantages like *in situ* treatments, it has one major disadvantage as it requires a long-term commitment of plants depending on its ability and growth in an unfavorable environment. Certain parameters must be carefully reviewed while designing an appropriate system of phytoremediation, including the type of contaminant to be treated, features of the contaminant, selection of plant species engaged in phytoremediation, and land type along with biotic and abiotic factors that affect phytoremediation (Eevers et al. 2017). Additionally, other factors such as temperature, soil organic matter contents, and pH also impact pesticide bioavailability along with degradation potential in the rhizosphere. Phytoremediation includes certain processes such as phytoextraction, rhizoremediation, phytodegradation, phytotransformation, phytovolatilization, phytostabilization, and phytodesalination. A detailed discussion of each one is given in the subsequent sub-sections.

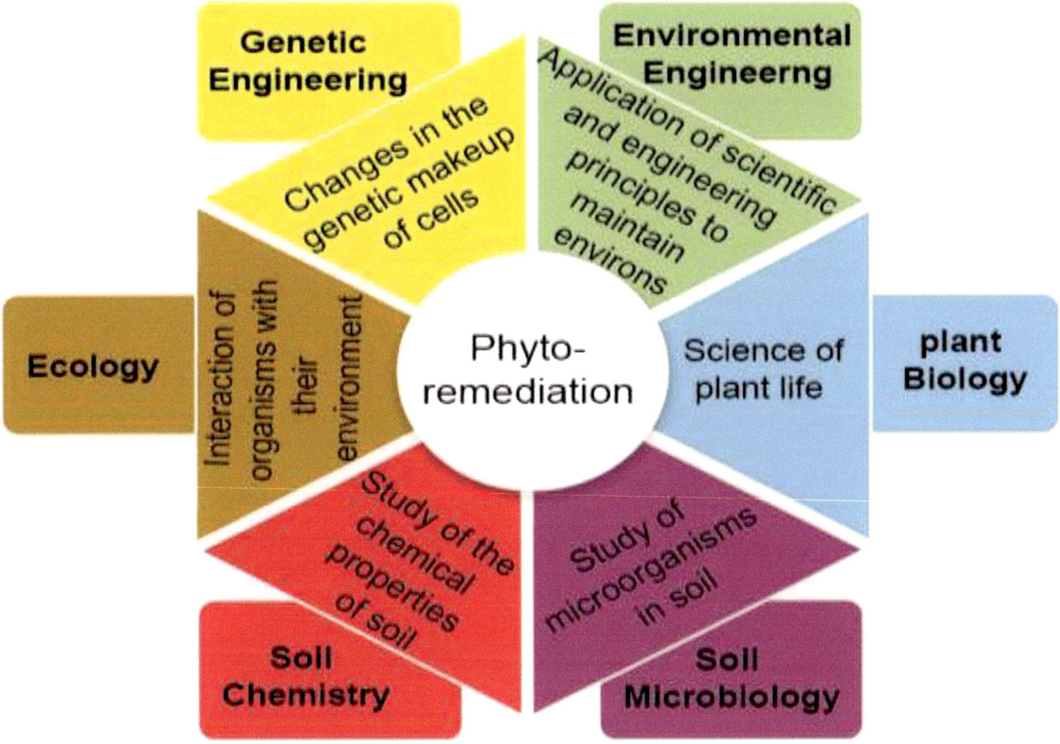

FIGURE 8.3 Phytoremediation research is a multidisciplinary approach.

8.3.1 Phytoextraction

Phytoextraction has emerged as a successful approach that clean up metal-polluted soil and water using ability of plants or algae to take up contaminants through their roots and accumulate them above ground into harvestable plant biomass. This technology is growing rapidly worldwide to overcome the adverse

TABLE 8.2

Important Plants Reported for Phytoremediation

Plant	Contaminant	Type	Reference
Salix babylonica	Cd	Phytoextractor	(Greger and Landberg 1999; Janssen et al. 2015)
Alpine pennycress	Cd and Zn	Hyper-accumulator	(Kirkham 2006)
Festuca arundinacea	Pb	Phytoextractor	(Khashij et al. 2018)
Sporobolus cryptandrus	Hydrocarbons	Phytoextractor	(Khrunyk et al. 2017)
Thlaspi arvense	Zn and Cu	Hyper-accumulator	(Pirzadah et al. 2015)
Medicago sativa	PAHs, Cd, Zn and Pb	Hyper-accumulator	(Li et al. 2020)
Typha latifolia	Cu and Pb	Sequester	(Xu et al. 2020)
Brassica juncea	Pb	Sequester	(Rathore et al. 2019; Raj et al. 2020)
Ambrosia artemisiifolia	Pb	Sequester	(Čudić et al. 2016)
Populus nigra	Cd, Cu and Pb	Hyper-accumulator	(Castro-Rodríguez et al. 2016)
Hordeum vulgare	NaCl	Phytoextractor	(Ali et al. 2017)
Solanum nigrum	Zn, Pb, Ni and Cd	Hyper-accumulator	(Saad-Allah and Elhaak 2017)
Beta vulgaris	NaCl	Hyper-accumulator	(Papazoglou and Fernando 2017)
Helianthus annuus	Pb, Sr-90, Cs-137	Hyper-accumulator	(Al-Jobori and Kadhim 2019)
Nicotiana tabacum	Cu, Cr, Pb, Zn	Hyper-accumulator	(Pourkhabbaz and Pourkhabbaz 2012)

impacts of pesticides and organic fertilizers. In general, phytoextraction is used for the elimination of heavy metals and other inorganics. In some cases, a conditioning fluid is added to the soil to stimulate metal mobilization so that the plants can absorb more contaminants, and is known as assisted phytoextraction. As such additives can increase plant's metal uptake, they can also raise metal availability in the soil beyond which the plants are able to translocate, causing potential leaching into the groundwater (Doumett et al. 2008). However, in both the cases, phytoextraction leads to the eradication of contaminants.

8.3.2 Rhizoremediation

It is the breakdown of organic contaminants, especially in the root zone, with the release of carbohydrates and acids stimulating microbial activity, which results in the biodegradation of contagions (Ahmed et al. 2015; Akbar et al. 2015; Jabeen et al. 2017). Rhizosphere is the volume of soil directly surrounded by plant roots and is heavily influenced by plant activities. Break down of pesticides is comparatively higher in rhizospheric soil than what it is in mass soil. Degradation in the rhizosphere occurs in a natural way, yet it can be improved by planting "fitting" plant species, or by adding pesticide-degrading microbes through inoculation (Gerhardt et al. 2009). This makes rhizosphere a more favorite habitat for microbes than bulk soil by providing a large amount of enzymatically active microbial networks. It is also called rhizofiltration when it comes to cleaning surface water or groundwater from contaminants.

8.3.3 Phytodegradation

On the contrary, phytodegradation is the complete breakdown of contaminants using plants or microorganisms in the plant body or soil through their enzymatic activity. Generally, it requires accumulation of pesticides which are then taken up by the plants and released through transpiration. In some cases, organic pollutants are made non-toxic by plant metabolism, while in other cases, microorganisms living in rhizosphere degrade these substances in soil or water (Ahmed et al. 2015; Jabeen et al. 2017).

8.3.4 Phytotransformation

Unlike phytodegradation, phytotransformation brings about chemical changes in contaminants as a direct result of plant digestion making them inactive. These complexes cannot be degraded to basic molecules by plants, and hence, the term phytotransformation signifies modifications in their chemical composition without a complete breakdown (Li et al. 2016).

8.3.5 Phytovolatilization

Phytovolatilization is the absorption of contaminants from soil or water and their transpiration into the atmosphere in the form of volatile molecules. When this process occurs in plant stem and leaves, it is called direct phytovolatilization; however, when contaminants volatilize from the root zone, it is termed as indirect phytovolatilization, as illustrated in Figure 8.4 (Limmer and Burken 2016). Poplar trees represented a successful example for phytoremediation through this process due to their high transpiration rate (Pilon-Smits 2015).

8.3.6 Phytostabilization

While phytoextraction removes metals from the soil, phytostabilization only reduces the mobility of toxic substances in the environments. Phytostabilization implicates reduction in the mobility of heavy metals in soil establishing a green-cover on contaminated sites through accumulation by roots, thereby reducing off-site contaminations. Phytostabilization generally includes transpiration and root absorption

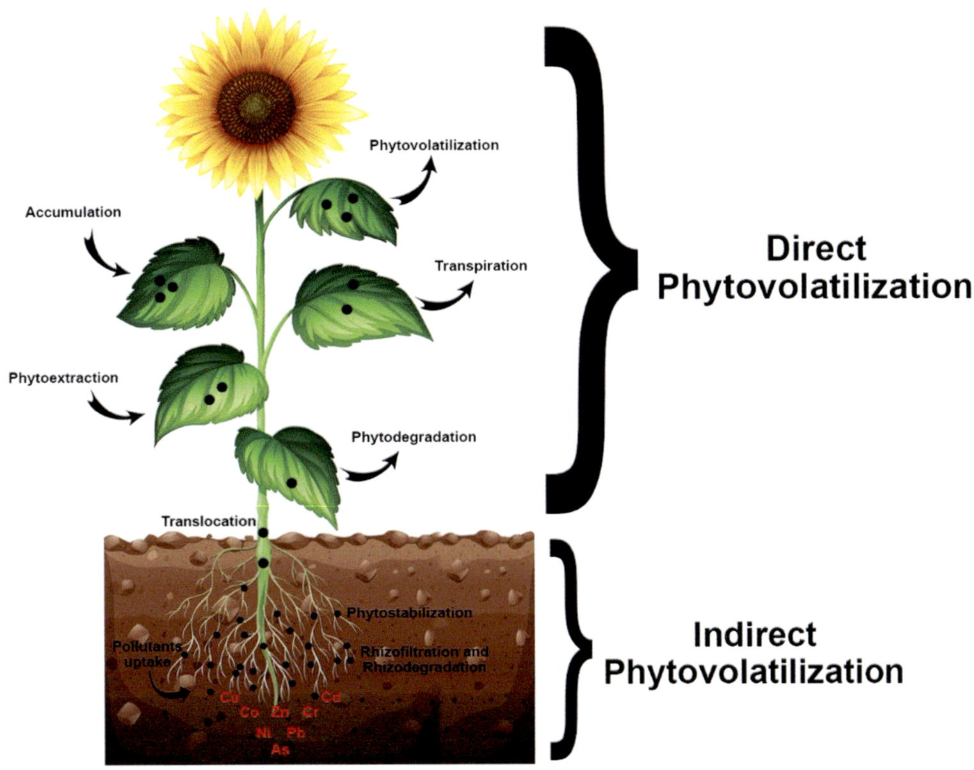

FIGURE 8.4 Illustration of different phytoremediation techniques in plants.

that ground contaminants by controlling erosion, adding organic matter to soil particles, and creating an aerobic environment in the root zone that binds contaminants, thereby reducing their bioavailability (Pilon-Smits 2015).

8.3.7 Phytodesalination

Phytodesalination is another recyclable biological approach that aims to assimilate sodic environments by halophytes (Ali et al. 2013; Islam et al. 2019). This new phytotechnique is based on the absorption and transportation of sodium ions from soil through roots to shoots of plants (Rabhi et al. 2015). All of the preceding phytotechniques can be impacted by plants and plant-associated microorganisms. Microbes, whether in the rhizospheric or endophytic region, have important roles in the remediation of contaminants through pesticide-degrading and plant growth promoting capacities (Glick 2003; 2010; Eevers et al. 2017). An illustration of different phytoremediation techniques is presented in model plant sunflower (Figure 8.4). Our world has a good experience of pilot-scale phytoremediation projects over the past two decades and these projects have made considerable progresses in accumulation of pesticides, fertilizers, heavy metals, and other contaminants, positively impacting our environments and ecosystem. Some of the large-scale projects conducted worldwide are listed in Table 8.3.

Nonetheless, this technology still have many limitations as some naturally originated plant species lack contaminants tolerance or clean-up capabilities, although researchers have discovered a phytomechanism that allows plants to grow in contaminated environments lethal to plant growth and development (Tangahu et al. 2011). Hereafter, it is obviously required to enhance the performance of the existing natural plant species, so as to acquire commercially-significant phytoremediation outputs.

TABLE 8.3
Model Phytoremediation Projects Implemented on Large Scale Worldwide

Plant Species	Pollutants	Location	Reference
Typha latifolia	Lignite and pyrolysis	Germany	(Kuschk et al. 2003)
Pteris vittata	Arsenic	USA	(Elless et al. 2005)
Phragmites australis	Nutrients and organic pollutants	UK	(Sun et al. 2006)
Persicaria, *Ludwigia peploides*, *Myriophyllum papillosum*, *Juncus usitatus* *Bolboschoenus medianus*, and *Typha domingensis*	Pesticide	Australia	(Rose et al. 2006)
Paulownia tomentosa	Metallic elements	Italy	(Doumett et al. 2008)
Phragmites australis	Organic pollutants	Germany	(Braeckevelt et al. 2008)
Freshwater algae	Nutrients (N, P)	USA	(Mulbry et al. 2008)
Phragmites australis *Typha latifolia*	Metallic elements	Taiwan	(Yeh et al. 2009)
Paulownia tomentosa	Metallic elements	Italy	(Doumett et al. 2010)
Phragmites australis	Organic pollutants	Germany	(Seeger et al. 2011)
Phragmites australis	Organic pollutants	Germany	(Braeckevelt et al. 2011)
Crown trees	Environmental pollutants	Hong Kong	(Ng et al. 2012)
Phragmites australis *Typha latifolia*	Organic pollutants	USA	(Ranieri et al. 2013)
Selaginella jacquemontii, *Rumex hastatus*, and *Plectranthus rugosus*	Metallic elements	Pakistan	(Muhammad et al. 2013)
Cyperus alopecuroides	Nutrients turbidity	India	(Devi and Jayakumar 2014)
Alternathera philoxeroides	Dye	India	(Rane et al. 2015)
Paspalum vaginatum, *P. vaginatum*, *Spartium junceum*, and *Tamarix gallica*	Metallic elements	Italy	(Doni et al. 2015)
Amorpha fruticosa	Petroleum	China	(Han et al. 2016)
Garcinia mangostana	CO_2 absorption	Indonesia	(Jaya 2018)
G. mangostana	Anti-oxidant	Indonesia	(Ayman et al. 2019)
Artocarpus heterophyllus, *Cyrtostachys lakka*, and *Carica papaya*	CO_2 absorption	Indonesia	(Ludang 2020)

8.4 Role of Biotechnology and Genetic Engineering (B & GE) in Phytoremediation

Unlike other strategies used for the enhancement of phytoremediation processes, the use of B & GE is the most significant approach for the improvement of plant's bioremediation capabilities. Transgenes, genetically engineered living organisms, generally have better performance like higher germination percentage and greater growth rate with enhanced accumulation of heavy metals than non-transformed plants (Koźmińska et al. 2018). Being the most powerful technique used to enhance natural phytoremediation capabilities, it can also introduce new understanding into the existing plants species. Moreover, genes may be transferred from one plant to another variety that suits best to the environmental conditions at the clean-up site using gene editing techniques like CRISPR-Cas9 (Buhari et al. 2016). Foreign gene may persuade plant's enzymatic activity by boosting up certain processes (up-regulated) or by depleting others (down-regulated) or it can also produce a new enzymatic activity. On the basis of this approach, enzymatic activities can be controlled all the time in every tissue (constitutive expression) or only in selected tissue (tissue-specific) or at certain times. With such management, it is probably possible to produce novel races of plants that would be more efficient to clean out certain waste of a particular area under defined conditions. It is also clear that bioremediation differs in plant capabilities depending on transporters, enzymes, root mass, microbial partners, and transpiration rate (Tangahu et al. 2011). Generally, a good decon species has properties like faster growth and high biomass, resilience, and resistance to impurities. Biotechnology, however, makes it possible to combine these properties in a single accumulator phenotype that could be more effective in phytoremediation. Additionally, if plant species have economic values, they can add more benefits to the ecosystem. All of the aforementioned properties have significant impacts on phytoremediation that may possibly be improved through B & GE (Dhankher et al. 2012).

8.4.1 Fates of Plant-Substrate Mechanism

Most contaminants have varied designations in plant-substrate mechanisms with altered rate-limiting factors for remediation that might be focused using GE. For instance, excretion of metal chelators and protons can increase bioavailability of certain metals in the rhizospheric region. Similarly, in case of rhizodegradation, secretion of degrading enzymes from roots might be up-regulated to stimulate microbial activity. Furthermore, absorption and transport of materials in plants depend on the consumption of transporter proteins that may be decreased by the number of transporter membranes, particularly for inorganic compounds. Similarly, transporters which mediate pollutants absorption and translocation might be overproduced in plants enhancing plants' assimilation capabilities (Ismael et al. 2019). However, organic compounds with moderate hydrophobic properties can normally move in and out irrespective of transporters.

8.4.2 GE-assisted Plant Tolerance

Plant tolerance, in turn, might be shortened by enzymes abundance to degrade pollutants, or by general antioxidants, however, depending on the expected limiting factors, any such enzymes may be overproduced to improve phytoremediation capacity using GE. In addition to up-regulate the existing genes, novel genes may be inserted from other plant species or any organism with the assistance of biotechnology to enhance gene expression for maximum resistance. In this way, an absolutely new remediation capacity might be brought into a suitable plant variety for phytoremediation. Moreover, several plant species have shown increased resistance under microbe's assistance; therefore, under such circumstances, biotechnology has made it possible to isolate such microorganisms and enrich mass soil to enhance phytoremediation through respective plant species (Buhari et al. 2016).

8.5 Microbial-Assisted Phytoremediation

Besides considering all other components for phytoremediation, plants grown on polluted soils might be able to resist some of the inhibitory effects of high concentration of metals within them; yet, synthesis of stress ethylene and depletion in the amount of iron could mostly result in a deceased plant growth and viability (Glick 2003). For this purpose, soil microorganisms play a substantial role in nutrients recycling, maintenance of soil structure, and detoxification of chemicals by either enhancing phytoremediation capacity of plants or reducing the phytotoxicity of the contaminated soils. Plants and their associated microbes can remediate these xenobiotics via stabilization, degradation in both rhizosphere and plants through extraction in harvestable plant parts, or volatilization. Currently, considerable research has been carried out on plants-associated microorganisms having potential implementation for the remediation of pesticide-contaminated environments

(Trognitz et al. 2016). It is also noteworthy that a wide range of microorganisms exist for the removal of many contaminants to handle environmental pollution. The bacterial catabolic pathways for the degradation of many compounds are also known (http://umbbd.ethz.ch/index.html) and their regulation has also been well-characterized (Kahlon 2016).

8.5.1 Microbial Accumulation and Remediation of Xenobiotics

Microorganisms have evolved to defend themselves against the toxicity of heavy metals through various mechanisms such as adsorption, methylation, oxidation, and reduction. Microorganisms absorb heavy metals actively through bioaccumulation or passively by adsorption, while microbial methylation assumes a significant role in the bioremediation of pesticides and fertilizers as methylated compounds are unstable and volatile in nature. For example, Mercury (Hg) (II), can be biomethylated by different bacterial species such as *Alcaligenes faecalis*, *Bacillus pumilus*, and *Brevibacterium iodinium* to gaseous methyl mercury (Igiri et al. 2018).

8.5.2 Use of Plant Growth Promoting Bacteria

However, to make the biodegradation of contaminants more efficient, microorganisms must enzymatically break down pollutants into less harmful products. Yet, it can happen only when ambient conditions allow microbial growth and development; sometimes applicability of these bacteria involves manipulation of environmental factors to allow their growth for the degradation of contaminants to proceed at a faster rate (Abatenh et al. 2017). Moreover, plant growth promoting bacteria may be used to relieve some of the toxicity of metals to plants. Likewise, for the purpose to assist plant growth promoting bacteria in their efficiency and multiplication, many techniques have been used where GE has played a vital role in manipulating microorganisms.

8.5.3 Genetically Engineered Microorganisms (GEMs)

B & GE have evolved microorganisms to utilize and eliminate hazardous materials under laboratory conditions by creating GEMs (Abatenh et al. 2017). These GEMs might be inspiration for some natural processes or artificial genetic exchange between microorganisms. GEMs have the potential of biological disposals in soil, groundwater, and other xenobiotic environments, showing enhanced degradative capabilities covering decontamination of a wide range of chemicals. The principal task of GEMs is to speed up biological disposition of wastes and increase substrate degradation with a small amount of cell mass by creating safer and neutralized environment.

8.6 Conclusion and Future Recommendations

In previous decades, due to the widespread application and utilization of pesticides, vast area is contaminated with pesticide residues. Thus, removal of these pollutants is crucially important for the safety of agricultural land, ecosystem, and human health. Therefore, phytoremediation seems to be an efficient and excellent approach for the elimination of these contaminants. Consecutively, native plant species that demonstrated excellent degradation potential in laboratory/field trails are promising candidates for achieving a clean environment free of contaminations. Efficiencies of plants can be boosted and developed using integrated approaches such as microbes and molecular techniques to improve the harvest time; shorten growth duration, and increase biomass production. However, phytoremediation alone cannot fight with these xenobiotics; so, it is very important to make a risk assessment of farm chemicals and identify other environment-friendly approaches to assist phytoremediation. Such approaches not only play a role in the current scenario but are also much needed for future generations where they will play a very active role in food security. Nonetheless, phytoremediation can promise a safe, better, and clean environment to our next generations.

REFERENCES

Abatenh, E., Gizaw, B., Tsegaye, Z. and Wassie, M. 2017. The role of microorganisms in bioremediation – a review. *J Environ Biol* 2:038–046.

Abhilash, P., Singh, B., Srivastava, P., Schaeffer, A. and Singh, N. 2013. Remediation of lindane by *Jatropha curcas*: utilization of multipurpose species for rhizoremediation. *Biomass Bioenerg* 51:189–193.

Abhilash, P. and Singh, N. 2010. Withania somnifera dunal-mediated dissipation of lindane from simulated soil: implications for rhizoremediation of contaminated soil. *J Soils Sediments* 10:272–282.

Ahmad, F., Iqbal, S., Anwar, S., Afzal, M., Islam, E., Mustafa, T. and Khan, Q. 2012. Enhanced remediation of chlorpyrifos from soil using ryegrass (*Lollium multiflorum*) and chlorpyrifos-degrading bacterium *Bacillus pumilus* C2A1. *J Hazard Mater* 237:110–115.

Ahmed, M., Jabeen, F., Ali, M., Ahmad, Z., Ahmed, F., Sarwar, B. S., Din, S. U., Hassan, M. and Jahan, S. 2015. Analysis of bifenthrin degrading bacteria from rhizosphere of plants growing at tannery solid waste. *Am J Plant Sci* 6:2042.

Akbar, S., Sultan, S. and Kertesz, M. 2015. Determination of cypermethrin degradation potential of soil bacteria along with plant growth-promoting characteristics. *Curr Microbiol* 70:75–84.

Akhtar, N. and Mannan, M. A. 2020. Mycoremediation: expunging environmental pollutants. *Biotechnol Rep* 26:e00452.

Al-Jobori, K. M. and Kadhim, A. K. 2019. Evaluation of sunflower (*Helianthus annuus*) for phytoremediation of lead contaminated soil. *J Pharm Sci Res* 11:847–854.

Ali, H., Khan, E. and Sajad, M. A. 2013. Phytoremediation of heavy metals—concepts and applications. *Chemosphere* 91:869–881.

Ali, M. B., Salem, E., and Sayed, M. A., 2017. Genetic variability of barley (*Hordeum vulgare*) genotypes in phytoremediation of heavy metals-contaminated soil. *Egypt J Agron* 39:383–399.

Alvarez, A., Benimeli, C. S., Sáez, J. M., Giuliano, A. and Amoroso, M. 2015. Lindane removal using Streptomyces strains and maize plants: a biological system for reducing pesticides in soils. *Plant Soil* 395:401–413.

Ayman, E., Hassan, S. M. and Osman, H. E. H. 2019. Mangosteen (*Garcinia mangostana*). In: *Nonvitamin and Nonmineral Nutritional Supplements*. Elsevier, Amsterdam, The Netherlands, pp:313–319.

Balázs, H. E., Schmid, C. A. O., Feher, I., Podar, D., Szatmari, P. M., Marincaş, O., Balázs, Z. R. and Schröder, P. 2018. HCH phytoremediation potential of native plant species from a contaminated urban site in Turda, Romania. *J Environ Manage* 223:286–296.

Becerra-Castro, C., Prieto-Fernandez, A., Kidd, P. S., Weyens, N. and Rodríguez-Garrido, B. 2013. Improving performance of *Cytisus striatus* on substrates contaminated with hexachlorocyclohexane (HCH) isomers using bacterial inoculants: developing a phytoremediation strategy. *Plant Soil* 362:247–260.

Bogdevich, O. and Cadocinicov, O. 2010. Elimination of acute risks from obsolete pesticides in Moldova: phytoremediation experiment at a former pesticide storehouse. In: *Application of Phytotechnologies for Cleanup of Industrial, Agricultural, and Wastewater Contamination*. Springer, Dordrecht, pp:61–85.

Braeckevelt, M., Mirschel, G., Wiessner, A., Rueckert, M., Reiche, N., Vogt, C., Schultz, A., Paschke, H., Kuschk, P. and Kaestner, M. 2008. Treatment of chlorobenzene-contaminated groundwater in a pilot-scale constructed wetland. *Ecol Eng* 33:45–53.

Braeckevelt, M., Reiche, N., Trapp, S., Wiessner, A., Paschke, H., Kuschk, P. and Kaestner, M. 2011. Chlorobenzene removal efficiencies and removal processes in a pilot-scale constructed wetland treating contaminated groundwater. *Ecol Eng* 37:903–913.

Buhari, M., Babura, S., Vyas, N., Badaru, S. and Harisu, O. 2016. Role of biotechnology in phytoremediation. *J Bioremediat Biodegrad* 7:330–339.

Cai, P., Ning, Z., Liu, Y., He, Z., Shi, J. and Niu, M. 2020. Diagnosing bioremediation of crude oil-contaminated soil and related geochemical processes at the field scale through microbial community and functional genes. *Ann Microbiol* 70:1–15.

Castro-Rodríguez, V., Gutiérrez, A., Canales, J., Cañas, R. A., Kirby, E. G., Avila, C. and Canovas, F. M. 2016. Poplar trees for phytoremediation of high levels of nitrate and applications in bioenergy. *Plant Biotechnol J* 14:299–312.

Chaudhry, Q., Blom-Zandstra, M., Gupta, S. K. and Joner, E. 2005. Utilising the synergy between plants and rhizosphere microorganisms to enhance breakdown of organic pollutants in the environment. *Environ Sci Pollut Res* 12:34–48.

Cheng, S. F., Huang, C. Y., Lin, Y. C., Lin, S. C. and Chen, K. L. 2015. Phytoremediation of lead using corn in contaminated agricultural land—An *in situ* study and benefit assessment. *Ecotoxicol Environ Safe* 111:72–77.

Chu, W., Wong, M. H. and Zhang, J. 2006. Accumulation, distribution and transformation of DDT and PCBs by *Phragmites australis* and *Oryza sativa*: II. Enzyme study. *Environ Geochem Health* 28:169–181.

Čudić, V., Stojiljković, D. and Jovović, A. 2016. Phytoremediation potential of wild plants growing on soil contaminated with heavy metals. *Arh Hig Rada Toksikol* 67:229–239.

Cui, L. E. and Yang, H. 2011. Accumulation and residue of napropamide in alfalfa (*Medicago sativa*) and soil involved in toxic response. *J Hazard Mater* 190:81–86.

da Conceição Gomes, M. A., Hauser-Davis, R. A., de Souza, A. N. and Vitória, A. P. 2016. Metal phytoremediation: general strategies, genetically modified plants and applications in metal nanoparticle contamination. *Ecotoxicol Environ Safe* 134:133–147.

Devi, B. M. and Jayakumar, K. 2014. Phytoremediation studies on tannery effluent using pilot scale constructed wetlands. In: Ramamohan K. (ed), *Hydrology and Watershed Management: Ecosystem Resilience-Rural and Urban Water Requirements*. Int J Environ Tech Sci 2:69–75.

Dhankher, O. P., Pilon-Smits, E. A., Meagher, R. B. and Doty, S. 2012. Biotechnological approaches for phytoremediation. In: *Plant Biotechnology and Agriculture*. Arie Altman, A., Hasegawa, P. M., Eds.; Academic Press: Oxford, UK, Elsevier, pp:309–328.

Doni, S., Macci, C., Peruzzi, E., Iannelli, R. and Masciandaro, G. 2015. Heavy metal distribution in a sediment phytoremediation system at pilot scale. *Ecol Eng* 81:146–157.

Dosnon-Olette, R., Trotel-Aziz, P., Couderchet, M. and Eullaffroy, P. 2010. Fungicides and herbicide removal in Scenedesmus cell suspensions. *Chemosphere* 79:117–123.

Doumett, S., Fibbi, D., Azzarello, E., Mancuso, S., Mugnai, S., Petruzzelli, G. and DelBubba, M. 2010. Influence of the application renewal of glutamate and tartrate on Cd, Cu, Pb and Zn distribution between contaminated soil and *Paulownia tomentosa* in a pilot-scale assisted phytoremediation study. *Int J Phytoremediation* 13:1–17.

Doumett, S., Lamperi, L., Checchini, L., Azzarello, E., Mugnai, S., Mancuso, S., Petruzzelli, G. and Del Bubba, M. 2008. Heavy metal distribution between contaminated soil and *Paulownia tomentosa*, in a pilot-scale assisted phytoremediation study: influence of different complexing agents. *Chemosphere* 72:1481–1490.

Durán-Lara, E. F., Valderrama, A. and Marican, A. 2020. Natural organic compounds for application in organic farming. *Agriculture* 10:41.

Eevers, N., Hawthorne, J., White, J., Vangronsveld, J. and Weyens, N. 2018. Endophyte-enhanced phytoremediation of DDE-contaminated using Cucurbita pepo: a field trial. *Int J Phytoremediation* 20:301–310.

Eevers, N., White, J. C., Vangronsveld, J. and Weyens, N. 2017. Bio- and phytoremediation of pesticide-contaminated environments: a review. *Adv Bot Res* 83:277–318.

Elless, M. P., Poynton, Y. C., Williams, A. C., Doyle, P. M., Lopez, C. A., Sokkary, A. D., Ferguson, B. W. and Balylock, M. J. 2005. Pilot-scale demonstration of phytofiltration for treatment of arsenic in New Mexico drinking water. *Water Res* 39:3863–3872.

Gent, M. P., White, J. C., Parrish, Z. D., Isleyen, M., Eitzer, B. D. and Mattina, M. I. 2007. Uptake and translocation of p, p′-dichlorodiphenyldichloroethylene supplied in hydroponics solution to Cucurbita. *Environ Toxicol Chem* 26:2467–2475.

Gerhardt, K. E., Huang, X. D., Glick, B. R. and Greenberg, B. M. 2009. Phytoremediation and rhizoremediation of organic soil contaminants: potential and challenges. *Plant Sci* 176:20–30.

Germaine, K. J., Liu, X., Cabellos, G. G., Hogan, J. P., Ryan, D. and Dowling, D. N. 2006. Bacterial endophyte-enhanced phytoremediation of the organochlorine herbicide 2, 4-dichlorophenoxyacetic acid. *FEMS Microbiol Ecol* 57:302–310.

Glick, B. R. 2003. Phytoremediation: synergistic use of plants and bacteria to clean up the environment. *Biotechnol Adv* 21:383–393.

Glick, B. R. 2010. Using soil bacteria to facilitate phytoremediation. *Biotechnol Adv* 28:367–374.

Greger, M. and Landberg, T. 1999. Use of willow in phytoextraction. *Int J Phytoremediation* 1:115–123.

Han, G., Cui, B., Zhang, X. and Li, K. 2016. The effects of petroleum-contaminated soil on photosynthesis of *Amorpha fruticosa* seedlings. *Int J Environ Sci Technol* 13:2383–2392.

Huan, Z., Xu, Z., Luo, J. and Xie, D. 2016. Monitoring and exposure assessment of pesticide residues in cowpea (*Vigna unguiculata*) from five provinces of southern China. *Regul Toxicol Pharmacol* 81:260–267.

Huang, M. T., Lu, Y. C., Zhang, S., Luo, F. and Yang, H. 2016. Rice (*Oryza sativa*) laccases involved in modification and detoxification of herbicides atrazine and isoproturon residues in plants. *J Agric Food Chem* 64:6397–6406.

Hussain, S., Siddique, T., Arshad, M. and Saleem, M. 2009. Bioremediation and phytoremediation of pesticides: recent advances. *Crit Rev Environ Sci Technol* 39:843–907.

Igiri, B. E., Okoduwa, S. I., Idoko, G. O., Akabuogu, E. P., Adeyi, A. O. and Ejiogu, I. K. 2018. Toxicity and bioremediation of heavy metals contaminated ecosystem from tannery wastewater: a review. *J Toxicol.* 2018:1–16.

Islam, M. S., Hosen, M. M. L. and Uddin, M. N. 2019. Phytodesalination of saline water using *Ipomoea aquatica*, *Alternanthera philoxeroides* and *Ludwigia adscendens*. *Int J Environ Sci Technol* 16:965–972.

Ismael, M. A., Elyamine, A. M., Moussa, M. G., Cai, M., Zhao, X. and Hu, C. 2019. Cadmium in plants: uptake, toxicity, and its interactions with selenium fertilizers. *Metallomics* 11:255–277.

Jabeen, F., Ahmed, M., Ahmed, F., Sarwar, M. B., Akhtar, S. and Shahid, A. A. 2017. Characterization of cypermethrin degrading bacteria: a hidden micro flora for biogeochemical cycling of xenobiotics. *Adv Life Sci* 4:97–107.

Jabeen, H., Iqbal, S., Ahmad, F., Afzal, M. and Firdous, S. 2016. Enhanced remediation of chlorpyrifos by ryegrass (*Lolium multiflorum*) and a chlorpyrifos degrading bacterial endophyte Mezorhizobium sp. HN3. *Int J Phytoremediation* 18:126–133.

Janssen, J., Weyens, N., Croes, S., Beckers, B., Meiresonne, L., Van Peteghem, P., Carleer, R. and Vangronsveld, J. 2015. Phytoremediation of metal contaminated soil using willow: exploiting plant-associated bacteria to improve biomass production and metal uptake. *Int J Phytoremediation* 17:1123–1136.

Kahlon, R. S. 2016. Biodegradation and bioremediation of organic chemical pollutants by Pseudomonas. *Pseudomonas: Mol Appl Biol* Springer, Cham, Switzerland, 343–417.

Khan, A., Nasir, I. A., Tabassum, B., Aaliya, K., Tariq, M. and Rao, A. Q. 2017. Expression studies of chitinase gene in transgenic potato against *Alternaria solani*. *Plant Cell Tiss Org Cult* 128:563–576.

Khashij, S., Karimi, B. and Makhdoumi, P. 2018. Phytoremediation with *Festuca arundinacea*: a mini review. *Int J Health Life Sci* 4:e86625.

Khrunyk, Y., Schiewer, S., Carstens, K. L., Hu, D. and Coats, J. R. 2017. Uptake of C14-atrazine by prairie grasses in a phytoremediation setting. *Int J Phytoremediation* 19:104–112.

Kidd, P. S., Prieto-Fernandez, A., Monterroso, C. and Acea, M. J. 2008. Rhizosphere microbial community and hexachlorocyclohexane degradative potential in contrasting plant species. *Plant Soil* 302:233–247.

Kirkham, M. B. 2006. Cadmium in plants on polluted soils: effects of soil factors, hyperaccumulation, and amendments. *Geoderma* 137:19–32.

Koźmińska, A., Wiszniewska, A., Hanus-Fajerska, E. and Muszyńska, E. 2018. Recent strategies of increasing metal tolerance and phytoremediation potential using genetic transformation of plants. *Plant Biotechnol Rep* 12:1–14.

Kuschk, P., Wiessner, A., Kappelmeyer, U., Weissbrodt, E., Kästner, M. and Stottmeister, U. 2003. Annual cycle of nitrogen removal by a pilot-scale subsurface horizontal flow in a constructed wetland under moderate climate. *Water Res* 37:4236–4242.

Leemans, M., Couderq, S. C. H., Demeneix, B. A. and Fini, J. B.2019. Pesticides with potential thyroid hormone-disrupting effects: a review of recent data. *Front Endocrinol* 10:743.

Li, L., Zhu, P., Wang, X. and Zhang, Z. 2020. Phytoremediation effect of *Medicago sativa* colonized by *Piriformospora indica* in the phenanthrene and cadmium co-contaminated soil. *BMC Biotechnol* 20:1–14.

Li, Y., Zhang, J., Zhu, G., Liu, Y., Wu, B., Ng, W. J., Appan, A. and Tan, S. K. 2016. Phytoextraction, phytotransformation and rhizodegradation of ibuprofen associated with *Typha angustifolia* in a horizontal subsurface flow constructed wetland. *Water Res* 102:294–304.

Limmer, M. and Burken, J. 2016. Phytovolatilization of organic contaminants. *Environ Sci Technol* 50:6632–6643.

Lone, M. I., He, Z. L., Stoffella, P. J. and Yang, X. 2008. Phytoremediation of heavy metal polluted soils and water: progresses and perspectives. *J Zhejiang Univ Sci B* 9:210–220.

Ludang, Y. 2020. Application of phytotechnology in determining plant species for greenspace in the city of palangka raya. *J Adv Res Eng Technol* 11:1–6.

Lushchak, V. I., Matviishyn, T. M., Husak, V. V., Storey, J. M. and Storey, K. B. 2018. Pesticide toxicity: a mechanistic approach. *EXCLI J* 17:1101.

McGuinness, M. and Dowling, D. 2009. Plant-associated bacterial degradation of toxic organic compounds in soil. *Int J Environ Res Public Health* 6:2226–2247.

Mercado-Borrayo, B., Cram, H. S., Rosas-Pérez, I., Hernández-Quiroz, M. and Ponce-De-León- Hill, C. 2015. Organophosphorus and organochlorine pesticides bioaccumulation by *Eichhornia crassipes* in irrigation canals in an urban agricultural system. *Int J Phytoremediation* 17:701–708.

Michalska, J., Piński, A., Żur, J. and Mrozik, A. 2020. Selecting bacteria candidates for the bioaugmentation of activated sludge to improve the aerobic treatment of landfill leachate. *Water* 12:140.

Mimmo, T., Bartucca, M. L., Del-Buono, D. and Cesco, S. 2015. Italian ryegrass for the phytoremediation of solutions polluted with terbuthylazine. *Chemosphere* 119:31–36.

Mitton, F. M., Miglioranza, K. S., Gonzalez, M., Shimabukuro, V. M. and Monserrat, J. M. 2014. Assessment of tolerance and efficiency of crop species in the phytoremediation of DDT polluted soils. *Ecol Eng* 71:501–508.

Mojiri, A., Ziyang, L., Tajuddin, R. M., Farraji, H. and Alifar, N. 2016. Co-treatment of landfill leachate and municipal wastewater using the ZELIAC/zeolite constructed wetland system. *J Environ Manage* 166:124–130.

Moklyachuk, L., Gorodiska, I., Slobodenyuk, O. and Petryshyna, V. 2010. Phytoremediation of soil polluted with obsolete pesticides in Ukraine. In: Kulakow, P. A., Pidlisnyuk, V. V., (eds.),*Application of Phytotechnologies for Cleanup of Industrial, Agricultural, and Wastewater Contamination*. Springer Science + Business Media: Berlin/Heidelberg, Germany, pp:113–124.

Moklyachuk, L., Petryshyna, V., Slobodenyuk, O. and Zatsarinna, Y. 2012. Sustainable strategies of phytoremediation of the sites polluted with obsolete pesticides. In: Vitale, K. (ed), *Environmental and Food Safety and Security for South-East Europe and Ukraine*. Springer-Verlag, New York, NY, pp:81–89.

Moore, M. and Locke, M. 2012. Phytotoxicity of atrazine, S-metolachlor, and permethrin to *Typha latifolia* germination and seedling growth. *B Environ Contam Toxicol* 89:292–295.

Morton, J. M., Blackburn, B. N., Bella, E., Steffy, M., Anderson, C., Massengill, R., Blackwell, L., Ka'aihue, R. and Chumley, J. 2014. Integrated pest management plan for eradicating *Elodea* from the Kenai Peninsula. In: *Elodea Subcommittee of the Kenai Peninsula Cooperative Weed Management Area Report*, Version June.

Muhammad, S., Shah, M. T., Khan, S., Siddique, U., Gul, N., Khan, M. U., Malik, R. N., Farooq, M. and Naz, A. 2013. Wild plant assessment for heavy metal phytoremediation potential along the mafic and ultramafic terrain in northern Pakistan. *BioMed Res Int* 1:9.

Muhammad, S., Tan, J., Deng, P., Li, T., He, H., Bian, J. and Wu, L. 2019. Pesticide application has little influence on coding and non-coding gene expressions in rice. *BMC Genomics* 20:1–13.

Mukherjee, I. and Kumar, A. 2012. Phytoextraction of endosulfan a remediation technique. *B Environ Contam Toxicol* 88:250–254.

Mulbry, W., Kondrad, S., Pizarro, C. and Kebede-Westhead, E. 2008. Treatment of dairy manure effluent using freshwater algae: algal productivity and recovery of manure nutrients using pilot-scale algal turf scrubbers. *Bioresour Technol* 99:8137–8142.

Ng, E., Chen, L., Wang, Y. and Yuan, C. 2012. A study on the cooling effects of greening in a high-density city: an experience from Hong Kong. *Build Environ* 47:256–271.

Nurzhanova, A., Kulakow, P., Rubin, E., Rakhimbayev, I., Sedlovskiy, A., Zhambakin, K., Kalugin, S., Kolysheva, E. and Erickson, L. 2010. Obsolete pesticides pollution and phytoremediation of contaminated soil in Kazakhstan. In: *Application of Phytotechnologies for Cleanup of Industrial, Agricultural, and Wastewater Contamination*. Springer, NATO Science for Peace and Security Series C: Environmental Security, pp:87–111.

Olivares, A. R., Carrillo-González, R., González-Chávez, M. C.A. and Hernández, R. M. S. 2013. Potential of castor bean (*Ricinus communis*) for phytoremediation of mine tailings and oil production. *J Environ Manage* 114:316–323.

Papazoglou, E. G. and Fernando, A. L. 2017. Preliminary studies on the growth, tolerance and phytoremediation ability of sugarbeet (*Beta vulgaris*) grown on heavy metal contaminated soil. *Ind Crops Prod* 107:463–471.

Parise, S. 2016. The effect of *Brassica juncea* phytoremediation using soil doped with concentrations of copper (II) sulfate solution on soil nitrate content. *South Carolina Junior Academy of Science* Paper 246.

Park, J. H., Lamb, D., Paneerselvam, P., Choppala, G., Bolan, N. and Chung, J. W. 2011. Role of organic amendments on enhanced bioremediation of heavy metal (loid) contaminated soils. *J Hazard Mater* 185:549–574.

Pascal-Lorber, S. and Laurent, F. 2011. Phytoremediation techniques for pesticide contaminations. In: Lichtfouse E. (ed), *Alternative Farming Systems, Biotechnology, Drought Stress and Ecological Fertilisation*. Springer, 6:77–105.

Pereira, R. C., Camps-Arbestain, M., Garrido, B. R., Macías, F. and Monterroso, C. 2006. Behaviour of α-, β-, γ-, and δ-hexachlorocyclohexane in the soil–plant system of a contaminated site. *Environ Pollut* 144:210–217.

Pilon-Smits, E. A. H. 2015. Selenium in plants. In: Luettge, U. (ed), *Progress in Botany*. Springer, Heidelberg, pp:93–107.

Pirzadah, T. B., Malik, B., Tahir, I., Kumar, M., Varma, A. and Rehman, R. U. 2015. Phytoremediation: an eco-friendly green technology for pollution prevention, control and remediation. *Soil Rem Plants Prospects Chall* 5:107–126.

Pourkhabbaz, A. and Pourkhabbaz, H. 2012. Investigation of toxic metals in the tobacco of different Iranian cigarette brands and related health issues. *Iran J Basic Med Sci* 15:636.

Rabhi, M., Atia, A., Abdelly, C. and Smaoui, A. 2015. New parameters for a better evaluation of vegetative bioremediation, leaching, and phytodesalination. *J Theor Biol* 383:7–11.

Raj, D., Kumar, A. and Maiti, S. K. 2020. *Brassica juncea* (Indian mustard): a putative plant species to facilitate the phytoremediation of mercury contaminated soils. *Int J Phytoremediation* 22:733–744. 10.1080/15226514.2019.1708861.

Rane, N. R., Chandanshive, V. V., Watharkar, A. D., Khandare, R. V., Patil, T. S., Pawar, P. K. and Govindwar, S. P. 2015. Phytoremediation of sulfonated Remazol Red dye and textile effluents by *Alternanthera philoxeroides*: an anatomical, enzymatic and pilot scale study. *Water Res* 83:271–281.

Ranieri, E., Gikas, P. and Tchobanoglous, G. 2013. BTEX removal in pilot-scale horizontal subsurface flow constructed wetlands. *Desalin Water Treat* 51:3032–3039.

Rathore, S., Shekhawat, K., Dass, A., Kandpal, B. and Singh, V. 2019. Phytoremediation mechanism in Indian mustard (*Brassica juncea*) and its enhancement through agronomic interventions. *Proc Natl Acad Sci India Sect B Biol Sci* 89:419–427.

Rissato, S. R., Galhiane, M. S., Fernandes, J. R., Gerenutti, M., Gomes, H. M., Ribeiro, R. and Almeida, M. V. 2015. Evaluation of *Ricinus communis* for the phytoremediation of polluted soil with organochlorine pesticides. *BioMed Res Int* 2:549863.

Romeh, A. A. 2015. Evaluation of the phytoremediation potential of three plant species for azoxystrobin-contaminated soil. *Int J Environ Sci Technol* 12:3509–3518.

Rose, M. T., Sanchez-Bayo, F., Crossan, A. N. and Kennedy, I. R. 2006. Pesticide removal from cotton farm tailwater by a pilot-scale ponded wetland. *Chemosphere* 63:1849–1858.

Saad-Allah, K. M. and Elhaak, M. 2017. Hyperaccumulation activity and metabolic responses of *Solanum nigrum* in two differentially polluted growth habitats. *J Saudi Soc Agric Sci* 16:227–235.

Salamanca, E. J. P., Madera-Parra, C. A., Avila-Williams, C. A., Rengifo-Gallego, A. L. and Ríos, D. A. 2015. Phytoremediation using terrestrial plants. In: Ansari AA, Gill SS, Gill R, Lanza GR, Newman L (eds.), *Phytoremediation: management of environmental contaminants, vol 2*. Springer, Cham, pp:305–319.

Seeger, E. M., Kuschk, P., Fazekas, H., Grathwohl, P. and Kaestner, M. 2011. Bioremediation of benzene-, MTBE- and ammonia-contaminated groundwater with pilot-scale constructed wetlands. *Environ Pollut* 159:3769–3776.

Shahid, M., Ahmed, B., Zaidi, A. and Khan, M. S. 2018. Toxicity of fungicides to *Pisum sativum*: a study of oxidative damage, growth suppression, cellular death and morpho-anatomical changes. *RSC Adv* 8:38483–38498.

Shi, X., Zhang, X., Chen, G., Chen, Y., Wang, L. and Shan, X. 2011. Seedling growth and metal accumulation of selected woody species in copper and lead/zinc mine tailings. *J Environ Sci* 23:266–274.

Singh, S., Sherkhane, P. D., Kale, S. P. and Eapen, S. 2011. Expression of a human cytochrome P4502E1 in *Nicotiana tabacum* enhances tolerance and remediation of γ-hexachlorocyclohexane. *New Biotechnol* 28:423–429.

Somtrakoon, K., Kruatrachue, M. and Lee, H. 2014. Phytoremediation of endosulfan sulfate-contaminated soil by single and mixed plant cultivations. *Water Air Soil Pollut* 225:1886.

Srichandan, H., Mohapatra, R. K., Parhi, P. K. and Mishra, S. 2019. Bioleaching approach for extraction of metal values from secondary solid wastes: a critical review. *Hydrometallurgy* 189:105122.

Sun, G., Zhang, X., Hu, Q., Zhang, H., Zhang, D. and Li, G. 2015. Biodegradation of dichlorodiphenyltrichloroethanes (DDTs) and hexachlorocyclohexanes (HCHs) with plant and nutrients and their effects on the microbial ecological kinetics. *Microb Ecol* 69:281–292.

Sun, G., Zhao, Y., Allen, S. and Cooper, D. 2006. Generating "tide" in pilot-scale constructed wetlands to enhance agricultural wastewater treatment. *Eng Life Sci* 6:560–565.

Sun, H., Xu, J., Yang, S., Liu, G. and Dai, S. 2004. Plant uptake of aldicarb from contaminated soil and its enhanced degradation in the rhizosphere. *Chemosphere* 54:569–574.

Tangahu, B. V., Abdullah, S., Rozaimah, S., Basri, H., Idris, M., Anuar, N. and Mukhlisin, M. 2011. A review on heavy metals (As, Pb, and Hg) uptake by plants through phytoremediation. *Int J Chem Eng* 2011:1155.

Tripathi, V., Fraceto, L. F. and Abhilash, P. 2015. Sustainable clean-up technologies for soils contaminated with multiple pollutants: plant-microbe-pollutant and climate nexus. *Ecol Eng* 82:330–335.

Trognitz, F., Hackl, E., Widhalm, S. and Sessitsch, A. 2016. The role of plant–microbiome interactions in weed establishment and control. *FEMS Microbiol Ecol* 92(10):fiw138.

Wang, F. Y., Tong, R. J., Shi, Z. Y., Xu, X. F. and He, X. H. 2011. Inoculations with arbuscular mycorrhizal fungi increase vegetable yields and decrease phoxim concentrations in carrot and green onion and their soils. *PLoS One* 6.

Wang, Q., Yang, J., Li, C., Xiao, B. and Que, X. 2013. Influence of initial pesticide concentrations in water on chlorpyrifos toxicity and removal by *Iris pseudacorus*. *Water Sci Technol* 67:1908–1915.

Wang, Q., Zhang, W., Li, C. and Xiao, B. 2012. Phytoremediation of atrazine by three emergent hydrophytes in a hydroponic system. *Water Sci Technol* 66:1282–1288.

Wehtje, G., Walker, R. H. and Shaw, J. N. 2000. Pesticide retention by inorganic soil amendments. *Weed Sci* 48:248–254.

White, J. C., Parrish, Z. D., Isleyen, M., Gent, M. P., Iannucci-Berger, W., Eitzer, B. D. and Mattina, M. J. I. 2005. Uptake of weathered p, p′-DDE by plant species effective at accumulating soil elements. *Microchem J* 81:148–155.

Willscher, S., Mirgorodsky, D., Jablonski, L., Ollivier, D., Merten, D., Buchel, G., Wittig, J. and Werner, P. 2013. Field scale phytoremediation experiments on a heavy metal and uranium contaminated site, and further utilization of the plant residues. *Hydrometallurgy* 131:46–53.

Xu, Q., Renault, S., Goltz, D. and Yuan, Q. 2020. Phytoremediation of waste dumping site soil and landfill leachate by using cattail (*Typha latifolia*). *Environ Technol* 41:1101–1106.

Xu, X. J., Sun, J. Q., Nie, Y. and Wu, X. L. 2015.*Spirodela polyrhiza* stimulates the growth of its endophytes but differentially increases their fenpropathrin-degradation capabilities. *Chemosphere* 125:33–40.

Yasin, M., Faisal, M., Mehdawi, A. and Pilon-Smits, E. 2015. Microbe-assisted selenium phytoremediation and phytomanagement of natural seleniferous areas. In: *Global Advances in Selenium Research from Theory to Application*: Proceedings of the 4th International Conference on Selenium in the Environment and Human Health 199.

Yeh, T., Chou, C. and Pan, C. 2009. Heavy metal removal within pilot-scale constructed wetlands receiving river water contaminated by confined swine operations. *Desalination* 249:368–373.

Zhang, C., Guanming, S. and Jian, S. H. U. 2015. Productivity effect and overuse of pesticide in crop production in China. *J Integr Agric* 14:1903–1910.

Zhou, Y., Tigane, T., Li, X., Truu, M., Truu, J. and Mander, U. 2013. Hexachlorobenzene dechlorination in constructed wetland mesocosms. *Water Res* 47:102–110.

Zhu, Z., Yang, X., Wang, K., Huang, H., Zhang, X., Fang, H., Li, T., Alva, A. K. and He, Z. 2012. Bioremediation of Cd-DDT co-contaminated soil using the Cd-hyperaccumulator *Sedum alfredii* and DDT-degrading microbes. *J Hazard Mater* 235:144–151.

9

Present Scenario of Plant Growth Promoting Microbes and Technologies Used for Biofertilizer Development

Nitu Rani, Rajinder Kaur, and Sukhminderjit Kaur
Chandigarh University, Punjab, India

CONTENTS

9.1 Introduction

Intensive farming with indiscriminate use of synthetic fertilizer and agrochemicals is adversely affecting the environment, soil health, and living organisms (Mulvaney et al. 2009). Globally, to overcome these deleterious effects, biofertilizers have emerged as an important priority, which is an eco-friendly and economical approach. Biofertilizers are natural fertilizers composed of soil microorganisms which increase crop growth and its productivity via direct and indirect mechanisms that include enhancing nutrients uptake, reducing pathogenic attack, and providing resistance against environmental stress. These soil microbes are plant growth promoting rhizospheric (PGPR) bacteria, fungi, and algae. In the year 1880, the discovery and application of nitrogen fixing bacteria accounted for the scope of several other rhizobial inoculations. Since then, many bacterial genera have evaluated as plant growth promoters, soil application of which augmented soil fertility, crop growth, and productivity. Microbes in soil are inhabited in the most active narrow region around plant root known as rhizosphere. Within rhizosphere, plant-microbe association predominates as the microorganisms for their proper growth depends on the presence of various amino acids, sugars, organic acids, vitamins, enzymes, etc. along with root exudates. The plant-microbe association practice had proved that microbes have tremendous capability to enhance nutrient bioavailability of nitrogen (N), phosphorus (P), potassium (K), iron (Fe), sulfur (S), and zinc (Zn) (Ahemad and Khan 2012; Haichar et al. 2014); they secrete growth-promoting substances like auxin and Gibberellic acid (Baca and Elmerich 2007; Spaepen and Vanderleyden 2011) and provide resistance to environmental stresses (Yao et al. 2010), protection against pathogens (Qi et al. 2016), and many others. Being promising and demonstrating popularity among the agriculturists, biofertilizers are the least utilized substances in the agricultural field and the reason being that agrochemicals have set standards of long shelf life, ease of use, and are easily accepted by the farmers. Furthermore, the success of biofertilizers primarily depends on the efficiency of isolates and the application technology, the results of which achieved till date are still with several drawbacks. For last 40 years, researchers have been focusing on the development of efficient strains and understanding their mode of action in the rhizosphere. Several microbial species have been identified as potential growth promoters and can be obtained from authenticated centers situated worldwide such as the American Type Culture Collection (ATCC), the Agricultural Research Service Culture Collection (Northern Regional Research Laboratory or NRRL), and many others.

Further, these potential microorganisms either in alive or latent state are employed for preparations and these preparations are termed as biofertilizers (Brahmaprakash and Sahu 2012; Kumar 2014). Numerous biofertilizers are available in the market with little or no negative effect. Still, the use of biofertilizer is low because of inconsistent performance, special care required during storage, high contamination, and unavailability of any specified standards. Apparently, acquiring more knowledge regarding formulation techniques, best carrier material, and delivery method would help to overcome various critical issues encountered by the inoculants industry. Nevertheless, researchers are extending their knowledge and focusing more toward the development of successful biofertilizers by considering several advantages and disadvantages gained through the literature. This chapter provides information about effective microbial inoculants, their characteristics, mode of actions along with carrier material, formulation development, and application in agriculture sector.

9.2 Potential Features of Microbes for Bioformulation

The presence of huge counts of microbes in regions around rhizosphere, at root surfaces, and in association with root surface enhances the growth and development of plants (Arora et al. 2010; Bhattacharyya and Jha 2012). The exact mechanism of their activity is not fully understood; however, various direct and indirect mechanisms are reported in the literature, which include symbiotic nitrogen fixation, nodule occupancy (Nehra and Choudhary 2015; Sahu and Brahmaprakash 2016), production of growth regulators, namely, auxin, gibberellic acid, cytokinin, and ethylene (Baca and Elmerich 2007; Spaepen and Vanderleyden 2011), increasing bioavailability of nutrients (Zhang et al. 2014), moisture availability (Marasco et al. 2012), and production of stress inhibitors (Tiwari et al. 2011). To demonstrate positive biofertilization activities, many microbial species and strains should possess the characteristics shown in Figure 9.1. Some of the well-studied plant-microbial associations which received a lot of attention include symbiotic association, mychorrhizal fungi association, and plant growth promoting rhizobacteria, namely, phosphate solubilizers, zinc solubilizers, potassium solubilizers, etc.

9.2.1 Nitrogen Fixing Bacteria

Among the nodulating species, nitrogen fixing bacteria, *Rhizobium* belonging to *Rhizobiaceae* family are the most important for inocula formulation. *Rhizobium* fix 50–100 kg/ha N with legumes alone. They are beneficial for pulses like pea, black gram, lentil, chickpea, etc., and oil seeds like groundnut and soyabean. *Rhizobium* strains colonize with the roots of their compatible legume crop forming nodules, which act as factories of ammonia production. Soil *Rhizobium* count increases with the increase in the legume crops present in the field. Every legume crop is associated with a particular species of *Rhizobium* forming effective nodules (Venkatashwarlu 2008).

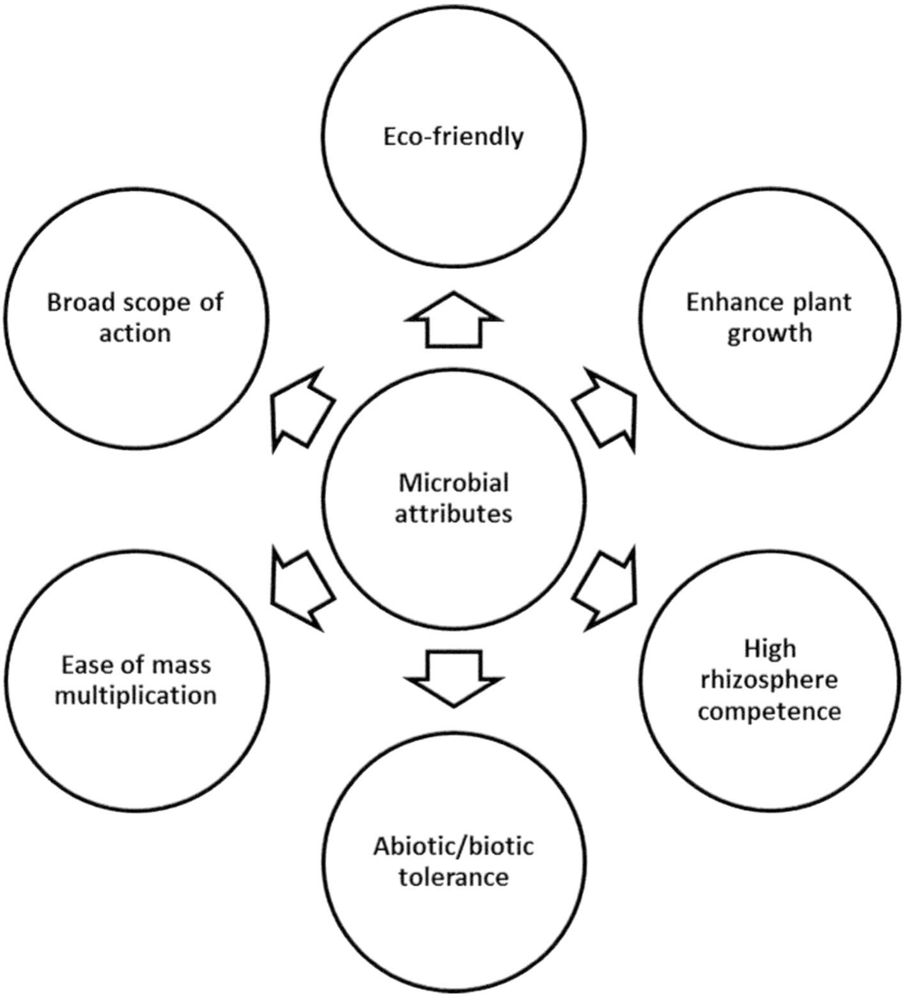

FIGURE 9.1 Characteristics of essential microbes.

9.2.1.1. *Azospirillum*

Azospirillum is a well-known genera belonging to *Spirilaceae* family. Apart from fixing nitrogen, they also produce growth regulating substances. Among the various species, only *Azospirillum lipoferum* and *A. brasilense* are widely distributed and employed as potential microbes for biofertilization globally. The bacteria of the genus *Azospirillum* form symbiotic association majorly with C4 plants, which include sugarcane, maize, sorghum, etc. Although the bacteria of the genus *Azospirillum* do not produce any visible nodule or outgrowth on roots after colonizing the roots, but a small proportion of them penetrates into the root tissues.

9.2.1.2 *Azotobacter*

Belonging to the family of *Azotobacteriaceae*, *Azotobacter* is a genus of free living, aerobic, and heterotroph bacteria generally found in alkaline soils. The *Azotobacter* count rarely exceeds 10^5cfu/g of soil because of lack of organic material and also presence of other microbes in the soil. Low count is generally found in uncultivated soils as well as near crop rhizosphere. The bacteria of the genus *Azotobacter* inhibit growth of many pathogenic fungi near the root region, and thus, have antifungal property.

Numerous species of *Azotobacter*, namely, *Azotobacter chroococcum*, *A. vinelandii*, *A. beijerinckii*, and *A. insigni* are reported in the rhizospheric region of rice, maize, sugarcane, vegetables, and plantation crops.

9.2.1.3 Cyanobacteria and Azolla

Cyanobacteria are often known as blue-green algae (BGA), and azolla are phototrophs enhancing plant growth through production of growth regulators – auxin and gibberllic acid. Also referred to as paddy organisms, these are abundant in paddy and fix 20–30 kg/ha N in waterlogged rice fields. Lowland rice production requires larger input of soil N. About 60% of N required is contributed by mineralization of soil organic N and through biological nitrogen fixation (BNF). The BGA form symbiotic relationship with ferns, fungi, and flowering plants. One of the best known cyanobacterium is *Anabaena azollae* that maintains beneficial symbiotic association with water fern Azolla and is being extensively utilized for rice. Azolla is considered of agronomical importance because of its increased growth rate, nitrogen fixation capability, and its quick soil decomposition. Thus, it has wide application as biofertilizer and as green manure in rice production. Apart from being nitrogen fixers, these biofertilizers also account for significant contributions in P, K, S, Zn, and other nutrients. Some important species of *Azolla* include *Azollapinnata*, *A. caroliniana*, *A. microphylla*, and *A. mexicana*.

9.2.1.4 Other Solubilizing Bacteria

Phosphate solubilizing bacteria are beneficial bacteria present in soil largely near the plant rhizosphere. The species classified as phosphate solubilizers belong to the genera *Acinetobacter*, *Arthrobacter*, *Bacillus*, *Erwinia*, *Enterobacter*, *Burkholderia*, *Achromobacter*, *Agrobacterium*, *Pseudomonas*, and *Serratia* (Hayat et al. 2010). The vital role of phosphate solubilizer is to solubilize organic and inorganic insoluble phosphate to soluble P, thereby enhancing its bioavailability to plants. Similarly, there are potassium solubilizing strains; the best known species include *Pseudomonas*, *Burkholderia*, *Acidothiobacillus ferrooxidans*, *Bacillus mucilaginosus*, *B. edaphicus*, *B. circulans*, and *Paenibacillus* sp. (Parmar and Sindhu 2013). Several microorganism species are also known to solubilize Zn, which include *Bacillus*, *Pseudomonas*, *Thiobacillus*, and *Saccharomyces*. They are capable of converting fixed insoluble Zn to soluble Zn, thereby enhancing their bioavailability to plants. They can be used with cheaper insoluble Zn source such as zinc oxide or zinc carbonate instead of costly chemical zinc sulfate (Rani et al. 2020). To develop an effective biofertilizer, formulation should contain either single strain or consortia of strains compatible with one another. Developing and utilizing consortium is more advantageous over a single strain. A few commercially available consortia are mentioned in Table 9.1.

TABLE 9.1

Commercial Microbial Consortia

S. No.	Commercially Available Microbial Consortia	Role as Growth Promoter	Reference
1	*Trichoderma* + *Brady rhizobium* (Excalibre-SA)	Enhances soya bean growth	Parnell et al. 2016
2	*Azospirillum* + *Azotobacter*	Has reduced 50% N-fertilizer application. Also known to enhance yield and oil quality in sesame.	Shakeri et al. 2016
3	*Bacillus cereus* PX35 + *B. subtilis* SM21 + *Serratia* sp. XY2	Reduces the incidence of root knot nematode, namely, Meloidogyne incognito in tomato, increases fruit yield and quality	Niu et al. 2016
4	*B. amyloliquefaciens* + *Tricodermavirens*	Enhances corn and tomato yield	Akladious and Abbas 2012
5	Arbuscular mycorrhizal fungi + *Trichoderma*	Enhances growth along with resistance against soil pathogen	Mwangi et al. 2011

Rhizobacteria applied alone or in consortia enhance the crop yield and nutritional quality, and thus, can be exploited in bioformulation, which can be further introduced into the market. Besides its role in nitrogen fixation, phosphorus and potassium solubilizers, as biocontrol agents or as PGPR, it is important to deliver these live inoculations to the right site in sufficient numbers, which is actually a hurdle for the industry and could only be achieved by selecting proper carrier material and formulation techniques.

9.3 Microbial Bioformulation

As per Xavier et al. (2004), formulation is an important aspect for the development of bio-inoculant of microbial strain. Direct application of microbes to soil is not easy as it becomes more susceptible to various factors like fluctuating temperature, pH, humidity, and soil conditions. Also for better survival, these uncovered and unprotected microbes need to compete with the diverse soil microflora. Formulation is a prime research area that affects the effectiveness of inoculants determining its success or failure rate. Formulation involves effective microbes with a suitable carrier and some additives. Currently, bioformulation technologies in the field of agriculture and environment are lagging, which can be overcome by targeting the various requirements of the potential bioformulation (Figure 9.2). The formulation is broadly divided into (1) conventional formulation, including liquid and solid formulation and (2) advanced formulation or encapsulated formulation (Figure 9.3). Among the various formulations,

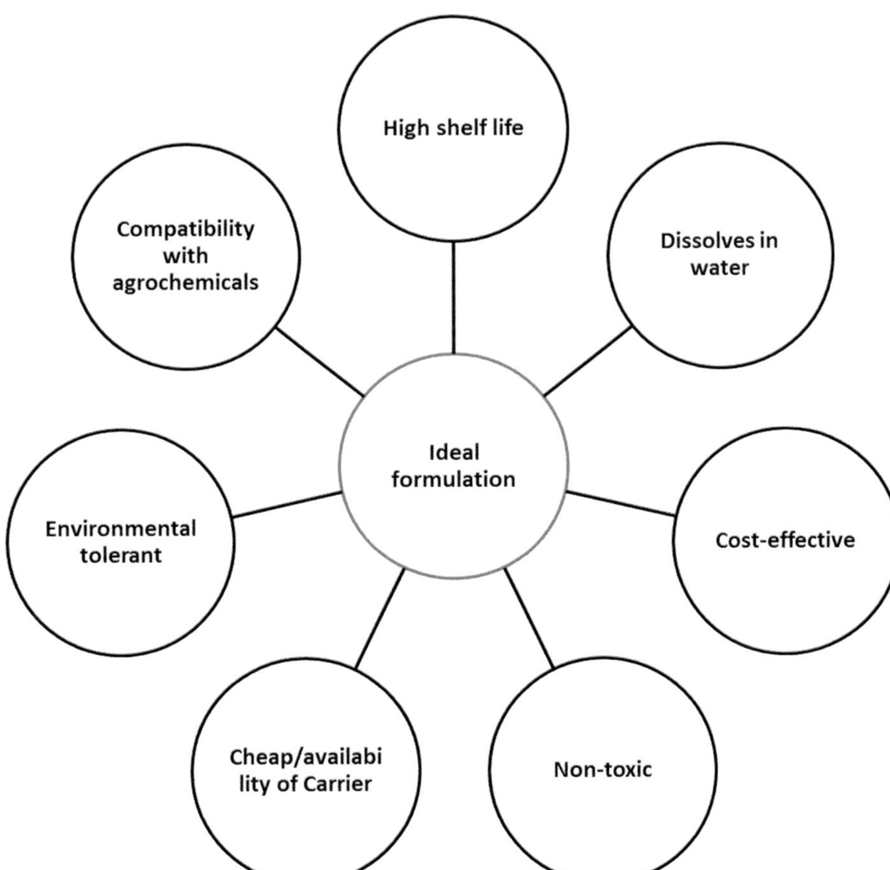

FIGURE 9.2 Characteristics of potential formulation.

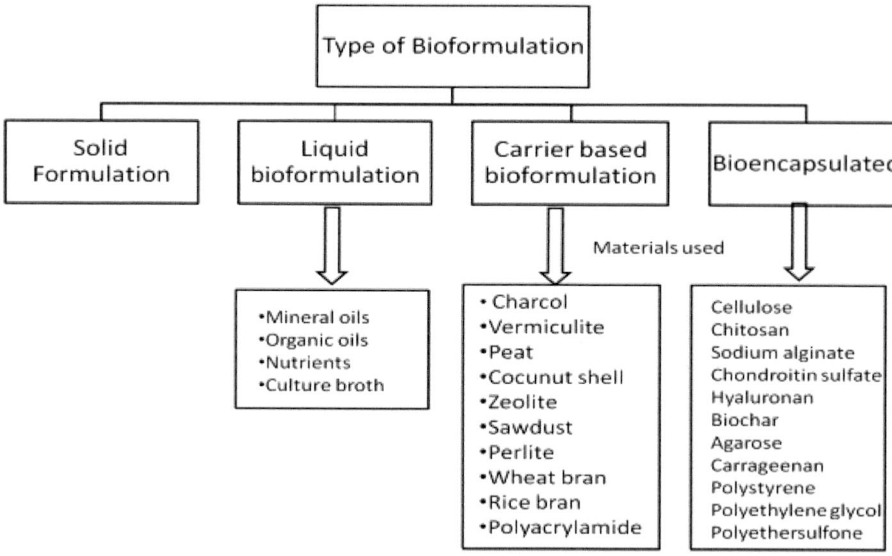

FIGURE 9.3 Types of bioformulations and the materials used (Kaur and Kaur 2018).

encapsulation is the newest and the most efficient technology which will not only provide covering and protection to the potential microbe but also provide micro-environment to the microbe to perform efficiently with high survival rate. Encapsulation and immobilization technique is employed to protect and cover microbe against adverse environmental stresses, to increase viability and efficiency, and maintain stability during production, storage, and handling.

9.3.1 Conventional Formulation

Conventional formulation of biofertilizers uses solid, liquid, or granulated forms of carrier. For seed application, generally powder and liquid inoculants are employed, whereas granular inoculants can be applied directly to the soil (Ben Rebah et al. 2007). The commercialization of inoculants is mostly in the form of either solid or liquid inoculants (Stephens and Rask 2000).

9.3.1.1 Solid Formulation

Solid carrier-based biofertilizers are mostly manufactured in India. Despite having benefits such as low cost of raw material and no requirement of sophisticated equipment, solid conventional formulation is less effective and thereby least accepted by the farmers. The reason behind this includes several disadvantages like unavailability of the best material, poor shelf life of up to six months, high contamination, intolerant to high temperature and UV rays, cell count of 10^8cfu/ml which gradually decreases with time, and high cost of bulk transportation.

9.3.1.2 Liquid Formulation

Liquid formulation is the preparation of effective microbes in nutrient media (water, mineral or organic oils), and there are two ways of its application that include sowing of seeds dipped into inoculant or inoculant being sprayed on the seeds (Bashan 1998). Ease of production, ease of handling, high shelf life, tolerance to temperature above 55°C and UV, and maintaining the cell count of 10^9cfu/ml are some of the advantages of liquid formulations. To improve inoculants performance, it allows addition of sufficient nutrients, cell protectants, and additives (Schulz and Thelen 2008). For instance, Singleton et al. (2002)

formulated a liquid inoculation containing *Rhizobium* with additives in order to improve the microbial viability. In another study by Taurian et al. (2010), they showed that the glycerol added to liquid formulation containing *Pseudomonas fluorescens* prolonged its viability during storage up to six months, and also, sucrose improved survival of phosphate solubilizing bacteria such as *Pantoea* spp. Glycerol was also used as cell protectant as it protected the cells from desiccation due to its high water holding capacity (Manikandan et al. 2010). Field application of liquid formulation is simple and can be done using hand/power sprayers, fertigation tanks, and can also be mixed with farm yard manure or other agrochemicals. It is also accepted by the farmers due to its compatibility with farm machinery like seed augers and air seeders. However, the formulation was not considered helpful due to its disadvantages, which included high risk of contamination, no protection from environmental stress, decreased viability with time, and cold condition required for its long-term storage (Singleton et al. 2002; Tittabutr et al. 2007; Albareda et al. 2008).

9.3.2 Carriers: A Potential Vehicle of Delivery

Carrier is a major portion of the inoculants that aids in enhancing the efficiency of biofertilizers. Being the transport vehicle, carrier should release the microbe at specific targeted site in controlled manner (Arora et al. 2008). Other criteria for selecting carrier material include ease of handling, increased shelf life, maintenance of viable cell count after formulation, high moisture holding capacity, good adhesive property, non-toxic nature, easy availability, low cost, sterilizable, and good pH buffering capacity; it has been also observed that carrier can be either sterile or non-sterile (Smith 1992). If the carrier is organic in nature, sterilization is must in order to maintain the sufficient count of the inoculants. Since autoclaving can alter the properties of the organic material and may produce some toxic substance to microbial strain, steaming is preferred. Steaming is a useful method of sterilization in case the carrier is made of soil-like material like peat and also for big batches as well. Depending on the system used, the time will vary with temperature range at 90–100°C and the cost will be several folds lower as compared to autoclaving. Another method of carrier sterilization without altering the properties of carrier material is gamma-irradiation. In this process, the carrier material is exposed to gamma radiation at 50 kGy to destroy bacteria and yeasts.

9.3.2.1 Different Types of Carriers in Biofertilizers

Carriers used in biofertilizers could be of organic, inorganic or synthetic origin. Soil material like peat, coal, clay, inorganic soil; organic material like vermicompost, wheat bran, rice bran, saw dust; or inert materials, namely, vermiculite, perlite, kaolin, bentonite are used in dry inoculants preparations. In case of liquid inoculants, broth cultures, oils, or oil-in water suspensions are used. Recently, organic polymers, biofilms, and nanoparticles are new areas studied and can be used in the formulation processes. Peat or turf is a carbonized vegetable tissue resulting from the decomposition of plants and mosses in wet condition, and due to its low cost, ease of availability, high moisture content, and high compressibility, it has been used for PGPR inoculants (Smith 1992). However, the composition and quality of peat is a major constraint that reduces the shelf life of the inoculants. Sterilization of peat is needed as it harbors a lot of unknown microbial contamination. Lignite is obtained from compressed peat and can be used as alternative to peat. *Rhizobium* increase viability at 28°C when lignite is used as carrier, especially when used in combination with soybean. Clay minerals are biocompatible, eco-friendly, and can be used in powder, granules, or suspension forms in agricultural formulation. Small size, high porosity, low permeability, and swelling properties associated with clay make it act as desiccant providing excellent storage for dried inoculants, and water holding capacity is high that enables high moisture content. Sawdust, an organic carrier, is useful for maintaining bacterial population of single strain or consortia due to its high porosity and water retention property (Arora et al. 2008). Talc, also known as steatite, is easily available from soapstone industries. Owing its small size, inert nature, and ease of availability, it is an excellent carrier for bio-inoculant formulation. Various scientists have used talc as carrier. For instance, Vidhyasekaran and Muthamilan (1995) used talc as carrier with 20% xanthum gum that ultimately enhanced the shelf life of *fluorescent Pseudomonads* during storage for the period of two months at 4°C. Another study by Bora et al. (2004) demonstrated that *P. putida* strain 30 and 180 survived a period of six months with population load of 10^8cfu/g of the product. In a study by Anitha and Rabeeth (2009), talc-based formulation of *Streptomyces griseus* when applied to the field protected tomato from *Fusarium* wilt. Vermiculite is an inert inorganic mineral with neutral pH, and it provides good aeration and moisture content in the formulation. A study by Saleh et al. (2001) showed that the use of vermiculite maintained the shelf life of *Azospirillum* with 1.3×10^7cfu/g for a period of ten months. Similarly, other inorganic materials such as bentonite and perlite are also available at different locations in several countries, but their uses are limited as their formulations are difficult, and also the microbial load of inorganic materials used as carrier is quite low, i.e., 10^2–10^3cfu/g compared to organic carrier. Combination of organic and inorganic carrier has been proved promising in enhancing the activity and viability of *Burkholderia* sp (Anandham et al. 2007).

9.4 Bioencapsulation: An Advanced Technology

Agricultural encapsulation or immobilization is entrapping active microbes into polymer matrix to retain their viability when introduced into the soil. The matrix will protect the inner cell from mechanical injuries as well as adverse environmental conditions. The two distinct features of bioencapsulation are to provide shelter against soil stress and to ensure the controlled gradual release of the desired microbe, thus making the microbes to survive and remain physiologically active for long period of time. Bioencapsulation is an alternative technology to conventional methods due to its advantages such as high cell viability, no cell leakage, and high microbial shelf life.

9.4.1 Bioencapsulation Materials

Both natural and synthetic polymers are considered for bioencapsulation of rhizobacteria. Naturally occurring polymers include alginate, carrageenan, agar-agar, and agarose, and synthetic polymers include polyacrylamides, polystyrene, and polyurethane. Polymers may be either homopolymers such as gelatine, or heteropolymers such as polylactic acid-polyglycolic acid. On the basis of monomer unit, the selection of the polymer is made and the molecular weight of the polymer plays a significant role. For example, in a study by Lu and Toy (2009), it has been revealed that the disadvantage was actually the use of polyethylene glycol, a high molecular weight compound that lowers the encapsulation process. The most popular polymer accounted for the microbial cell encapsulation is the alginate. It is mainly derived from *Macrocystis pyrifera* (kelp) and *Sargassum sinicola*, and is composed of D-mannuronic acid and L-glucuronic acid. In the presence of a multivalent cation (Ca^{2+}), alginate forms a dense three-dimensional lattice having pore-size range of 0.005–0.2 mm in diameter. The key role of alginate matrix in encapsulation is to provide protection and enhance the shelf life of microbial strain. Several studies had proven the high survival of rhizobacteria in alginate inoculant over long period of storage. In a study by Trejo et al. (2012), *A. lipoferum* immobilized using alginate beads showed high survival of 10^{10}cfu/g when stored for a year at 37°C. Similarly, rhizobacteria *Bacillus subtilis* and *Pseudomonas corrugata* immobilized in sodium alginate beads retained their potential and demonstrated a high value of 10^8cfu/g after three years when stored at 4°C (Trivedi et al. 2008). Extending the storage period to 14 years, strain *A. brasilense* Cd and *P. fluorescens*313 immobilized in alginate beads with 10^5–10^6cfu/g demonstrated high effectiveness toward the positive growth of wheat plants. Schoebitz et al. (2013) confirmed that the most critical step for microbial survival in encapsulation technique is drying. Alginate alone cannot overcome this issue and requires addition of filler, which will further improve mechanical strength and lead to prolonged release of cells to the target. Starch is a cheap and abundant natural polymer which is known to improve the survival rate by reducing 50% water content and can, thus, acts as a good candidate for filler material in alginate bead.

9.4.2 Techniques of Bioencapsulation

With a wide range of applications in probiotics formulation and other medicinal industries, encapsulation is gaining attention in the field of agriculture and encourages researchers to work enthusiastically to develop encapsulated biofertilizers. In agricultural bioencapsulation, several aspects need to be considered in order to guarantee the bacterial survival. Bioencapsulation is achieved in mainly three stages:

1. Incorporation of an active ingredient into a solid or liquid matrix
2. Dispersion or spraying a solution onto solid particles with constant mechanical stirring.
3. Stabilization by chemical or physiochemical on a droplet.

Broadly, the techniques are categorized as physical methods (spray drying, spray chilling/cooling, extrusion, or fluidized bed); chemical methods (co-crystallization, molecular inclusion, or interfacial polymerization); and physicochemical processes (coacervation, liposomes, and gelation/inverse gelation).

9.4.2.1 Physical Methods

9.4.2.1.1 Spray Drying

Spray drying is an economical method used in large-scale production of microbial formulation. Development of the microcapsule through spray drying involves preparation of dispersed bacterial cell in carrier with homogenization. The dispersion is then fed to spray dryer and atomized with the nozzle. The capsules are collected from the bottom of dryer after dehydration by hot air contacting the atomized material. Despite being simple continuous process with large production, the technique is not regarded as the method of choice because of high temperature and rapid drying that would result in high microbial mortality, especially for temperature sensitive microbes. In a study, *Pseudomonas fluorescens-putida* were encapsulated using the method of spraydrying. At time zero, with inlet temperature 80°C, no bacterial cell was found in live condition (Amiet-Charpentier et al. 1998); however, at 60°C, bacterial survival was approximately 10^7cfu/g after two months of storage, thus making it clear that desiccation due to high temperature lowers bacterial survival, and hence, protective measures are needed for better survival.

9.4.2.1.2 Extrusion

Extrusion is a simple convenient process used commercially to produce large volume of material. In this method, a microbe is mixed with hydrocolloids, followed by extrusion into a hardening solution resulting in the formation of capsule (Krasaekoopt et al. 2003). The extruder die plays an important role in determining the shape and size of beads, which vary according to the size of needle pore and the distance of free fall. Extruder die, in turn, depends on the relationship between flow, temperature, and deformation (Sokhey et al. 1997). Moreover, introducing the pump to the system would result in development of small-sized and uniform beads. Although it has very low loading capacity, the main feature of this method is its stability over oxidation since the glassy carbohydrate matrices has good barrier properties.

9.4.2.1.3 Fluidized Bed Spray Method

To overcome the problems of spraydrying method, fluidized bed spray method was introduced. The method involves passing the particle (to be coated) into the coating chamber, wherein the particle is fluidized with hot air. Soon after formation of coated material, it is sprayed through a nozzle onto the particles, and it results in the formation of a film, which is further followed by a series of wetting and drying stages. The minute droplets out of the sprayed liquid fall onto the surface of the particles and bind them together. Finally, hot air evaporates the solvent out allowing the coating material to remain on the particles (Jacquot and Pernetti 2004). Advantage of fluidized bed spray include particular sized particle formation, reduced dehydration and temperature gradients (Morgan et al. 2006), fast drying, smaller flow area, low porosities, high thermal efficacy, low cost, and easily controlled process. Besides so many advantages, this method lowers the viability of microbial inoculants, and hence, is not used commonly (Deaker et al. 2004).

9.4.2.2 Chemical Methods

9.4.2.2.1 Molecular Inclusion

In this method, inclusion complexes are formed when the host compound has a cavity to accommodate small guest molecules. An example is cyclodextrins, an enzymatic modification of starch molecules and beta cyclodextrin with hydrophobic interior and hydrophilic exterior that enable conformational inclusion for a wide range of compounds. Once the guest compound is complexed, presence of water or high temperature is required for its release. One of the main limitations of using cyclodextrins is that they are not cost-effective. Success of the method has been seen in several medical applications, and thus, could be a promising approach in agriculture.

9.4.2.2.2 Interfacial Polymerization

It is a polymerization occurring at the interface between two immiscible liquids, one involving suspension of the cells as the discontinuous phase and another is organic solvent as continuous phase. Further, a biocompatible reagent is also added to the emulsion suspension to enhance the reaction. This method allows bulk loading, but involves high pH and toxic chemicals (Yeo et al. 2001). The toxicity can be avoided using vegetable oil as dispersant, and by reducing cell exposure to water insoluble cross-linking agent during the formation of membrane.

9.4.2.3 Physiochemical Process

9.4.2.3.1 Coacervation

It involves the formation of polymer droplets in suspension based on the liquid-liquid phase separation, liquid rich in polymer phase, and another phase called coacervate. Coacervation is a phenomenon where core particle and polymer should be compatible with each other, equilibrium should exist between the two phases, and core particle should be insoluble in coacervation medium. In the aqueous phase with more than one polymer, change in pH, ionic strength, and temperature induced by the core material would form second polymer acting as wall material. Coacervation can either be simple or complex. In case of simple coacervation, there is interaction between dissolved polymer and a low molecular weight substance such as gelatine with alcohol or sodium sulfate. Complex coacervation involves interaction between two charged polymers, one cationic and another anionic. Polymer containing rhizobacteria can be obtained with complex type of coacervation method. Disadvantage with coacervation is its high cost.

9.4.2.3.2 Liposomes

Liposomes are small spherical vesicles where the outer membrane has at least one hydrated bilayer surrounding an encapsulated active material (Immordino et al. 2006). Currently, the process is costlier and has low loading capacity; hence, it is considered only at laboratory scale.

9.4.2.3.3 Ionic Gelation

Ionic gelation is a low-cost conventional method which involves formation of capsule. An aqueous drop containing the active material and sodium alginate is dropped into the solution of divalent cation calcium chloride, instantaneously forming calcium alginate, a permanent film over the entire drop. The capsule thus formed is placed in sodium citrate that will slowly solubilize calcium through the formation of soluble citrate complexes, ungelling the internal portion of the drop. In a study by Covarrubias et al. (2012), the alginate beads keep cells inside the beads and prevent inside entry of microbes even under extreme conditions. The survival of rhizobacteria can be further improved by adding materials like starch to the alginate beads (Schoebitz et al. 2013).

9.4.2.3.4 Inverse Gelation

It is an inverse of ionic gelation where the calcium suspension containing the calcium chloride and bacterial cells is dropped into alginate solution forming liquid core calcium alginate capsules

(Abang et al. 2012). Inverse gelation using starch is performed so to increases the solid content of the membrane. In a study by Abang et al. (2012), alginate was used to form film covering core material, and different ways of incorporating the calcium source in oil were tested. The variables measured during the study included calcium chloride concentrations, sodium alginate concentration, and curing time along with properties such as thickness and elastic modulus.

9.4.2.3.5 Bead by Oil-Entrapped Emulsion

This emulsification involves formation of beads by dipping alginate contained with a suspension of cells and glycerol into a calcium solution. Diffusion of the calcium in alginate droplets leads to gelation (Lopez et al. 2012). Addition of starch increases the solid content, and addition of glycerol increases the viscosity, cell viability, and controlled release of microbes, thereby overcoming the drawbacks of alginate beads. Thus, this method is assumed to be an effective technology for storing and delivering microbes.

9.5 Conclusion

The rapidly increasing population of the world is imprinting adverse impact on our environment. One of the contributors includes the chemical fertilizers applied to the agriculture field to overcome the food need. Biofertilizers, in turn, have emerged as a substitute to the chemical fertilizers, which is an eco-friendly and economical approach. Since 1980s, biofertilizers are accounted as vital component of integrated nutrient management approach. Further, production of biofertilizer was accelerated with the emergence of organic farming. Several *in vitro* studies have led to the isolation of numerous microbes known as PGPR, which enhance crop growth and productivity, thus acting as potential microbes for biofertilization. However, achieving potential microbe *in vitro* will not guarantee its success until used in the field. Conventional methods like liquid and solid formulations have been developed for the application of biofertilizers in agriculture field. The disadvantages with solid inoculants include temperature sensitivity, decreased cell viability, and decreased shelf life during storage condition. On the other hand, no protection to PGPR against soil stress is the major drawback associated with liquid formulation. To overcome the drawbacks, carrier material was introduced as the key factor to the bioformulation. Besides many carriers as suitable vehicle and with numerous encapsulation techniques introduced, no encapsulated formulation has yet shown any success in the field and not been approved by the industry for field application. New and modified encapsulation techniques with modified properties and functions are emerging worldwide. Agriculture-based industry and research fraternity should consider the upgraded methods and work together in order to overcome the various constraints and offer suitable inputs to develop fail-proof biofertilizers for the farmers.

REFERENCES

Abang, S., Chan, E. S. and Poncelet, D. 2012. Effects of process variables on the encapsulation of oil in ca-alginate capsules using an inverse gelation technique. *J Microencapsul* 29:417–428.

Ahemad, M. and Khan, M. S. 2012. Effect of fungicides on plant growth promoting activities of phosphate solubilizing *Pseudomonas putida* isolated from mustard (*Brassica compestris*) rhizosphere. *Chemosphere* 86(9):945–950.

Akladious, S. A. and Abbas, S. M. 2012. Application of *Trichoderma harziunum* T22 as a biofertilizer supporting maize growth. *Afri J Biotechnol* 11(35):8672–8683.

Albareda, M., Rodriguez-Navarro, D. N., Camacho, M. and Temprano, F. J. 2008. Alternatives to peat as a carrier for rhizobia inoculant: solid and liquid formulations. *Soil Biol Biochem* 40:2771–2779.

Amiet-Charpentier, C., Gadille, P., Digat, B. and Benoit, J. P.1998. Microencapsulation of rhizobacteria by spray-drying: formulation and survival studies. *J Microencaps* 15:639–659.

Anandham, R., Choi, K. H., Gandhi, P. I., Yim, W. J., Park, S. J., Kim, K. A. and Sa, T. M. 2007. Evaluation of shelf life and rock phosphate solubilization of *Burkholderia* sp. in nutrient-amended clay, rice bran and rock phosphate-based granular formulation. *World J Microbiol Biotechnol* 23(8):1121–1129.

Anitha, A. and Rabeeth, M.2009. Control of Fusarium wilt of tomato by bioformulation of *Streptomyces griseus* in green house condition. *Afr J Basic Appl Sci* 1:9–14.

Arora, N. K., Khare, E., Naraian, R. and Maheshwar, D. K. 2008. Sawdust as a superior carrier for production of multipurpose bioinoculant using plant growth promoting rhizobial and pseudomonad strains and their impact on productivity of *Trifolium repense*. *Curr Sci* 95:90–94.

Arora, N. K., Khare, E. and Maheshwari, D. K. 2010. Plant growth promoting rhizobacteria: constraints in bioformulation, commercialization, and future strategies. In: *Plant Growth and Health Promoting Bacteria*. Springer, Berlin/Heidelberg, pp:97–116.

Baca, B. E. and Elmerich, C. 2007. Microbial production of plant hormones. In: *Associative and Endophytic Nitrogen-Fixing Bacteria and Cyanobacterial Associations*. Springer, Dordrecht, pp:113–143.

Bashan, Y. 1998. Inoculants of plant growth-promoting bacteria for use in agriculture. *Biotechnol Adv* 16(4):729–770.

Ben Rebah, F., Prévost, D., Yezza, A. and Tyagi, R. D. 2007. Agro-industrial waste materials and wastewater sludge for rhizobial inoculant production: a review. *Bioresour Technol* 98:3535–3546.

Bhattacharyya, P. N. and Jha, D. K. 2012. Plant growth promoting rhizobacteria (PGPR): emergence in agriculture. *World J Microbiol Biotechnol* 28(4):1327–1350.

Bora, T., Ozaktan, H., Gore, E. and Aslan, E. 2004. Biological control of *Fusarium oxysporum* sp. Melonis by wettable powder formulations of the two strains of *Pseudomonas putida*. *J Phytopathol* 152:471–475.

Brahmaprakash, G. P. and Sahu, P. K. 2012. Biofertilizers for sustainability. *J Ind Inst Sci* 92(1):37–62.

Covarrubias, S. A., de-Bashan, L. E., Moreno, M. and Bashan, Y. 2012. Alginate beads provide a beneficial physical barrier against native microorganisms in wastewater treated with immobilized bacteria and microalgae. *Appl Microbial Biotechnol* 93(6):2669–2680.

Deaker, R., Roughley, R. J. and Kennedy, I. R. 2004. Legume seed inoculation technology –a review. *Soil Biol Biochem* 36(8):1275–1288.

Haichar, F. E. Z., Santaella, C., Heulin, T. and Achouak, W. 2014. Root exudates mediated interactions belowground. *Soil Biol Biochem* 77:69–80.

Hayat, Q., Hayat, S., Irfan, M. and Ahmad, A. 2010. Effect of exogenous salicylic acid under changing environment: a review. *Environ Exp Bot* 68:14–25.

Immordino, M. L., Dosio, F. and Cattel, L. 2006. Stealth liposomes: review of the basic science, rationale, and clinical applications, existing and potential. *Inter Nanomed* 1(3):297–299.

Jacquot, M. and Pernetti, M. 2004. Spray coating and drying processes. In: Nedovic V. and Willaert R. (eds), *Fundamentals of Cell Immobilisation Biotechnology*. Kluwer Academic Publishers, Dordrecht, pp:343–356.

Kaur, R. and Kaur, R. 2018. Biological alternates to synthetic fertilizers: efficiency and future scopes. *Ind J Agr Res* 52(6):587–595.

Krasaekoopt, W., Bhandari, B. and Deeth, H. 2003. Evaluation of encapsulation techniques of probiotics for yoghurt. *Int Dairy J* 13:3–13.

Kumar, V. 2014. Characterization, bio-formulation development and shelf-life studies of locally isolated biofertilizer strains. *Octa J Environ Res* 2(1):32–37.

Lopez, M. D., Maudhuit, A., Pascual-Villalobos, M. J. and Poncelet, D. 2012. Development of formulations to improve the controlled-release of linalool to be applied as an insecticide. *J Agri Food Chem* 60(5):1187–1192.

Lu, J. and Toy, P. H. 2009. Organic polymer supports for synthesis and for reagent and catalyst immobilization. *Chem Rev* 109:815–838.

Manikandan, R., Saravanakumar, D., Rajendran, L., Raguchander, T. and Samiyappan, R. 2010. Standardization of liquid formulation of *Pseudomonas fluorescens* Pf1 for its efficacy against Fusarium wilt of tomato. *Biol Control* 54:83–89.

Marasco, R., Rolli, E., Ettoumi, B., Vigani, G., Mapelli, F., Borin, S., Abou-Hadid, A. F., El-Behairy, U. A., Sorlini, C., Cherif, A., Zocchi, G. and Daffonchio, D. 2012. A drought resistance promoting microbiome is selected by root system under desert farming. *PLoS One* 7:e48479.

Morgan, C. A., Herman, N., White, P. A. and Vesey, G. 2006. Preservation of microorganisms by drying: a review. *J Microbiol Methods* 66(2):183–193.

Mulvaney, R. L., Khan, S. A. and Ellsworth, T. R. 2009. Synthetic nitrogen fertilizers deplete soil nitrogen: a global dilemma for sustainable cereal production. *J Environ Qual* 38:2295–2314.

Mwangi, M. W., Monda, E. O., Okoth, S. A.andJefwa, J. M. 2011. Inoculation of tomato seedlings with *Trichoderma harzianum* and arbuscular mycorrhizal fungi and their effect on growth and control of wilt in tomato seedlings. *Braz J Microbio* 42(2):508–513.

Nehra, V. and Choudhary, M. 2015. A review on plant growth promoting rhizobacteria acting as bioinoculants and their biological approach towards the production of sustainable agriculture. *J Appl Nat Sci* 7(1):540–556.

Niu, D. D., Zheng, Y., Zheng, L., Jiang, C.-H., Zhou, D.-M. and Guo, J.-H. 2016. Application of PSX biocontrol preparation confers root-knot nematode management and increased fruit quality in tomato under field conditions. *Biocontrol Sci Technol* 26:174–180.

Parmar, P. and Sindhu, S. S. 2013. Potassium solubilization by rhizosphere bacteria: influence of nutritional and environmental conditions. *J Microbiol Res* 3(1):25–31.

Parnell, J. J., Berka, R., Young, H. A., Sturino, J. M., Kang, Y., Barnhart, D. M. and DiLeo, M. V. 2016. From the lab to the farm: an industrial perspective of plant beneficial microorganisms. *Front Plant Sci* 7:1110.

Qi, X., Xiaofeng, S., Guo, H., Qi, J. and Cheng, H. 2016. VdThit, a thiamine transport protein, is required for pathogenicity of the vascular pathogen *Verticillium dahlia*. *MPMI* 29(7):545–559.

Rani, N., Kaur, R. and Kaur, S. 2020. Zinc solubilising bacteria to augment soil fertility – a comprehensive review. *Int J AgriSci Vet Med* 8(1):38–44.

Sahu, P. K. and Brahmaprakash, G. P. 2016. Formulations of biofertilizers–approaches and advances. In: *Microbial Inoculants in Sustainable Agricultural Productivity*. Springer, New Delhi, pp:179–198.

Saleh, S. A., Mekhemar, G. A. A., El-Soud, A., Ragab, A. A. and Mikhaeel, A. A. 2001. Survival of Azorhizobium and Azospirillum in different carrier materials: inoculation of wheat and *Sesbania rostrata*. *Bull Fac Agric Univ Cairo* 52:319–338.

Schoebitz, M., Osman, J. and Ciampi, L. 2013. Effect of immobilized *Serratia* sp. by spray-drying technology on plant growth and phosphate uptake. *Chil J Agric Anim Sci* 29:111–119.

Schulz, T. J. and Thelen, K. D. 2008. Soybean seed inoculants and fungicidal seed treatment effects on soybean. *Crop Sci* 48:1975–1983.

Shakeri, E., Modarres-Sanavy, S. A. M., Amini Dehaghi, M., Tabatabaei, S. A. and Moradi-Ghahderijani, M. 2016. Improvement of yield, yield components and oil quality in sesame (*Sesamum indicum* L.) by N-fixing bacteria fertilizers and urea. *Arch Agro Soil Sci* 62(4):547–560.

Singleton, P., Keyser, H. and Sande, E. 2002. Development and evaluation of liquid inoculants, In: Herridge D. (ed), *Inoculants and Nitrogen Fixation of Legumes in Vietnam*. ACIAR Proceedings 109e, Canberra, pp:52–66.

Smith, R. S. 1992. Legume inoculant formulation and application. *Canadian Microbiol* 38(6):485–492.

Sokhey, A. S., Ali, Y. and Hanna, M. A. 1997. Effects of die dimensions on extruder performance. *J Food Eng* 31:251–261.

Spaepen, S. and Vanderleyden, J. 2011. Auxin and plant-microbe interactions. *Cold Spring Harb Perspect Biol* 3(4):a001438.

Stephens, J. H. G. and Rask, H. M. 2000. Inoculant production and formulation. *Field Crops Res* 65:249–258.

Taurian, T., Anzuay, M. S., Angelini, J. G., Tonelli, M. L., Ludueña, L., Pena, D., Ibáñez, F. and Fabra, A. 2010. Phosphate-solubilizing peanut associated bacteria: screening for plant growth promoting activities. *Plant Soil* 329:421–431.

Tittabutr, P., Payakapong, W., Teaungroong, N., Singleton, P. W. and Boonkerd, N. 2007. Growth survival and field performance of bradyrhizobial liquid inoculant formulations with polymeric additives. *Sci Asia* 33:69–77.

Tiwari, S., Singh, P., Tiwari, R., Meena, K. K., Yandigeri, M., Singh, D. P. and Arora, D. K. 2011. Salt-tolerant rhizobacteria-mediated induced tolerance in wheat (*Triticum aestivum*) and chemical diversity in rhizosphere enhance plant growth. *Biol Fertil Soil* 47(8):907–916.

Trejo, A., de-Bashan, L. E., Hartmann, A., Hernandez, J. P., Rothballer, M., Schmid, M. and Bashan, Y. 2012. Recycling waste debris of immobilized microalgae and plant growth-promoting bacteria from wastewater treatment as a resource to improve fertility of eroded desert soil. *Environ Exp Bot* 75:65–73.

Trivedi, P., Pandey, A. and Palni, L. M. S. 2008. In vitro evaluation of antagonistic properties of *Pseudomonas corrugata*. *Microbiol Res* 163(3):329–336.

Venkatashwarlu, B. 2008. *Role of Bio-Fertilizers in Organic Farming: Organic Farming in Rain Fed Agriculture*.Central Institute for Dry Land Agriculture, Hyderabad, pp:85–95.

Vidhyasekaran, P. and Muthamilan, M. 1995. Development of formulations of *Pseudomonas fluorescens* for control of chickpea wilt. *Plant Dis* 79:782–786.

Xavier, I. J., Holloway, G. and Legget, M. 2004. Development of rhizobial inoculant formulations. *Crop Manag* 3(1):318–322.

Yao, L., Wu, Z., Zheng, Y., Kaleem, I. and Li, C. 2010. Growth promotion and protection against salt stress by *Pseudomonas putida* Rs-198 on cotton. *Eur J Soil Biol* 46:49–54.

Yeo, Y., Baek, N.and Park, K. 2001. Microencapsulation methods for delivery of protein drugs. *Biotechnol Bioprocess Eng* 6(4):213–230.

Zhang, L., Fan, J., Ding, X., He, X., Zhang, F. and Feng, G. 2014. Hyphosphere interactions between an arbuscular mycorrhizal fungus and a phosphate solubilizing bacterium promote phytate mineralization in soil. *Soil Biol Biochem* 74:177–183.

10

Recent Biotechnological Applications of Archaeal Domain

Ruth Gill, Rajinder Kaur, Nitu Rani, and Sukhminderjit Kaur
Chandigarh University, Punjab, India

CONTENTS

10.1 Introduction

The existing life forms on the earth are majorly divided into three domains: Archaea, Eubacteria, and Eukarya. The domain Archaea consists of single-celled prokaryotic organisms that lack nucleus and show combined characteristics of other two domains of Bacteria and Eukarya. The word "archaea" comes from Greek word "archaios," which means "primitive" or "ancient," and in fact, some of the archaea express characteristics that resemble that of the ancient times. The exceptional feature of archaea is their survival potential in extreme conditions of scorching heat, high acidity, and saline environments (Leuko and Rettberg 2016). The archaeal members such as *Pyrolobus fumarii* can grow in the hydrothermal vents where temperature reaches about 113°C (Blöchl et al. 1997). *Picrophilus* sp. can survive in acidic conditions where pH is low or even zero (Rampelotto 2013). The Dead Sea also harbors many *Haloferax volcanii* spp. which can grow in high salt concentration (Haque et al. 2020). In the year 1977, the whole scientific community was amazed when Carl-Woese et al. (1977) examined the prokaryotic DNA sequences and observed the variations within prokaryotes (Carl-Woese et al. 1977). They reported a group of bacteria was methane producing and able to proliferate at high temperature; therefore, they divided those bacteria in a separate group. The life was divided into three domains: Eubacteria, Archaebacteria, and Eukaryota. Later, the domain Archaebacteria was renamed as Archaea (Blohs et al. 2019). Panawala (2017) described the differences between archea and bacteria based on the cell wall, shape, lipid membrane, and growth conditions, which are shown in Table 10.1. According to the Universal Tree of Life, the deepest and the shortest branches are of hyperthermophilic organisms that grow at high temperature. This indicates that hyperthermophiles may be the extension of the last life present on the earth and most

TABLE 10.1

Major Differential Characteristics between Archaea and Bacteria

S. No.	Characteristics	Archaea	Bacteria
1.	Environment	Extreme	Present everywhere
2.	Cell wall	Lacks peptidoglycan	Peptidoglycan is present
3.	Shapes	Spiral, rod, sphere, flat, square, plate in shape	Vibrio, rods, filamentous, cocci, bacilli, spirochetes shaped
4.	Lipid membrane	Ether-linked	Ester-linked
5.	Major types	Halophiles, thermophiles, methanogens	Gram-positive and Gram-negative
6.	RNA polymerase	Several	One
7.	Initiation tRNA	Methionine	Formyl methionine
8.	Examples	*Halobacterium salinarum, Haloferax volcanii, Sulfolobusacido caldarius, Thermoplasma volcanium, Methanogenium frigidum*	*Staphylococcus aureus, Pseudomonas aeruginosa, Helicobacter pylori, Klebsiella pneumoniae, Bacillus cereus*

of the hyperthermophiles belong to domain Archaea. Various studies showed that eukaryotes are derived from the fusions of archaeal and bacterial cells (Koonin 2015). The Archaeal species have many industrial applications because their metabolite products are stable in extreme environmental conditions. The enzymes (xylanases) of *Pyrodictium abyssi* and *Thermococcus zilliggi* AN1 show stability at temperature more than 100°C and such archaea are used in paper and pulp industries (Andrade et al. 2001). The DNA polymerase isolated from *Thermococcus Kodak araensis* KOD1 has the capability to show high extension rates and low errors in the process of amplification of DNA sequence (Egorova and Antranikian 2005). Archaeal spp. like *Thermococcus litoralis*, *Sulfolobus solfataricus*, *Thermoanaerobacter brockii*, and *Thermococcus guaymasensis* have the ability to produce active alcoholic compounds in pharmaceutical industries (Boehmwald et al. 2016). The enzyme "superoxide dismutases HB27/Recombinant" from the psychrophilic *Thermusther mophilus* is being used in medical treatments (Liu et al. 2011). Thermophilic and psychrophilic alkaline lipases from *Geobacillus* sp. and proteases secreted by *Aeropyrum pernix* are utilized in detergent production (Boehmwald et al. 2016; Munoz et al. 2013). According to molecular analysis, Archaeal domain has two major subdivisions: the Euryarchaeota and the Crenarchaeota, one minor subdivision: the Korarchaeota, and other subdivisions: Thaumarchaeota and Nanoarchaeota. The further classification is based on the type of growing conditions of these organisms and is shown in in Figure 10.1 (Guanghong et al. 2015).

10.2 Cultivation Strategies of Archaea

The cultivation of archaea requires similar basic procedure like bacteria, including sample collection, enrichment, and isolation. But unlike bacteria, the isolation process of archaea is difficult with low success rate. Slow *in vitro* growth rate of archaea and their low abundance result in the outgrowth of bacteria. Limited availability of nutrients and growth factors along with oxidative stress conditions leads to the problems related to cultural procedures of archaea. To combat this situation, appropriate growth conditions, nutrients requirements, trace elements, and other additives need to be characterized. Presently, advanced cultivation methods, including Hungate method for anaerobic archaea and continuous cultivation methods in fermentors are available for archaeal isolation (Sun et al. 2020). Genetic and transcriptional information is used for the formulations of selective media for archaeal isolation, and genetic encoding provides information regarding genes for metabolism of carbon, sulfur, nitrogen, ammonia, and may suggest potential substrates for archaeal cultivation. On the similar note, the genetic information results in the tolerance of

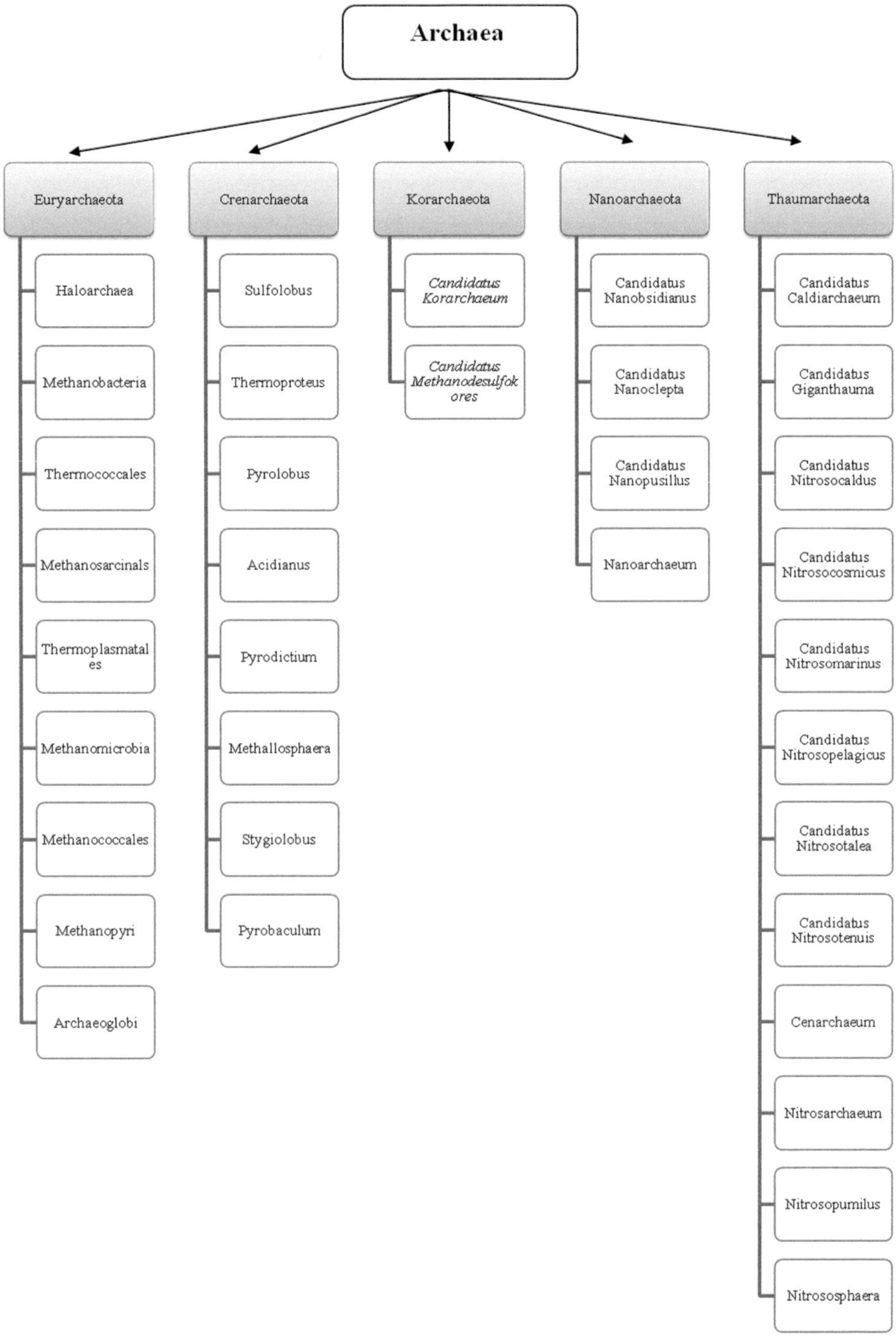

FIGURE 10.1 Classification of archaea according to growth conditions.

anaerobic/aerobic conditions, metal stress, carbon dioxide stress, which may reveal the ideal conditions for its cultivation. Innovation techniques such as co-culturing, direct interspecies electron transfer (DIET), high throughput culturing (HTC), single-cell isolation, and simulation of the natural habitat are some possible ways to enhance the cultivation of archaea for industrial applications (Sun et al. 2020).

10.3 Biotechnological Applications of Archaea

Archaeal species and the metabolites secreted by them have shown unusual stability at extreme environmental conditions. These metabolites have the ability to work at extreme high and low temperature, varied pH range, high pressure, and high salt concentrations at which most of the bacterial enzymes and metabolites usually get inactivated. This ability of archaea has gained momentum and captured the interest of many researchers throughout the scientific community. A number of industrially important metabolites, including biosurfactants, extremoenzymes, halocins, cationic antimicrobial peptides (CAMPs), polyhydroxyalkanoates (PHAs), retinal proteins, compatible solutes, heat stress proteins, and exopolysaccharides are secreted by Archaea (Charlesworth and Burns 2016). Archaea are also involved in the important biological processes like bio-mining, methane production, and bioremediation processes (Elisabetta 2010). Some of the important applications of the archaea in the field of biotechnology are explained in the subsequent sections.

10.3.1 Archaea in Biomethane Production

Biomethane is odorless, colorless, and non-toxic gas used as a domestic and commercial biofuel. Methanogenic archaea are the important group of archaea that can produce biomethane by the anaerobic fermentation using dead plants and animals, manure, organic waste, and sewage. Methanogenic archaea use nutrient-rich substrates, which are fermented to produce H_2, CO_2, and acetate using complex anaerobic reactions, and finally, undergo methanogenesis to produce the biogas (Figure 10.2) (Balat and Balat 2009; Kassim et al. 2017). Methanogenic archaea, based on the morphology and growth conditions, are classified into five orders: Methanobacterials, Methanococcales, Methanomicrobiales, Methanosarcinales, and Methanopyrales (Enzmann et al. 2018). They have used different mechanisms for methane production and are given in Table 10.2. *Methanosarcina soligelidi* and *M. barkeri*, the halotolerant methanogenic archaea, can potentially act as a biological source of methane (Maus et al. 2020). *Methanobacterium arbophilicum* isolated from tree wood tissue can degrade the pectin and cellulose to produce methane by using H_2 in methanogenesis reaction. *M. barkeri* and *Methanothermobacter thermautotrophicus* can convert four molecules of CO_2 to methane

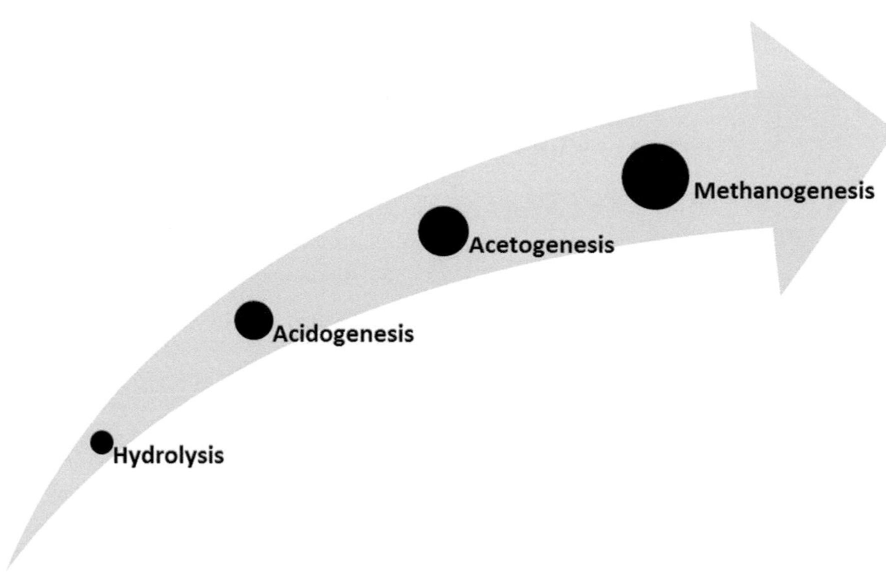

FIGURE 10.2 Different stages of methane production in methanogens.

TABLE 10.2

Methanogenic Archaea Help in Methanogenesis (Liu and Whitman 2008)

S. No.	Methanogenic Archaea	Morphology	Growth Conditions	Mechanism of Methane Production
1.	Methanobacterials	Rod-shaped, non-motile	pH range 6–9	Produce methane gas using CO_2 (electron acceptor) and H_2 (electron donor)
2.	Methanococcales	Cocci and contains flagella	High salt concentration	Take up CO_2 and H_2 for methane production
3.	Methanomicrobiales	Regular or irregular coccus-shaped	30–55°C	Use acetate for methane production
4.	Methanosarcinales	Rod-shaped	Low temperature	Split acetate or disproportionate the methyl group in the organic compound to produce methane
5.	Methanopyrales	Rod-shaped and motile	84–110°C	Reduce H_2 and CO_2 for methanogenesis

by using CO dehydrogenase (CODH), and H_2 as electron donor (Enzmann 2018). *Methanosarcina Acetivorans* can also utilize carbonmonoxide (CO) to produce a large amount of formate and acetate, and the group of *Methanosarcina* and *Methanotrix* can produce methane by using acetate. The methylotrophic group of *Methanosarcinales* can utilize methylamines, methanol/methanethiol, or dimethyl sulfide for the production of methane (Franziska et al. 2018).

10.3.2 Biosurfactant

Many Archaeal species are known to produce biosurfactants, and different archaea secreting industrially important biosurfactants are depicted in Table 10.3. These amphiphilic compounds produced by archaea have fatty acids or glycolipids as hydrophobic moiety, and sugars, proteins, amino acids, carboxylic acids as hydrophilic moiety (Kaur and Kaur 2019; Hamme and Urban 2009). Biosurfactant functions by lowering down the surface tension (ST) and portal tensions which occur between molecules at the surface and interface. In comparison to traditional chemically derived surfactants, biosurfactant are non-toxic, renewable, and biodegradable, and are known to exhibit activity in extreme conditions (Cameotra et al. 2010). Archeal biosurfactants also manifested applications in bioremediation of oil in water samples and soil (Van et al. 2009), cosmetic, pharmaceutical, and food industries (Nitschke and Costa 2007). Halophilic archaea, viz. *Halovivax* strain A21, *Haloarcula strain* D21, *Haloferax mediterranei*, and *Haloarcula japonica* have been known to produce biosurfactant at high salt concentration (Kebbouche et al. 2009; Kassim et al. 2017). Elazzazy et al. (2015) reported the production of anionic lipopeptide biosurfactant from alkali-halo-thermophilic bacteria (*Virgibacillus salarius*). The biosurfactant produced by *V. salarius* exhibited the emulsifying activity of 80% and decreased surface tension by 30 mN/m under wide temperature and pH range.

10.3.3 Antibiotics

The antibiotics or antimicrobial compounds present in the market currently are either derived from mesophilic bacteria or semi-synthetically derived by chemical processes. The emergence of antibiotic resistance era demands for novel antibiotics and the isolation of antibiotics from the unexplored archaea can prove as a milestone toward this application. Because of their adaptations to diverse range of habitats, they have the potential to produce antimicrobial compounds with potent activities. One such antimicrobial peptide secreted by *Halobacteriaceae* is halocin which is the proteinous bacteriocin (more than

TABLE 10.3
Industrially Important Biosurfactant Produced by Different Archaea

S. No.	Archaea	Biosurfactant	Archaeal Type	Application	Reference
1.	*Achromobacter* sp. PS1, *Bacillus* sp. SLDB1	Glycolipids	Thermophiles	Bioremediation	Joy et al. 2017
2.	*Ochrobactrum* sp. GREW1, *Bacillus* sp. SB2	Lipopeptides	Thermophiles	Bioremediation	Joy et al. 2017
3.	*Aneurinibacillus aneurinilyticus* SBP-11	Aneurinifactin	Halophiles	Pharmaceutical industries	Balan et al. 2017
4.	*Bacillus licheniformis*	Lipopeptide	Thermophiles	Microbial enhanced oil recovery (MEOR)	El-Sheshtawy 2015
5.	*B. licheniformis* AL 1.1	Lichenysin	Thermophiles and Psychrophiles	Cosmetics industry	Coronel-Leon et al. 2015
6.	*Rhodococcus* sp., *Pseudomonas* sp., *Pseudoalteromonas* sp. and *Idiomarina* sp.	Glycolipids	Psychrophiles	Not known	Malavenda et al. 2015
7.	*Kocuria marina* BS-15	Lipopeptide	Halophiles	Pharmaceutical Industries	Sarafin et al. 2014
8.	*Oceanobacillus* sp. BRI 10	Glycolipoprotein	Psychrophiles	Food, pharmaceutical and cosmetics industry, and bioremediation processes	Jadhav et al. 2013
9.	*Halomonas* sp. BS4	Glycolipids, polymeric Substances, and lipopeptides	Halophiles	Pharmaceutical Industries	Donio et al. 2013
10.	*Natrialba* sp.	Glycoproteins, glycolipids or lipopeptides	Halophiles	Bioremediation	Kebbouche-Gana et al. 2013
11.	*Thermusaqu aticus* CCM 3488, *Thermus* sp. CCM 2842, *Meiothermus ruber* CCM 4212	Ramnolipids	Thermophiles	Not known	Rezanka et al. 2011
12.	*Pseudomonas fluorescens* BD5	Pseudofactin II	Psychrophiles	Not known	Janek et al. 2010
13.	*Halovivax* sp. A21, *Haloarcula* sp. D212	Peptidoglycolipid and glycoproteins	Halophiles	Oil industry	Kebbouche-Gana et al. 2009
14.	*B. licheniformis* BAS 50	Lipopeptide	Thermophiles and Halophiles	MEOR	Yakimov et al. 1997
15.	*Rhodococcus* sp. GA-05	Trehalose lipids	Psychrophiles and Halophiles	Bioremediation	Yakimov et al. 1999
16.	*Bacillus subtilis* BBK-1	Lipopeptides: bacillomycin L, plipastatin, and surfactin	Halophiles	Not known	Roongsawang et al. 2002

TABLE 10.4

Different Types of Halocins Secreted by Archea

S. No.	Halophilic Archaea	Type of Halocin	Spectrum of Action	Reference
1	Haloferax mediterranei R4 (ATCC33500)	H4	Narrow	Kumar and Tiwari 2019
2	*H. larsenii* HA1	HA1	Narrow	Kumar et al. 2016
3	*H. volcanii* KPSl	KPS1	Broad	Kavitha et al. 2011
4	*H. mediterranei* Sech7a	Sech7a	Narrow	Pasic et al. 2008
5	*Halobacterium* strain GRB	G1	Broad	Sun et al. 2005
6.	*H.* strain AS7092	C8	Broad	Li et al. 2003
7.	*Haloarchaeon* Gla2.2	H2	Broad	Platas et al. 2002
8.	*H. mediterranei*	H1	Wide	Platas et al. 2002

10 kDA) made up of two parts: large protein part (35 kDa) and peptide microhalocins (3.6 kDa). Based on the antagonistic studies, different types of halocins have been reported to be produced by archaea (Table 10.4). Due to the presence of hydrophobic microhalocins, a 311 residue pro-protein halocin can withstand the process of desaltation without destroying its antimicrobial activity. Moreover, it is comparatively insensitive to heat and can be stored for a period of seven years (Shand and Leyva 2007). Halocin secreted by *Natrinema* spp. BTSH10 is found to be stable in the pH range of 5.0–9.0 and can tolerate the temperature up to 100°C (Karthikeyan et al. 2013). These archaea can survive in the high saline environment because they secrete organic solutes, which balance the ion concentration outside as well as inside the cell and keep the cell undamaged (Gonzalo et al. 2002; Jan et al. 2007; Torsten et al. 2007). Extreme halophillic archaea such as *Natrinema* (Karthikeyan et al. 2013), *Haloferax larsenii* (Kumar et al. 2016) have been reported to produce halocins. Halocin producing archaea find extensive applications in textile industry; tanning requires the use of high salt concentration which allows the growth of trivial halophilic organisms, and halocin secreting archaea can be employed to hinder their growth (Shandand Leyva 2007). Similarly, halocin producing archaea such as *Haloarcula*, *Halobacterium*, and *Halorubrum* outgrow the pathogenic haloarchaea in salted food, therefore stimulating significant prospective in food processing industry (Olivier et al. 2014). Another class of antimicrobial compounds secreted by archaea are broad-spectrum amphipathic CAMPs, which show antimicrobial activity against a wide range of Gram positive, Gram negative bacteria, fungi as well as protozoa (Lei et al. 2019). Archaea *Sulfolobusis landicus* can be regarded as the cellular factories for the secretion of CAMP like antimicrobial compounds (Notomista et al. 2015).

10.3.4 Extremozymes

Microbial enzymes are in demand in industries as they exhibit many advantages such as better catalytic activity, low risk of contamination, increased solubility of substrates, and lower viscosity. Most of the enzymes are obtained from fungi and bacteria, but presently, archaeal enzymes are in spotlight because of their survival in extreme conditions where the fungal and bacterial enzymes normally get denatured. Extremophilic enzymes exhibit the great potential to be used in biotechnological field (Elleuche et al. 2014) in which proteomics, metablomics, and genomics play a major role in the enhancement of novel extremozymes, which are currently in high demands by biotechnological industries. Extremophiles have been broadly classified as thermophiles, psychrophiles, halophiles, and piezophiles (Table 10.5), and they secrete different extermozymes. Thermophilus archaea have two categories – thermophiles and hyperthermophiles – in which thermophiles showed their optimum growth in the temperature range of 50–70°C, but hyperthermophiles can grow up to the temperature of 115°C (Egorova and Antranikian 2005). Both groups of thermophiles belong to Euryarchaeota, Nanoarchaeota, Crenarchaeota, and Korarchaeota (Egorova and Antranikian 2005). The archaea were also found in geothermal systems, hot springs, and in volcanic areas where the temperature is extremely high. The structural studies of protein confirms higher hydrogen bonds, low thermo-labile amino acid concentration, shorter loop

TABLE 10.5
Different Enzymes Secreted by Extermophiles

S. No.	Extremophiles	Habitat	Growth Requirements	Archeal Strains	Enzymes	Reference
1.	Thermophiles	Volcanic areas, hot springs	Temperature range:–50°C to 115°C	*Thermosphaera aggregans*, *Pyrococcus furiosus*, *Thermococcus zilligii* (Tzi)	Glycoside hydrolases, DNA polymerase	Ausili et al. 2007; Griffiths 2007
2.	Psychrophiles	Sea ice, polar, glacier	Temperature:0°C to–45°C	*Serratiarubidaea*, *Stenotrophomonas maltophilia*	Proteases	Kuddus and Ramteke 2009
3.	Halophiles	Dead Sea and Great Salt Lake	Salt Concentration:–3.4 M to 0.8 M	*Haloarcula* sp.	Amylase	Fukushima et al. 2005
4.	Piezophiles	Deep sea and hydrothermal vents	Pressure:–100 MPa	*Pyrococcusyayanosii*	Amidase and amidase nitrilase	Fu et al. 2014

structure, and increased hydrophobicity and surface charge as the major reasons underlining the stability of these enzymes (Reed et al. 2013). Glycoside hydrolase enzymes isolated from thermophiles are used for the hydrolysis of carbohydrates and can tolerate higher temperatures. Enzymes like α-glycosidase, α-pullulanases, and α-amylases also reported to perform glucose hydrolysis at high temperature of 60–105°C and pH of 4.5–6.0 (Leveque et al. 2000). These enzymes serve as environment-friendly and cost-effective catalytic units in industries. *Bacillus stearothermophilus* and *B. licheniformis* are the two bacterial strains which are widely used in the industries for the production of hydrolytic enzymes (Souza and Magalhães, 2010). However, various studies revealed that the hydrolase enzymes produced by archaea have higher activity compared to bacteria (Vieille and Zeikus 2001). The archaeal group of *Thermophilum*, *Desulfurococcus*, *Sulfolobus*, and *Staphylothermus* are also found to produce enzymes that are active at temperature range of 40–100°C (Bertoldo and Antranikian 2002). Various archaeal strains such as *Thermococcus fumicolans* (Tfu) (Cambon-Bonavita et al. 2000), *Pyrococcus abyssi* (IsisTM) (Dietrich et al. 2002), *Thermococcus celericrescens* (TceI) (Kim et al. 2011), and *T. zilligii* (Tzi) (Griffiths et al. 2007) have the capability to produce thermostable DNA polymerases of commercial grade. Psychrophiles are the bacteria that are able to multiply at temperature ranges from 0°C to –10°C, are found to show activity at subzero temperatures, and can survive very low temperatures of up to –45°C (Danilovich et al. 2018). Psychrophiles are found naturally in polar areas, glacier, permafrost, sea ice, and in artificial environment like refrigerator. For their survival, psychrophiles generate enzymes that enable them to survive in cold temperatures. Cold adapted enzymes show distinctive properties like higher structural flexibility and more catalytic sites (Pearce, 2017). The genomic and proteomic studies revealed that psychrophiles like *Methanococcoides burtonii*, *Methanogenium frigidum*, *Serratia rubidaea*, and *Stenotrophomonas maltophilia* (Kuddus and Ramteke 2009) have the ability to produce commercial grade proteases (Allen et al. 2009). The microorganisms that can survive under high salt concentration (at least 2M) such as in Dead Sea and Great Salt Lake are known as extreme halophiles. Enzymes from halophiles have specific ion binding sites, less hydrophobic surface contact, and high acidic amino acid content that help them to remain stable and active at high salt concentrations (Siglioccolo et al. 2011), and probably, these were the first-ever enzymes used for industrial purposes. The amylase from *Haloarcula* spp. was stable and functional up to high salt concentration of 4.3 M, temperature 50°C, and showed stability in the presence of chloroform, toluene, and benzene (Fukushima et al. 2005). The enzyme halophilic galactosidases secreted by *Haloferax alicante* was active at higher salt concentration of 4M NaCl. It has been reported that *Haloferax mediterranei* can produce 38% Poly-β-hydroxyalkanoate (PHA) grown at salt concentration of 150 g/L (Don et al. 2006). PHA is biodegradable in nature, and thus, can be used in the synthesis of bioplastic. Many industries are now interested in PHA for the production of plastic extracted from non-fossil source, i.e., archaea. Apart from bioplastic, PHA is also used by pharmaceutical, medical, and agricultural industries. Microorganisms that are able to survive at high pressure are called piezophiles, and can be isolated from hydrothermal vents and deep sea. They have the characteristics for the adaptation to high temperature as they exhibit small hydrogen bonding aminoacids, multimers of proteins and packed hydrophobic core (Reed et al. 2013). Amidase nitrilase is the enzyme produced by *Pyrococcus yayanosii* that catalyzes the hydrolysis of diethylamide (Fu et al. 2014). Diethylamide is used for preparing synthetic fibers and solvent in organic synthesis and is utilized as a solvent for resins, plastic, and other chemical reactions. Formamide is used as solvent in pharmaceutical industries and act as a softening agent in the production of glues and paper. In many other organic compounds such as plasticizers, acetamide was used as a solvent and for denaturing alcohol (Heinrich 2019).

10.3.5 Biomining

Mining sector is regarded as one of the largest industries involved in extraction of minerals and metals from their ores from the earth's crust. The extraction of minerals is done by chemical processing and employing high temperature conditions, which ultimately results in environmental pollution. However, biomining is an alternative method to overcome such problems. Biomining refers to the extraction of minerals and metals of economic interest from their ores utilizing microorganisms. Biomining is involved in the extraction of copper, gold, cobalt, zinc, uranium, and nickel; it is comparatively simple and cheaper; involves metal recovery from low-grade ores; and increases quantity and reduces environmental pollution

(Kumar and Yaashikaa 2020). Various microorganisms such as bacteria and archaea can be used to carry out the process of biomining that involves bioleaching and biooxidation. Bioleaching is accomplished by using microorganisms to convert insoluble form of metal into soluble form, whereas biooxidation is mainly applied for the extraction of gold from its ore (Faramarzi et al. 2020). In this process, microorganisms breakdown the arsenic and insoluble pyrite compounds, and remove metal sulfides from the ores using two pathways considered for metal sulfides dissolution, viz. polysulfate pathway and thiosulfate pathway. In case of thiosulfate pathway, the ferric ions produced during microbial processing help in the solubilization of metals and thiosulfate is the predominant intermediate compound produced (Srichandan et al. 2019). Electron extraction by ferric ions helps in the oxidation of acid insoluble metal sulfides such as tungstenite; on contrary, in case of polysulphate pathway, both proton and ferric ions solubilize the acid soluble metal sulfides such as chalcopyrite and the sulfide cation is the primary intermediate product (Fomchenko and Muravyov 2019). Archaea have the ability to demerize the sulfate to free disulfide in order to produce sulfur element and different species are involved in the process of biomining, bioleaching and biooxidation. *Metallosphaere sedula* is a single-celled microbe that carries out the process of biomining by secreting enzyme similar to polyphosphate/NAD kinase and changes the metal sulfur compound to its elemental form (Kassim et al. 2017). The bioleaching of copper and molybdenum from chalcopyrite and molybdenite ores is accomplished by employing thermophilic and acidophilic archaea, including *Sulfolobus brierleyi* and *S. acidocaldarius* (Abhilash and Pandey 2015). *Ferroplasma acidiphilum* grows at an optimum temperature of 40–60°C is known to oxidize ferrous ions (Golyshina et al. 2000). *Acidianus infernus* belong to thermophilic archaea that catalyze the oxidation and reduction of sulfur (Ding et al. 2011). Furthermore, *Stygiolobus azoricus* are thermophilic obligate anaerobic archaea that utilize sulfur for respiration process; however, *Sulfolobus shibatae* is a species of thermophilic facultative chemolithotroph involved in oxidation of sulfur (Johnson 2014). Hence, the process of biomining could be carried out efficiently and effectively with the utilization of different archaeal species.

10.3.6 Bioremediation

Bioremediation is carried out with the help of natural or genetically engineered living organisms such as bacteria, fungi, and archaea which convert contaminants and pollutants into less harmful compounds. Archaea have the capability to break down aromatic compounds as well as hydrocarbons present in water and soil. This process relies on the capability of archaea to grow and multiply in the contaminated environment. It has been observed that the various parameters like temperature, pH, and salinity play a vital role in the microbial growth and production of metabolites that ultimately degrade the compounds (Hamzah et al. 2010); however, extreme conditions such as high temperature and salt concentrations limit the bioremediation process. This is because many microorganisms do not have the ability to grow in harsh conditions, but archaea have the capability to remove the environmental contaminants even in these conditions. Archaea such as Methanomicrobiales, Methanosarcinales, Methanobacteriales and Thermoplasmatales have been reported to degrade the petroleum hydrocarbons in oil spills (Zhang et al. 2012), and various researchers have revealed that archaea have the ability to act as a biocatalyst in bioremediation processes (Erdoğmuş et al. 2013). Majorly, archaea have prominent catabolic activity in comparison to bacteria and fungi because archaea have been used as a primary feasible biocatalyst applicant (Fathepure 2014), as archaea have the capability to grow in either extreme pH condition or under high or low temperatures, where usually bacteria are used for bioremediation, but cannot survive in these extreme harsh conditions (Lin et al. 2014). It has also been observed that some of the archaea showed high resistance to antibiotics such as streptomycin, penicillin, and cycloheximide and due to this ability, they have been used in the bioremediation of environment containing antibiotics (Al-Mailem et al. 2010).

10.3.7 Archeal Metabolites

Archaea secrete various metabolites such as osmotically active solutes/compatible solutes, proteins, special lipids, and exopolysaccharides that have ability to be used in various industrial applications. They also secrete some unique organic solutes (osmolytes) that are not produced by other organisms such as osmotically active solutes secreted by *Methanococcus thermolithotrophicus* have the properties of tackling high osmotic pressure and prevent denaturation of other macromolecules (Roberts 2000). The compatible solutes produced by halophillic bacteria can be used as stabilizers for biomolecules as they help in maintaining the positive water balance inside the cell and also suitable for cellular metabolism (Santosand da-Costa 2001). Ectoine (1, 4, 5, 6-tetrahydro-2-methyl-4-pyrimidine carboxylic acid) was produced by halophiles and halotolerant archaea to survive in osmotic stress in high-salt environment (Alexander et al. 2011). Ectoine and its derivatives have been proven to be useful as moisturizers in cosmetics (Montitsche et al. 2000) and stabilizers in the polymerase chain reaction. Mannosylglycerate (MG) and di-gylycerol phosphate are the components of compatible solutes produced by halophilic archaea such as *Pyrococcus woesei*, *P. furiosus*, and *Archaeoglobus fulgidus*. MG showed more stabilizing effect of heating or freeze-drying on several enzymes than other stabilizers (Lamosa et al. 2000). Heat stress proteins (e.g., chaperonins) secreted by archaea are important for the regulation of key transcriptional enzymes (An et al., 2017). Membrane-bound retinal pigments such as halorhodopsin and bacteriorhodopsin (BR) are secreted by halophillic archaea and are useful to perform bioenergetics processes by using light energy (Kassim et al. 2017). Because of thermodynamic and photochemical stability of BR, it can be used in technical applications such as spatial light modulators, artificial retina, holography, and associative and volumetric optical memories. COBEL (Barcelona, Spain) and Munich Innovative Biomaterial (Munich, Germany) companies are using BR isolated from *Halobacterium salinarum* (Margesin and Schinner 2001) and can also be utilized to convert adenosine diphosphate (ADP) to adenosine triphosphate (ATP) for the continuation of biochemical energy (Kassim et al. 2017).

10.3.8 Application in Biowaste

Archaea as methanogens are applied in the treatment of sewage sludge, agricultural wastes, municipal waste, and industrial wastes under anaerobic conditions (Arora and Panosyan 2019. Under the harsh environmental conditions, halophiles archaea have also been utilized for bioremediation, degradation of several organic pollutants and in the treatment of waste waters from the textile, especially in the first dye bath liquor from dying process containing a high load of salt (Margesin and Schinner 2001), as well as have decreased the time required for remediation of polluted region due to biosurfactant properties. Banat et al. (2000) carried out a study on oil spills in marine environments and revealed positive results against hydrocarbon pollution. Another study carried out by Lovley (2001) has also reported hydrocarbon degradation by the archaeal microbes. Archaeal sulfate-reducing microorganisms are isolated and have an ability to degrade alkanes anaerobically. It has been reported that some archaeal microbes utilize sulfate as electron acceptor for oxidation of methane anaerobically and hypothesized that this involves the reversal of methanogenesis catalyzed scavenging of an electron-carrying metabolite by sulfate-reducing bacteria (Spormann and Widdel 2000).

10.4 Conclusion

Due to their unique genetic makeup, archaea can withstand diverse environmental conditions which make their metabolites resilient to degradation during various industrial processes. Various metabolites secreted by archaea exhibit wide potential applications in industrial processes. The role of archaea in methane production and biowaste degradation is well-known. Enzymes secreted by archea have shown better catalytic activity compared to enzymes isolated from fungi and bacteria. The antimicrobial peptides produced by archaea are extensively used in textile industries. The biosurfactants of archaea origin are non-toxic, renewable, and biodegradable and are known to exhibit activity in extreme conditions. Biomining and bioleaching also employ the use of archaea to carry out the process of biooxidation. In a nutshell, it can be said that archaea can be seen as the source of various metabolites with potential industrial applications, and in future, we can expect more use of archaea-based enzymes, biosurfactants, antibiotics, and other products in biotechnology.

REFERENCES

Abhilash, G. A. and Pandey, B. D. 2015. Bioleaching of low grade granitic chalcopyrite ore by hyperthermophiles elucidation of kinetics-mechanism. *Metallur Res Technol* 112(5):1–14.

Alexander, S., Reshetnikov, V. N., Khmelenina, Ildar I., Mustakhimov, Y. A. and Trotsenko, 2011. Genes and enzymes of ectoine biosynthesis in halotolerant methanotrophs. *Meth Enzymol* 495:15–30.

Allen, M. A., Lauro, F. M., Williams, T. J., Burg, D., Siddiqui, K. S., De-Francisci, D., Chong, K. W. Y., Pilak, O., Chew, H. H., De-Maere, M. Z., Ting, L., Katrib, M., Ng, C., Sowers, K. R., Galperin, M. Y., Anderson, I. J., Ivanova, N., Dalin, E., Martinez, M., Lapidus, A., Hauser, L., Land, M., Thomas, T. and Cavicchioli, R. 2009. The genome sequence of the psychrophilic archaeon, *Methanococcoides burtonii*: the role of genome evolution in cold adaptation. *ISME J* 3(9):1012–1035.

Al-Mailem, D. M., Sorkhoh, N. A., Al-Awadhi, H., Eliyas, M. and Radwan, S. S. 2010. Biodegradation of crude oil and pure hydrocarbons by extreme halophilic archaea from hypersaline coasts of the Arabian Gulf. *Extremophiles* 14(3):321–328.

An, Y. J., Rowland, S. E., Na, J. H., Spigolon, D., Hong, S. K., Yoon, Y. J., Lee, J. H., Robb, F. T. and Cha, S. S. 2017. Structural and mechanistic characterization of an archaeal-like chaperonin from a thermophilic bacterium. *Nat Commun* 8(1):1–11.

Andrade, C. M., Aguiar, W. and Antranikian, B. 2001. Physiological aspects involved in production of xylanolytic enzymes by deep-sea hyperthermophilic archaeon *Pyrodictium abyssi. Appl Biochem Biotechnol* 91–93:655–669.

Arora, N. K. and Panosyan, H. 2019. Extremophiles: applications and roles in environmental sustainability. *Environ Sustain* 2:217–218.

Ausili, A., Cobucci-Ponzano, B., Di Lauro, B., D'Avino, R., Perugino, G., Bertoli, E., Scirè, V, Rossi, M., Tanfani, F. and Moracci, M. 2007. A comparative infrared spectroscopic study of glycoside hydrolases from extremophilic archaea revealed different molecular mechanisms of adaptation to high temperatures. *Proteins* 67(4): 991–1001.

Balan, S. S., Kumar, C. G. and Jayalakshmi, S. 2017. Aneurinifactin, a new lipopeptidebiosurfactant produced by a marine *Aneurinibacillus aneurinilyticus* SBP-11 isolated from Gulf of Mannar: purification, characterization and its biological evaluation. *Microbiol Res* 194:1–9.

Balat, M. and Balat, H. 2009. Biogas as a renewable energy source –a review. *Energ Sourc A*31(14):1280–1293.

Banat, I. M., Makkar, R. S.and Cameotra, S. S. 2000. Potential commercial applications of microbial surfactants. *Appl Microbiol Biotechnol* 53:495–508.

Bertoldo, C. and Antranikian, G. 2002. Starch-hydrolyzing enzymes from thermophilic archaea and bacteria. *Curr Opin Chem Biol* 6(2):151–160.

Blöchl, E, Rachel, R, Burggraf, S, Hafenbradl, D, Jannasch, H.W., Stetter, K.O. 1997. *Pyrolobus fumarii,* gen. and sp. nov., represents a novel group of archaea, extending the upper temperature limit for life to 113 degree C. *Extremophiles* 1:14–21.

Blohs, M., Mahnert, A., Spang, A., Dombrowski, N., Krupovic, M. and Klingl, A. 2019. Archaea –an introduction. *Encyclo Microbiol* 4:243–252.

Boehmwald, F., Muñoz, T., Patricio, F.and Patricio, B. 2016. Functional screening for the discovery of new extremophilic enzymes. In: Rampelotto P. (ed) *Biotechnol Extremophiles*. Springer, Cham, pp:321–350.

Cambon-Bonavita, M. A., Schmitt, P., Zieger, M., Flaman, J. M., Lesongeur, F., Gerard, R., Bindel, D., Frisch, N., Lakkis, Z., Dupret, D., Barbier, G. and Quérellou, J.2000. Cloning, expression, and characterization of DNA polymerase I from the hyperthermophilic archaea *Thermococcus fumicolans. Extremophiles* 4(4):215–225.

Cameotra, S. S., Makkar, R. S., Kaur, J. and Mehta, S. K. 2010. Synthesis of biosurfactants and their advantages to microorganisms and mankind. In: Sen R. (ed), *Biosurfactants*. Springer, New York, pp:261–280.

Charlesworth, J. and Burns, B. P. 2016. Extremophilic adaptations and biotechnological applications in diverse environments. *AIMS Microbiol* 2(3):251–261.

Coronel-Leon, J., de Grau, G., Grau-Campistany, A., Farfan, M., Rabanal, F., Manresa, A.and Marqués, A. M. 2015. Biosurfactant production by AL 1.1, a *Bacillus licheniformis* strain isolated from Antarctica: production, chemical characterization and properties. *Ann Microbiol* 65:2065–2078.

Danilovich, M. E., Sánchez, L. A., Acosta, F. and Delgado, O. D. 2018. Antarctic bioprospecting: in pursuit of microorganisms producing new antimicrobials and enzymes. *Polar Biol* 41(7):1417–1433.

Dietrich, J., Schmitt, P., Zieger, M., Preve, B., Rolland, J. L., Chaabihi, H. and Gueguen, Y. 2002. PCR performance of the highly thermostable proof-reading B-type DNA polymerase from *Pyrococcus abyssi*. *FEMS Microbiol Lett* 217(1):89–94.

Ding, J., Zhang, R., Yu, Y., Jin, D., Liang, C., Yi, Y., Zhu, W. and Xia, J. 2011. A novel acidophilic, thermophilic iron and sulfur-oxidizing archae on isolated from a hot spring of Tengchong, Yunnan, China. *Brazilian J Microbiol* 42(2):514–525.

Don, T. M., Chen, C. W. and Chan, T. H. 2006. Preparation and characterization of poly (hydroxyalkanoate) from the fermentation of *Haloferax mediterranei*. *J Biomater Sci Polym Ed* 17 (12):1425–1438.

Donio, M. B. S., Ronica, F. A., Thanga- Viji, V., Velmurugan, S., Adlin, J. J., Michaelbabu, M., Dhar, P., Citarasu, T. 2013. Halomonas sp. BS4, a biosurfactant producing halophilic bacterium isolated from solar salt works in India and their biomedical importance. Springer Plus 2:149.

Egorova, K. and Antranikian, G. 2005. Industrial relevance of thermophilic archaea. *Curr Opin Microbiol* 8(6):649–655.

Elisabetta, B. 2010. Archaeal transformation of metals in the environment, *FEMS Microbiol Ecol* 73(1):1–16.

El-Sheshtawy, H. S., Aiad, I., Osman, M. E., Abo-ELnasr, A. A., Kobisy, A. S. 2015. Production of biosurfactant from Bacillus licheniformis for microbial enhanced oil recovery and inhibition the growth of sulphate reducing bacteria. *Egypt J Pet* 24:155–162.

Elleuche, S., Schröder, C., Sahm, K. and Antranikian, G. 2014. Extremozymes — biocatalysts with unique properties from extremophilic microorganisms. *Curr Opin Biotechnol* 29:116–123.

Elazzazy, A. M., Abdelmoneim, T. S. and Almaghrabi, O. A. 2015. Isolation and characterization of biosurfactant production under extreme environmental conditions by alkali-halo-thermophilic bacteria from Saudi Arabia. *Saudi J Biol Sci* 22(4):466–475.

Enzmann, F., Mayer, F., Rother, M. and Holtmann, D. 2018. Methanogens: biochemical background and biotechnological applications. *AMB Express* 8(1):1–10.

Erdoğmuş, S. F., Mutlu, B., Korcan, S. E., Güven, K. and Konuk, M. 2013. Aromatic hydrocarbon degradation by halophilic archaea isolated from Çamaltı Saltern, Turkey. *Water Air Soil Pollut* 224(3):1449.

Faramarzi, M. A., Mogharabi-Manzari, M. and Brandl, H. 2020. Bioleaching of metals from wastes and low-grade sources by HCN-forming microorganisms. *Hydrometallurgy* 191:105228.

Fathepure, B. Z. 2014. Recent studies in microbial degradation of petroleum hydrocarbons in hypersaline environments. *Front Microbiol* 5(173):1–16.

Fomchenko, N. V. and Muravyov, M. I. 2019. Effect of sulfide mineral content in copper-zinc concentrates on the rate of leaching of non-ferrous metals by biogenic ferric iron. *Hydrometallurgy* 185:82–87.

Franziska, E., Florian, M., Michael, R. and Dirk, H. 2018. Methanogens: biochemical background and biotechnological applications. *AMB Express* 8:1–12.

Fu, L., Li, X., Xiao, X. and Xu. 2014. Purification and characterization of a thermostable aliphatic amidase from the hyperthermophilicarchae on *Pyrococcus yayanosii* CH1. *Extremophiles* 18:429–440.

Fukushima, T., Mizuki, T., Echigo, A., Inoue, A. and Usami, R. 2005. Organic solvent tolerance of halophilic α-amylase from a haloarchae on, *Haloarcula* sp. Strain S-1. *Extremophiles* 9(1):85–89.

Golyshina, O. V., Pivovarova, T. A., Karavaiko, G. I., Kondratéva, T. F., Moore, E. R., Abraham, W. R., Lünsdorf, H., Timmis, K. N., Yakimov, M. M. and Golyshin, P. N.2000. Ferroplasma acidiphilum gen. nov., sp. nov., an acidophilic, autotrophic, ferrous-iron-oxidizing, cell-wall-lacking, mesophilic member of the *Ferroplasmaceae* fam. nov., comprising a distinct lineage of the archaea. *Inter J Syst Evol Microbial* 50(3):997–1006.

Gonzalo, P., Inmaculada, M. and Ricardo, A. 2002. Purification and biological characterization of halocin H1 from *Haloferax mediterranei* M2a. *Int Microbiol* 5:15–19.

Griffiths, K., Nayak, S., Park, K., Mandelman, D., Modrell, B., Lee, J., Ng, B., Gibbs, M. D. and Bergquist, P. L. 2007. New high fidelity polymerases from *Thermococcus* species. *Protein Expr Purif* 52(1):19–30.

Guanghong, Z., Zhao, X. and Hao. 2015. Phylogeny and taxonomy of archaea: a comparison of the whole-genome-based CVTree approach with 16S rRNA sequence analysis. *Life (Basel)* 5(1):949–968.

Hamme, J. D. and Urban, J. 2009. Biosurfactants in bioremediation. In: Singh A., Kuhad R.C. and Ward O.P. (eds), *Advances in Applied Bioremediation*. Springer, Berlin/Heidelberg, pp:73–89.

Hamzah, A., Rabu, A., Azmy, R. F. H. R. and Yussoff, N. A. 2010. Isolation and characterization of bacteria degrading Sumandak and South Angsi oils. *Sains Malays* 39(2):161–168.

Haque, R. U., Paradisi, F. and Allers, T. 2020. Haloferax volcanii for biotechnology applications: challenges, current state and perspectives. *Appl Microbiol Biotechnol* 104(4):1371–1382.

Heinrich, L. A. 2019. Future opportunities for bio-based adhesives–advantages beyond renewability. *Green Chem* 21(8):1866–1888.

Jan, B., Antonio, J. P., Nathalie, P. and Erhard, B. 2007. Osmotically induced synthesis of the compatible solute hydroxyectoine is mediated by an evolutionarily conserved ectoine hydroxylase. *J Biol Chem* 282:31147–31155.

Janek, T., Łukaszewicz, M., Rezanka, T. and Krasowska, A.2010. Isolation and characterization of two new lipopeptide biosurfactants produced by *Pseudomonas fluorescens* BD5 isolated from water from the Arctic Archipelago of Svalbard. *Bioresour Technol* 101:6118–6123.

Jadhav, V. V., Yadav, A., Shouche, Y. S., Aphale, S., Moghe, A., Pillai, S., Arora, A. and Bhadekar, R. B. 2013. Studies on biosurfactant from Oceanobacillus sp. BRI 10 isolated from Antarctic sea water. *Desalination* 318:64–71.

Johnson, D. B. 2014. Biomining biotechnologies for extracting and recovering metals from ores and waste materials. *Curr Opin Biotechnol* 30:24–31.

Joy, S., Rahman, P. K. S. and Sharma, S. 2017. Biosurfactant production and concomitant hydrocarbon degradation potentials of bacteria isolated from extreme and hydrocarbon contaminated environments. *Chem Eng J* 317:232–241.

Karthikeyan, P., Bhat, S. G. and Chandrasekaran, M. 2013. Halocin SH10 production by an extreme haloarchaeon *Natrinema* sp. BTSH10 isolated from salt pans of South India. *Saudi J Boil Sci* 20(2):205–212.

Kassim, M. A., Hashim, N. H. F. and Yusof, N. A.2017. Archaea domain as biocatalyst in environmental biotechnology and industrial applications. *J Adv Microbiol* 5(2):1–21.

Kaur, S. and Kaur, R.2019. Biosurfactant from *Lactobacillus* sp. as an antibiofilm agent. *Biotechnol* 100(3):335–43.

Kavitha, P., Lipton, A. P., Sarika, A. R. and Aishwarya, M. S. 2011. Growth characteristics and halocin production by a new isolate, *H. volcanii* KPSl from Kovalam Solar Saltern (India). *Res J Biol Sci* 6:257–262.

Kebbouche-Gana, S., Gana, M. L., Khemili, S., Fazouane-Naimi, F., Bouanane, N. A., Penninckx, M. and Hacene, H. 2009. Isolation and characterization of halophilic Archaea able to produce biosurfactants. *J Ind Microbiol Biotechnol* 36(5):727–738.

Kebbouche-Gana, S., Gana, M. L., Ferrioune, I., Khemili, S., Lenchi, N., Akmouci-Toumi, S., Bouanane-Darenfed, N. A. and Djelali, N. D. 2013. Production of biosurfactant on crude date syrup under saline conditions by entrapped cells of *Natrialba* sp. strain E21, an extremely halophilic bacterium isolated from a solar saltern (Ain Salah, Algeria). *Extremophiles* 17:981–993.

Kim, K. P., Bae, H., Kim, I. H. and Kwon, S. T. 2011. Cloning, expression, and PCR application of DNA polymerase from the hyperthermophilic archaeon, *Thermococcus celer. Biotechnol Lett* 33(2):339–346.

Koonin, E. V. 2015. Origin of eukaryotes from within archaea, archaeal eukaryome and bursts of gene gain: eukaryogenesis just made easier? *Philos Trans R Soc Lond B Biol Sci* 370(1678):20140333.

Kuddus, M. and Ramteke, P. W. 2009. Cold-active extracellular alkaline protease from an alkaliphilic *Stenotrophomonas maltophilia*: production of enzyme and its industrial applications. *Can J Microbiol* 55(11):1294–301.

Kumar, V. and Tiwari, S. K.2019. Halocin diversity among Halophilic Archaea and their applications. In: Kumar Das (Eds.) *Microbial Diversity in Ecosystem Sustainability and Biotechnological Applications.* Springer, Singapore, pp:497–532.

Kumar, V., Saxena, J. and Tiwari, S. K. 2016. Description of a halocin-producing *Haloferax larsenii* HA1 isolated from Pachpadra salt lake in Rajasthan. *Arch Microbiol* 198(2):181–192.

Kumar, P. S. and Yaashikaa, P. R. 2020. Recent trends and challenges in bioleaching technologies. In: Navanietha K. R. and Sani R. K. (eds), *Biovalorisation of Wastes to Renewable Chemicals and Biofuels.* Elsevier, pp:373–388.

Lamosa, P., Burke, A., Peist, R., Huber, R., Liu, M. Y., Silva, G., Rodrigues-Pousada, R., Gall. J., Maycock, C. and Santos, H. 2000. Thermo-stabilization of proteins by diglycerol phosphate, a new compatible solute from the hyperthermophile *Archaeoglobus fulgidus. Appl Environ Microbiol* 66:1974–1979.

Lei, J., Sun, L. Huang, S., Zhu, C., Li, P., He, J., Mackey, V., Coy, D. H. and He, Q. 2019. The antimicrobial peptides and their potential clinical applications. *Am J Transl Res* 11(7):3919–3931.

Leuko, S. and Rettberg, P. 2016. The astrobiological potential of halophilic archaea. 41st COSPAR Scientific Assembly.

Leveque, E., Janecek, S., Haye, B. and Belarbi, A. 2000. Thermophilic archaeal amylolytic enzymes. *Enzyme Microb Technol* 26(1):3–14.

Li, Y., Xiang, H., Liu, J., Zhou, M. and Tan, H. 2003. Purification and bio- logical characterization of halocin C8, a novel peptide antibiotic from Halobacterium strain AS7092. *Extremophiles* 7:401–407.

Lin, J., Hao, B., Cao, G., Wang, J., Feng, Y., Tan, X. and Wang, W. 2014. A study on the microbial community structure in oil reservoirs developed by water flooding. *J Petroleum Sci and Eng* 122:354–359.

Liu, Y. and Whitman, W. B. 2008. Metabolic, phylogenetic, and ecological diversity of the methanogenic archaea. *Ann NY Acad Sci* 1125(1):171–189.

Liu, J., Yin, M., Zhu, H., Lu, J. and Cui, Z. 2011. Purification and characterization of a hyperthermostable Mn-superoxide dismutase from *Thermus thermophilus* HB27. *Extremophiles* 15:221–226.

Lovley, D. R. 2001. Bioremediation: anaerobes to the rescue. *Science* 239:1444–1446.

Malavenda, R., Rizzo, C., Michaud, L., Gerce, B., Bruni, V. and Syldatk, C., Hausmann, R. and Giudice, A. L. 2015. Biosurfactant production by Arctic and Antarctic bacteria growing on hydrocarbons. *Polar Biol* 38:1565–1574.

Margesin, R. and Schinner, F. 2001. Potential of halotolerant and halophilic microorganisms for biotechnology. *Extremophiles* 5:73–83.

Maus, D., Heinz, J., Schirmack, J., Airo, A., Kounaves, S. P., Wagner, D. and Schulze-Makuch, D. 2020. Methanogenic archaea can produce methane in deliquescence-driven mars analog environments. *Sci Rep* 10(1):1–7.

Montitsche, L. H. and Galinski, E. 2000. Ectoine and ectoine derivatives as moisturizers in cosmetics. Patent US060071.

Munoz, P., Correa-Llantén, D. and Blamey, J. 2013. Production, purification and partial characterization of four lipases from a thermophile isolated from Deception Island. *Lipids* 48:527–533.

Nitschke, M. and Costa, S. 2007. Biosurfactants in food industry. *Trends Food Sci Technol* 18(5):252–259.

Notomista, E., Falanga, A., Fusco, S., Pirone, L., Zanfardino, A., Galdiero, S. and Contursi, P. 2015. The identification of a novel *Sulfolobus islandicus* CAMP-like peptide points to archaeal microorganisms as cell factories for the production of antimicrobial molecules. *Microb Cell Fact 14*(1): 126–132.

Olivier, H., Jeanne, F., Bruno, D. and Jacques, M. 2014. Exploring the diversity of extremely halophilic archaea in food-grade salts. *Inter J Food Microbiol* 191:36–44.

Pasic, L., Velikonja, B. H. and Ulrih, N. P. 2008. Optimization of the culture conditions for the production of a bacteriocin from halophilic archaeon Sech7a. *Prep Biochem Biotechnol* 38:229–245.

Panawala, L. 2017. Difference between archaebacteria and eubacteria. https://pediaa.com/difference-between-archaebacteria-and-eubacteria/.

Pearce, D. A. 2017. Extremophiles in Antarctica: life at low temperatures. In: Stan-Lotter H., Fendrihan S. (eds) *Adaption of Microbial Life to Environmental Extremes*. Springer, Cham, pp:99–131.

Platas, G., Meseguer, I. and Amils, R. 2002. Purification and biological characterization of halocin H1 from Haloferax mediterranei M2a. *Int Microbiol* 5:15–19.

Rampelotto, P. H. 2013. Extremophiles and extreme environments. *Life (Basel)* 3(3):482–485.

Reed, C. J., Lewis, H., Trejo, E., Winston, V. and Evilia, C. 2013. Protein adaptation in archeal extremophiles. *Archaea* 1–14.

Rezanka, T., Siristova, L. and Sigler, K. 2011. Rhamnolipid-producing thermophilic bacteria of species Thermus and Meiothermus. *Extremophiles* 15:697–709.

Roberts, M. F. 2000. Osmoadaptation and osmoregulation in archaea. *Front Bio Sci* 5(362):796–812.

Roongsawang, N., Thaniyavarn, J., Thaniyavarn, S., Kameyama, T., Haruki, M., Imanaka, T., Morikawa, M. and Kanaya, S. 2002. Isolation and characterization of a halotolerant Bacillus subtilis BBK-1 which produces three kinds of lipopeptides: bacillomycin L, plipastatin, and surfactin. *Extremophiles* 6:499–506.

Santos, H. and da-Costa, M. S. 2001. Organic solutes from thermophiles and hyperthermophiles. *Method Enzymol* 334:302–315.

Sarafin, Y., Donio, M. B. S., Velmurugan, S., Michaelbabu, M. and Citarasu, T. 2014. *Kocuria marina* BS-15 a biosurfactant producing halophilic bacteria isolated from solar salt works in India. *Saudi J Biol Sci* 21:511–519.

Shand, R. F. and Leyva, K. J. 2007. Peptide and protein antibiotics from the domain archaea: halocins and sulfolobicins. In: Riley M. A. and Chavan M. A. (eds), *Bacteriocins: Ecology and Evolution*. Springer, Berlin/Heidelberg, pp:93–109.

Siglioccolo, A., Paiardini, A., Piscitelli, M. and Pascarella, S. 2011. Structural adaptation of extreme halophilic proteins through decrease of conserved hydrophobic contact surface. *BMC Struct Biol* 11(50):1–12.

Souza, P. M. D. and Magalhães, P. O. 2010. Application of microbial α-amylase in industry –a review. *Braz J Microbiol* 41(4):850–861.

Spormann, A. M. and Widdel, F. 2000. Metabolism of alkylbenzenes, alkenes, and other hydrocarbons in anaerobic bacteria. *Biodegradation* 11:85–105.

Srichandan, H., Mohapatra, R. K., Parhi, P. K. and Mishra, S. 2019. Bioleaching approach for extraction of metal values from secondary solid wastes: a critical review. *Hydrometallurgy* 189:105122.

Sun, C., Li, Y., Mei, S., Lu, Q., Zhou, L. and Xiang, H. 2005. A single gene directs both production and immunity of halocin C8 in a haloarchaeal strain AS7092. *Mol Microbiol* 57:537–549.

Sun, Y., Liu, Y. and Pan, J. F. Wang and M. Li 2020. Perspectives on cultivation strategies of archaea. *Microb Ecol* 79:770–784.

Torsten, S., Thomas, M., Dirk, B., Hauke, H. and Uta, B. 2007. Continuous synthesis and excretion of the compatible solute ectoine by a transgenic, nonhalophilic bacterium. *Appl Environ Microbiol* 73:3343–3347.

Vieille, C. and Zeikus, G. J. 2001. Hyperthermophilic enzymes: sources, uses, and molecular mechanisms for thermostability. *Microbiol Mol Biol Rev* 65(1):1–43.

Yakimov, M. M., Amro, M. M., Bock, M., Boseker, K., Fredrickson, H. L., Kessel, D. G. and Timmis, K. N. 1997. The potential of Bacillus licheniformis strains for in situ enhanced oil recovery. *J Pet Sci Eng* 18:147–160.

Yakimov, M. M., Abraham, W. R., Meyer, H., Giuliano, L., Golyshin, P. N. 1999. Structural characterization of lichenysin A components by fast atom bombardment tandem mass spectrometry. *Biochim Biophys Acta* 1438:273–280.

Zhang, D. C., Mörtelmaier, C. and Margesin, R. 2012. Characterization of the bacterial archaeal diversity in hydrocarbon-contaminated soil. *Sci Total Environ* 421–422:184–196.

11

Role of Fungal Endophytes in Improving Abiotic Stress Tolerance in Plants

Bushra Jan[1], **Zafar A. Reshi**[1], **and F. A. Mohiddin**[2]
[1]*University of Kashmir, Hazratbal, Jammu and Kashmir, India*
[2]*Sher-e-Kashmir University of Agricultural Sciences and Technology, Jammu and Kashmir, India*

CONTENTS

11.1 Introduction

Endophytic fungus is a fungal microorganism that spends all or part of its lifespan colonizing in the stable tissues of the host plant, intercellularly and intracellularly, without causing diseases in the host. These fungal endophytes infect plant tissues asymptomatically and have been isolated from all kinds of plants, including both woody and non-woody plant species and even algae (Khare et al. 2018; Chhipa and Deshmukh 2019; Nouh et al. 2020). Fungal endophytes have been found to profoundly affect plant ecology, fitness, evolution (Zanne et al. 2020), shaping of plant communities (Dastogeer et al. 2020), and the community structure and diversity of associated organisms such as insects, bacteria, and nematodes (Omacini et al. 2001; Van Overbeek and Saikkonen 2016; Zhou et al. 2018). Mycorrhizal fungi only colonize plant roots and grow into the rhizosphere while endophytes completely reside inside plant tissues and may grow within roots, stems, and/or leaves (Stone et al. 2004; Heinonsalo et al. 2017). Endophytic fungi have been largely identified in temperate plants, but recently studies have also focused on tropical plants for the presence of these fungi (El Mansy et al. 2020). Fungal endophytes can invade almost every part of the plant, including root, branch, twig, bark, leaves, petals, bulbs, fruits, and seed tissues (Nouh et al. 2020). Endophytes, in addition to bacteria and other microbes, also contain a number of fungi, including sleeping fungi and those fungi that may cause latent infections Endophytic fungal life cycles may involve vertical transmission of fungal propagules through host seeds and/or horizontal transmission from one plant to other plants of the same species. While vertical transmission has been well studied, relatively little work has been done on horizontal dissemination (Tadych et al. 2014). *Piriformospora indica, Trichoderma* spp., *Epicoccum nigrum, Penicillium* spp., *Alternaria, Cladosporium, Colletotrichum graminicola, Fusarium* spp., *Aspergillus flavus, Fusarium oxysporum, Aspergillus niger, Chaetomium globosum, Saccharomyces cerevisiae, Cladosporium cladosporioides, Aspergillus, Curvularia, Trichoderma koningii, Gilmaniella, Arthrobotrys foliicola, Acremonium zeae, F. verticillioides, Beauveria bassiana, Alternaria alternata, Paecilomyces* spp., and others are the most widespread endophytic fungi that are isolated from many plants and classified (Nouh et al. 2020). Such endophytic fungi are known to affect the ecology of the host plants, often by raising their capacity to survive and tolerate environmental and biological stresses; the mechanism involved therein are, however, only partially understood. Endophytes are also considered important in the plant defense against insect pests, host pathogens, and other herbivore animals and these can also improve tolerance of plants to pathogenic agents, stresses like biotic and abiotic ones (Kaur 2020).

11.2 Classification of Endophytic Fungi

Based on taxonomy, ecological functions, plant host range, tissue specificity, patterns of colonization, and evolutionary relatedness two different groups of endophytic fungi have been identified (Schaechter 2011): C-endophytes (clavicipitaceous endophytes) and NC-endophytes (non-clavicipitaceous endophytes). Most C-endophytes (*Epichloe/Neotyphodium*) are generally indicated to be spread systemically and vertically by plants, and infect grass exclusively. Whereas NC-endophytes are taxonomically diverse, transmitted horizontally from plants to plants and colonize almost all plants in ecosystems viz., angiosperms, conifers, ferns, and allies (Mane and Vedamurthy 2018). Besides these two major groups, there are further four classes of endophytic fungi (Rodriguez et al. 2009) (Table 11.1).

11.2.1 Class I Endophytic Fungi or Clavicipitaceous Fungi

The class I endophytic fungi comprise a limited group of phylogenetically related clavicipitaceous species that are difficult to cultivate and limited to some cold areas and warm-season grasses (Bischoff and White 2005). Such endophytes specifically live their complete life cycle inside the aerial portion of the host plant, establishing non-pathogenic, systemic, and typically intercellular connections (Khiralla et al. 2016). Class I endophytes have often been found to enhance the productivity of plants, produce

TABLE 11.1

Different Classes of Endophytic Fungi and Their Role in Stress Management with Examples

Class of Endophytic Fungi	Stress Management	Examples
Class I	Increase plant biomass, confer drought tolerance, produce chemicals that are toxic to animals, reduce herbivory, overcome mineral imbalance and soil acidity	*Neotyphodium* endophytes, clavicipitaceous endophytic fungi
Class II	Herbivore resistance, resistance to heat, salt, disease and drought, and increase biomass below and above ground	Non-clavicipitaceous endophytic fungi
Class III	Increase in host fitness	Includes both members of Ascomycota and Basidiomycota
Class IV	Trap and eliminate oxygen radicals generated during abiotic stress	Dark septate endophytes

toxic chemicals against animals, impart drought resistance, and, in effect, reduce herbivory (Clay 1988; Khiralla et al. 2016). The benefits of these fungi, however, appear to depend on the host species, the host genotype, and environmental conditions (Hidayat 2019). Over time, the magnitude of endophyte infection also increases in grass populations which suggests that endophytes offer an adaptive advantage to their hosts, although they grow in the host at the cost of metabolism. Colonization by endophytes enhances the growth rate of perennial ryegrass and tall fescue (Clay 1988). *Neotyphodium* endophytes also enhance drought resistance in grasses through osmoregulation and stomatal regulation, and also protect plants against nitrogen starvation or water stress (Chandran et al. 2020). These mechanisms have enabled perennial ryegrass to colonize large areas of south-eastern US that are otherwise too hot and dry for the plant. Endophyte-infected grasses express adaptations to a variety of biotic (González-Teuber et al. 2020) and abiotic stress including drought (Singh et al. 2020), mineral imbalance (Azad and Kaminskyj 2016), and soil acidity (Govindasamy et al. 2018). Endophyte-infected grasses are also more competitive and show vigorous growth than noninfected grasses in resource-limited environments (Lakshmanan et al. 2017).

11.2.2 Class II Endophytes or Non-Clavicipitaceous Fungi

Class II or NC-endophytic fungi colonize leaves, roots, and stems and are capable of developing substantial plant infections that are spread by seed coats and/or rhizomes; have low rhizosphere abundance; habitat-adapted fitness benefits in addition to non-habitat-adapted benefits; and usually have high infection rates (90–100%) in plants growing in high-stress habitats (Rodriguez et al. 2009). Nonetheless, fungal endophytes have been known to confer fitness benefits on host plants including herbivore resistance; resistance against drought, salt, disease, and heat; and increased biomass both above and below the ground (Rodriguez and Redman 2008; Shanmugapriya and Balasubramanian 2017). Non-embryonic tissues of *Dichanthelium lanuginosum*, a geothermal plant, are colonized by *Curvularia protuberate* (Redman et al. 2002). Neither the plant nor the fungus is able to survive temperatures above 40°C if grown nonsymbiotically. The symbiotic association between the two renders both the partners ability to withstand temperatures up to 65°C. Similar correlation between fungal symbiosis and fitness value was observed in *Fusarium culmorum* which colonizes all non-embryonic tissues of coastal dunegrass (*Leymus mollis*); if grown nonsymbiotically, the host plant does not survive and the growth of the endophyte is reduced when exposed to increased salinity levels in their native habitat (Rodriguez et al. 2008). Nonetheless, when symbiotically grown, both partners tolerate salinity levels of the sea water (300–500 mM NaCl). Evidently, *C. protuberate* and *F. culmorum* residing in plant tissues are able to prevent the detrimental effects of temperature and salt stress. On the basis of aforementioned observations, it was hypothesized that some class II endophytes are mutual; giving hosts positive fitness benefits, especially enhanced abiotic stress tolerance in reciprocation of the nutrients obtained from host tissues (Rodriguez et al. 2009). Field experiments conducted on geothermal soils and coastal beaches are in agreement with the experimental observations that nonsymbiotic plants are able to thwart the influence of abiotic stresses in their natural habitat only after colonization by specific endophytes. Further research has shown that the efficiency of endophytes to confer habitat-specific stress tolerance is an adaptive mechanism that is established at the subspecies level (Rodriguez et al. 2008).

11.2.3 Class III Endophytic Fungi

The endophytic fungi of class III are separated according to their role in horizontal transmission, above-ground tissue occurrence, localized infection formation, increase in host fitness, and richness in biodiversity. This class of endophytes involves highly diverse fungi and are often associated with aerial tissues of nonvascular and tropical plants, seedless vascular plants, conifers, angiosperms (both woody and non woody), boreal and artic flora, and flowers and fruits (Rodriguez et al. 2008). This class includes both members of *Ascomycota* and *Basidiomycota* (*Dikaryomycota*), where *Ascomycota* dominates with only few members of *Basidiomycota* i.e. *Agaricomycotina*, *Pucciniomycotina*, and *Ustilaginomycotina* (Tejesvi and Pirttilä 2018).

11.2.4 Class IV Endophytic Fungi or Dark Septate Endophytes

The class IV endophytic fungi are usually highly pigmented (brown to black in color) and are usually associated with mycorrhizal fungi, also referred as pseudo-mycorrhizal fungi (Rodriguez et al. 2008). Such fungi are generally referred to as dark septate endophytes (DSE) and are grouped as class IV endophytic fungi. DSEs are usually categorized on the basis of their restriction to plant roots and the presence of dark melanized septa; these fungal endophytes have been found essential for the survival of plant host under stress environments as cell wall melanin can capture and remove the oxygen radicals formed during abiotic stress (Richier et al. 2005). However, other role of these fungal endophytes is still underway (Knapp et al. 2018).

11.3 Strategies of Fungal Endophytes to Combat Stress and Improve Plant Health

Great efforts are being steered in studying endophytes with respect to their diversity, ecology, evolutionary biology, and their role in plant defense mechanism elicited against biotic and abiotic stresses. Endophytic biotechnology is seen as a promising instrument in harnessing plants for important agricultural, industrial, and economic products. The resident endophytes in plants have been found to effectively improve productivity of plant metabolites and products of agricultural importance and also the level of tolerance to biotic and abiotic stresses (Wani et al. 2015; Phurailatpam and Mishra 2020) (Figure 11.1). A number of important novel bioactive substances have been obtained from different endophytic species. New biotechnology applications such as phytoremediation and bioremediation are gaining considerable attention (Nandy et al. 2020; Pietro-Souza et al. 2020) and in healthy plants, endophytes play important roles in bioremediation through three different mechanisms known as phytostimulation, biofertilization, and biocontrol (de Carvalho et al. 2020; Guerrero-Zúñiga et al. 2020).

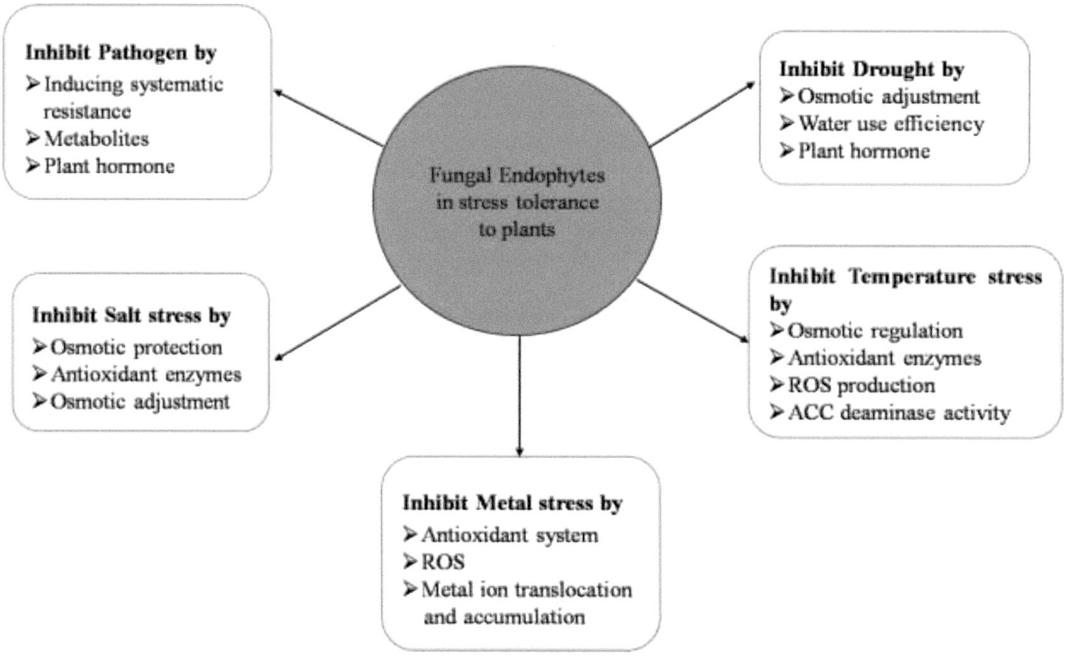

FIGURE 11.1 Flow chart representing the various methods adopted by fungal endophytes to mitigate abiotic stress tolerance in plants.

11.3.1 Endophytes: A Source for Bioactive and Novel Compounds

It has been found that specific bioactive substances are exuded by endophytic microbes without causing any noticeable damage to their host tissues (Ancheeva et al. 2020). Different endophytic microbes have been found to produce or secrete bioactive substances which improve plant resistance against plant pathogens. These substances are being employed as antimicrobial, anticancer, antidiabetic, and antiviral agents in the field of pharmaceutical and extra bioactive substances (Rustamova et al. 2020). Other important bioactive substances have also been found to be secreted by endophytic microbes in the form of phenolics, alkaloids, terpenes and terpenoids, polyketones, flavonoids, peptides, quinols, steroids, and toxic compounds which act as insecticides such as azadirachtin which have been successfully examined to be applicable in the field of medicine, agriculture, and industry (Zinniel et al. 2002; Meena et al. 2019). Many of these bioactive compounds have shown antimicrobial (i.e., antifungal, antibacterial, and antiviral), antioxidant, antineoplastic, anti-leishmanial, cytotoxicity, antiproliferative, and fuel production activities (Wang and Dai 2011; Rustamova et al. 2020). Endophytes also produce several antifungal compounds such as pestaloside, cryptocandin, cryptocin, pestalopyrone, ecomycins, and pseudomycins (Yu et al. 2010). It has been observed that fungal endophytic species are native source for terpenoids, phenolics, flavonoids, sugars, alkaloids, tannins, saponin, and round worm opposing compounds also (Bogner et al. 2017; Vasundhara et al. 2019).

11.3.2 Extracellular Enzyme Production

Endophytic fungi play a pivotal role in plant health via extracellular enzyme production, which has been considered as inevitable mechanism for endophyte colonization in plant kingdom. Different types of enzymes having extracellular activity, such as oxidoreductases, transferases, lyases, and hydrolases, are produced by microorganisms including fungi (Traving et al. 2015). It has been found that numerous macromolecules which include sugar-based polymers, lignin, carbohydrates, and proteins are catabolized by these enzymes into micromolecules, organic phosphate, and are then transported all through the cells, where they are constantly metabolized and play a pivotal role to inculcate the host symbiosis process (Strong and Claus 2011). Also, it has been observed that extracellular hydro-lyase enzymes combat several pathogenic infections (Leo et al. 2016). Different endophytic microbes produce enzymes, which include chitinase, protease, lipase, and β-1,3-glucanase, that are extracellular in nature and lyse the pathogen cell walls (both fungi and bacteria) and hence, can be used as biocontrol agents (Fouda et al. 2015). It has been studied that there is a correlation between the increase in growth of endophytic microbes within the tissues of host plants and diminution in the disease-causing pathogens. This correlation is due to increased production of enzymes like cellulases, xylanases, pectinases, lipases, phosphatases, amylase, proteases, and glucosidases by these microbial endophytes (Ayob and Simarani 2016). A promoting effect on the growth of a number of medicinal plants with improvement in their ability to survive in adverse environmental conditions has been found due to the extracellular enzymes which are produced by endophytic fungi (Chathurdevi and Gowrie 2016). Moreover, isolation and identification of 50 endophytic strains of fungi which include laccase, lipase, amylase, pectinase, cellulase, and protease activity has been achieved (Sunitha et al. 2013; Mishra et al. 2019).

11.3.3 Plant Growth-Promoting Activity

One of the viable and practical strategies that can be used for reducing the effect of stress without compromising plant growth is to harness the potential of plant-microbe (endophyte) relationship. Endophytic fungi produce a diverse range of metabolite substances which are used by plants to improve plant growth (Saleem et al. 2020). Fungal endophytes augment growth of plants by the production of enzymes and variety of bioactive compounds. Under unfavorable conditions, an improvement in plant growth-promotion has been found due to the endophytic organisms particularly fungi which involve *Sebacina vermifera, P. indica and* other species of *Penicillium* and *Colletotrichum* (Hamilton and Bauerle 2012). Many plant species associated with plant growth-promoting endophytic fungi tend to have evident improved qualities along with increased growth of plants and relative disease resistance against phytopathogenic fungi, viruses, bacteria, and plant parasitic nematodes by the synthesis of phytohormone like cytokinins, indole-3-acetic acid (IAA), gibberellins, phosphate solubilization, siderophores, uptake of nutrients, and plant pathogen antagonism. In addition, induction of physical or chemical changes in relation to plant protection by endophytic fungi has been signified through a process known as induced systemic resistance (ISR). Fungal endophytes have been proven to be very beneficial for the plants which are continuously facing abiotic stresses. Increase in the tolerance of plants against various stresses, such as salinity, heat, cold, heavy metals, and drought, by plant growth-promoting fungi (PGPF) has been confirmed by several reports (Khan et al. 2012; Rana et al. 2019).

11.3.4 Bioremediation/Biodegradation Activity

Endophytic fungi have a potent capability to augment growth of plants in contaminated soil through breakdown of hazardous compounds. Elimination of hazardous wastes and pollutants from contaminated environments with help of biological processes is known as bioremediation. It occurs by break down of said pollutants and wastes into smaller components (Yakop et al. 2019). Induction of bioremediation ability has been reported in *Nicotiana tabacum* plants through inoculation of their seeds with endophytes (Mastretta et al. 2009). This inoculation has shown an improvement in the production of plant biomass under the stress of Cadmium (Cd) and it has also been found that tissues of inoculated plants were having higher concentration of Cd as compared to non-inoculated plants. This suggests the important role of inoculation of seeds with endophytes to assimilate and accumulate heavy metals. Various strains of endophytic fungi were allowed to grow in broth and agar media containing polyester polyurethane (PUR) as a carbon source in order to enumerate the role played by these fungi in the itemization of contaminants like plastics (Russell et al. 2011). It has been found that two isolates of *Pestalotiopsis microspora* among endophytic fungi have the capability to utilize PUR both in aerobic and anaerobic situations and also produce enzyme serine hydrolase for the degradation of PUR.

11.4 Stress Tolerance Mechanisms

At least two mechanisms are found to be involved in symbiotically associated stress tolerance: (1) activation of stress response systems in host very shortly after exposure to a stress that leads the plants in a direction of mitigating or avoiding the effects of stress (Schulz et al. 1999); and (2) biosynthesis of some anti-stress biochemicals by endophytes (Miller et al. 2002). Besides anti-stress chemical biosynthesis, over evolutionary time, the sustenance of the plant-fungal mutualisms lies in the fact that the host stress response system is regulated by endophytes as these act as bio-triggers (Rodriguez et al. 2004; Eid et al. 2019) (Figure 11.1).

11.4.1 Osmotic Adjustment

Plant-water relationships get affected due to stress conditions like heat, drought, and salinity which triggers plants to respond in complex ways such as stressed plants produce increased concentration of osmolytes (Wang et al. 2000). Solute potential and matrix potential are the two primary components which determine the osmotic potential in plants and it is believed that fungi, which are symbiotic associates, mainly add to the matrix potential that can enhance drought tolerance in plants by water retention. It has been experimentally demonstrated that when nonsymbiotic panic grass and tomato were exposed to heat stress, these plants produced higher osmolyte concentrations as predicted, and this correlated to the development of successive desiccation and wilting symptoms before the death of plants (Rodriguez et al. 2004). Whereas symbiotic plants produced either lowered (Tiger-Like tomato) or similar (Rutgers tomato and panic grass) concentrations of osmolytes as compared to non-stressed controls (Marquez et al. 2007). Mostly *Epichloe* effect investigations on stress tolerance focus on water relations, drought recovery and osmotic adjustment (Cheplick et al. 2000), rate of photosynthesis under heat or water stress (Amalric et al. 1999), and drought protective osmolyte accumulation in tissues of grass (Richardson et al. 1992). During water stress, it was demonstrated that the tall fescue endophyte showed a momentous increment in the elasticity of cell wall which was measured by ratio of turgid weight to dry weight (TW/DW) and bulk modulus tissue elasticity (White et al. 1992). Regardless of the underlying cause, the general pattern of osmotic concentrations in plants that are tolerant to heat stress differs from plants that succumb due to heat stress (nonsymbiotic), signifying the use of different approaches by symbiotic plants other than mounting concentrations of osmolytes to alleviate the heat stress impacts (Rodriguez et al. 2008).

11.4.2 Osmotic Protection

In tall fescue, the stomatal conductance involved in drought protection is least expected than osmotic protection (Richardson et al. 1993), but in *Neotyphodium* sp-*Festuca arizonica* interactions, it has been found that reduced stomatal conductance may be significant (Morse et al. 2002). It is speculated that the fungal loline alkaloids are centered around by osmoprotectants, which are sufficient in those symbiota for which the endophyte has a known and consistent positive effect on the tolerance of drought (Malinowski and Beleskey 2000). It has been observed that lolines qualify for several prerequisite criteria like highly water soluble, in general enhancing the drought or heat response, and being nonhazardous to plant cells. However, it is not known significantly whether lolines on reaching sufficient levels can affect water osmotic balance. Involvement of these alkaloids can be expected to shield macromolecules from denaturation and/or help in scavenging of reactive oxygen species (ROS) associated with drought stress, possibilities have not yet been tested (Schardl et al. 2004). Other potential protectants and osmoregulators are sugar alcohols and soluble sugars which are formed by plants or endophytes or both (Richardson et al. 1992).

11.4.3 Water Use Efficiency

Regardless of the colonizing endophytes, the consumption of water by nonsymbiotic plants is more significant than symbiotic ones. It has also been enumerated that lesser fluid is used by dunegrass plants, panic grass, tomato, and rice, than nonsymbiotic plants. These symbiotic plants suggested efficient water usage by reduced consumption of water and attainment of enhanced biomass. Increased water use efficiency and decreased water consumption can be presented as a sole mechanism of drought tolerance to symbiotic plants (Rodriguez et al. 2008). After the removal of endophytes from tall fescue grass, a substantial (>50%) stand loss has been reported (Read and Camp 1986). This loss is being characteristically connected to drought periods, and in comparison to endophyte-free tall fescue grass, the endophyte-infected tall fescue has been found to exhibit better recovery from drought stress (Arachevaleta et al. 1989). It has been proposed that in tall fescue, the grass endophytes particularly *Neotyphodium coenophialum* affect plant water relations, nutrient acquisition, as well as assimilation and allocation of photosynthesis (Bacon 1993). Overall, the physiological responses of endophyte-infected grasses appear to be improved vis-a-vis adverse environmental conditions (Schardl et al. 2004). Nevertheless, studies have been carried out to demonstrate that numerous species of grass ascertain complex connections between plant genotype, status of endophyte, nutrient and water availability, and spatial competition (Faeth and Sullivan 2003).

11.4.4 Reactive Oxygen Species

It has been suggested that plants maintain relationships with fungi, which can either be pathogens or symbionts and it was recently reported that both host and fungi produces ROS in many symbiotic and pathogenic systems (Tanaka et al. 2006; Tsai et al. 2020). Therefore, in a symbiotic association or pathogenic interaction, ROS which comes from both interactions as a multipart mixture must be dealt by both the plant and the microbe. ROS are not inevitably dangerous for the partners, they can be helpful in controlling the interaction and facilitate signal (Rouhier et al. 2008). For example, the development of a symbiotic association between the grass *Lolium perenne* and *Epichloë festucae*, there is a requisition for the hydrogen peroxide or superoxide production by a fungal NADPH oxidase for a fungal endophyte, but once this gene gets inactivated, the interaction changes from mutualistic to antagonistic (Tanaka et al. 2006). Whatever the case may be, both microbe and the plant have developed a remarkable array of enzymatic and non-enzymatic antioxidant systems in order to maintain suitable concentrations of ROS within their own cells. In contrast, if a plant is exposed to any stress, it has been seen that nonsymbiotic tissues experienced bleaching and indicated the generation of ROS while the tissues which are symbiotic in nature remained green as such. Hence this suggested that endophytes either directly scavenge ROS or induces plants to do, or they prevent the production of ROS when plant is exposed to abiotic stresses (Ma et al. 2019). It has been presumed that the role of ROS may differ in plant having symbioses with endophytes of class I and class II (Rodriguez et al. 2008). Class I endophyte *E. festucae* seems to produce ROS to maintain mutualisms and limit colonization of host (Tanaka et al. 2006), while class II endophytes retard the production of ROS to probably reduce down the abiotic stress impacts. Further studies may elucidate whether the differences between endophytes of class I and class II are general or an expression of individual isolates (Rodriguez et al. 2008).

11.4.5 Antioxidant Enzymes

A pivotal role is played by antioxidant enzymes in fungal symbiosis, which confer the tolerance against abiotic stress conditions (Rouhier et al. 2008). Antioxidants comprise the compounds of lower molecular weight and include ascorbate, tocopherol, and glutathione and the enzymes superoxide dismutases (SOD), thiol- or ascorbate-dependent peroxidases, catalases, glutathione reductases, monodehydro ascorbate reductases, and dehydro ascorbate reductases (Rouhier et al. 2008). These enzymes play an important role in removing the ROS either directly (catalases, ascorbate, SOD or thiol-dependent peroxidases) or indirectly via regeneration of the two major redox molecules in the cell, i.e., glutathione and ascorbate (dehydro ascorbate reductases, glutathione reductases, and monodehydro ascorbate reductases) (Rouhier et al. 2008). The studies have shown that there is a strong connection between alleviation of antioxidant enzyme and plant tolerance to salt stress (Sekmen et al. 2007). The fungal endophyte induces tolerance to *Nicotiana benthamiana* by elevating the antioxidant enzyme expression (Dastogeer et al. 2019).

11.4.6 ACC Deaminase

The production of 1-aminocyclopropane-1-carboxylate (ACC) deaminase enzyme is common in bacterial endophytes, whereas some of the fungal endophytic strains also possessed genes for ACC deaminase synthesis. Among fungal strains, some species of *Penicillium, Mortierella*, and *Caralluma* have been found to produce ACC (Ozimek et al. 2018; Ali et al. 2019). The endophytic microbial communities living inside the plants and synthesizing ACC deaminase use ethylene and cleave the ethylene precursor ACC to α-ketobutyrate and ammonia. In this way, the ethylene level is lowered in developing or stressed plants. The endophytes which are producing ACC deaminase help in plant growth under stress conditions and also perform an important role in nodulation. This function of ACC deaminase is variable in different plant species (Ali et al. 2012).

11.5 Conclusion

Almost all the species around the globe harbor one or more endophytes which protect the plant from diverse biotic and abiotic stresses and improve their growth. Several researches demonstrated that fungal endophytes increase plant growth and provide plant tolerance from adverse stress conditions thereby increasing host fitness in stress conditions such as extreme temperature (cold/heat), drought, heavy metal pollution, salinity etc. These fungal endophytes can shield hosts from phytopathogens and help in diminution of impacts of toxicity of heavy metals, herbicides, and insecticides. These could be adopted as an alternate solution to support plants in adjustment to varied stress conditions such as salinity, drought, nutrient stress, temperature stress, and heavy metals pollution. It is now commonly held that almost all the indigenous/native plants successfully survive the ecosystems under abiotic stress due to endophytic microbes that have co-evolved with their specific hosts (Barrow et al. 2008). In addition, several studies suggested that endophytic fungi typically have a broad range of hosts and exhibit non-mutualistic behaviors in specific hosts, and can form mutual symbiotic relationships with genetically different plant species and confer tolerance for stress (disease and/or drought). Capitalizing on this property of fungal endophytes may enable us to utilize the host endophytic fungi isolated from abiotically stressed ecosystems to impart desirable traits, such as stress resistance, to genetically distant stress-sensitive plant species. This approach may engender stress tolerance in a wide variety of agricultural crops and native plants and, in effect, may help boost their agricultural and economic value. In order to materialize these objectives, the primary requirement is to evaluate the effectiveness of known endophytic fungi in varied environments and discover the hitherto unknown or unidentified fungal endophytes and their ecological role. However, most of the research on endophytes is confined to greenhouse experiments under controlled conditions. The successful utilization of their potential, therefore, necessitates the need for rigorous field trials and acclimatization experiments in order to evaluate their effectiveness in natural environments. Through evolutionary adjustments, microbes have become almost inalienable components of plant microbiomes, and almost all plants have entered into ecological interactions with microbes of one kind or the other. Therefore, it is imperative to study plant growth and development vis-a-vis these microbial interactions in order to gain more comprehensive and inclusive picture of the phenomena involved therein. The comprehensive knowledge about fungal endosymbionts may help us to enter into new era of sustainable development, relying almost exclusively on the organic routes of agricultural crop production. The unraveling of the intricacies of such plant-microbe interactions and their possible role in improving crop production, qualitatively as well as quantitatively, may be a great leap forward toward attaining Sustainable Development Goals.

REFERENCES

Ali S., Charles, T. C. and Glick, B. R. 2012. Delay of flower senescence by bacterial endophytes expressing 1-aminocyclopropane-1-carboxylate deaminase. *J Appl Microbiol* 113(5):1139–1144.

Ali, S., Khan, S. A., Hamayun, M., Iqbal, A., Khan, A. L., Hussain, A. and Shah, M. 2019. Endophytic fungi from *Caralluma acutangula* can secrete plant growth promoting enzymes. *Fresenius Environ Bull* 28:2688–2696.

Amalric, C., Sallanon, H., Monnet, F., Hitmi, A. and Coudret, A. 1999. Gas exchange and chlorophyll fluorescence in symbiotic and non-symbiotic ryegrass under water stress. *Photosynthetica* 37(1):107–112.

Ancheeva, E., Daletos, G. and Proksch, P. 2020. Bioactive secondary metabolites from endophytic fungi. *Curr Med Chem* 27(11):1836–1854.

Arachevaleta M., Bacon, C. W., Hoveland, C. S. and Radcliffe, D. E. 1989. Effect of the tall fescue endophyte on plant response to environmental stress. *Agron J* 81(1):83–90.

Ayob, F.W. and Simarani, K. 2016. Endophytic filamentous fungi from a *Catharanthus roseus*: identification and its hydrolytic enzymes. *Saudi Pharm J* 24(3):273–278.

Azad, K. and Kaminskyj, S. 2016. A fungal endophyte strategy for mitigating the effect of salt and drought stress on plant growth. *Symbiosis* 68 (1-3):73–78.

Bacon, C. W. 1993. Abiotic stress tolerances (moisture, nutrients) and photosynthesis in endophyte infected tall fescue. *Agric Ecosys Environ* 44 (1-4):123–141.

Barrow, J. R., Lucero, M. E., Reyes-Vera, I. and Havstad, K. M. 2008. Do symbiotic microbes have a role in plant evolution, performance and response to stress? *Commun Integr Biol* 1(1):69–73.

Bischoff, J. F. and White, J. F. 2005. Evolutionary development of the Clavicipitaceae. *Mycol Ser* 23:505.

Bogner, C. W., Kamdem, R. S., Sichtermann, G., Mattheaus, C., Heolscher, D., Popp, J., Proksch, P., Grundler, F. M. and Schouten, A. 2017. Bioactive secondary metabolites with multiple activities from a fungal endophyte. *Microb Biotechnol* 10(1):175–188.

Chandran, V., Shaji, H. and Mathew, L. 2020. Endophytic microbial influence on plant stress responses. In: Ajay Kumar and Radhakrishnan E. K (eds), *Microbial Endophytes*. Woodhead Publishing, Elsevier, pp:161–193.

Chathurdevi, G. and Gowrie, S. U. 2016. Endophytic fungi isolated from medicinal plant—a source of potential bioactive metabolites. *Int J Curr Pharm Res* 8:50–56.

Cheplick, G. P., Perera, A. and Koulouris, K. 2000. Effect of drought on the growth of *Lolium perenne* genotypes with and without fungal endophytes. *Funct Ecol* 14:657–667.

Chhipa, H. and Deshmukh, S. K. 2019. Fungal endophytes: rising tools in sustainable agriculture production. *Endophytes Second Metab* 1–24.

Clay, K. 1988. Fungal endophytes of grasses: a defensive mutualism between plants and fungi. *Ecology* 69:10–16.

Dastogeer, K. M., Tumpa, F. H., Sultana, A., Akter, M. A. and Chakraborty, A. 2020. Plant microbiome—an account of the factors that shape community composition and diversity. *Curr Plant Biol* 23:100161.

Dastogeer, K.M.G., Li, H., Wylie, S. and Jones, M.G.K. 2019. Fungal endophytes mediate plant drought tolerance through alteration of host metabolism, enzymatic activities and gene expression. *Mol Plant-Microbe Interact* 32(10S):S192.

de Carvalho, J.O., Broll, V., Martinelli, A.H.S. and Lopes, F.C. 2020. Endophytic fungi: positive association with plants. In: Sharma Vivek, Salwan Richa and Khalil Laith, Al-Ani Tawfeeq (eds), *Molecular Aspects of Plant Beneficial Microbes in Agriculture.* Academic Press, Elsevier, Cambridge, Massachusetts, pp:321–332.

Eid, A. M., Salim, S. S., Hassan, S. E. D., Ismail, M. A. and Fouda, A. 2019. Role of endophytes in plant health and abiotic stress management. In: Kumar Vivek, Prasad Ram, Kumar Manoj and Choudhary Devendra K. (eds), *Microbiome in Plant Health and Disease.* Springer, Singapore, pp:119–144

El Mansy, S. M., Nouh, F. A. A., Mousa, M. K. and Abdel-Azeem, A. M., 2020. Endophytic fungi: diversity, abundance, and plant growth-promoting attributes. In: Yadav Ajar Nath, Mishra Shashank, Kour Divjot, Yadav Neelam and Kumar Anil (eds), *Agriculturally Important Fungi for Sustainable Agriculture.* Springer, Cham, pp:21–59.

Faeth, S. H. and Sullivan, T. J. 2003. Mutualistic, asexual endophytes in a native grass are usually parasitic. *Am Nat* 161:310–325.

Fouda, A. H., Hassan, S. E. D., Eid, A. M., and Ewais, E. E. D. 2015. Biotechnological applications of fungal endophytes associated with medicinal plant Asclepias sinaica (Bioss.). *Ann Agricul Sci* 60(1), 95–104.

González-Teuber, M., Vilo, C., Guevara-Araya, M. J., Salgado-Luarte, C. and Gianoli, E. 2020. Leaf resistance traits influence endophytic fungi colonization and community composition in a South American temperate rainforest. *J Ecol* 108(3):1019–1029.

Govindasamy, V., George, P., Raina, S. K., Kumar, M., Rane, J. and Annapurna, K. 2018. Plant-associated microbial interactions in the soil environment: role of endophytes in imparting abiotic stress tolerance to crops. In: Kumar Bal Santanu, Mukherjee Joydeep, Choudhury Burhan Uddin, Kumar Dhawan Ashok (eds), *Advances in Crop Environment Interaction.* Springer, Singapore, pp:245–284.

Guerrero-Zúñiga, A. L., López-López, E., Rodríguez-Tovar, A. V. and Rodríguez-Dorantes, A. 2020. Functional diversity of plant endophytes and their role in assisted phytoremediation. In: Bharagava Ram Naresh and Saxena Gaurav (eds), *Bioremediation of Industrial Waste for Environmental Safety,* Volume II. Springer, Singapore, pp:237–255.

Hamilton, C. E. and Bauerle, T. L. 2012. A new currency for mutualism? Fungal endophytes alter antioxidant activity in hosts responding to drought. *Fungal Diversity* 54(1), 39–49.

Heinonsalo, J., Buée, M. and Vaario, L. M. 2017. Root-endophytic fungi cause morphological and functional differences in scots pine roots in contrast to ectomycorrhizal fungi. *Bot* 95(2):203–210.

Hidayat, I., 2019. Dark septate endophytes and their role in enhancing plant resistance to abiotic and biotic stresses. In: Sayyed R. Z., Arora Naveen Kumar and Reddy M. S. (eds), *Plant Growth Promoting Rhizobacteria for Sustainable Stress Management,* Volume I. Springer, Singapore, pp:35–63.

Kaur, T., 2020. Fungal endophyte-host plant interactions: role in sustainable agriculture. In: Hasanuzzaman Mirza, Fujita Masayuki, Carvalho Marcelo, Filho Minhoto Teixeira, Nogueira Thiago Assis Rodrigues, Galindo Fernando Shintate (eds), *Sustainable Crop Production.* In tech Open, London, UK.

Khan, A. L., Hamayun, M., Hussain, J., Kang, S. M. and Lee, I. J. 2012. The newly isolated endophytic fungus *Paraconiothyrium* sp. LK1 produces ascotoxin. *Molecules* 17:1103–1112.

Khare, E., Mishra, J. and Arora, N. K. 2018. Multifaceted interactions between endophytes and plant: developments and prospects. *Front Microbiol* 9:2732.

Khiralla, A., Spina, R., Yagi, S., Mohamed, L. and Laurain-Mattar, D. 2016. Endophytic fungi: occurrence, classification, function and natural products. In Hughes Evelyn (ed), *Endophytic Fungi: Diversity, Characterization and Biocontrol.* Nova publishers, New York, pp:1–19.

Knapp, D. G., Németh, J. B., Barry, K., Hainaut, M., Henrissat, B., Johnson, J., Kuo, A., Lim, J. H. P., Lipzen, A., Nolan, M. and Ohm, R. A. 2018. Comparative genomics provides insights into the lifestyle and reveals functional heterogeneity of dark septate endophytic fungi. *Sci Rep* 8(1):1–13.

Lakshmanan, V., Ray, P. and Craven, K. D. 2017. Toward a resilient, functional microbiome: drought tolerance-alleviating microbes for sustainable agriculture. In: *Plant Stress Tolerance*. Humana Press, New York, pp:69–84.

Leo, V. V., Passari, A. K., Joshi, J. B., Mishra, V. K., Uthandi, S., Ramesh, N., Gupta, V. K., Saikia, R., Sonawane, V. C. and Singh, B. P. 2016. A novel triculture system (CC3) for simultaneous enzyme production and hydrolysis of common grasses through submerged fermentation. *Front Microbiol* 7:447.

Ma, L., Li, X., Wang, L., Li, Y., Bu, N. and Yu, C., 2019. Endophytic infection modulates ROS-scavenging systems and modifies cadmium distribution in rice seedlings exposed to cadmium stress. *Theor Exp Plant Physiol* 31(4):463–474.

Malinowski, C. P. and Beleskey, D. P. 2000. Adaptations of endophyte-infected cool-season grasses to environmental stresses: mechanisms of drought and mineral stress tolerance. *Crop Sci* 40:923–940.

Mane, R. and Vedamurthy, A. 2018. The fungal endophytes: sources and future prospects. *J Med Plants Stud 6*:121–126.

Márquez, L. M., Redman, R. S., Rodriguez, R. J. and Roossinck, M. J. 2007. A virus in a fungus in a plant: three-way symbiosis required for thermal tolerance. *Science* 315:513–515.

Mastretta, C., Taghavi, S., van der Lelie, D., Mengoni, A., Galardi, F., Gonnelli, C., Barac, T., Boulet, J., Weyens, N. and Vangronsveld, J. 2009. Endophytic bacteria from seeds of *Nicotiana tabacum* can reduce cadmium phytotoxicity. *Int J Phytoremediation* 11(3):251–267.

Meena, H., Hnamte, S. and Siddhardha, B., 2019. Secondary metabolites from endophytic fungi: chemical diversity and application. In: Singh Bhim Pratap (ed), *Advances in Endophytic Fungal Research*. Springer, Cham, pp:145–169.

Miller, J. D., Mackenzie, S., Foto, M., Adams, G. W. and Findlay, J. A. 2002. Needles of white spruce inoculated with rugulosin-producing endophytes contain rugulosin reducing spruce budworm growth rate. *Mycol Res* 106:471–479.

Mishra, R., Kushveer, J. S., Revanthbabu, P. and Sarma, V. V. 2019. Endophytic fungi and their enzymatic potential. In: Singh Bhim Pratap (ed), *Advances in Endophytic Fungal Research*. Springer, Cham, pp:283–337.

Morse, L. J., Day, T. A. and Faeth, S. H. 2002. Effect of *Neotyphodium* endophyte infection on growth and leaf gas exchange of Arizona fescue under contrasting water availability regimes. *Environ Exp Bot* 48:257–268.

Nandy, S., Das, T., Tudu, C. K., Pandey, D. K., Dey, A. and Ray, P. 2020. Fungal endophytes: futuristic tool in recent research area of phytoremediation. *S Afr J Bot* 134:285–295.

Nouh, F. A. A., Nahas, H. H. A. and Abdel-Azeem, A. M. 2020. Agriculturally important fungi: plant-microbe association for mutual benefits. In: Yadav Ajar Nath, Mishra Shashank, Kour Divjot, Yadav Neelam and Kumar Anil (eds), *Agriculturally Important Fungi for Sustainable Agriculture*. Springer, Cham, pp:1–20.

Omacini, M., Chaneton, E. J., Ghersa, C. M. and Müller, C. B. 2001. Symbiotic fungal endophytes control insect host-parasite interaction webs. *Nature* 409:78–81.

Ozimek, E., Jaroszuk-Ściseł, J., Bohacz, J., Korniłłowicz-Kowalska, T., Tyśkiewicz, R., Słomka, A., Nowak, A. and Hanaka, A., 2018. Synthesis of indoleacetic acid, gibberellic acid and ACC-deaminase by *Mortierella* strains promote winter wheat seedlings growth under different conditions. *Int J Mol Sci* 19(10):3218.

Phurailatpam, L. and Mishra, S., 2020. Role of plant endophytes in conferring abiotic stress tolerance. In: Hasanuzzaman Mirza (ed), *Plant Ecophysiology and Adaptation under Climate Change: Mechanisms and Perspectives II*. Springer, Singapore, pp: 603–628.

Pietro-Souza, W., de Campos Pereira, F., Mello, I. S., Stachack, F. F. F., Terezo, A. J., da Cunha, C. N., White, J. F., Li, H. and Soares, M. A. 2020. Mercury resistance and bioremediation mediated by endophytic fungi. *Chemosphere* 240:24874.

Rana, K. L., Kour, D., Sheikh, I., Yadav, N., Yadav, A. N., Kumar, V., Singh, B. P., Dhaliwal, H. S. and Saxena, A. K. 2019. Biodiversity of endophytic fungi from diverse niches and their biotechnological applications. In: Singh Bhim Pratap (ed), *Advances in Endophytic Fungal Research*. Springer, Cham, pp:105–144.

Read, J. C. and Camp, B. J. 1986. The effect of the fungal endophyte *Acremonium coenophialum* in tall fescue on animal performance, toxicity and stand maintenance. *Agron J* 78:848–850.

Redman, R. S., Sheehan, K. B., Stout, R. G., Rodriguez, R. J. and Henson, J. M. 2002. Thermotolerance generated by plant/fungal symbiosis. *Science* 298(5598), 1581–1581.

Richardson, M. D., Chapman, G. W., Hoveland, C. S. and Bacon, C. W. 1992. Sugar alcohols in endophyte-infected tall fescue under drought. *Crop Sci* 32:1060–1061.

Richardson, M. D., Hoveland, C. S. and Bacon, C. W. 1993. Photosynthesis and stomatal conductance of symbiotic and nonsymbiotic tall fescue. *Crop Sci* 33:145–149.

Richier, S., Furla, P., Plantivaux, A., Merle, P. L. and Allemand, D. 2005. Symbiosis-induced adaptation to oxidative stress. *J Exp Biol* 208(2):277–285.

Rodriguez, R. J., Henson, J., Van Volkenburgh, E., Hoy, M., Wright, L., Beckwith, F., Kim, Y. O. and Redman, R. S. 2008. Stress tolerance in plants via habitat-adapted symbiosis. *ISME J* 2(4):404–416.

Rodriguez, R. and Redman, R. 2008. More than 400 million years of evolution and some plants still can't make it on their own: plant stress tolerance via fungal symbiosis. *J Exp Bot* 59(5), 1109–1114.

Rodriguez, R. J., Redman, R. S. and Henson, J. M. 2004. The role of fungal symbioses in the adaptation of plants to high stress environments. *Mitigat Adapt Strateg Global Change* 9(3), 261–272.

Rodriguez, R. J., White, J. F., Arnold, A. E. and Redman, R. S. 2009. Fungal endophytes: diversity and functional roles. *New Phytol* 182:314–330.

Rouhier, N., San Koh, C., Gelhaye, E., Corbier, C., Favier, F., Didierjean, C. and Jacquot, J. P. 2008. Redox based anti-oxidant systems in plants: biochemical and structural analyses. *Biochim Biophys Acta* 1780:1249–1260.

Russell, J. R., Huang, J., Anand, P., Kucera, K., Sandoval, A. G., Dantzler, K. W., Hickman, D., Jee, J., Kimovec, F. M., Koppstein, D., Marks, D. H., Mittermiller, P. A., Núñez, S. J., Santiago, M., Townes, M. A., Vishnevetsky, M., Williams, N. E., Vargas, M. P., Boulanger, L. A., Bascom-Slack, C. and Strobel, S. A. 2011. Biodegradation of polyester polyurethane by endophytic fungi. *Appl Environ Microbiol* 77(17):6076–6084.

Rustamova, N., Bozorov, K., Efferth, T. and Yili, A. 2020. Novel secondary metabolites from endophytic fungi: synthesis and biological properties. *Phytochem Rev* 1–24.

Saleem, H., Mohsin, H., Tanvir, R. and Rehman, Y. 2020. Culturable endophytic fungal communities associated with cereal crops and their role in plant growth promotion. In: *Plant Microbiomes for Sustainable Agriculture*. Springer, Cham, pp:53–77.

Schaechter, M. ed., 2011. *Eukaryotic Microbes*. Academic Press, Cambridge, Massachusetts.

Schardl, S. L., Leuchtmann, A. and Spiering, M. J. 2004. Symbioses of grasses with seedborne fungal endophytes. *Annu Rev Plant Biol* 55:315–340.

Schulz, B., Rommert, A. K., Dammann, U., Aust, H. J. and Strack, D. 1999. The endophyte-host interaction: a balanced antagonism? *Mycol Res* 10:1275–1283.

Sekmen, A. H, Turkan, I. and Takio S. 2007. Differential responses of antioxidative enzymes and lipid peroxidation to salt stress in salt-tolerant *Plantago maritima* and salt-sensitive plantago media. *Physiol Plant* 131:399–411.

Shanmugapriya, T. and Balasubramanian, V. 2017. Biological interactions of endophytic fungi and their applications in various fields. *Indian J Public Health Res Dev* 8(3):416–420.

Singh, S., Kumar, V., Dhanjal, D. S., Sidhu, G. K., Datta, S., Kumar, S. and Singh, J. 2020. Endophytic microbes in abiotic stress management. In: Kumar Ajay and Singh Vipin Kumar (eds), *Microbial Endophytes*. Woodhead Publishing, Cambridge, England, pp:91–123.

Stone, J. K., Polishook, J. D. and White, J. R. J. 2004. Endophytic fungi. In: Mueller G., Bills G. F. and Foster M. S. (eds), *Biodiversity of Fungi: Inventory and Monitoring Methods*. Elsevier, Burlington, pp:241–270.

Strong, P. J. and Claus, H. 2011. Laccase: a review of its past and its future in bioremediation. *Crit Rev Environ Sci Technol* 41(4):373.

Sunitha, V. H, Devi, D. N. and Srinivas, C. 2013. Extracellular enzymatic activity of endophytic fungal strains isolated from medicinal plants. *WJAS* 9:1–9.

Tadych, M., Bergen, M. and White, J. F. 2014. *Epichloë* spp. associated with grasses: new insights on life cycles, dissemination and evolution. *Mycologia* 106(2):181–201.

Tanaka, A., Christensen, M. J., Takemoto, D., Park, P. and Scott, B. 2006. Reactive oxygen species play a role in regulating a fungus-perennial ryegrass mutualistic interaction. *Plant Cell* 18:1052–1066.

Tejesvi, M. V. and Pirttilä, A. M. 2018. Endophytic fungi, occurrence, and metabolites. In: Anke Timm and Schüffler Anja (ed), *Physiology and Genetics*. Springer, Cham, Switzerland, pp:213–230.

Traving, S. J, Thygesen, U. H, Riemann, L. and Stedmon, C. A. 2015. A model of extracellular enzymes in free-living microbes: which strategy pays off? *Appl Environ Microbiol* 81:7385–7393.

Tsai, H. J., Shao, K. H., Chan, M. T., Cheng, C. P., Yeh, K. W., Oelmüller, R. and Wang, S. J. 2020. *Piriformospora indica* symbiosis improves water stress tolerance of rice through regulating stomata behavior and ROS scavenging systems. *Plant Signal Behav* 15(2):1722447.

Van Overbeek, L.S. and Saikkonen, K., 2016. Impact of bacterial–fungal interactions on the colonization of the endosphere. *Trends Plant Sci* 21(3):230–242.

Vasundhara, M., Reddy, M. S. and Kumar, A., 2019. Secondary metabolites from endophytic fungi and their biological activities. In: Gupta Vijai Kumar and Pandey Anita (eds), *New and Future Developments in Microbial Biotechnology and Bioengineering*. Elsevier, pp:237–258.

Wang, W. X., Barak, T., Vinocur, B., Shoseyov, O. and Altman, A. 2000. Abiotic resistance and chaperones: possible physiological role of SP1, a stable and stabilizing protein from Populus. In: Vasil Indra K. (ed), *Plant Biotechnology 2000 and Beyond*. Springer, Dordrecht, pp: 439–443.

Wang, Y. and Dai, C. C. 2011. Endophytes: a potential source for biosynthesis transformation and biodegradation. *Ann Microbiol* 61:207–215.

Wani, Z. A, Ashraf, N., Mohiuddin, T. and Riyaz-Ul-Hassan, S. 2015. Plant-endophyte symbiosis, an ecological perspective. *Appl Microbiol Biotechnol* 99:2955–2965.

White, R. H., Engelke, M. C., Morton, S. J., Johnson-Cicalese, J. M. and Ruemmele, B. A. 1992. *Acremonium* endophyte effects on tall fescue drought tolerance. *Crop Sci* 32:1392–1396.

Yakop, F., Taha, H. and Shivanand, P. 2019. Isolation of fungi from various habitats and their possible bioremediation. *Curr Sci* 116(5):733.

Yu, H., Zhang, L., Li, L., Zheng, C., Guo, L., Li, W., Sun, P. and Qin, L. 2010. Recent developments and future prospects of antimicrobial metabolites produced endophytes. *Microbiol Res* 165:437–449.

Zanne, A. E., Abarenkov, K., Afkhami, M. E., Aguilar-Trigueros, C. A., Bates, S., Bhatnagar, J. M., Busby, P. E., Christian, N., Cornwell, W. K., Crowther, T. W. and Flores-Moreno, H. 2020. Fungal functional ecology: bringing a trait-based approach to plant-associated fungi. *Biol Rev* 95(2):409–433.

Zhou, W., Wheeler, T. A., Starr, J. L., Valencia, C. U. and Sword, G. A. 2018. A fungal endophyte defensive symbiosis affects plant-nematode interactions in cotton. *Plant Soil* 422(1-2):251–266.

Zinniel, D. K., Lambrecht, P., Harris, N. B., Feng, Z., Kuczmarski, D. and Higley, P. 2002. Isolation and characterization of endophytic colonizing bacteria from agronomic crops and prairie plants. *Appl Environ Microbiol* 68:2198–2208.

12

Microbes and Their Role in Abiotic Stress Tolerance in Food Crops

Anjali Pathak[1], Mir Sajad Rabani[1], Mahendra K. Gupta[1], Shivani Tripathi[1], and Sadhna Pandey[2]
[1]Jiwaji University, Madhya Pradesh, India
[2]Govt. K.R.G. PG Autonomous College, Madhya Pradesh, India

CONTENTS

12.1 Introduction

Agricultural productivity all over the world is hampered due to changing climatic conditions that lead to various abiotic stresses (Padgham 2009; Grayson 2013). These stress factors have increased in the environment due to anthropogenic and natural factors which ultimately affect the principal crop productivity. Due to adverse changes in climate, extreme factors including prolonged drought, frost damages, high alkalinity and temperature, severe rain and floods etc. are more likely to increase in future. Thus, to combat these stresses different approaches are sought where microbes can play a significant role and provide suitable models for understanding the stress tolerance and mitigation mechanisms in the plants (Grover et al. 2011). At present, only about 3.5% of land area is said to be unaffected by different environmental stresses as per the reports of FAO (2007). Various abiotic stresses that encompass drought, high salinity, acidic conditions, high light intensity, and high/low temperature etc. have been reported by various scientists (Chaves and Oliveira 2004; Agrawal and Grover 2006; Nakashima and Yamaguchi-Shinozaki 2006; Hirel et al. 2007). Globally, the drought stress or water deficit has been reported to affect about 64% of the land area; about 13% of land cover is found to be affected by floods (anoxia), 9% by mineral deficit, 6% by salinity, 72% by acidity and cold (Mittler 2006; Cramer et al. 2011). Riadh et al. (2010) had reported that out of the world's 5.2 billion ha of agricultural dry land area, 3.6 billion ha has been found to be affected by serious problems of salinity, soil erosion, and soil degradation. The estimated loss in crop production has been about US$27.3 billion as 50% of total cultivated land consists of salt affected soils (Qadir et al. 2014). The malleable effect of brininess or saltiness on farming is start ball rolling. There is over 34 million ha of irrigated land that has been continuously affected by the alkalinity (FAO 2012). Therefore, agriculture sectors is said to be one of the most vulnerable sector for the damage due to climatic change. It has been demonstrated in various studies that increase in water stress (reduction in rainy days as well as rise in temperature) has led to decline in the wheat and paddy crop productivity in South Asia. It has also been reported that the average temperature in Indian subcontinent increased by about 0.570°C in past 100 years. Further, the agro-ecological disturbances caused by the ignorance of stresses and agricultural losses (like crop production and reduction in soil health) have not been accurately estimated and studied. It is obvious that such stresses affect larger land area and also the quality and quantity of crop production (Cramer et al. 2011). Different technological experts and communities around the globe are searching for the solutions of plant stress tolerance strategies. Without compromising the adequate development to meet the food requirements under available or limited resources, the scientific community around the world is seriously involved in plant stress management practices. A number of eco-friendly schemes are in operation to improve crop production under various environmental restrictions. Plant microbe integration is one of those techniques; such an application involves nourishing the plants natal protection system by engaging favorable microbes. Analysis based on forecasting plant-microbe interactivity at biochemical, molecular, and physiological levels has shown that involvement of microbes help plants to exhibit positive responses toward abiotic stresses on larger scale (Farrar et al. 2014). Studies have revealed that the microorganisms found in the rhizospheric region of crop plants, degraded soils, alkaline and infertile soils as well as endophytes inhabiting plant tissues are well identified sources of microbe-mediated stress tolerance in crop plants (Kaushik and Subhashini 1985; Barka et al. 2006; Rodriguez et al. 2008; Yang et al. 2009; Singh et al. 2013). Among several beneficial microbes in abiotic stress mitigation, the plant growth-promoting bacteria (PGPB) or rhizobacteria (PGPR); mycorrhizal fungi, actinomycetes, and nitrogen-fixing cynobacteria are the most studied groups (Glick 2012; Azcon et al. 2013; Singh et al. 2013; Grover et al. 2016; Elhindi et al. 2017). Hence, microorganisms can play a pivotal role in the mitigation of different kinds of abiotic stresses, if extraction of their traits for tolerance, genetic diversity with their ubiquity, and role of interaction with food crops would be used in agricultural sectors. Moreover, by influencing the physiochemical characteristics of rhizospheric soil through production of various chemical enzymes like exopolysaccharides and biofilm formation, microbes can act upon high response to abiotic stresses such as chilling injury, alkalinity, prolonged drought, metal toxicity, and high/low temperature. For this, microbes use different mechanisms such as heat shock proteins (HSP) and induction of osmo-protectants (e.g., induced systemic tolerance, IST) (Grover et al. 2011). The ability of some microbes to confer plant stress tolerance may open a new opportunity for mitigating the adverse effects of global climatic change on agricultural production sector. It is well known that plant roots are colonized by several soil-borne microbes mainly bacteria and fungi. These may have significant effects on crop production by inducing crop alteration to abiotic stresses (Grover et al. 2011). Among different soil-borne microbes, arbuscular mycorrhizal fungi (AMF) play a vital role in alleviation of abiotic environmental stresses. They are mainly microscopic and filamentous microorganisms which colonize the cortical tissues by extending their hyphae deep into the roots (Shokri and Maadi, 2009). The symbiotic relationship of AMF with plants helps plants to survive in highly extreme conditions by protecting their roots against various soil-borne pathogens. Further, symbiotic association of AMF plays several other beneficial roles such as improving the rhizosphere as well as soil quality and the photosynthesis process, enhancing the fastidious nutrient uptake and protein synthesis, modifying the microbial population, etc. It also retains the integrity of membrane, prevents nutritional disorders, and governs the reactive oxygen species (ROS) accumulation by increasing the activity of antioxidant enzymes and the changes in gene transcript levels in signaling pathways or stress response (Maya and Matsubara 2013; Talaat and Shawky 2014a; b; 2015; Shabani et al. 2016). There are certain bacteria known as diazotrophs that have the ability to fix atmospheric nitrogen (N_2) for crops in nitrogen-limiting conditions. These N_2-fixing bacteria, which act as free-living bacteria, perform biological nitrogen fixation through symbiotic relationship with legumes by root nodulation. This association enhances the production, growth, and development of crops under various stressful conditions (Gomez-Sagasti and Marino 2015). Also, there are some other beneficial soil microbes—PGPBs—which colonize the rhizospheric or endorhizospheric regions of crops, enhance growth, and increase tolerance against various abiotic stresses. The activity of PGPRs is regulated by the secretion and production of various substances because of the alteration of the microbial community present in the rhizospheric niche (Naik and Patil 2004). These rhizobacteria are also helpful in improving the growth and conferring stress tolerance in certain crops and may serve as sustainable tools for attaining sustainable agriculture (Shrivastava and Kumar 2015; Gouda et al. 2018; Nagargade et al. 2018). There are several ways through which different PGPRs may facilitate the growth and proliferation of crops. These can also act as biocontrol agents and inhibit the growth of various plant pathogens directly or indirectly by secreting some bioactive compounds (Ahemad and Kibret 2014; Gupta and Kaushal 2017). The word "induced systemic tolerance (IST)" is used for PGPR-induced physical and chemical changes in crops. IST is exhibited by certain microorganisms which leads to increase in tolerance to certain abiotic stresses (Belimov and Wenzel 2009; Cohen et al. 2015; Singh and Jha 2016). There are numerous bacteria that belong to different genera such as, *Klebsiella, Microbacterium, Pantoea, Pseudomonas, Azospirillum, Azotobacter, Serratia*, etc. that confer stress tolerance and other beneficial traits to plants to develop efficiently under stressful conditions (Grover et al. 2011; Nadeem et al. 2016). Meanwhile, the use of these microbial bioinoculants in management of different biotic and abiotic stresses is currently gaining significance worldwide. Therefore, the current chapter highlights the role of beneficial soil microbes in plant tolerance against major abiotic stresses, such as salinity, drought, heat, freezing injury, and heavy metal poisoning. In addition, the need and importance of using stress-tolerant microbial bioinoculants in agriculture are discussed and future directions have also been highlighted.

12.2 Plant-Microbe Interaction through Rhizospheric and Root-Associated Soil

Abiotic stresses can be alleviated by the use of microbes in different extreme environmental conditions. The plant-microbe interaction in the rhizospheric or root-associated region found in extreme environmental conditions are helpful against different abiotic stresses. Rhizosphere is defined as an immediate zone in the soil which surrounds the roots of a plant where intense chemical and biological parameters occur in a narrow sleeve around the axes of the root hairs (Garcia et al. 2001). Free-living rhizospheric microbes or mutualistic associations of these with plants (e.g., Eubacteria, Actinobacteria, archaea, viruses, and fungi) (Philippot et al. 2013) can be beneficial for the deterioration of abiotic factors present in the environment. The soil and plant microbiome are found to be a valuable assets for improving food crop yield and these help in sustaining fertility of soil (Gopal and Gupta 2016). The exudates produced from plant roots are primary factors that exhibit the specificity in selecting microbes for root colonization from the bulk of soil population, hence, the rhizospheric microbiome can be used against abiotic stresses (Grayston et al. 1998). Herewith, microorganisms found in plant soil exhibit various metabolic activities which make them capable to cope up with severe abiotic stresses. Studies revealed that various biomolecules, such as trehalose, present in microbes are involved in providing stress tolerance against different environmental factors, stabilizing the cell membrane, and preventing denaturation of proteins in bacteria like *Escherichia coli* thus, protecting it from cold stress. In yeasts these biomolecules provide protection against osmotic stress, heat, and desiccation by acting as ROS scavenger (Hottiger et al. 1987; Hounsa et al. 1998; Benaroudj et al. 2001). Plant-microbe interactions generally facilitate a number of

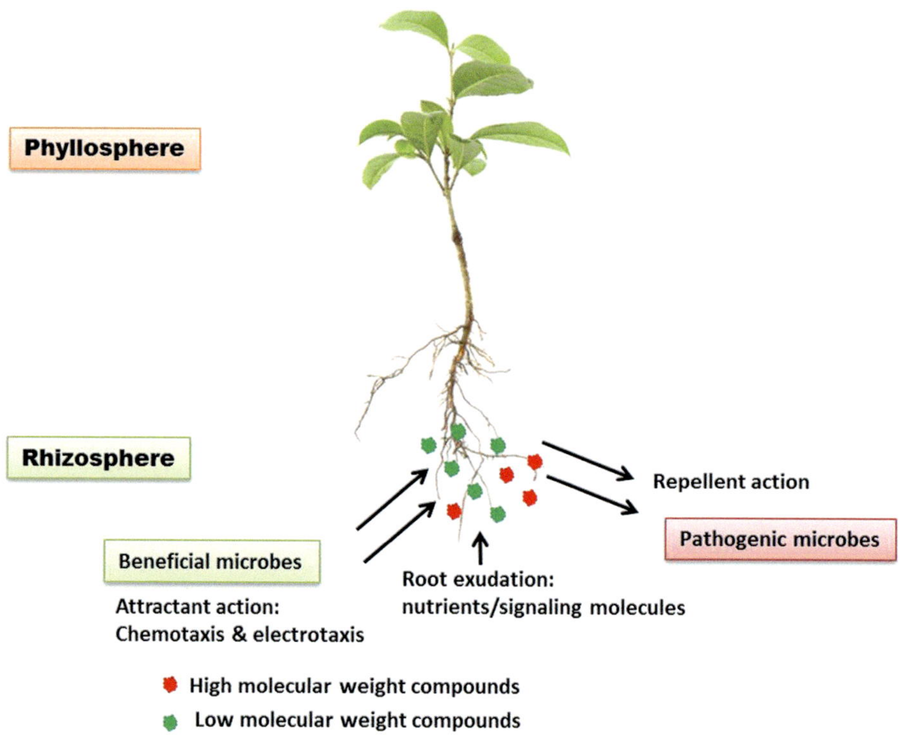

FIGURE 12.1 Plant-microbe interaction in the rhizospheric region of plant soil (Chagas et al. 2018).

transformations in the rhizospheric region. These transformations influence several facets of ecosystem functioning which include, the nutrient cycling particularly, nitrogen and carbon sequestration. Different studies have revealed the plant-microbes interaction with other fauna and flora in the environment; these interactions are generally modulated by abiotic factors (Figure 12.1). The zone of rhizosphere is one of the most active site consisting of several microbes known to facilitate nutrient transformations. The plant-microbe interaction found in the rhizospheric region manifest the soil fertility and plant yield growth and improvement. Several processes take place in the rhizosphere, where microbes interact with the help of different mechanisms like quoram sensing, volatilization, immobilization and mobilization of macro and micronutrients, etc. (Velmourougane et al. 2017). Therefore, the regions associated or surrounded by the roots are found to be microbially active (Hiltner 1904). Moreover, the rhizosphere is the most important niche for plant and microbial interactions in terms of exudation, nutrient uptake, and transformation (White 2003; biogeochemical cycles). For abiotic stress conditions, various omics approaches have been introduced (e.g., genomics, proteomics, transcriptomics, and phenomics) for understanding the mechanisms behind plant-microbe interactions (Luan et al. 2015).

12.3 Effect of Abiotic Stresses on Crops and Mitigation by Microbes

The effects of abiotic stresses in environment can be reduced by actions of the beneficial microbes. Food crops are subjected to different environmental tensions such as abiotic and biotic stress conditions. Abiotic stress conditions have a major impact on crop yield all over the world, particularly reducing the yield of most important food crops. Generally, the abiotic stresses are interlinked as osmotic tensions resulting in ion distribution and homeostasis disturbances in the cell. This is mainly because of the changes found in expression of gene pattern, which ultimately affects the development of crops and its productivity (Kumar 2013). To understand the stress response mechanism in food crops, it is very necessary to identify the genes which are responsible for abiotic stresses, as the microbial activity is influenced by various stress factors. Coordinated research and technology development is needed to select and establish microorganisms in stressed ecosystem. The functional diversity of agriculture and important microbes in a stressed environment has been critically studied by many authors (Grahm 1992; Zahran 1999; Venkateswarlu et al. 2008). It has been reported that the different species like *Bradyrhizobium, Azospirillium, Pseudomonas*, and *Bacillus* are found in Indian desert ecosystems, alkaline and saline areas, and in acidic soils including highly eroded hill slopes (Tilak et al. 2005; Selvakumar et al. 2009). The microorganisms found in these ecosystems are helpful in mitigating various abiotic stresses. Some of the abiotic stresses and their associated microbial attributes are discussed in subsequent text.

12.3.1 Drought Stress

Among abiotic stresses, drought is the major stress factor known and most tremendously growing problem in agricultural practices. Overcrowding is the limiting factor for plant growth, yield, and productivity. Between different extreme conditions in environment, drought stress limits crops proliferation and growth (Shankar et al. 2014). In a study conducted by Lim and Kims (2013), it was found that the growth and productivity of pepper plant in presence of *Bacillus licheniformis* K11 strain has enhanced the crop growth under drought stress. Drought stress tolerance helps plants to proliferate and maintain themselves and as a result of this plant crops combat the drought stressed conditions. The growth arrest is first response of crop plants when they are subjected to drought stress. Under drought stress, shoot growth reduces and osmotic adjustments take place for the metabolic activities, mobilization of compounds, and other cell functions (Kumar 2013). 1-Aminocyclopropane-1carboxylate (ACC) deaminase-producing bacteria have been found to be the best alternative for reducing drought effects on plants. Mayank et al. (2004) worked on ACC deaminase-producing bacteria and revealed that the bacteria facilitated the growth of plants under drought stress and the strain used was obtained from the rhizospheric soil of plant *Lycium shawii* (grown in a dry river bed in the Arava region of the Southern Negev desert in Israel). Previous studies have reported that the efficacy and enzymatic action of ACC deaminase in protecting plants against yield loss induced by drought stress (Barnawal et al. 2013) and also co-inoculation of many other species of rhizobium (*Rhizobium tropici* and *Paenibacillus polymyxa*) resulted in increased plant shoot, height, and nodule formation (Figueiredo et al. 2008).

12.3.2 Heavy Metal Stress

Nowadays, rapid industrialization, irrigation with sewage wastewater, and the usage of chemical fertilizers has increased, ultimately making agricultural soil heavy metal toxic and therefore, causing dangerous effects on plant and soil ecosystem. Cadmium (Cd) is one of the heavy metal which is considered as major environmental concern in agriculture ecosystem. It was described from different studies that this metal can remain in agricultural soil over thousands of years (Kumar 2013). A number of reports demonstrated that heavy metals, such as Cd (Cadmium), Zn (Zinc), Pb (Lead), Cu (Copper), and Cr (Chromium), when present in soil affect the growth, development, and productivity of plants (Chatterjee et al. 2008; Tsonev and Lidon 2012; Vezzaa et al. 2018). Various microorganisms release a metal-chelating compound that influences the uptake of heavy metals like, Zn, Fe, and Cu. It has been reported that when Cd-contaminated soil of barley field was inoculated with the species *Klebsiella mobilis* CIAM 880, it showed high grain productivity with reduction in Cd contents (Barka et al. 2006). These results were obtained due to great potential of bacterial species which binds the free Cd ions into complex forms thus reducing their availability to crop plants (Pishchik et al. 2002).

12.3.3 Salinity

Soil salinity is a growing problem in worldwide agricultural practices. It has been found to be a crucial limiting factor for agricultural food crops especially in hypersaline, arid, or semiarid environments in the world. At least 20% of the world's cultivated land area is affected by salinity thereby reducing the crop yield (Flowers 2004). Different physiological damages of plants are seen due to soil salinity stress in ecosystem. These damages generally include nutrient imbalance, sodium (Na^+) and chloride (Cl^-) toxicity, ethylene production of ROS, and interference with photosynthesis. Other effects are inhibition of seed germination, seedling formation, hybrid vigor, flowering, and fruiting (Sairam and Tyagi 2004). It has been noticed that salinity affects the plants by two different primary factors, i.e., through osmotic stress and ionic toxicity (Kumar 2013). The osmotic pressure of plant cells has been found higher with respect to soil solution under normal conditions as plant cells always use higher osmotic pressure to take up essential mineral, salts, and water from soil solution to their parts of root cells. However, under salt stress conditions the osmotic pressure of soil solution increases than the osmotic pressure of plant cells because of the presence of high salt concentration and consequently, this limits the capacity of crop plants in uptake of minerals (e.g., potassium (K^+) and calcium (Ca^{2+})) and water. At the same time, Na^+ and Cl^- ions move into cells and disrupt metabolic functions in the cytosol and the cell membrane of crop plants (Kumar 2013). It has been demonstrated by the scientists that in the presence of up to 172mM NaCl, the species *Acromobacter piechaudii* ARV8 shows high capability for growth of tomato plant (Mayank et al. 2004). In addition to this, other researchers observed that ACC deaminase-producing bacterial strains can also facilitate the growth of food crops in presence of inhibitory salt levels (Siddikee et al. 2010; Bal et al. 2013).

12.3.4 Flooding Stress and Anoxia

Flooding and anoxia is also considered as an imperative abiotic factor which imposes huge decline in crop productivity. Different studies have revealed that flooding can affect the uptake of soil nutrients as well as it is responsible for the oxygen depletion in the environment. There are numerous known microorganisms present in the rhizospheric region (soil surrounded by roots) of plants that have been found very helpful when inoculated in soil under flooding or anoxia conditions perceived by the environment (Table 12.1). By inoculation, microbes, such as strains of AMF, *Bradyrhizobium*, and the strain of *Glomus intraradices*, made noteworthy contributions to flood tolerance of *P. officinalis* seedling crop through improving their growth and phosphorus (P) acquirement in leaves (Fougnies et al. 2007). Increased ethylene concentration in plants generally induces the expression of necrosis, leaf chlorosis, epinasty (wilting), and reduced biomass production (Li et al. 2013). However, by the use of ACC deaminase-producing bacteria from rhizospheric soil it is easy to limit the high rate of ethylene and also it can be very useful for supporting crop growth development in much adverse conditions (Barnawal et al. 2012). It has been reported that in hypoxic and normoxic (no oxygen limitation) conditions, the cucumber plant roots inoculated with *Pseudomonas putida* UW4 were able to synthesize ACC deaminase (Li et al. 2013).

12.3.5 High Temperature

High temperature is also an important abiotic factor which seriously affects the plant growth and yield, predominantly in crops. In coming days, the increased variation in high temperature may lead to the major reduction in crop plants. Thus, a lot of work needs to be done in the direction of heat stress response processes in crop plants as well as toward the cultivation of heat-alleviating breeds (Kumar 2013). There are many microbes in soil which are considered as phosphorus (P) and potassium (K) solubilizers. These microbes are associated with various rhizospheric plants like epiphytic and endophytic. P-solubilizing microorganisms have potential and can solubilize inorganic phosphate compounds ultimately producing various organic acids (Vyas et al. 2009; Yadav et al. 2015). Several groups of microorganisms are reported which have ability to solubilize K at very high temperatures such as *Bacillus, Delftia, Providencia, Pseudomonas, Rhodobacter, Alcaligenes, and Salmonella* (Verma et al. 2014; 2015; 2016).

12.3.6 Cold

The cold stress phenomenon is a primary abiotic factor that constraints crop productivity by affecting the food quality. For surviving in stress conditions, crops immobilize and modify themselves through their metabolic activities (Kumar 2013). The microorganisms found in the cold regions play a pivotal role in agricultural fields. These cold-adapted microorganisms have great significant applications in biotechnological fields. Microorganisms protect plants in low temperature by secreting anti-freezing

TABLE 12.1

Mitigation of Abiotic Stresses in Different Food Crops by Beneficial Microbes

Inoculants	Stress Type	Beneficial Microbe	Crop Plant	Tolerance Mechanism
Bacteria	Salt	*Pseudomonas fluorescens*	*Arachis hypogaea*	By the synthesis of ACC deaminase
		Azospirillum	*Lactuca sativa*	Promotes higher biomass, ascorbic acid content, antioxidant capacity, and a lower browning intensity
		Burkholderia, Arthrobacter, and Bacillus	*Vitis vinifera, Capsicum annuum*	Increased accumulation of proline
	Drought	*Pseudomonas putida H-2-3*	*Glycine max*	Lowers the level of abscisic acid and salicylic acid and raises the level of jasmonic acid content. Modulates antioxidants by declining superoxide dismutase, flavonoids, and radical scavenging activity
		Bacillus megaterium and Glomus sp.	*Trifolium*	Indole acetic acid and proline production
		Pantoea agglomerans	*Triticum aestivum*	Rhizosphere soil aggregation through exopolysaccharide
	Heavy metals (Cd, Ni, As, Cu, Pb and Zn) Cd and Pb	*Bacllus thuringiensis GDB-1*	*Alnus firma*	Production of phytohormones, i.e., siderophore and ACC deaminase, and solubilization of phosphorus. Increased biomass, chlorophyll content, nodule number, and heavy metal (As, Cu, Pb, Ni, and Zn) accumulation in *Abies firma* seedling
		Thiobacillus thiooxidans and P. putida	*Gladiolus grandiflorus* L.	Promote root length, plant height, dry biomass of the plant and enhanced accumulation of Cd and Pb
	Zn toxicity	*Pseudomonas aeruginosa*	*Triticum aestivum*	Improved biomass, N and P uptake and total soluble protein
	Flood	*P. putida, Enterobacter cloacae*	*Lycopersicon esculentum*	Synthesis of ACC deaminase
	Nutrient deficiency	*Azotobacter, Chroococcum, and Piriformospora indica*	*Triticum aestivum*	Improved mineral and nutrient uptake (exclusively zinc)
Fungi	Coastalsalinity	*Scytonema*	*Oryza sativa*	Gibberellic acid and extra cellular products
	Drought, salinity	*Arbuscular mycorrhizal fungi*	*Sorghum bicolor*	Improved water retention
Virus	Drought	Brome mosaic virus	*Oryza sativa*	-

Adapted from Ilangumaran and Smith 2017.

compounds in freezing circumstances. These low-temperature-resistant microbes could be used in agricultural sectors as bioinoculants and fertilizers for crop plant improvement. Also, it has been found in various studies that these cold-resistant microbes have the ability of solubilization of various nutrients, and these fix atmospheric nitrogen and also secrete growth hormones (e.g., indole-acetic acid [IAA], cytokinin, gibberellin, and iron-chelating substances) mainly in hilly areas (Shukla et al. 2016; Singh et al. 2016). The cold-adapted enzyme-secreting microbes have been isolated from the different regions where temperature was found very low. Some of these microbes include *Arthrobacter, Bacillus, Burkholderia, Brevundimonas, Pseudomonas, Citricoccus, Exiguobacterium, Flavobacterium, Kocuria,*

Microbacterium, Janthinobacterium, Lysinibacillus, Methylobacterium, Providencia, Paenibacillus, and *Serratia* (Mishra et al. 2011; Selvakumar et al. 2013; Verma et al. 2015; Yadav et al. 2014; 2015a; 2015b; 2016; 2017).

12.4 Rhizospheric Microbes and Their Role in Stress Tolerance

Microorganisms are found beneficial to the crop plants through their beneficial associations that occur in rhizospheric region (symbiotic or free-living). These associations are majorly found in region of rhizospheric soil in plants, for example, presence of micro-algae associated bacteria (Gomez et al. 2012). There are various valuable microorganisms that are helpful in the alleviation of abiotic factors; some of them are described in subsequent subsections.

12.4.1 Plant Growth-Promoting Rhizobacteria

Bacteria found in rhizosphere are called as rhizobacteria and these are classified by their role in symbiotic relationship with crop plants but are not relatively involved in soil profile. These are free-living bacteria that are directly linked with the root surface and reside inside the roots of plants, for example, in endophytes (Kloepper and Beauchamp 1992). In natural ecosystem, PGPRs follow different mechanisms for growth and enhancement of food crops. Therefore, certain PGPR strains are known which not only help in abiotic stress tolerance but also have the capability to promote plant growth under high-stress conditions. Rhizobacteria are found around or on the root surface and promote growth of the crops, hence, are also involved directly/indirectly in modulating crop plant hormones. Also, these are helpful in resource acquisition, transformation, and mobilization of different nutrients like nitrogen, phosphorus, and other necessary minerals. Moreover, these rhizobacteria also have inhibitory effects and are involved indirectly in the inhibition of several pathogens so that the plant growth and development could be sustained (Hayat et al. 2010). On the basis of rhizobacterial existence in spaces found between the cells of cortex, rhizosphere or rhizoplane region, they are categorized as extracellular PGPR (ePGPR) (e.g., *Azotobacter, Azospirillum, Bacillus, Burkholderia, Agrobacterium, Caulobacter, Micrococcus, Pseudomonas, Serratia Chromobacterium, Erwinia, Flavobacterium,* etc.; Bhattacharyya and Jha 2012) and the intracellular PGPR (iPGPRs) that exist specifically in nodular forms inside the root cells, these includes *Azorhizobium, Bradyrhizobium, Allorhizobium, Mesorhizobium*, and *Rhizobium* (Figueiredo et al. 2011). Studies demonstrated that the microbial communities found in rhizosphere exhibits a unique plant growth developmental trait due to the existence of actinomycetes. Some of the biocontrol agents are worth to be revealed because they have shown great potential inhibitory effects against several root fungal pathogens such as *Thermobifida* sp., *Micromonospora* sp., *Streptomyces* spp., etc. (Bhattacharyya and Jha 2012). A lot of research work has been carried out by several scientists all over the world for exploring novel rhizobacterial traits to combat different abiotic stress tolerance such as temperature, drought and salinity (Mayank et al. 2004), flooding and anoxia (Li et al. 2013), heavy metal toxicity (Ma et al. 2011). Besides this, several plant growth promoting characteristics, for instance phytohormones (e.g., siderophore, ACC, hydrogen cyanide (HCN)), ammonia production, and nitrogenase activities (Ahemad and Khan 2012; Glick 2012; Jahanian et al. 2012) have been investigated.

12.4.2 Vesicular or Arbuscular Mycorrhizal Fungi

AMF are generally used as biofertilizers, hence, these are energy-saving, cost-effective as well as environment friendly. Mycorrhizae are the root symbionts and are merely the product of relationship between fungi with roots of plant. Different scientists have also revealed that AMFs have positive results on those crops which were found inadequate in transporting nutrients like Mg, N, Ca, Zn, Fe, Cu, K, and P due to exposure to unfavorable and abiotic environmental factors (De Andrade et al. 2008; Talaat and shawky 2011;2013; Bagheri et al. 2012). These mycorrhizae also provide mineral elements like Ca, K, N, P, S, and Zn to host plants and also in return obtain their nutrients from plant. The colonization mechanism of vesicular mycorrhizal fungi (VAM) or AMF improves the water absorption capability in several maize crops when they get exposed to high saline conditions. AMF increases the root hydraulic conductivity and also adjusts the osmotic balance in food crops to cope up with high saline conditions (Sheng et al. 2008). Scientists reported the fossil evidences as well as analyses DNA sequences of this mutualism and it showed that this association had appeared around 400–460 million years ago (Simon et al. 1993; Remy et al. 1994). VAM belong to the class Zygomycetes, order Endogonales, family Endogonaceae (Benjamin 1979). The VAM have far and wide significance in the food sector; hence they can be used to obtain a sustainable agriculture. The fungus performs colonization in the root cortex by forming network of mycelial (i.e., septate or aseptate) and arbuscules (branched finger-like) hyphae. Mycorrhizal fungi play a vital role in improving the growth of host plants by increasing nutrient uptake, producing several growth-promoting substances, and helping in mitigation of abiotic stress (i.e., heavy metal toxicity, salinity, drought, high temperature, and flooding); in addition to this, it shows synergistic cooperation with other useful microbes (Sreenivasa and Bagyaraj 1989). Moreover, the mycorrhizae have great significant value in agriculture field as it improves yield of cereals, vegetables, and fruit crops. It suppresses the action of several pathogens like nematodes and fungal infestations. Using the combination of microbes like AMF can reduce different environmental problems whether it is associated with abiotic stress tolerance or with excessive usage of harmful pesticides. Hence, high prompted research is going on in this area for the reduction and exclusion of abiotic stresses and pathogenicity so that increasing consumer demands could be met for organic and sustainable food products.

12.4.3 Some Other Useful Microbes

Several other microorganism, such as cynobacteria, various soil fungi, algal species, etc., are also found in nature, which show high potential against abiotic environmental stresses. Photosynthetic and nitrogen-fixing cynobacteria (also known as blue green algae, BGA) found as a part of tropical field ecosystem (e.g., paddy) can be used as good alternatives of agrochemicals, with their significant benefits for environment and economy (Blaak et al. 1993). Several soil fungal species are known, for example, *mycorrhizal* and *Trichoderma* spp. These colonize the diverse forms of micro-inhabitants in soil and their purposeful or functional diversity enhances the pathogen levels and therefore, plays an essential role in the crop improvement programs (Smith and Read 2008). Crawford et al. (1993) reported that the cynobacteria (BGA) when associate with the red and brown algae, provide a better approach for the maintenance of food crops; hence, these can be used as a potential biofertilizers in agricultural fields. There are large number of algal biofertilizers used in agricultural fields of paddy that have capability to fix atmospheric nitrogen, for example, BGA, *Nostoc* sp., *Anabaena* sp., *Tolypothrix* sp., etc. Other well-known biofertilizers include *Rhizobium* spp. (known as, symbiotic nitrogen fixers). These species follow the association with leguminous crop plants. Likewise, the actinomycetes help in the growth of rhizosphere. It is stimulated by the root exudates of plants and has been found antagonistic against several fungal pathogens and is also responsible for the growth and synthesis of antimicrobial compounds.

12.5 Mechanisms of Microbes in Mitigation of Abiotic Stress

Microorganisms help crops to cope with abiotic stresses and influence nutrient uptake, disease management, and plant growth promotion. A long list of studies has demonstrated the beneficial functions and the effects of various microorganisms in promotion of crop yield and development under stress conditions. Saxena et al. (2005) reported that the beneficial microbes colonize the rhizospheric/endorhizospheric regions of plants thereby promoting crop growth via direct or indirect mechanisms. Meena et al. (2017) revealed that the cellular processes underlying plant-microbe interactions have been progressively studied at the different levels, such as physiological, biochemical, and molecular, to get the proper understanding of their mutual relationships. Plant growth-enhancing bacterial strains have been isolated from various stress-tolerant wild plant species that are known to serve as booming inoculants for the essential agricultural crops (Coleman-Derr and Tringe 2014). Abiotic stresses are known to cause drastic changes in the food crop production and growth by halting the normal functioning of crop plants. Therefore, growth-promoting microbes are utilized through several means and strategies to combat abiotic stresses by production of phytohormones (such as, gibberellins, auxins, IAA, ethylene, and cytokinins) under stressed environments (Fahad et al. 2015). Hence, several mechanisms are followed by the microorganism to mitigate abiotic stresses in agricultural fields. Some of the mechanisms have been described in following subsections.

12.5.1 Synthesis of Phytohormone and Their Modulation

Beneficial microorganisms colonize around the rhizosphere and thereby promote crop plant growth and development by two different modes, i.e., direct and indirect. PGPRs have capability to produce plant hormones for crop improvement; along with this, the phytohormones produced by them critically take part in abiotic stress mitigation (Goswami et al. 2015; Wani et al. 2016). Studies demonstrated that many phytohormones (i.e., IAA, ethylene, and cytokinins) are produced by the phytostimulators. Moreover, auxins (IAA) stimulate long-term and instant responses in developing crop plantlets (Shi et al. 2009; Kapoor et al. 2012). Phytohormones are responsible for the physiological responses of the crop plants and when a plant senses any stress stimuli, it immediately exhibits an effective response against the stimuli in the form of stress-specific signaling cascades (Chinnusamy et al. 2004; Andreasson and Ellis 2010). Other useful defense pathways are also known such as salicyclic acid, ethylene, or jasmonic acid pathways that trigger the crop plants for mitigation against abiotic stress responses (Zabala et al. 2009). There are several direct or indirect mechanisms of microbes that can mitigate the effects of abiotic stresses in plants (Figure 12.2).

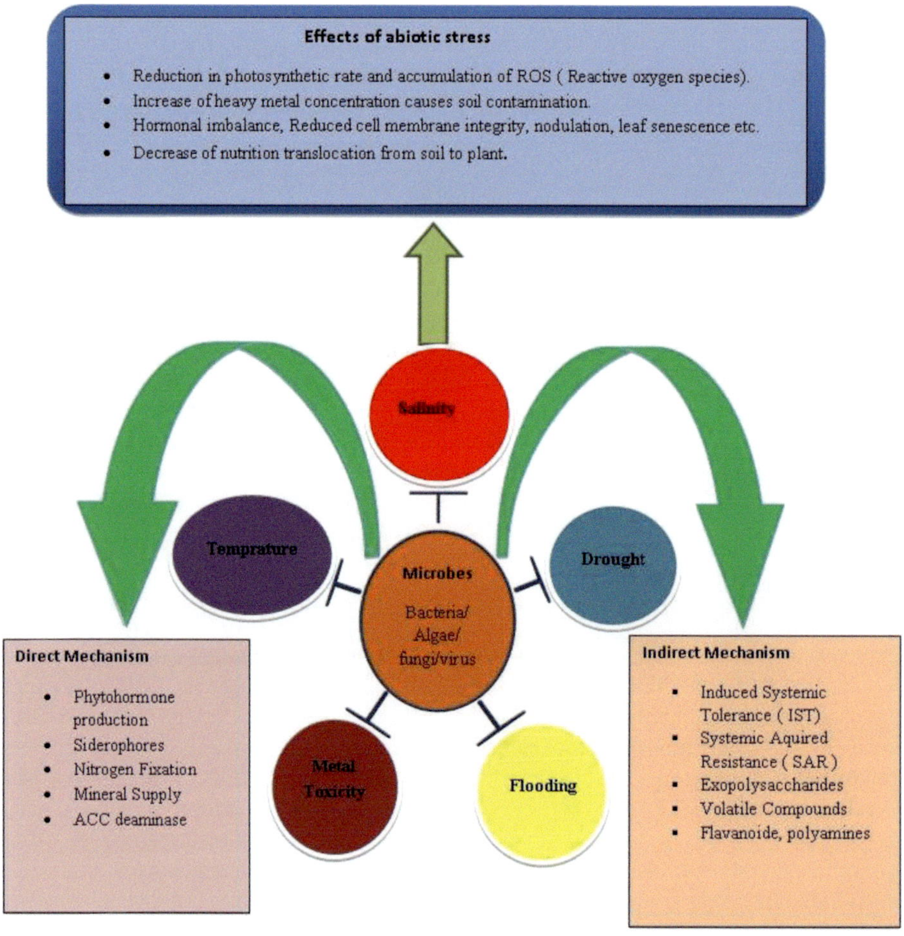

FIGURE 12.2 Schematic diagram showing effects of abiotic stresses thereby microorganisms inhibiting various abiotic stresses by their direct and indirect mechanisms.

12.5.2 Ethylene and ACC Deaminase

Ethylene is a gaseous plant hormone which is endogenously regulated by crop homeostasis thus reducing root and shoot development under high stress conditions. Production and response of ethylene in stressed environment can raise crop plant tolerance or accelerate senescence (Morgan and Drew 1997). Therefore, ethylene regulates the adaptation of crops to stress at the outlay of growth and proliferation. When ethylene level rises under high stress condition, the transcription mechanism of auxin response factors gets halted and it restraints the crop expansion. PGPRs or PGPB are also known to be very good alternatives for modulating ethylene levels in plants with the help of ACC deaminase-synthesizing bacteria (Figure 12.3). PGPRs synthesize ACC deaminase enzyme that restricts the biosynthesis of ethylene hormone in plants. ACC deaminase-producing rhizobacteria sequester and degrade the plant ACC and therefore provide nitrogen and energy to the plants. ACC deaminase enzyme breaks down ACC into ammonia and α-ketobutyrate and these compounds are further metabolized by the bacteria (Holguin and Glick 2001). PGPB supply unique nitrogen sources in the form of ACC to the plants that enables them to proliferate/grow under extreme stress conditions (Hontzeas et al. 2006). The enzyme ACC deaminase acts like a sink as it reduces the level of ethylene within the plants subsequently, as a result of which crop plants that grow in presence of ACC deaminase-containing PGPB generally have longer roots, shoots, and greater biomass with more resistance quality against different physiological abiotic stresses (Glick et al. 2007). Zahir et al. (2008) had also reported that the inoculation of ACC deaminase-producing bacteria can be helpful under drought stress conditions, as it induces the uptake of more water from deep soil and increases the water use efficiency in crop plants.

12.5.3 Auxin

Auxin is also an important plant hormone and its biosynthesis takes place via multiple number of pathways in rhizobacteria. Its biosynthesis includes the utilization of tryptophan found in exudates of plant roots and also its conversion into IAA. IAA is absorbed by the roots of plants. Auxin synthesis also helps in the maintenance of plant growth and proliferation by their signaling cascade (followed by plant's endogeneous IAA pool) (Dodd et al. 2010; Spaepen and Vanderleyden 2011).

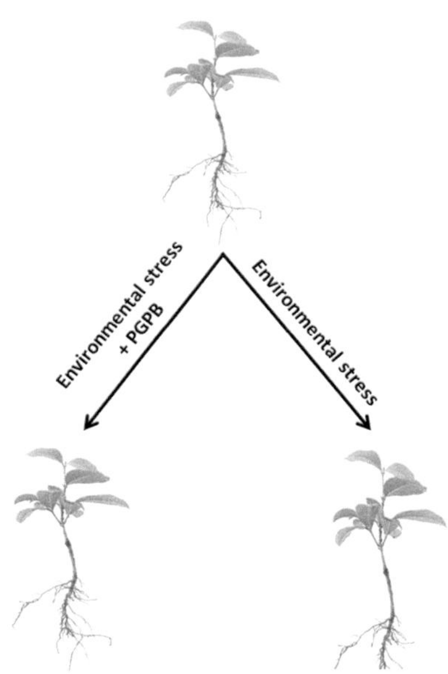

FIGURE 12.3 Shows reduction in the environmental stress and enhancement of crop plant in presence of ACC deaminase plant growth-promoting bacteria (Glick 2014).

12.5.4 Accumulation of Protective Compounds

Several scientists and their co-workers have reported that PGPRs secrete extracellular compounds like proteins, volatiles, polyamines, and hormones, which are found helpful in many signaling cascades and regulatory functionalities of plants leading to positive impact on crop development (Kang et al. 2014; Bhattacharyya et al. 2015; Smith et al. 2015). Accumulation of various compounds such as phenols, flavonoids, antioxidants, osmolytes as well as the initiation of transcription factors (TFs) in the crop plants is influenced by the essential stress-specific genes thereby increasing the suitable defense mechanisms against abiotic stresses (Koussevitzky et al. 2008; Atkinson et al. 2013; Prasch and Sonnewald 2013). Crop plants modulate certain activities in a parallel manner such as accretion of antioxidants, compatible solutes, signaling of chaperones, and transforming molecular processes to survive in heat stress conditions (Wahid et al. 2007).

12.5.4.1 Exopolysaccharides

Among several microbial species, bacteria are the one that take part in the stabilization and protection of crops indirectly or directly through various mechanisms. Many bacterial species are known to secrete exopolysaccharides (EPS) that are important for the attachment along with other bacterial species within the root zone surfaces and their associated soil particles. The secreted EPS bind to soil particles and performs numerous functions, for example, it stabilizes the soil structure and formation, mounts water holding capability of soil thereby improving its cation exchanging capacity (Upadhyay et al. 2011). EPS also play an important role in the protection of crop plants against environmental fluctuations, nutrient retention by forming an enclosed matrix of micro colonies for protecting crops (Balsanelli et al. 2014). EPS are also important for useful nodules formation by legume-rhizobia symbiotic relationship and for mature biofilm formation (Stoodley et al. 2002; Skorupska et al. 2006).

12.5.4.2 Polyamines

Polyamines (PA) are the compounds that are found naturally but also exist as synthetic growth regulators in plants. PAs are the aliphatic amines having low molecular weight with marked antioxidant property. PAs are small and positively charged organic compounds having ubiquitous nature, found in all living species. A large number of PAs (e.g., spermine, spermidine, and putrescine) are important for various plant physiological functions like cell proliferation, growth and development, as well as mitigation of abiotic stress (Gupta et al. 2013).

12.5.4.3 Volatile Compounds

PGPRs are responsible for the production of volatile organic compounds (VOC) that are very well known to stimulate crop plant enhancement. VOCs are also helpful in the modulation of stress responses by resulting in better shoot and root biomass in crop plants (Bailly and Weisskopf 2012).

12.6 Fungi- and Algae-Mediated Stress Response

Different fungal and algal species are known to stimulate plant growth under extreme environments. The AMFs perform symbiotic association with the other roots of advanced plants and play a pivotal role in stress mitigation along with nutrient recycling in crops (Al-Karaki and Al-Raddad 1997). The mycorrhizal fungi are used to enhance the stress tolerance capability in crops under salinity and drought stresses. AMFs also play an essential role in the nutrient uptake of mainly P, which helps in the reduction of other compounds like Na and Cl ions (mainly responsible for the destruction of aerial parts in crops), thus enhancing the uptake of other useful compounds such as, K, Mg, Ca ions for improving ionic balances among K/Na^+ and Ca/Na (Evelin et al. 2009; Daei et al. 2009).

Apart from this, algal species (e.g., cynobacteria) are involved in symbiotic relationship with a variety of host plants, thereby supporting their growth and imitation under stressed conditions (Adams et al. 2013). Cynobacteria, prior well-known as BGA, come under Cynophyceae. It augments the fertility of soil by N_2fixtion (by diazotrophs) and help in the solubilization of phosphate compound as well as releases many other organic compounds/metabolites. Moreover, several studies on cyanobacteria demonstrated their ability to withstand advantageously under different stressful environments viz. drought, salinity, etc. (Asthana et al. 2005; Sen et al. 2017).

12.7 Climate Change: A Major Factor Faced by Plant and Microbes for Crop Improvement

The challenges of abiotic stress on plant growth and proliferation are apparent in the midst of the emerging ecological impact of climate change. The growing world population also restraints crop production due to competition for ecological resources (Wallace et al. 2003; Bellard et al. 2012). However, the climatic change has predicted to affect large-scale agricultural yield mainly at low altitudes of populated developing countries. This may be due to the unpleasant effects of escalating carbon dioxide (CO_2) and high temperature conditions. And this problem of climatic factors in environment is leading researchers to a prompted and challenging investigation toward adaptation strategies (Rosenzweig et al. 2014). The restraints to global food supply have encouraged researchers to develop smart crops that may be resilient to extreme climatic changes (Wheeler and Von Braun 2013). A numerous variety of plant species are exposed to plant abiotic stresses, thus rapid focus on the studies of abiotic factors or stressors found in ecosystem is required (Sulmon et al. 2015). The stressors generally include the high and low level of light, UV radiations (UV-B and UV-A), extreme temperatures, drought and flooding (due to water), chemical substances (includes heavy metals and pH range), salinity (due to excess or deficient amount of Na^+) etc. Among these stressors, the combination of heat and drought is most commonly accompanied in nature and causes a serious effect on plant growth and development (Suzuki et al. 2014). Yoshida et al. (2014) revealed that a major challenge in the abiotic stress biology is to understand how plants experience different stresses and how they get crossed through initial signals, the type of response pathway they produced, and how they are genetically determined. Thus, abiotic stresses will remain a challenge to the natural ecosystem and agriculture. Currently, competing uses of landfarms and worldwide growing population has become a challenge for us to produce more crop in less area with deteriorating resources (e.g., water and landfarms).

12.8 Conclusion and Future Prospects

Climate change has increased the concurrence of various abiotic and biotic stresses which affect the crop production considerably. Different techniques and methods have been adapted on large scale to combat these types of stresses in order to meet the economic and sustainable development. Thus, plant growth-promoting microbes have received a considerable attention of agronomists and environmentalists and it is the rapidly growing research area. The beneficial microbes including PGPB, rhizobia, and AMF may act as candidate organisms for effective, eco-friendly, and sustainable approach to combat these environmental stresses. They have a great potential to manage various agricultural and environmental problems, with the ultimate aim to enhance the crop production and maintain soil structure. The composite and vigorous interactions between microorganisms and plant roots under the various abiotic stresses not only affect the crops but also the physical, chemical, and structural properties of soil. The mitigation of abiotic stresses in crop plants opens a new chapter for the application of microorganisms in the field of agriculture. Selection of microorganisms from stressful environments may contribute significantly in the agriculture sector. Therefore, taking the current available resources and technologies, there is a requirement of intensive research work in the field of agriculture and assessment of more potent microorganisms.

REFERENCES

Adams, D. G., Bergman, B., Nierzwicki-Bauer, S. A., Duggan, P. S., Rai, A. N. and Schussler, A. 2013. Cyanobacterial plant symbioses. In: Rosenberg E., DeLong E. F., Lory S., Stackebrandt E., Thompson F. (eds), *The Prokaryotes springer.* Springer, Heidelberg–Berlin, Germany, pp:359–400.

Agarwal, S. and Grover, A. 2006. Molecular biology, biotechnology and genomics of flooding-associated low O_2 stress response in plants. *Crit Rev Plant Sci* 25:1–21.

Ahemad, M. and Khan, M. S. 2012. Evaluation of plant growth promoting activities of rhizobacterium *Pseudomonas putida* under herbicide stress. *Ann Microbiol* 62(4):1531–1540.

Ahemad, M. and Kibret, M. 2014. Mechanisms and applications of plant growth promoting rhizobacteria: current perspective. *J King Saud Uni-Sci* 26(1):1–20.

Al-Karaki, G. N. and Al-Raddad, A. 1997. Effect of arbuscular mycorrhizal fungi and drought stress on growth and nutrient uptake of two wheat genotypes differing in drought resistance. *Mycorrhiza* 7(2):83–88.

Andreasson, E. and Ellis, B. 2010. Convergence and specificity in the *Arabidopsis* MAPK nexus. *Trends Plant Sci* 15:106–113.

Asthana, R. K., Srivastava, S., Singh, A. P., Kayastha, A. M. and Singh, S. P. 2005. Identification of maltooligosyltrehalose synthase and maltooligosyltrehalose trehalohydrolase enzymes catalysing trehalose biosynthesis in *Anabaena* 7120 exposed to NaCl stress. *J Plant Physiol* 162(9):1030–1037.

Atkinson, N. J., Lilley, C. J. and Urwin, P. E. 2013. Identification of genes involved in the response of *Arabidopsis* to simultaneous biotic and abiotic stresses. *Plant Physiol* 162:2028–2041.

Azcon, R., Medina, A., Aroca, R. and Ruiz-Lozano, J. M. 2013. Abiotic stress remediation by the arbuscular mycorrhizal symbiosis and rhizosphere bacteria/yeast interactions. In: de Brujin F. J. (ed), *Molecular Microbial Ecology of the Rhizosphere.* John Wiley & Sons Inc., New Jersey, pp:991–1002.

Bagheri, V., Shamshiri, M. H., Shirani, H. and Roosta, H. 2012. Nutrient uptake and distribution in mycorrhizalpistachio seedlings under drought stress. *J Agric Sci Technol* 14:1591–1604.

Bailly, A. and Weisskopf, L. 2012. The modulating effect of bacterial volatiles on plant growth: current knowledge and future challenges. *Plant Signal Behav* 7:79–85.

Bal, H. B., Nayak, L., Das, S. and Adhya, T. K. 2013. Isolation of ACC deaminase PGPR from rice rhizosphere and evaluation their plant growth promoting activity under salt stress. *Plant Soil* 366:93–105.

Balsanelli, E., de Baura, V. A., Pedrosa, F. D., de Souza, E. M. and Monteiro, R. A. 2014. Exopolysaccharide biosynthesis enables mature biofilm formation on abiotic surfaces by *Herbaspirillum seropedicae. PLoS One* 9:e110392.

Barka, E. A., Nowak, J. and Clement, C. 2006. Enhancement of chilling resistance of inoculated grapevine plantlets with a plant growth-promoting rhizobacterium, *Burkholderia phytofirmans* strain PsJN. *App Environ Microbiol* 72(11): 7246–7252.

Barnawal, D., Bharti, N., Maji, D., Chanotiya, C. S. and Kaira, A. 2012. 1-aminocyclopropane-1-carboxylic acid (ACC) deaminase-containing rhizobacteria protect *Ocimum sanctum* plants during water logging stress via reduced ethylene generation. *Plant Physiol Biochem* 58:27–35.

Barnawal, D., Maji, D., Bharti, N., Chanotiya, C. S. and Kalra, A. 2013. ACC deaminase-containing *Bacillus subtilis* reduces stress ethylene-induced damage and improves mycorrhizal colonization and rhizobial nodulation in *Trigonella foenum-graecum* under drought stress. *J Plant Growth Regul* 32:809–822.

Belimov, A. A. and Wenzel, W. W. 2009. The role of rhizosphere microorganisms in heavy metal tolerance of higher plants. *Asp Appl Biol* 98:81–90.

Bellard, C., Bertelsmeier, C., Leadley, P., Thuiller, W. and Courchamp, F. 2012. Impacts of climate change on the future of biodiversity. *Ecol Lett* 15:365–377.

Benaroudj, N., Lee, D. H. and Goldberg, A. L. 2001. Trehalose accumulation during cellular stress protects cells and cellular proteins from damage by oxygen radicals. *J Biol Chem* 276:24261–24267.

Benjamin, R. K. 1979. Zygomycetes and their spores. In: Kendrick B. (ed), *The Whole Fungus: The Sexual Asexual Synthesis.* National Museums of Natural Sciences, Ottawa, Canada, pp:573–622.

Bhattacharyya, D., Garladinne, M. and Lee, Y. H. 2015. Volatile indole produced by *Rhizobacterium proteus* vulgaris JBLS202 stimulates growth of *Arabidopsis thaliana* through auxin, cytokinin, and brassinosteroid pathways. *J Plant Growth Regul* 34:158–168.

Bhattacharyya, P. N. and Jha, D. K. 2012. Plant growth-promoting rhizobacteria (PGPR): emergence in agriculture. *World J Microbiol Biotechnol* 28(4):1327–1350.

Blaak, H., Schnellmann, J., Walter, S., Henrissat, B. and Schrempf, H. 1993. Characteristics of an exochitinase from *Streptomyces olivaceoviridis*, its corresponding gene, putative protein domains and relationship to other chitinases. *Eur J Biochem* 214(3):659–669.

Chagas, F. O., Pessotti, R. C., Caraballo-Rodríguez, A. M. and Pupo, M. T. 2018. Chemical signaling involved in plant–microbe interactions. *Chem Soc Rev* 47:1652–1704.

Chaves, M. M. and Oliveira, M. M. 2004. Mechanisms underlying plant resilience to water deficits: prospects for water-saving agriculture. *J Exp Bot* 55:2365–2384.

Chinnusamy, V., Schumaker, K. and Zhu, J. K. 2004. Molecular genetics perspectives on cross-talk and specificity in abiotic stress signalling in plants. *J Exp Bot* 55:225–236.

Cohen, A. C., Bottini, R., Pontin, M., Berli, F. J., Moreno, D., Boccanlandro, H. and Piccoli, P. N. 2015. *Azospirillum brasilense* ameliorates the response of *Arabidopsis thaliana* to drought mainly via enhancement of ABA levels. *Physiol Plant* 153:79–90.

Coleman-Derr, D. and Tringe, S. G. 2014. Building the crops of tomorrow: advantages of symbiont-based approaches to improving abiotic stress tolerance. *Front Microbiol* 5:283.

Cramer, G. R., Urano, K., Delrot, S., Pezzotti, M. and Shinozaki, K. 2011. Effects of abiotic stress on plants: a systems biology perspective. *BMC Plant Biol* 11:163.

Crawford, D. L., Lynch, J. M., Whipps, J. M. and Ousley, M. A. 1993. Isolation and characterization of actinomycete antagonists of a fungal root pathogen. *Appl Environ Microbiol* 59(11):3899–3905.

Daei, G., Ardekani, M. R., Rejali, F., Teimuri, S. and Miransari, M. 2009. Alleviation of salinity stress on wheat yield, yield components, and nutrient uptake using arbuscular mycorrhizal fungi under field conditions. *J Plant Physiol* 166(6):617–625.

De Andrade, S. A., de Silviera, A. P., Jorge, R. A. and de Abreu, M. F. 2008. Cadmium accumulation in sunflower plants influenced by arbuscular mycorrhiza. *Inter J Phytorem* 10:1–13.

Dodd, I. C., Zinovkina, N. Y., Safronova, V. I. and Belimov, A. A. 2010. Rhizobacterial mediation of plant hormone status. *Ann Appl Biol* 157:361–379.

Elhindi, K. M., El-Din, A. S. and Elgorban, A. M. 2017. The impact of arbuscular mycorrhizal fungi in mitigating salt-induced adverse effects in sweet basil (*Ocimum basilicum* L.). *Saudi J Biol Sci* 24(1):170–179.

Evelin, H., Kapoor, R. and Giri, B. 2009. Arbuscular mycorrhizal fungi in alleviation of salt stress: a review. *Ann Bot* 104(7):1263–1280.

Fahad, S., Hussain, S., Bano, A., Saud, S., Hassan, S., Shan, D., Khan, F. A., Khan, F., Chen, Y., Wu, C. and Tabassum, M. A. 2015. Potential role of phytohormones and plant growth-promoting rhizobacteria in abiotic stresses: consequences for changing environment. *Environ Sci Pollut Res* 22(7):4907–4921.

FAO. 2007. Remediation of arsenic for agriculture sustainability, food security and health in Bangladesh. FAO Working paper. Rome, FAO. (Available at: http://www.fao.org/nr/water/docs/FAOWATER_ARSENIC.pdf).

Farrar, K., Bryant, D. and Cope-Selby, N. 2014. Understanding and engineering beneficial plant–microbe interactions: plant growth promotion in energy crops. *Plant Biotechnol J* 12(9):1193–1206.

Figueiredo, M. V. B., Burity, H. A. and Martinez, C. R. et al. 2008. Alleviation of water stress effectsin common bean (*Phaseolus vulgaris* L.) by co-inoculation *Paenibacillus Rhizobium tropici*. *Appl Soil Ecol* 40:182–188.

Figueiredo, M. V. B., Seldin, L., Araujo, F. F. and Mariano, R. L. R. 2011. Plant growth promoting rhizobacteria: fundamentals and applications. In: Maheshwari D. K. (ed), *Plant Growth and Health Promoting Bacteria*. Springer-Verlag, Berlin–Heidelberg, Germany, pp:21–43.

Flowers, T. 2004. Improving crop salt tolerance. *J Exp Bot* 55:307–319.

Fougnies, L., Renciot, S., Mulle, F., Plenchette, C., Prin, Y., de Faria, S. M., Bouvet, J. M., Sylla, S. N., Dreyfus, B. and Ba, A. M. 2007. Arbuscular mycorrhizal colonization and nodulation improve flooding tolerance in *Pterocarpus officinalis* Jacq. seedlings. *Mycorrhiza* 17(3):159–166.

Garcia, J. L., Probanza, A., Ramos, B. and Manero, F. J. 2001. Ecology, genetic diversity and screening strategies of plant growth promoting rhizobacteria. *J Plant Nutr Soil Sci* 164:1–7.

Glick, B. R. 2012. Plant growth-promoting bacteria: mechanisms and applications. *Scientifica* 963401.

Glick, B. R., Cheng, Z., Czarny, J. and Duan, J. 2007. Promotion of plant growth by ACC deaminase-producing soil bacteria. *Eur J Plant Pathol* 119:329–339.

Gomez, C. G., Valero, N. V. and Brigard, R. C. 2012. Halotolerant/alkalophilic bacteria associated with the cyanobacterium *Arthrospira platensis* (nordstedt) gomont that promote early growth in *Sorghum bicolor* (L.). *Agronomía Colombiana* 30(1):111–115.

Gomez-Sagasti, M. and Marino, D. 2015. PGPRs and nitrogen fixing legumes: a perfect team for efficient Cd phytoremediation? *Front Plant Sci* 81:1–9.

Gopal, M. and Gupta, A. 2016. Microbiome selection could spur next-generation plant breeding strategies. *Front Microbiol* 7:1971.

Goswami, D., Thakker, J. N. and Dhandhukia, P. C. 2015. Simultaneous detection and quantification of indole-3-acetic acid (IAA) and indole-3-butyric acid (IBA) produced byrhizobacteria from l-tryptophan (trp) using HPTLC. *J Microbiol Method* 110:7–14.

Gouda, S., Kerry, R. G., Das, G., Paramithiotis, S., Shin, H. S. and Patra, J. K. 2018. Revitalization of plant growth promoting rhizobacteria for sustainable development in agriculture. *Microbiol Res* 206:131–140.

Grahm, P. H. 1992. Stress tolerance in rhizobium, bradyrhizobium and nodulation under adverse soil conditions. *Can J Microbial* 38:475–484.

Grayson, M. 2013. Agriculture and drought. *Nature* 501:S1.

Grayston, S. J., Wang, S. Q., Campbell, C. D. and Edwards, A. C. 1998. Selective influence of plant species on microbial diversity in the rhizosphere. *Soil Biol Biochem* 30:369–378.

Grover, M., Ali, S. Z., Sandhya, V., Rasul, A. and Venkateswarlu, B. 2011. Role of microorganisms in adaptation of agriculture crops to abiotic stresses. *World J Microbiol Biotechnol* 27:1231–1240.

Grover, M., Bodhankar, S., Maheswari, M. and Srinivasarao, C.2016. Actinomycetes as mitigators of climate change and abiotic stress. In: Subramaniam G., Arumugam S. and Rajendran V. (eds), *Growth Promoting Actinobacteria*. Springer, Singapore, pp:203–212.

Gupta, K., Dey, A. and Gupta, B. 2013. Plant polyamines in abiotic stress responses. *Acta Physiol Plant* 35:2015–2036.

Gupta, S. and Kaushal, R. 2017. Plant growth promoting rhizobacteria: bioresouce for enhanced productivity of Solanaceous vegetable crops. *Acta Sci Agric* 1(3):10–15.

Hayat, R., Ali, S., Amara, U., Khalid, R. and Ahmed, I. 2010. Soil beneficial bacteria and their role in plant growth promotion: a review. *Ann Microbiol* 60(4):579–598.

Hiltner, L.1904. Über neue Erfarungen und Problem eauf dem Gebieteder Bodenbakteriologie. *Arbeiten der Deutschen land wirtschaftl Gesellschaft* 98:59–78.

Hirel, B., Gouis, J., Ney, B. and Gallais, A. 2007. The challenge of improving nitrogen use efficiency in crop plants: towards a more central role for genetic variability and quantitative genetics within integrated approaches. *J Exp Bot* 58:2369–2387.

Holguin, G. and Glick, B. R. 2001. Expression of the ACC deaminasegene from *Enterobacter cloacae* UW4 in *Azospirillum brasilense*. *Microb Ecol* 41:281–288.

Hontzeas, N., Hontzeas, C. E. and Glick, B. R. 2006. Reaction mechanisms of the bacterial enzyme 1-aminocyclopropane-1-carboxylate deaminase. *Biotechnol Adv* 24:420–426.

Hottiger, T., Boller, T. and Wiemken, A. 1987. Rapid changes of heat and desiccation tolerance correlated with changes of trehalose content in *Saccharomyces cerevisiae* cells subjected to temperature shifts. *FEBS Lett* 220(1):113–115.

Hounsa, C. G., Brandt, E. V., Thevelein, J., Hohmann, S. and Prior, B. A. 1998. Role of trehalose in survival of *Saccharomyces cerevisiae* under osmotic stress. *Microbiol* 144(3):671–680.

Ilangumaran, G. and Smith, D. L. 2017. Plant growth promoting rhizobacteria in amelioration of salinity stress: a systems biology perspective. *Front Plant Sci* 8:1768.

Jahanian, A., Chaichi, M. R., Rezaei, K., Rezayazdi, K. and Khavazi, K. 2012. The effect of plant growth promoting rhizobacteria (PGPR) on germination and primary growth of artichoke (*Cynara scolymus*). *Inter J Agric Crop Sci* 4:923–929.

Kang, S. M., Radhakrishnan, R., Khan, A. L., Kim, M. J., Park, J. M., Kim, B. R., Shin, D. H. and Lee,I. J. 2014. Gibberellin secreting rhizobacterium, *Pseudomonas putida* H-2-3 modulates the hormonal and stress physiology of soybean to improve the plant growth under saline and drought conditions. *Plant Physiol Biochem* 84:115–124.

Kapoor, R., Kumar, R. A., Kumar, A., Patil, S., Pratush, A. and Kaur, M. 2012. Indole acetic acid production by fluorescent *Pseudomonas* isolated from the rhizospheric soils of malus and pyrus. *Recent Res Sci Technol* 4(1):06–09.

Kaushik, B. D. and Subhashini, D. 1985. Amelioration of salt affected soils with blue green algae: II, improvement soil properties. *Proc Ind Natl Sci Acad Sec B* 51(3):386–389.

Kloepper, J. W. and Beauchamp, C. J. 1992. A review of issues related to measuring colonization of plant roots by bacteria. *Can J Microbiol* 38(12):1219–1232.

Kumar, M. 2013. Crop plants and abiotic stress. *J Biomol Res. Ther* 3:e125.

Li, J., McConkey, B. J., Cheng, Z., Guo, S. and Glick, B. R. 2013. Identification of plant growth-promoting rhizobacteria-responsive proteins in cucumber roots under hypoxic stress using a proteomic approach. *J Proteomics* 84:119–131.

Lim, J. H. and Kim, S. D. 2013. Induction of drought stress resistance by multi-functional PGPR Bacillus licheniformis K11 in pepper. *Plant Pathol J* 29:201–208.

Luan, Y., Cui, J., Zhai, J., Li, J., Han, L. and Meng, J. 2015. High throughput sequencing reveals differential expression of miRNAs in tomato inoculated with *Phytophthora infestans*. *Planta* 241(6):1405–1416.

Ma, Y., Rajkumar, M., Luo,Y. and Freitas, H. 2011. Inoculation of endophytic bacteria on host and non-host plants effects on plant growth and Ni uptake. *J Hazard Matr* 195:230–237.

Maya, M. A. and Matsubara, Y. 2013. Influence of arbuscular mycorrhiza on the growth and antioxidative activity in cyclamen under heat stress. *Mycorrhiza* 23:381–390.

Mayank, S., Tirosh, T. and Glick, B. R. 2004. Plant growth promoting bacteria confer resistance in tomato plants to salt stress. *Plant Physiol Biochem* 42(6):565–572.

Meena, K. K., Sorty, A. M., Bitla, U. M., Choudhary, K., Gupta, P., Pareek, A., Singh, D. P., Prabha, R., Sahu, P. K., Gupta, V. K., Singh, H. B., Krishanani, K. K. and Minhas, P. S. 2017. Abiotic stress responses and microbe-mediated mitigation in plants: the omics strategies. *Front Plant Sci* 8:172.

Mishra, P. K., Bisht, S. C., Ruwari, P., Selvakumar, G., Joshi, G. K., Bisht, J. K., Bhatt, J. C. and Gupta, H. S. 2011. Alleviation of cold stress in inoculated wheat (*Triticum aestivum* L.) seedlings with psychrotolerant Pseudomonas from NW Himalayas. *Arch Microbiol* 193:497–513.

Mittler, R. 2006. Abiotic stress, the field environment and stress combination. *Trends Plant Sci* 11:15–19.

Morgan, P. W. and Drew, M. C. 1997. Ethylene and plant responses to stress. *Physiol Plant* 100:620–630.

Nadeem, S. M., Ahmad, M., Naveed, M., Imran, M., Zahir, Z. A. and Crowley, D. E. 2016. Relationship between in vitro characterization and comparative efficacy of plant growth-promoting rhizobacteria for improving cucumber salt tolerance. *Arch Microbiol* 198:1–9.

Nagargade, M., Tyagi, V. and Singh, M. K. 2018. Plant growth-promoting rhizobacteria: a biological approach toward the production of sustainable agriculture, in role of rhizospheric microbes in soil. In: Meena V. S. (ed), *Stress Management and Agriculture Sustainability*. Springer Singapore, Singapore, pp:205–223.

Naik, N. M. and Patil, A. B. 2004. Effect of growth promoting rhizobacteria on soybean (Glycine max L.) under pot culture. Microbes: Wheels of Organic Farming, 13th Southern Regional Conference on Microbial Inoculants. 33–34.

Nakashima, K. and Yamaguchi-Shinozaki, K. 2006. Regulons involve dinosmotic stress-responsive and cold stress-responsive gene expression in plants. *Physiol Plant* 126:62–71.

Padgham, J. 2009. *Agricultural Development under a Changing Climate: Opportunities and Challenges for Adaptation*. Agriculture and Rural Development & Environmental Departments, The World Bank, Washington, DC.

Philippot, L., Raaijmakers, J. M., Lemanceau, P. and van der Putten, W. H. 2013. Going back to the roots: the microbial ecology of the rhizosphere. *Nat Rev Microbiol* 11:789–799.

Pishchik, V. N., Vorobyev, N. I., Chernyaeva, I. I., Timofeeva, S. V., Kozhemyakov, A. P., Alexeev, Y. V. and Lukin, S. M. 2002. Experimental and mathematical simulation of plant growth promoting rhizobacteria and plant interaction under cadmium stress. *Plant Soil* 243(2):173–186.

Prasch, C. M. and Sonnewald, U. 2013. Simultaneous application of heat, drought, and virus to *Arabidopsis* plants reveals significant shifts in signaling networks. *Plant Physiol* 162:1849–1866.

Qadir, M., Quillérou, E., Nangia, V., Murtaza, G., Singh, M., Thomas, R. J., Drechsel, P. and Noble, A. D. 2014. Economics of salt-induced land degradation and restoration. *Nat Resour Forum* 38:282–295.

Remy, W., Taylor, T. N., Hass, H. and Kerp, H. 1994. Four hundred-million-year-old vesicular arbuscular mycorrhizae. *Proc Natl Acad Sci U S A* 91(25):11841–11843.

Riadh, K., Wided, M., Hans-Werner, K. and Chedly, A. 2010. Responses of halophytes to environmental stresses with special emphasis to salinity. *Adv Bot Res* 53:117–145.

Rodriguez, R. J., Henson, J., Van Volkenburgh, E., Hoy, M., Wright, L., Beckwith, F., Kim, Y. O. and Redman, R. S. 2008. Stress tolerance in plants via habitat-adapted symbiosis. *ISME J* 2(4):404–416.

Rosenzweig, C., Elliott, J., Deryng, D., Ruane, A. C., Müller, C., Arneth, A., Boote, K. J., Folberth, C., Glotter, M., Khabarov, N., Neumann, K., Piontek, F., Pugh, T. A. M., Schmid, E., Stehfest, E., Yang, H. and Jones, J. W. 2014. Assessing agricultural risks of climate change in the 21st century in a global gridded crop model intercomparison. *Proc Natl Acad Sci USA* 111:3268–3273.

Sairam, R. K. and Tyagi, A. 2004. Physiology and molecular biology of salinity stress tolerance in plants. *Curr Sci* 86:407–421.

Saxena, A. K., Shende, L.R. and Pandey, A. K. 2005. Culturing of plant growth promoting rhizobacteria. In: Gopi K. P. and Varma A. (eds), *Basic Research Applications of Mycorrhizae*. I K International Pvt Ltd, New Delhi, pp:453–474.

Selvakumar, G., Joshi, P., Mishra, P. K., Bisht, J. P. and Gupta, H. S. 2009. Mountain aspects influence the genetic clustering of psychrotolerant phosphate solubilizing Pseudomonas in the Uttarkhand Himalayas. *Curr Microbiol* 59:432–438.

Selvakumar, G., Joshi, P., Suyal, P., Mishra, P. K., Joshi, G. K., Venugopalan, R., Bisht, J. K., Bhatt, J. C. and Gupta, H. S. 2013. Rock phosphate solubilization by psychrotolerant *Pseudomonas* spp. and their effect on lentil growth and nutrient uptake under polyhouse conditions. *Ann Microbiol* 63:1353–1362.

Sen, S., Rai, S., Yadav, S., Agrawal, C., Rai, R., Chatterjee, A. and Rai, L. C. 2017. Dehydration and rehydration—induced temporal changes in cytosolic and membrane proteome of the nitrogen fixing cyanobacterium *Anabaena* sp. PCC 7120. *Algal Res* 27: 244–258.

Shabani, L., Sabzalian, M. R. and Pour, S. M. 2016. Arbuscular mycorrhiza affects nickel translocation and expression of ABC transporter and metallothionein genes in *Festuca arundinacea. Mycorrhiza* 26:67–76.

Shankar, A. K., Maheswari, M., Yadav, S., Desai, S., Bhanu, D., Attal.,N. B. and Venkateswarlu, B. 2014. Drought stress responses in crops. *Funtion Integ Genom* 14:11–22.

Sheng, M., Tang, M., Chen, H., Yang, B., Zhang, F. and Huang, Y. 2008. Influence of arbuscular mycorrhizae on photosynthesis and water status of maize plants under salt stress. *Mycorrhiza* 18:287–296.

Shi, Y., Lou, K. and Li, C. 2009. Promotion of plant growth by phytohormone-producing endophytic microbes of sugar beet. *Biol Fertil Soils* 45:645–653.

Shokri, S. and Maadi, B. 2009. Effects of arbuscular mycorrhizal fungus on the mineral nutrition and yield of *Trifolium alexandrium* plants under salinity stress. *J Agron* 8:79–83.

Shrivastava, P. and Kumar, R. 2015. Soil salinity: a serious environment issue and plant growth promoting bacteria as one of the tools for its alleviation. *Saudi J Biol Sci* 22:123–131.

Shukla, L., Suman, A., Yadav, A. N., Verma, P. and Saxena, A. K. 2016. Syntrophic microbial system for exsitude gradation of paddy straw at low temperature under controlled and natural environment. *J Appl Biol Biotechnol* 4:30–37.

Siddikee, M. A., Chauhan, P. S., Anandham, R., Han, G. H. and Sa, T. 2010. Isolation characterization and use of plant growth promotion under salt stress of ACC deaminase-producing halotolerant bacteria derived from coastal soil. *J Microbiol Biotechnol* 20:1577–1584.

Simon, L., Bousquet, J., Levesque, R. C. and Lalonde, M. 1993. Origin and diversification of endomycorrhizal fungi and coincidence with vascular land plants. *Nature* 363(6424):67–69.

Singh, P., Tiwari, A., Singh, S. P. and Asthana, R. K. 2013. Desiccation induced changes in osmolytes production and the antioxidative defence in the cyanobacterium *Anabaena* sp. PCC 7120. *Physiol Mol Biol Plants: Intl J Funct Plant Biol* 19(1): 61–68.

Singh, R. N., Gaba, S., Yadav, A. N., Gaur, P., Gulati, S., Kaushik, R. and Saxena, A. K. 2016. First high quality draft genome sequence of a plant growth promoting and cold active enzymes producing psychrotrophic *Arthrobacter agilis* strain L77. *Stand Genom Sci* 11:54.

Singh, R. P. and Jha, P. N. 2016. Mitigation of salt stress in wheat plant (*Triticum aestivum*) by ACC deaminase bacterium *Enterobacter* sp. SBP-6 isolated from sorghum bicolor. *Acta Physiol Plant* 38:110.

Skorupska, A., Janczarek, M., Marczak, M., Mazur, A. and Krol, J. 2006. Rhizobial exopolysaccharides: genetic control and symbiotic functions. *Microb Cell Fact* 5:7.

Smith, D. L., Subramanian, S., Lamont, J. R. and Bywater-Ekegard, M. 2015. Signaling in the phytomicrobiome: breadth and potential. *Front Plant Sci* 6:709.

Smith, S. E. and Read, D. J. 2008. *Mycorrhizal Symbiosis*. In: Smith S. E. and Read D. J. (eds). Academic Press, London, pp:1–800.

Spaepen, S. and Vanderleyden, J. 2011. Auxin and plant-microbe interactions. *Cold Spring Harb Perspect Biol* 3:a001438.

Sreenivasa, M. N. and Bagyaraj, D. J. 1989. Use of pesticides for mass production of vesicular-arbuscular mycorrhizal inoculum. *Plant Soil* 119(1):127–132.

Stoodley, P., Sauer, K., Davies, D. G. and Costerton, J. W. 2002. Biofilms as complex differentiated communities. *Annu Rev Microbiol* 56:187–209.

Sulmon, C., VanBaaren, J., Cabello-Hurtado, F., Gouesbet, G., Hennion, F., Mony, C., Renault, D., Amrani, M., Wiegand, A. E. and Gérard, C. 2015. Abiotic stressors and stress responses: what commonalities appear between species across biological organization levels? *Environ Pollut* 202:66–77.

Suzuki,N., Rivero, R. M., Shulaev, V., Blumwald, E. and Mittler, R. 2014. Abiotic and biotic stress combinations. *New Phytologist* 203:32–43.

Talaat, N. B. and Shawky, B. T. 2011. Influence of arbuscular mycorrhizae on yield, nutrients, organicsolutes, and antioxidant enzymes of two wheat cultivars under salt stress. *J Plant Nutr Soil Sci* 174:283–291.

Talaat, N. B. and Shawky, B. T. 2013. Modulation of nutrient acquisition and polyamine pool in salt-stressed wheat (*Triticum aestivum* L.) plants inoculated with arbuscular mycorrhizal fungi. *Acta Physiol Plant* 35:2601–2610.

Talaat, N. B. and Shawky, B. T. 2014a. Protective effects of arbuscular mycorrhizal fungi on wheat (*Triticum aestivum* L.) plants exposed to salinity. *Environ Exp Bot* 98:20–31.

Talaat, N. B. and Shawky, B. T. 2014b. Modulation of the ROS-scavenging system in salt-stressed wheat plants inoculated with arbuscular mycorrhizal fungi. *J Plant Nutr Soil Sci* 177:199–207.

Talaat, N. B. and Shawky, B. T. 2015. Plant-microbe interaction and salt stress tolerance in plants. In: Wani S. H. and Hossain M. A. (eds), *Managing Salt Tolerance in Plants: Molecular and Genomic Perspectives*. CRC Press/Taylor & Francis Group, Boca Raton, pp:267–289.

Tilak, K. V. B. R., Ranganayaki, N., Pal, K. K., De, R., Saxena, A. K., Nautiyal, C. S., Mittal, S., Tripathi, A. K. and Johri, B. N. 2005. Diversity of plant growth and soil health supporting bacteria. *Curr Sci* 89(1):135–150.

Tsonev, T. and Lidon, F. J. C. 2012. Zinc in plants—an overview. *Emir J Food Agric* 24(4):322–333.

Upadhyay, S. K., Singh, J. S. and Singh, D. P. 2011. Exopolysaccharide-producing plant growth-promoting rhizobacteria under salinity condition. *Pedosphere* 21:214–222.

Velmourougane, K., Saxena, G. and Prasanna, R. 2017. Plant-microbe interactions in the rhizosphere: mechanisms and their ecological benefits.In: *Plant-Microbe Interactions in Agro-Ecological Perspectives: Microbial Interactions and Agro-Ecological Impacts*. Springer, Singapore, pp:193–214.

Venkateswarlu, B., Desai, S. and Prasad, Y. G. 2008. Agriculturally important microorganisms for stressed ecosystems: challenges in technology development and application. In: Khachatourians G. G., Arora D. K., Rajendran T. P. and Srivastava A. K. (eds), *Agriculturally Important Microorganisms*, vol. 1. Academic World, Bhopal, pp:225–246.

Verma, P., Yadav, A. N., Kazy, S. K., Saxena, A. K. and Suman, A. 2014. Evaluating the diversity and phylogeny of plant growth promoting bacteria associated with wheat (*Triticum aestivum*) growing in central zone of India. *Int J Curr Microbiol Appl Sci* 3:432–447.

Verma, P., Yadav, A. N., Khannam, K. S., Panjiar, N., Kumar, S., Saxena, A. K. and Suman, A. 2015. Assessment of genetic diversity and plant growth promoting attributes of psychrotolerant bacteria allied with wheat (*Triticum aestivum*) from the northern hills zone of India. *Ann Microbiol* 65:1885–1899.

Verma, P., Yadav, A. N., Khannam, K. S., Kumar, S., Saxena, A. K. and Suman, A.2016. Molecular diversity and multifarious plant growth promoting attributes of bacilli associated with wheat (*Triticum aestivum* L.) rhizosphere from six diverse agro-ecological zones of India. *J Basic Microbiol* 56:44–58.

Vezzaa, M. E., Llanesb, A., Travagliab, C., Agostinia, E. and Talanoa, M. A. 2018. Arsenic stress effects on root water absorption in soybean plants: physiological and morphological aspects. *Plant Physiol Biochem* 123:8–17.

Vyas, P., Rahi, P. and Gulati, A. 2009. Stress tolerance and genetic variability of phosphate-solubilizing fluorescent *Pseudomonas* from the cold deserts of the trans-Himalayas. *Microb Ecol* 58:425–434.

Wahid, A., Gelani, S., Ashraf, M. and Foolad, M. R. 2007. Heat tolerance in plants: an overview. *Environ Exp Bot* 61:199–223.

Wallace, J. S., Acreman, M. C. and Sullivan, C. A. 2003. The sharing of water between society and ecosystems: from conflict to catchment-based co-management. *Philos Trans R Soc Lond B Biol Sci* 358:2011–2026.

Wani, S. H., Kumar, V., Shriram, V. and Sah, S. K. 2016. Phytohormones and their metabolic engineering for abiotic stress tolerance in crop plants. *Crop J* 4:62–176.

Wheeler, T. and Von Braun, J. 2013. Climate change impacts on global food security. *Science* 341:508–513.

White, P. J. 2003. Ion transport. In: Thomas B., Murphy D. J. and Murray D. J. (eds), *Encyclopedia of Applied Plant Sciences*. Academic press, London, pp:625–634.

Yadav, A. N., Sachan, S. G., Kaushik, R. and Saxena, A. K. 2014. Cold tolerant microbes from Indian Himalayas: diversity and potential applications in biotechnology and agriculture. In: Proceeding of 84th annual session of NASI and symposium on "Desert Science-Opportunities and Challenges" p:17.

Yadav, A. N., Sharma, D., Gulati, S., Singh, S., Dey, R., Pal, K. K., ... and Saxena, A. K. 2015. Haloarchaea endowed with phosphorus solubilization attribute implicated in phosphorus cycle. *Scie Rep* 5(1), 1–10.

Yadav, A. N., Sachan, S. G., Verma, P. and Saxena, A. K. 2015a. Prospecting cold deserts of north western Himalayas for microbial diversity and plant growth promoting attributes. *J Biosci Bioeng* 119:683–693.

Yadav, A. N., Verma, P., Sachan, S. G., Kaushik, R. and Saxena, A. K. 2015b. Microbes mediated alleviation of cold stress for growth and yield of wheat (*Triticum aestivum* L.). In: Proceeding of international conference on "Low Temperature Science and Biotechnological Advances." doi:10.13140/RG.2.1.2374.2883.

Yadav, A. N., Verma, P., Sachan, S. G., Kaushik, R. and Saxena, A. K. 2016. Microbiome of Indian Himalayan regions: molecular diversity, phylogenetic profiling and biotechnological applications. In: Proceeding of 86th annual session of NASI & the symposium on "Science, Technology and Entrepreneurship for Human Welfare in The Himalayan Region" p:58.

Yadav, A. N., Verma, P., Kumar, V., Sachan, S. G. and Saxena, A. K. 2017. Extreme cold environments: a suitable niche for selection of novel psychrotrophic microbes for biotechnological applications. *Adv Biotechnol Microbiol* 2:1–4.

Yang, J., Kloepper, J. W. and Ryu, C. M. 2009. Rhizosphere bacteria help plants tolerate abiotic stress. *Trends Plant Sci* 14(1):1–4.

Yoshida, T., Mogami, J. and Yamaguchi-Shinozaki, K. 2014. ABA-dependent and ABA-independent signaling in response to osmotic stress in plants. *Curr Opin Plant Biol* 21:133–139.

Zabala, M. D., Bennett, M. H., Truman, W. H. and Grant, M. R. 2009. Antagonism between salicylic and abscisic acid reflects early host-pathogen conflict and moulds plant defence responses. *Plant J* 59:375–386.

Zahir, Z. A., Munir, A., Asghar, H. N., Arshad, M. and Shaharoona, B. 2008.Effectiveness of rhizobacteria containing ACC-deaminase for growth promotion of pea (*Pisum sativum*) under drought conditions. *J Microbiol Biotechnol* 18(5):958–963.

Zahran, H. H. 1999. Rhizobium-legume symbiosis and nitrogen fixation under severe conditions and in an arid climate. *Microbiol Molec Biol Rev* 3(4):968–989.

13

Plant Growth Promoting Rhizobacteria (PGPR): Eco-Friendly Approach for Sustainable Agriculture

Bushra Jan[1], Sheikh Sajad[1], Zafar A. Reshi[1], and F. A. Mohiddin[2]
[1]University of Kashmir, Jammu and Kashmir, India
[2]Sher-e-Kashmir University of Agricultural Sciences and Technology, Jammu and Kashmir, India

CONTENTS

13.1 Introduction

The region of the soil in close contact with roots of plants is called rhizosphere; it is rather a dynamic ecosystem inhabited by remarkable diversity of microorganisms, including bacteria and fungi, and is characterized by an array of complex interactions taking place between the microorganisms and the roots of the plant. Most of the bacterial species inhabiting rhizosphere have been reported to exhibit a stimulatory effect on the plant development via a plethora of mechanisms and are collectively called plant growth promoting rhizobacteria (PGPR) (Majeed et al. 2018). A wide variety of taxa such as *Proteobacteria*, *Cyanobacteria*, *Firmicutes actinobacteria*, and *Bacteroidetes* inhabit plant rhizosphere and have been shown to possess the characteristics typical of PGPR (Tilak et al. 2005).

The PGPR comprise both symbiotic bacteria living in a mutualistic relationship with the plant in specialized structures or nodules present on the roots of the host species and saprophytes living freely in the rhizosphere, often found near, on, or inside the roots of the host plant. The former group of rhizobacteria includes mainly *Rhizobium* and other genera, while the latter belong to different genera and species, notably *Agrobacterium radiobacter, Azospirillum* spp., *Bacillus* spp., and *Pseudomonas* spp. (Benaissa 2019). The plant-rhizobacterial interactions are mediated by the root exudates of the host plant which usually act as signaling molecules to chemotactically draw and adhere rhizospheric bacteria to rhizosphere (Kour et al. 2019). The diversity and abundance of rhizospheric microbiome communities are largely determined, among many other factors, by host genotype, host species, type of soil, etc. An interplay of myriad of abiotic and biotic factors influences rhizosphere ecology in natural and agricultural soils, which in turn depends on the nature and role of PGPR in plant growth and development. PGPR are characteristically known for their growth-promoting effects on the host plant and as a result have emerged as an important and promising tool in the realization of sustainable agricultural and economic development (Kumar et al. 2020). From a functional perspective, Somers et al. (2004) have categorized PGPR as biofertilizers that promote plant growth, directly or indirectly, via a number of activities including enhanced production of plant growth hormones and siderophores (Dilnashin et al. 2020), solubilization of phosphorus and nutrient transformation (Khan et al. 2019), reduction in heavy metal toxicity (Ojuederie and Babalola 2017; Khanna et al. 2019), biological nitrogen fixation (Gouda et al. 2018), and elicitation of the production of secondary metabolites that are antagonistic to phytopathogens (Backer et al. 2018). Due to the positive effects of PGPR on plant growth promotion as well as on their eco-friendly nature, the scientists have lately recognized the potential role of PGPRs as substitute to chemical fertilizers. The PGPRs are now gaining momentum for their use as biofertilizers/bio-inoculants in sustainable agriculture in place of ecologically hostile chemical fertilizers that are infamous for their harmful repercussions such as loss of soil fertility, low yield index, and decline in diversity of indigenous microbes (Kudoyarova et al. 2019). PGPR are now widely employed in such diverse fields as agriculture, horticulture, silviculture, and environmental cleanup activities. In view of the role of PGPR, especially in relation to sustainable development across diverse sectors, advanced innovative technologies are now employed to gather comprehensive information about the nature and activity of PGPR. For instance, rhizo-engineering is used to change the microbiome of rhizosphere aiming to decipher the structural, functional, and ecological behavior of PGPR populations (Ahmad and Zaib 2020). Albeit there are enormous literary evidence for the function of PGPR in plant growth promotion, the full realization of their potential is limited by a number of challenges, especially the inadequacy of the information and wide gaps in the realization of their potential in controlled conditions (greenhouse) and the field. One of the proposed ways to bridge this gap is the adoption of new biotechnological methods along with the use of nano-encapsulation and micro-encapsulation approaches. Some of the novel strategies need to be adopted to harness the potential of PGPR as a technique for disease management and enhancement of crop production system. This chapter is structured into four sections. In the first section, the role of PGPR in sustainable agriculture highlighting various mechanisms involved in PGPR-mediated plant growth promotion, such as biological nitrogen fixation, phosphate solubilization, and phytohormone (auxin, cytokinins, gibberellin, and ethylene) production, has been presented along with a review of our current understanding of their impact on the plant growth promotion. The second section brings forth recent findings on the role and application of PGPR as biocontrol agents, with a focus on modes of action vis-à-vis production of antibiotics, hydrogen cyanide (HCN), and other hydrolytic enzymes as well as induction of systemic resistance through various induced systemic resistance (ISR) and systemic acquired resistance (SAR) pathways. The third section highlights the role of PGPR in the induction of stress tolerance in plants to a variety of stress factors, including drought, salinity, and temperature. In the fourth section, a critical analysis of the possible role of PGPR in sustainable agriculture and its allied sectors is presented. Finally, a note on the importance of an integrated approach for the use of PGPR in agriculture is advanced, with a special emphasis on the adoption of modern developments in the area of biotechnology and genetic engineering in an attempt to improve and maximize the potential applications of PGPR, and to understand the modes of action and functional diversity of PGPR strains.

13.2 Role of PGPR in Sustainable Agriculture

The role of PGPR in the promotion of plant growth and development involves direct and indirect mechanisms (Glick 2012). Direct mechanisms include phytohormone production (e.g., gibberellins, cytokinins, and auxins), biological nitrogen fixation, production of enzymes and siderophores, mineral solubilization (e.g., iron and phosphorus), systemic resistance induction; and the indirect mechanisms include biocontrol, production of antibiotics, chelation of usable iron in the rhizosphere, secretion of hydrolytic enzymes, and rhizosphere niche competition (Ahmad and Zaib 2020). In addition, nitrogen transformation, improved phosphate bioavailability, an exhibition of unique enzymatic activity, and plant defense against phytopathogens with antibiotic development mediated by PGPR can also successfully enhance crop quality in agriculture (Vejan et al. 2016). For a better understanding and appreciation of the role of PGPR in sustainable agriculture, each of the aforementioned direct and indirect growth-promoting mechanisms is explained and explicated in Figure 13.1.

13.2.1 Biological Nitrogen Fixation

Nitrogen fixation is one of the most important mechanisms employed by PGPR for their stimulatory effect on plant growth and development. PGPR fix nitrogen both symbiotically and nonsymbiotically. The symbiotic nitrogen fixation is a mutualistic association between the microorganism and plants. Usually, the microorganisms form a nodule in plant roots where nitrogen fixation takes place. The nitrogen-fixing symbiotic relationship between the rhizobia and legume is most widely studied. *Rhizobia* is

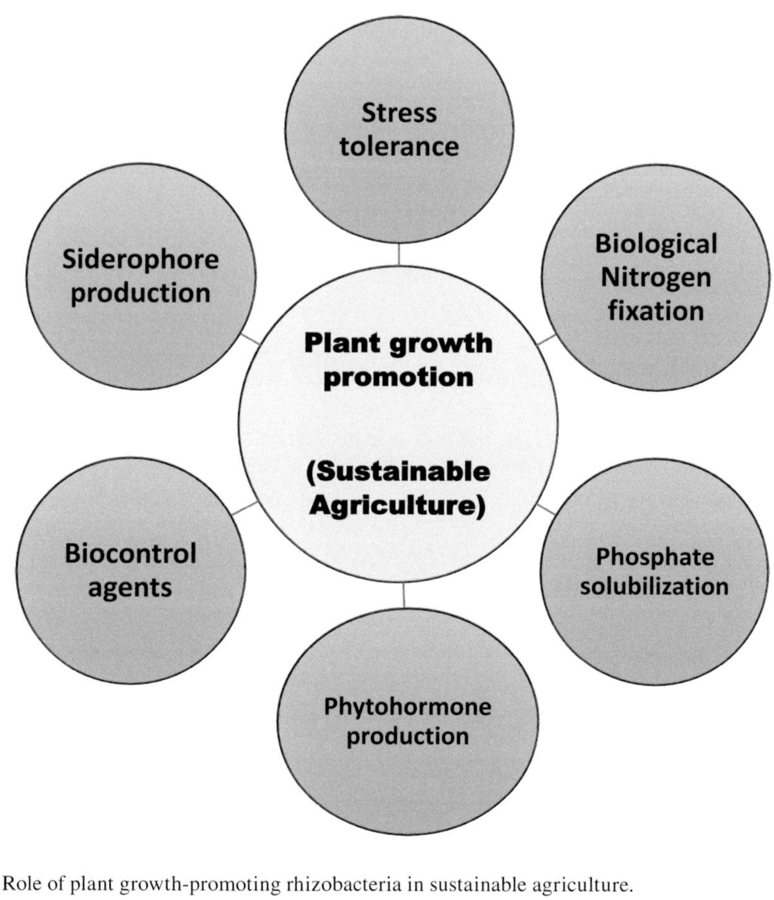

FIGURE 13.1 Role of plant growth-promoting rhizobacteria in sustainable agriculture.

a large community of rhizobacteria capable of living in symbiotic association with leguminous plants, invading their roots, and inducing the formation of root nodules where nitrogen is fixed into ammonia and made available to the plant (El-Rahman et al. 2020). Besides *Rhizobium*, many other common rhizobacteria that live as symbionts in the leguminous plants are *Bradyrhizobium*, *Sinorhizobium*, and *Mesorhizobium*. *Frankia*, a Gram-positive member of actinomycete, fixes nitrogen in symbiotic association with a broad spectrum of plants, particularly, nonleguminous plants. Free-living diazotrophs, on the contrary, carry out nonsymbiotic fixation of nitrogen and can be used to stimulate growth in rice and radish (nonleguminous plants). Some of the most common genera of free-living diazotrophs include *Rhodospirillum*, *Clostridium*, *Achromobacter*, *Herbaspirillum*, *Pseudomonas*, *Arthrobacter*, *Azospirillum*, *Azotobacter*, *Acetobacter*, *Enterobacter*, *Azomonas*, *Klebsiella*, *Bacillus*, *Beijerinckia*, etc. (Miransari 2013). It has been reported that application of diazotrophs has enhanced the production of grasses (perennial as well as annual) (Tilak et al. 2005). Many species of the genus *Azospirillum* are associated with rhizopshere of crops and cereals and can fix nitrogen under microaerophilic conditions as well. Numerous rapidly dividing rhizospheric strains of *Azospirilla* include seven species, viz. *A. largimobile*, *A. dobereinerae*, *A. irakense*, *A. halopraeferens*, *A. amazonense*, *A. lipoferum*, and *A. brasilense*. These PGPR involved in biological nitrogen fixation are an important tool for an integrated approach toward retention of the nitrogen level in agricultural soil, plant growth promotion, and disease control.

13.2.2 Phosphate Solubilization

Phosphorus (P) is the second most important element required for plant growth and sustenance. It plays a significant role in nearly all major metabolic processes of plants, including respiration, signal transduction, photosynthesis, biosynthesis of macromolecules, and energy transfer (Billah et al. 2019; Alaylar et al. 2020). Soil phosphorus mineral is abundantly available in organic as well as inorganic forms but more than 95% phosphate present in soil is insoluble, immobilized, and in precipitated form(Alaylar et al. 2020). Plants can utilize phosphate only in two soluble forms: either monobasic ($H_2PO_4^{-1}$) or dibasic (HPO_4^{-2}) ions (Awais et al. 2019). PGPR and phosphate-solubilizing bacteria (PSB) present in the soil adopt multiple strategies and mechanisms to utilize unavailable forms of phosphorus thus making phosphorus available to plants (Sharma et al. 2013). The major mechanisms used by PSB for inorganic phosphorus solubilization are: (1) release of protons (Illmer and Schinner 1992), production of chelating substances (Awais et al. 2019) and inorganic acids (Awais et al. 2019); (2) production of citric and gluconic acids (low-molecular weight organic acids) (Rodriguez et al. 2004) that bind phosphate to its carboxyl and hydroxyl groups while chelating cations and causing soil acidification, all leading to the release of soluble phosphate (Kpomblekou and Tabatabai 1994); and (3) exopolysaccharides formation which solubilizes tricalcium phosphates in growth medium, thereby releasing free phosphorus and thus affecting homeostasis solubilization (Yi et al. 2008). These PSB have been characterized as members of various genera such as *Rhizobium*, *Arthrobacter*, *Flavobacterium*, *Erwinia*, *Microbacterium*, *Beijerinckia*, *Enterobacter*, *Bacillus*, *Rhodococcus*, *Pseudomonas*, *Burkholderia*, *Serratia*, etc. (Oliveira et al. 2009; Billah et al. 2019; Alaylar et al. 2020). The enzymes like phosphatases, including phosphodiesterase, phosphotriesterase, and phosphomonoesterase, are responsible for accumulation of organic phosphorus by catalyzing the hydrolysis of phosphoric esters (Rodriguez and Fraga 1999). Among the rhizobacteria capable of producing phytase, the genera *Staphylococcus*, *Enterobacter*, *Pseudomonas*, *Serratia*, *Burkholderia*, and *Bacillus* are the most common cultured bacteria (Shedova et al. 2008). It has been studied that many phosphate-solubilizing rhizobacteria are extremely effective and their efficacy can be enhanced when co-inoculated with other plant growth promoting (Matias et al. 2009), or with mycorrhizal (Babana and Antoun 2006; Matias et al. 2009) or non-mycorrhizal fungi (Babana and Antoun 2006). Therefore, the combination of various efficient microbes with plant growth-promoting properties seems to be a potential strategy for sustainable agriculture. All these PSB have garnered considerable attention for their use as biofertilizers to enhance plant development and production (Munir et al. 2019; Ben Zineb et al. 2020), alone or in consortium with other PGPR (Pahari et al. 2020).

13.2.3 Phytohormone Production

13.2.3.1 Indole Acetic Acid (IAA)

The capability of most PGPR to produce plant hormones such as IAA has been considered to be one of their most important traits involved in plant growth promotion. The synthesis of auxins is at the center stage of several primary plant-microbial interactions. Almost all the microbes in rhizosphere are capable of producing IAA and perform in combination with endogenous IAA in plants to stimulate cell division and increase mineral and nutrient uptake from soil (Vessey 2003). IAA mainly affects plant proliferation, expansion and differentiation of plant cells; stimulates root elongation; initiates lateral and adventitious root formation; increases the rate of xylem and root development, seed and tuber germination; controls processes of vegetative growth; mediates responses to gravity, light intensity, and fluorescence; influences pigment formation, photosynthesis, biosynthesis of different metabolites, and stress resistance (Spaepen and Vanderleyden 2011). At present, auxin-synthesizing rhizobacteria are the most well-studied phytohormone producers (Tsavkelova et al. 2006; Spaepen et al. 2007; Khan et al. 2020) and these synthesize IAA both via tryptophan and tryptophan-independent pathways; the former pathway employs diverse mechanism and tryptophan is usually found in exudates of roots while lower quantities of IAA are synthesized via tryptophan-independent pathways (Spaepen et al. 2007). Among various PGPR strains, *Azospirillum* is reported to be one of the best studied IAA producers (Asiloglu et al. 2020; Ejaz et al. 2020; Kargapolova et al. 2020) and other IAA-producing bacteria belonging to genera *Azotobacter* (Nosheen et al. 2019; Latef et al. 2020), *Aeromonas* (Halda-Alija 2003), *Bacillus* (Chakraborty et al. 2010; Guo et al. 2019), *Pseudomonas* (Hariprasad and Niranjana 2009; Khan et al. 2020), *Burkholderia* (Halda-Alija2003), *Enterobacter* (Shoebitz et al. 2009; Zafar-Ul-Hye et al. 2019), and *Rhizobium* (Ghosh et al. 2008) have been isolated from different rhizosphere soils. Root-growth promotion by the free-living PGPR, e.g. *Enterobacter cloacae*, *Acetobacter dizotrophicous*, *Alkaligenes faecalis*, and species of *Pseudomonas*, *Azospirillum*, and *Xanthomonas* has been related to low level of IAA secretion. However, microbial hormones are known to be effective for their slow and continuous release qualifying for the low threshold levels between inhibitory and stimulatory effects of chemically produced hormones. PGPR producing IAA are being used as inoculums to stimulate germination, enable rapid root growth and alter the root system architecture, and significantly improve the root biomass (Boominathan 2020; Prakash et al. 2020).

13.2.3.2 Cytokinins, Gibberellin, and Ethylene

Besides IAA, other phytohormones involved in PGPR action are cytokinins (Ortíz-Castro et al. 2008) and gibberellic acid (GA_3) (Naz et al. 2009). These are very essential plant hormones present in low concentrations in plants and microbes that increase root initiation and development, cell proliferation, and are associated with photosynthesis and chlorophyll differentiation (Vessey 2003). It is also known that these induce stomatal opening, suppression of apical dominance induced by auxin, and inhibition of plant senescence, especially in leaves (Baca and Elmerich 2007). Some PGPR strains such as *Pseudomonas fluorescens*, *Paenibacillus polymyxa*, and *Rhizobium leguminosarum* were reported to produce cytokinins (Timmusk and Wagner 1999). Several gibberellins (GAs)-producing PGPR have been reported which most commonly produce GA_3; however, in plants GA_1 is the most active form primarily responsible for stem elongation (Davies 2010). There are almost 89 GAs known so far (Amara et al. 2015), and PGPR genera such as *Rhizobia*, *Herbaspirillum*, *Bacillus*, *Acetobacter*, and *Azospirillum* were reported to produce GAs (Janzen et al. 1992). Ethylene (gaseous hormone) is another key phytohormone that participates in many biological activities such as promotion of root formation, inhibition of root elongation, leaf abscission, seed germination, fruit ripening, activation of other phytohormone production, etc. (Glick et al. 2007). It is produced by various biotic and abiotic mechanisms in soil (Glick et al. 2007). The prerequisite for the production of ethylene is enzyme 1-aminocyclopropane-1-carboxylic acid oxidase (ACCO). The inoculation of plant growth-promoting strains of *Pseudomonas* spp. in combination with *R. leguminosarum* has resulted in lowering of ethylene production, thereby enhancing fresh weight, nitrogen content in grain of lentil, nodule dry weight, nodule number, and grain and straw yield (Iqbal et al. 2012). 1-aminocyclopropane-1-carboxylic acid (ACC) deaminase activity was extensively reported in numerous strains of PGPR such as *Enterobacter*

(Penrose and Glick 2001), *Ralstonia*, *Azospirillum*, *Agrobacterium*, *Burkholderia* (Sessitsch et al. 2005; Blaha et al. 2006), *Pseudomonas* (Belimov et al. 2001; Blaha et al. 2006), *Methylobacterium fujisawaense* (Madhaiyan et al. 2006), *Variovorax paradoxus*, *Bacillus*, *Alcaligenes* (Belimov et al. 2001), *Rhizobium* (Ma et al. 2003), *Rhodococcus* (Safronova et al. 2012), and *Sinorhizobium meliloti* (Belimov et al. 2005). Many bacterial genera having ACC deaminase activity have been identified so far. Several studies demonstrated a greater resistance to environmental stress by host plants inoculated with ACC deaminase-producing rhizobacteria (Danish et al. 2020).

13.2.4 Siderophore Production

Siderophores are low-molecular weight iron-chelating compounds that are produced by many microbes for trapping iron from the soil (Schwyn and Neilands 1987; Singh et al. 2019). This condition usually occurs in neutral to alkaline soils due to low iron solubility at elevated pH. (Sharma and Johri 2003). Based on functional groups, these siderophores are grouped into three categories: hydroxamates, catecholates, and carboxylates. There are presently over 500 different forms of siderophores identified, of which 270 have been structurally identified (Cornelis 2010). Iron acquisition through siderophore production plays an important role in defining the effectiveness of bacteria in colonizing host roots and rhizosphere (Crowley 2006). Siderophores have been involved both directly and indirectly in enhancing PGPR-mediated host growth and can inhibit the development of pathogens by trapping Fe^{3+}ion in rhizosphere (Siddiqui 2005). Plants also use their own mechanisms to accumulate iron, e.g. dicots use a protein that transforms insoluble Fe^{3+}ion into more soluble Fe^{2+} ion or root membrane reductase ion; whereas monocots achieve this by production of phytosiderophores (Crowley 2006). Some common PGPR producing siderophores are: *Aeromonas* spp., *Bacillus* spp., *Azadirachta* spp., *Azotobacter* spp., *Burkholderia* spp., *Pseudomonas* spp., *Rhizobium* spp., *Serratia* spp., and *Streptomyces* spp. (Sujatha and Ammani 2013). The possible role of siderophores can be demonstrated in certain disease suppressive soils where phytopathogens, such as *Fusarium* and *Pythium*, are inhibited by *pseudomonads*. Siderophore-producing PGPR are recognized as efficient biocontrol agents (BCAs) against pathogens. Nevertheless, in most instances, siderophores tend to be more effective in enhancing the competence of disease-suppressive bacteria in the rhizosphere. But the capability of biocontrol PGPR can be improved by inoculating the soil with the mixture of other heterologous siderophore-producing rhizobacteria (Crowley 2006). Also, heavy metal pollution is largely influenced by siderophores. Siderophore-producing PGPR are gaining commercial importance for their safe and self-replicating behavior; the latter circumvents the repeated application, but the target organisms do not develop resistance (Schalk et al. 2011).

13.3 PGPR as BCAs

Many effective bacterial strains are known to be potential BCAs due to their rapid colonization and fast proliferation around plant rhizosphere. The use of PGPR as BCAs against soil-borne fungal pathogens is an alternative or complementary strategy to mitigate physical and chemical disease management practices (Singh et al. 2019). PGPR indirectly enhance plant growth via suppressing the pathogen by producing chemicals that inhibit the growth and invasion of phytopathogens. Rhizobacteria, due to their ability to perform biocontrol of disease or insect pests (biopesticides), can be utilized as substitutes to chemical pesticides for disease control (Bajracharya 2019; Pahari et al. 2020). These PGPR offer reliable alternatives for sustainable and eco-friendly agriculture, inter alia, with their ability to enhance the fertility of soil and plant growth indirectly. PGPR have also been reported to confer drought resistance to plants, such as tomato and pepper, grown in water-deficient soils (Aroca and Ruiz-Lozano 2009) as well as disease control. For sustainable biocontrol, the mode of action of PGPR strains is a determining factor in the type of disease management strategy implemented. The PGPR render disease control resistance by producing one or several kinds of antibiotics which check the pathogen growth (Kenawy et al. 2019). These antibiotic-producing PGPR have been reported to minimize or remove plant-bound infections in many crops such as wheat, rice, maize, chickpea, barley, and tomato (Raaijmakers et al. 2002; Karthika et al. 2020). Genetic study of many *Pseudomonas* biocontrol strains indicates that a positive correlation exists between disease elimination and

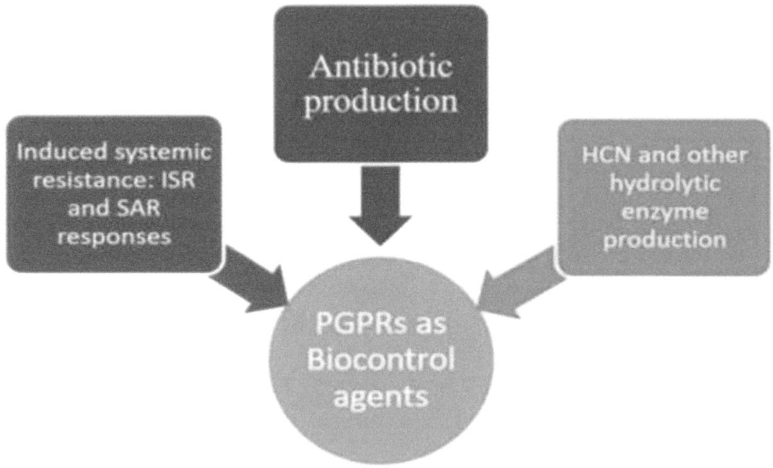

FIGURE 13.2 Plant growth-promoting rhizobacteria as biocontrol agents by mediating various responses.

antibiotic production (Sahni and Prasad 2020). Most biocontrol mechanisms involve a direct antagonism by developing antibiotics, siderophores, HCNs, hydrolytic enzymes (e.g., chitinases, proteases, lipases, etc.) or an indirect involvement by competing with the pathogen for a niche which can become infectious site and nutrient site (Lugtenberg and Kamilova2009; Tariq et al. 2014). The capability of PGPR to reduce the growth of phytopathogens is largely attributed to the synthesis of a range of different types of antibiotics (Kenawy et al. 2019). The mechanisms such as SAR, ISR in plants and regulation of hormonal levels can activate the biocontrol system (Van Loon 2007; Kannojia et al. 2019) in the plant tissues (Figure 13.2).

13.3.1 Antibiotic Production

Antibiotic production is the most powerful and the most studied biological control mechanism of PGPR against plant disease-causing pathogens (Kenawy et al. 2019). It has been proved under *in vitro* conditions that various antibiotics produced by PGPR strains are effective against various plant pathogens (Tariq et al. 2014). Many antibiotics produced by PGPR have been identified and include butyrolactones, zwittermicin A, kanosamine, oligomycin A, oomycin A, xanthobaccin, 2,4-diacetylphloroglucinol (2,4-DAPG), and many others (Fernando et al. 2005). The 2,4-DAPG is produced by some strains of *Pseudomonas* which is one of the most common PGPR strain (Rezzonico et al. 2007). The 2,4-DAPG has a broad range of antibiotic activities such as antihelmintic (Cronin et al. 1997), antibacterial (Velusamy et al. 2006), and antifungal (Loper and Gross 2007; Rezzonico et al. 2007). Some antibiotics such as xanthobaccin, zwittermicin A, kanosamine, and oligomycin A produced by *Stenotrophomonas* spp., *Bacillus* spp., and *Streptomyces* spp. have been reported to check the spread of phytopathogens (Compant et al. 2005). The detailed study of the specificity and mode of action of these antibiotics has been done and many of these strains with biocontrol activities are commercially available. However, the continuous application of these antibiotic-producing bacteria is beset with the complications of the development of antibiotic resistance in some pathogens. As a remedial measure thereof, a few researchers have suggested the utilization of biocontrol strains that synthesize HCN as well as one or more antibiotics (Glick 2012). *Bacillus amyloliquefaciens* is well-recognized for lipopeptide and polyketide secretion for biocontrol activity and is a recommended PGPR against soil-borne pathogens (Gupta and Vakhlu 2015).

13.3.2 Hydrogen Cyanide Production and Other Hydrolytic Enzymes

Apart from antibiotic production, many PGPR also generate inhibitory volatile compounds, such as HCN and other lytic enzymes (Rezzonico et al. 2007). HCN is a volatile, secondary metabolite that increases life span of many soil microbes by killing competing microbes, thereby reducing possible nutritional competition (Siddiqui 2005). Glycine, the precursor of HCN, in association with the action of HCN synthetase enzyme forms HCN, which is associated with the plasma membrane of certain rhizobacteria (Blumer and Haas 2000). Currently, it has been shown that various bacterial genera can produce HCN, including species of *Rhizobium*, *Bacillus*, *Alcaligenes*, *Pseudomonas*, and *Aeromonas* (Devi et al. 2007; Ahemad and Kibret 2014). Some biocontrol bacteria produce enzymes, including lipases, chitinases, proteases, cellulases, and β-1,3 glucanases, that can break cell walls of various fungal pathogens (Amara et al. 2015; Pahari et al. 2020). These enzymes with biocontrol activity are effective against a number of fungal pathogens, including *Fusarium oxysporum*, *Sclerotium rolfsii*, *Botrytis cinerea*, *Pythium ultimum*, *Rhizoctonia solani*, and *Phytophthora* spp. (Nadeem et al. 2013). Many researchers have claimed the effectiveness of PGPR as efficient BCAs, including *Azotobacter chroococcum* against *F. oxysporum* in *Sesamum indicum*, and *Pseudomonas putida* and *P. fluorescens* (Voisard et al. 1989) against *Macrophomina phaseolina* in chickpea under the field conditions (Maheswari and Rajagopal 2013).

13.3.3 Induction of Systemic Resistance in Host

13.3.3.1 ISR

ISR is characterized by improved defense capability of plant in response to adverse environmental stimuli, thereby strengthening plant's innate defense system against subsequent biotic challenges (Kannojia et al. 2019). ISR does not target specific pathogens, but includes signaling pathways for jasmonic acid and ethylene within the plant to enhance the response of host plant to a variety of phytopathogens (Glick 2012). ISR does not involve close contact between the biocontrol-stimulating PGPR and the phytopathogens (Bakker et al. 2007). ISR agents such as flamma, lipopolysaccharides (LPs), siderophores, homoserine lactones, cyclic lipopeptides, 2,4-DAPG, and volatiles such as 2,3-butanediol are induced by a number of different individual bacterial components (Doornbos et al. 2012).

13.3.3.2 SAR

PGPR can also provide systemic resistance to the host plants through activation of their defense mechanisms in response to infection by a pathogenic agent (Pieterse et al. 1998). SAR develops when plants initiate their defense system and successfully induce a hypersensitive response to primary infection by a pathogen which becomes limited to small gray, desiccated tissue in local necrotic damage. A striking difference between two defense systems is the dependence of SAR response system on salicylic acid (SA) (Pettersson and Baath 2004). Certain PGPR in their rhizospheric habitat produce relatively lesser quantity of SA to trigger SA-dependent signaling pathway response (Dekkers et al. 2000). Furthermore, a significant proportion of PGPR microbes triggers defense response via an SA-independent pathway which involves jasmonic acid and ethylene signals (Pieterse et al. 1998). Besides, findings on mechanism of biocontrol pathways by various PGPR have also revealed that several strains inhibit the pathogen growth by adding extra callose, lignin, and phenolic depositions on epidermal and cortical cell walls (Singh et al. 2019) and also by activating defense genes encoding chitinase, peroxidase, polyphenol oxidase, and phenylalanine ammonia lyase (Benhamou and Paulitz 2000) and enzymes that are responsible for the phytoalexins production (Ganeshamoorthi et al. 2008).

13.4 PGPRs in Stress Tolerance

Salinity in agricultural soil is a major issue in the world and a significant environmental factor that decreases agricultural plant development and production. The rise in the salt levels in soil can affect the physiological and metabolic activities of crops, often resulting in the declivity in both qualitative and quantitative growth and productivity of crops. Of late, several recombinant salt-resistant plants have been created to mitigate this crisis (salt stress) in the agricultural crop productivity but little success has been achieved. A viable and environment-friendly alternative has been the use of PGPR which not only reduce soil salinity and a load of pesticides and chemical fertilizers but also improve soil nutrition, seed germination, plant development, and crop yield under the saline soil conditions. PGPR are now recognized as potential bacteria that stand different environmental conditions for wider ranges of their fluctuation such as higher temperatures, pH, and salt content in soils (Subiramani et al. 2020). It has been confirmed that salinity-adapted microbes isolated from rhizosphere and bulk soil of halophytes increase plant development directly by phytohormone production, biological nitrogen fixation, phosphate solubilization, siderophore production, ACC deaminase activity; or indirectly by biocontrol activity (Verma et al. 2017). From an ecological perspective, these halophilic microbes facilitate the adaptability of halophytic plants living in the saline environments. Knowledge of plant-microbe interactions may serve as a potential guiding tool for the protection of crops, soil reclamation, and redressal of many other issues confronting the agriculture sector. Drought is a potential ecological challenge that affects plant development and yield. Cost-effective and environment-friendly approaches are imperative for the mitigation and compensation of yield losses incurred due to drought. Drought stress tolerance is a complex characteristic including groups of genes; hence, application of recombinant technology to generate drought-tolerant crops is a challenging task. In this background, the application of PGPRs to mitigate drought stress is gaining attention as a desirable and economically viable alternative approach (Kour et al. 2019). In natural environments, microorganisms that are able to cope with low temperatures are always the dominant flora, and hence, can be considered as the most effective colonizers on the earth. Cold-adapted microbes thrive well at sub-zero temperatures (Yadav et al. 2015). Microbial activities of cold-adapted microbes have been reported even at the sub-zero temperatures. Generally, the growth rates and microbial activity of the psychrophilic microbes are higher at low temperatures than at the highest temperature of the growth, and more often are explained as the basis for the successful microbial adjustment to the natural cold environment(Subiramani et al. 2020). Prospecting the cold habitats has resulted in a wide variety of microorganisms being isolated. In agriculture and pharmaceutical industry, the psychrophiles have numerous biotechnological applications. The microbial biodiversity from temperate climates can be used as a database for selecting the bioinoculants capable of producing plant growth promoting activities and improving crop growth and yield in high altitudes with low temperatures (Yadav et al. 2019).

13.5 Applications in Agriculture

The essential precondition for introducing microorganisms into the field is to predict beforehand the nature of their interaction with the resident microbial community. The choice of suitable strains for improving the efficacy of PGPR under various conditions is of great importance (Prasad et al. 2019). Various strains of PGPR can increase the supply of various nutrients such as nitrogen or phosphorous. Species of genus *Azospirillum* and *Rhizobium* can fix nitrogen and are, therefore, more effective along with nitrogen fertilizers. Arbuscular mycorrhiza (AM) fungi and bacteria capable of solubilizing phosphorus in the soil can increase phosphorus by producing enzymes such as phosphatase. The synergistic associations between AM fungi and PGPR such as nitrogen-fixing *Rhizobium* can enhance the efficacy of soil microorganisms (Artursson et al. 2006). This is of great importance in enhancing biofertilizer efficiency by producing the biofertilizers with an appropriate combination of microbes (Verma et al. 2018). In the recent past, the use of biofertilizers in agriculture has attracted considerable attention from the research community and is now widely accepted as an environmentally sustainable practice owing much to the negative fallouts of use of chemical inorganic fertilizers. Bacterial formulations can be species-specific or a consortium of microbes. Soil microbes are capable of generating growth-enhancing substances that suppress the impact of numerous abiotic stress conditions, such as water, salt, pH stress, temperature stress, nutrient deficiency, and pollutants (Kumar et al. 2019; Singh et al. 2019). The past few decades have witnessed a significant rise in the human population that has taken a heavy toll on the food security of the world. The issue is further exacerbated by the limited availability of land for crop production, more so in the wake of accelerating rate of land diversion for nonagricultural purposes. In order to fulfill the increasing food demand amidst such constraints as land availability, farmers across the globe have largely relied upon the use of chemical fertilizers for a speedy increase in crop production. This approach to agriculture proved to be a mixed blessing as crop production significantly showed an upsurge, but the ecological and health repercussions largely out-shadowed the achievements in crop productivity on account of the use of chemical fertilizers. The detrimental consequences of the massive chemical inputs into farm systems have put the viability of crop production as well as the protection of environmental health at risk. The scientific and the farming communities have now roped in biofertilizers as an environment-friendly and cost-effective substitute to address this menace. The burden of the ever-growing population and the consequent pressures mounted on the environmental health warrant a thorough research on the applicability and potential prospects of these microbial organisms over a large scale.

13.6 Future Prospects

The potential benefits of PGPR can be increased and improved upon in light of the latest developments in the area of biotechnology and genetic engineering. Genetic engineering has opened new windows for the application of biodiversity for the benefit of humans, in addition to many other useful applications; PGPR are no exception to the class of organisms being genetically engineered for extended services to the mankind. In both laboratory and greenhouse studies, the applications of rhizosphere soil, with suitable bacterial communities, to crops have yielded promising results (Gouda et al. 2018). Application of genetically engineered PGPR in the remediation of complex contaminated soils (Etesami and Maheshwari 2018) and expected increase in agricultural productivity has emerged as another focal area for research in the recent past. However, an area of concern in relation to PGPR is that their application as biofertilizers and biopesticides shows a significant gap between laboratory and glasshouse results and the results obtained in the field. This is primarily due to the rhizospheric conditions into which the bacteria are introduced in the field and the environmental stress they are subjected to therein. With further advancement in the understanding mechanisms of action of these bacteria, it becomes progressively clear that PGPR employ an array of mechanisms for promoting plant growth. However, most studies are still based on a single pathway and have still been unable to address the relative influence of the various mechanisms that are involved for promoting development of crops. The selection of the most viable strain of PGPR warrants comprehensive knowledge about its mechanisms of action and ecological aspects of its introduction vis-à-vis resident PGPR and other plant rhizosphere microbial community (Singh et al. 2019). Although the effects of different management procedures or soil modifications on indigenous bacteria's plant growth promoting activity remain unknown, inoculation is now considered to increase the crop productivity. The impact of PGPR inoculums on improving crop nutrition, crop production, disease management in agriculture, and more importantly, on sustainable development is well-documented now. Nonetheless, along with further refinements in the understanding of their ecology, a more holistic account of their mechanisms of action and the conditions favorable for their role in bioaugmentation of crop soil needs considerable attention. Another area relating to PGPR warranting special attention is the means of making such bacteria available to the crop rhizosphere, taking into account suitable carriers, shelf life, and the survival of these rhizobacteria. Additional optimization is needed to introduce effective PGPR strains into agriculture in the better fermentation and formulation processes.

Acknowledgments

The authors are grateful to the University Grants Commission for its support under the scheme, "Centre with Potential for Excellence in Particular Areas," CSIR for Senior Research Fellowship in favor of Sheikh Sajad, and the Department of Botany, University of Kashmir, Hazratbal Srinagar, Jammu and Kashmir, for laboratory facilities. There are no conflicts of interest.

REFERENCES

Ahemad, M. and Kibret, M. 2014. Mechanisms and applications of plant growth promoting rhizobacteria: current perspective. *J King Saud Univ Sci* 26(1):1–20.

Ahmad, I. and Zaib, S. 2020. Mighty microbes: plant growth promoting microbes in soil health and sustainable agriculture. In: Giri Bhoopander and Varma Ajit (eds), *Soil Health*. Springer, Cham, pp:243–264.

Alaylar, B., Egamberdieva, D., Gulluce, M., Karadayi, M. and Arora, N. K. 2020. Integration of molecular tools in microbial phosphate solubilization research in agriculture perspective. *World J Microbiol Biotechnol* 36(7):1–12.

Amara, U., Khalid, R. and Hayat, R. 2015. Soil bacteria and phytohormones for sustainable crop production. In: Maheshwari Dinesh K. (ed), *Bacterial Metabolites in Sustainable Agroecosystem*. Springer, Cham, pp:87–103.

Aroca, R. and Ruiz-Lozano, J. M. 2009. Induction of plant tolerance to semi-arid environments by beneficial soil microorganisms—a review. In: Lichtfouse Eric (ed), *Climate Change, Intercropping, Pest Control and Beneficial Microorganisms*. Springer, Dordrecht, pp:121–135.

Artursson, V., Finlay, R. D. and Jansson, J. K. 2006. Interactions between arbuscular mycorrhizal fungi and bacteria and their potential for stimulating plant growth. *Environ Microbiol* 8(1):1–10.

Asiloglu, R., Shiroishi, K., Suzuki, K., Turgay, O.C., Murase, J. and Harada, N. 2020. Protist-enhanced survival of a plant growth promoting rhizobacteria, *Azospirillum* sp. B510, and the growth of rice (*Oryza sativa* L.) plants. *Appl Soil Ecol* 154:103599.

Awais, M., Tariq, M., Ali, Q., Khan, A., Ali, A., Nasir, I. A. and Husnain, T. 2019. Isolation, characterization and association among phosphate solubilizing bacteria from sugarcane rhizosphere. *Cytol Genet* 53 (1):86–95.

Babana, A. H. and Antoun, H. 2006. Effect of tilemsi phosphate rock-solubilizing microorganisms on phosphorus uptake and yield of field-grown wheat (*Triticum aestivum* L.) in Mali. *Plant Soil* 287(1-2):51–58.

Baca, B. E. and Elmerich, C. 2007. Microbial production of plant hormones. In: Elmerich Claudine and Newton William E. (eds), *Associative and Endophytic Nitrogen-Fixing Bacteria and Cyanobacterial Associations*. Springer, Dordrecht, pp:113–143.

Backer, R., Rokem, J. S., Ilangumaran, G., Lamont, J., Praslickova, D., Ricci, E., Subramanian, S. and Smith, D. L. 2018. Plant growth-promoting rhizobacteria: context, mechanisms of action, and roadmap to commercialization of bio-stimulants for sustainable agriculture. *Front Plant Sci* 9:1473.

Bajracharya, A. M. 2019. Plant growth promoting rhizobacteria (PGPR): biofertiliser and biocontrol agent—review article. *J Balkumari Col* 8:42–45.

Bakker, P. A., Pieterse, C. M. and Van Loon, L. C. 2007. Induced systemic resistance by fluorescent *Pseudomonas* spp. *Phytopathol* 97(2):239–243.

Belimov, A. A., Hontzeas, N., Safronova, V. I., Demchinskaya, S. V., Piluzza, G., Bullitta, S. and Glick, B. R. 2005. Cadmium-tolerant plant growth-promoting bacteria associated with the roots of Indian mustard (*Brassica juncea* L. Czern.). *Soil Biol Biochem* 37(2):241–250.

Belimov, A. A., Safronova, V. I., Sergeyeva, T. A., Egorova, T. N., Matveyeva, V. A., Tsyganov, V. E., Borisov, A. Y., Tikhonovich, I. A., Kluge, C., Preisfeld, A. and Dietz, K. J. 2001. Characterization of plant growth promoting rhizobacteria isolated from polluted soils and containing 1-aminocyclopropane-1-carboxylate deaminase. *Can J Microbiol* 47(7):642–652.

Ben Zineb, A., Trabelsi, D., Ayachi, I., Barhoumi, F., Aroca, R. and Mhamdi, R. 2020. Inoculation with elite strains of phosphate-solubilizing bacteria enhances the effectiveness of fertilization with Rock phosphates. *Geomicrobiol J* 37(1):22–30.

Benaissa, A. 2019. Plant growth promoting rhizobacteria—A review. *Algerian J Environ Sci Technol* 5:1.

Benhamou, N. and Paulitz, T. C. 2000. Defense enzymes induced in cucumber roots by treatment with plant growth-promoting rhizobacteria (PGPR) and *Pythium aphanidermatum*. *Physiol Mol Plant Pathol* 56:13–23.

Billah, M., Khan, M., Bano, A., Hassan, T. U., Munir, A. and Gurmani, A. R. 2019. Phosphorus and phosphate solubilizing bacteria: keys for sustainable agriculture. *Geomicrobiol J* 36(10):904–916.

Blaha, D., Prigent-Combaret, C., Mirza, M. S. and Moënne-Loccoz, Y. 2006. Phylogeny of the 1-aminocyclopropane-1-carboxylic acid deaminase-encoding gene ACDS in phytobeneficial and pathogenic proteobacteria and relation with strain biogeography. *FEMS Microbiol Ecol* 56(3):455–470.

Blumer, C. and Haas, D. 2000. Mechanism, regulation, and ecological role of bacterial cyanide biosynthesis. *Arch Microbiol* 173(3):170–177.

Boominathan, U. 2020. Screening PGPR for improving seed germination, seedling growth and yield of maize (*Zea mays* L.). *Studies Indian Place Names* 40(20):425–430.

Chakraborty, U., Chakraborty, B. N. and Chakraborty, A. P. 2010. Influence of *Serratia marcescens* TRS-1 on growth promotion and induction of resistance in *Camellia sinensis* against *Fomes lamaoensis*. *J Plant Interact* 5(4):261–272.

Compant, S., Reiter, B., Sessitsch, A., Nowak, J., Clément, C. and Barka, E.A. 2005. Endophytic colonization of *Vitisvinifera* L. by plant growth-promoting bacterium *Burkholderia* sp. strain PsJN. *Appl Environ Microbiol* 71(4):1685–1693.

Cornelis, P. 2010. Iron uptake and metabolism in pseudomonads. *Appl Microbiol Biotechnol* 86(6):1637–1645.

Cronin, D., Moenne-Loccoz, Y., Fenton, A., Dunne, C., Dowling, D.N. and O'gara, F. 1997. Role of 2, 4-diacetylphloroglucinol in the interactions of the biocontrol pseudomonad strain F113 with the potato cyst nematode *Globodera rostochiensis*. *Appl Environ Microbiol* 63(4):1357–1361.

Crowley, D.E.2006. Microbial siderophores in the plant rhizosphere. In: Barton Larry L. and Abadia Javier (eds), *Iron Nutrition in Plants and Rhizospheric Microorganisms*. Springer, Dordrecht, pp:169–198.

Danish, S., Zafar-ul-Hye, M., Mohsin, F. and Hussain, M. 2020. ACC-deaminase producing plant growth promoting rhizobacteria and biochar mitigate adverse effects of drought stress on maize growth. *PLoS One* 15(4):0230615.

Davies, P. J. 2010. The plant hormones: their nature, occurrence, and functions. In: Davies Peter J. (ed), *Plant Hormones*. Springer, Dordrecht, pp:1–15.

Dekkers, L. C., Mulders, I. H., Phoelich, C. C., Chin-A-Woeng, T. F., Wijfjes, A. H. and Lugtenberg, B. J. 2000. The sss colonization gene of the tomato-*Fusarium oxysporum* f. sp. radicis-lycopersici biocontrol strain *Pseudomonas fluorescens* WCS365 can improve root colonization of other wild-type *Pseudomonas* spp. bacteria. *Mol Plant Microbe Interact* 13(11):1177–1183.

Devi, K. K., Seth, N., Kothamasi, S. and Kothamasi, D. 2007. Hydrogen cyanide-producing rhizobacteria kill subterranean termite *Odontotermes obesus* (Rambur) by cyanide poisoning under in vitro conditions. *Curr Microbiol* 54(1):74–78.

Dilnashin, H., Birla, H., Hoat, T. X., Singh, H. B., Singh, S. P. and Keswani, C. 2020. Applications of agriculturally important microorganisms for sustainable crop production. In: Sharma Vivek, Salwan Richa, Khalil Laith and Al-Ani Tawfeeq (eds), *Molecular Aspects of Plant Beneficial Microbes in Agriculture*. Academic Press, Cambridge, Massachusetts, United States, pp:403–415.

Doornbos, R. F., van Loon, L. C. and Bakker, P. A. 2012. Impact of root exudates and plant defense signaling on bacterial communities in the rhizosphere. A review. *Agron Sustain Dev* 32(1):227–243.

Ejaz, S., Batool, S., Anjum, M. A., Naz, S., Qayyum, M. F., Naqqash, T., Shah, K. H. and Ali, S. 2020. Effects of inoculation of root-associative *Azospirillum* and *Agrobacterium* strains on growth, yield and quality of pea (*Pisum sativum* L.) grown under different nitrogen and phosphorus regimes. *Sci Hortic* 270:109401.

El-Rahman, A., Rehab, A. M. and Abo Taleb, H. H. 2020. Response of some new chickpea genotypes to rhizobial inoculation and foliar application with plant growth promoting rhizobacteria (PGPR). *J Plant Prod* 11(2):89–94.

Etesami, H. and Maheshwari, D. K. 2018. Use of plant growth promoting rhizobacteria (PGPRs) with multiple plant growth promoting traits in stress agriculture: action mechanisms and future prospects. *Ecotoxicol Environ Saf* 156:225–246.

Fernando, W. D., Nakkeeran, S. and Zhang, Y. 2005. Biosynthesis of antibiotics by PGPR and its relation in biocontrol of plant diseases. In: Siddiqui Zaki A. (ed), *PGPR: Biocontrol and Biofertilization*. Springer, Dordrecht, pp:67–109.

Ganeshamoorthi, P., Anand, T., Prakasam, V., Bharani, M., Ragupathi, N. and Samiyappan, R. 2008. Plant growth promoting rhizobacterial (PGPR) bioconsortia mediates induction of defense-related proteins against infection of root rot pathogen in mulberry plants. *J Plant Interact* 3(4):233–244.

Ghosh, S., Sengupta, C., Maiti, T. K. and Basu, P. S. 2008. Production of 3-indolylacetic acid in root nodules and culture by a *Rhizobium* species isolated from root nodules of the leguminous pulse *Phaseolusmungo*. *Folia Microbiol* 53(4):351.

Glick, B. R. 2012. Plant growth-promoting bacteria: mechanisms and applications. *Scientifica* 1:15.

Glick, B. R., Todorovic, B., Czarny, J., Cheng, Z., Duan, J. and McConkey, B. 2007. Promotion of plant growth by bacterial ACC deaminase. *CRC Crit Rev Plant Sci* 26(5-6):227–242.

Gouda, S., Kerry, R. G., Das, G., Paramithiotis, S., Shin, H. S. and Patra, J. K. 2018. Revitalization of plant growth promoting rhizobacteria for sustainable development in agriculture. *Microbiol Res* 206:131–140.

Guo, D., Yuan, C., Luo, Y., Chen, Y., Lu, M., Chen, G., Ren, G., Cui, C., Zhang, J. and An, D. 2019. Biocontrol of PGPR strain *Bacillus amyloliquefaciens* Ba168 against *Phytophthora nicotianae* on tobacco. *Bio Rxiv* pp:700–757.

Gupta, R. and Vakhlu, J. 2015. Native *Bacillus amyloliquefaciens* W2 as a potential biocontrol for *Fusarium oxysporum* R1 causing corm rot of *Crocus sativus. Eur J Plant Pathol* 143(1):123–131.

Halda-Alija, L. 2003. Identification of indole-3-acetic acid producing freshwater wetland rhizosphere bacteria associated with *Juncus effusus* L. *Can J Microbiol* 49(12):781–787.

Hariprasad, P. and Niranjana, S. R. 2009. Isolation and characterization of phosphate solubilizing rhizobacteria to improve plant health of tomato. *Plant Soil* 316(1–2):13–24.

Illmer, P. and Schinner, F. 1992. Solubilization of inorganic phosphates by microorganisms isolated from forest soils. *Soil Biol Biochem* 24(4):389–395.

Iqbal, M. A., Khalid, M., Shahzad, S. M., Ahmad, M., Soleman, N. and Akhtar, N. 2012. Integrated use of *Rhizobium leguminosarum*, plant growth promoting rhizobacteria and enriched compost for improving growth, nodulation and yield of lentil (*Lens culinaris* medik.). *Chil J Agric Res* 72(1):104.

Janzen, R. A., Rood, S. B., Dormaar, J. F. and McGill, W. B. 1992. *Azospirillum brasilense* produces gibberellin in pure culture on chemically-defined medium and in co-culture on straw. *Soil Biol Biochem* 24 (10):1061–1064.

Kannojia, P., Choudhary, K. K., Srivastava, A. K. and Singh, A. K. 2019. PGPR bioelicitors: induced systemic resistance (ISR) and proteomic perspective on biocontrol. In: Singh Amit Kishore, Kumar Ajay and Singh Pawan Kumar (eds), *PGPR Amelioration in Sustainable Agriculture.* Woodhead Publishing, Sawston, United Kingdom. pp:67–84.

Kargapolova, K. Y., Burygin, G. L., Tkachenko, O. V., Evseeva, N. V., Pukhalskiy, Y. V. and Belimov, A. A. 2020. Effectiveness of inoculation of in vitro-grown potato microplants with rhizosphere bacteria of the genus *Azospirillum. Plant Cell Tissue Organ Cult* 1–9.

Karthika, S., Varghese, S. and Jisha, M. S. 2020. Exploring the efficacy of antagonistic rhizobacteria as native biocontrol agents against tomato plant diseases. *3 Biotech* 10(7):1–17.

Kenawy, A., Dailin, D. J., Abo-Zaid, G. A., AbdMalek, R., Ambehabati, K. K., Zakaria, K. H. N., Sayyed, R. Z. and El Enshasy, H. A. 2019. Biosynthesis of antibiotics by PGPR and their roles in biocontrol of plant diseases. In: Sayyed R. Z. (ed), *Plant Growth Promoting Rhizobacteria for Sustainable Stress Management.* Springer, Singapore, pp:1–35.

Khan, N., Bano, A. and Babar, M. A. 2019. The stimulatory effects of plant growth promoting rhizobacteria and plant growth regulators on wheat physiology grown in sandy soil. *Arch Microbiol* 201(6):769–785.

Khan, N., Bano, A., Ali, S. and Babar, M. A. 2020. Crosstalk amongst phytohormones from planta and PGPR under biotic and abiotic stresses. *Plant Growth Regul* 90:189–203.

Khanna, K., Jamwal, V. L., Gandhi, S. G., Ohri, P. and Bhardwaj, R. 2019. Metal resistant PGPR lowered Cd uptake and expression of metal transporter genes with improved growth and photosynthetic pigments in *Lycopersicon esculentum* under metal toxicity. *Sci Rep* 9(1):1–14.

Kour, D., Rana, K. L., Yadav, N., Yadav, A. N., Kumar, A., Meena, V. S., Singh, B., Chauhan, V. S., Dhaliwal, H. S. and Saxena, A. K. 2019. Rhizospheric microbiomes: biodiversity, mechanisms of plant growth promotion, and biotechnological applications for sustainable agriculture. In: *Plant Growth Promoting Rhizobacteria for Agricultural Sustainability.* Springer, Singapore, pp:19–65.

Kpomblekoua, K. and Tabatabai, M. A. 1994. Effect of organic acids on release of phosphorus from phosphate rocks. *Soil Sci* 158(6):442–453.

Kudoyarova, G., Arkhipova, T. N., Korshunova, T., Bakaeva, M., Loginov, O. and Dodd, I. C. 2019. Phytohormone mediation of interactions between plants and non-symbiotic growth promoting bacteria under edaphic stresses. *Front Plant Sci* 10:1368.

Kumar, A., Kumar, R., Kumari, M. and Goldar, S. 2020. Enhancement of plant growth by using PGPR for a sustainable agriculture: a review. *Int J Curr Microbiol Appl Sci* 9(2):152–165.

Kumar, A., Patel, J. S., Meena, V. S. and Srivastava, R. 2019. Recent advances of PGPR based approaches for stress tolerance in plants for sustainable agriculture. *Biocatal Agric Biotechnol* 20:101271.

Latef, A. A. H. A., Alhmad, M. F. A., Kordrostami, M., Abo–Baker, A. B. A. E. and Zakir, A. 2020. Inoculation with *Azospirillum lipoferum* or *Azotobacter chroococcum* reinforces maize growth by improving physiological activities under saline conditions. *J Plant Growth Regul* 1–14.

Loper, J. E. and Gross, H. 2007. Genomic analysis of antifungal metabolite production by *Pseudomonas fluorescens* Pf-5. In: Bakker P. A. H. M., Raaijmakers J. M., Bloemberg G., Höfte M., Lemanceau P. and Cooke B. M. (eds), *New Perspectives and Approaches in Plant Growth-Promoting Rhizobacteria Research.* Springer, Dordrecht, pp:265–278.

Lugtenberg, B. and Kamilova, F. 2009. Plant-growth-promoting rhizobacteria. *Annu Rev Microbiol* 63:541–556.

Ma, W., Sebestianova, S. B., Sebestian, J., Burd, G. I., Guinel, F. C. and Glick, B. R. 2003. Prevalence of 1-aminocyclopropane-1-carboxylate deaminase in *Rhizobium* spp. *Antonie Van Leeuwenhoek* 83(3):285–291.

Madhaiyan, M., Poonguzhali, S., Ryu, J. and Sa, T. 2006. Regulation of ethylene levels in canola (*Brassica campestris*) by 1-aminocyclopropane-1-carboxylate deaminase-containing *Methylobacterium fujisawaense*. *Planta* 224(2):268–278.

Maheswari, S. and Rajagopal, K. 2013. Biodiversity of endophytic fungi in *Kigeliapinnata* during two different seasons. *Current Sci* 104:515–518.

Majeed, A., Muhammad, Z. and Ahmad, H. 2018. Plant growth promoting bacteria: role in soil improvement, abiotic and biotic stress management of crops. *Plant Cell Rep* 37(12):1599–1609.

Matias, S. R., Pagano, M. C., Muzzi, F. C., Oliveira, C. A., Carneiro, A. A., Horta, S. N. and Scotti, M. R. 2009. Effect of rhizobia, mycorrhizal fungi and phosphate-solubilizing microorganisms in the rhizosphere of native plants used to recover an iron ore area in Brazil. *EurJ Soil Biol* 45(3):259–266.

Miransari, M. 2013. Soil microbes and the availability of soil nutrients. *Acta Physiol Plant* 35(11):3075–3084.

Munir, I., Bano, A. and Faisal, M. 2019. Impact of phosphate solubilizing bacteria on wheat (*Triticum aestivum*) in the presence of pesticides. *Braz J Biol* 79(1):29–37.

Nadeem, S. M., Naveed, M., Zahir, Z. A. and Asghar, H. N. 2013. Plant–microbe interactions for sustainable agriculture: fundamentals and recent advances. In: Arora Naveen Kumar (ed), *Plant Microbe Symbiosis: Fundamentals and Advances*. Springer, New Delhi, pp:51–103.

Naz, I., Bano, A. and Ul-Hassan, T. 2009. Isolation of phytohormones producing plant growth promoting rhizobacteria from weeds growing in Khewra salt range, Pakistan and their implication in providing salt tolerance to glycine max L. *Afr J Biotechnol* 8:21.

Nosheen, A., Bano, A., Naz, R., Yasmin, H., Hussain, I., Ullah, F., Keyani, R., Hassan, M. N. and Tahir, A. T.2019. Nutritional value of *Sesamum indicum* L. was improved by *Azospirillum* and *Azotobacter* under low input of NP fertilizers. *BMC Plant Biol* 19(1):466.

Ojuederie, O. B. and Babalola, O. O. 2017. Microbial and plant-assisted bioremediation of heavy metal polluted environments: a review. *Int J Environ Res Public* 14(12):1504.

Oliveira, C. A., Alves, V. M. C., Marriel, I. E., Gomes, E. A., Scotti, M. R., Carneiro, N. P., Guimaraes, C. T., Schaffert, R. E. and Sá, N. M. H. 2009. Phosphate solubilizing microorganisms isolated from rhizosphere of maize cultivated in an oxisol of the Brazilian cerrado biome. *Soil Biol Biochem* 41(9):1782–1787.

Ortíz-Castro, R., Valencia-Cantero, E. and López-Bucio, J. 2008. Plant growth promotion by *Bacillus megaterium* involves cytokinin signaling. *Plant Signal Behav* 3(4):263–265.

Pahari, A., Pradhan, A., Nayak, S. K. and Mishra, B. B. 2020. Plant growth promoting rhizobacteria (PGPR): prospects and application. *Front Soil Environ Microbiol* 47.

Penrose, D. M. and Glick, B. R. 2001. Levels of ACC and related compounds in exudate and extracts of canola seeds treated with ACC deaminase-containing plant growth-promoting bacteria. *Can J Microbiol* 47(4):368–372.

Pettersson, M. and Baath, E. 2004. Effects of the properties of the bacterial community on pH adaptation during recolonisation of a humus soil. *Soil Biol Biochem* 36(9):1383–1388.

Pieterse, C. M., Van Wees, S. C., Van Pelt, J. A., Knoester, M., Laan, R., Gerrits, H., Weisbeek, P. J. and Van Loon, L. C. 1998. A novel signaling pathway controlling induced systemic resistance in *Arabidopsis*. *Plant Cell* 10(9):1571–1580.

Prakash, V., Khan, M. Y., Rai, P., Prasad, R., Tripathi, D. K. and Sharma, S. 2020. Exploring plant rhizobacteria synergy to mitigate abiotic stress: a new dimension toward sustainable agriculture. In: Tripathi Durgesh Kumar, Singh Vijay Pratap, Chauhan Devendra Kumar, Sharma Shivesh, Prasad Sheo Mohan, Dubey Nawal Kishore and Ramawat Naleeni (eds), *Plant Life Under Changing Environment.*. Academic Press, Cambridge, Massachusetts, United States, pp:861–882.

Prasad, M., Srinivasan, R., Chaudhary, M., Choudhary, M. and Jat, L. K. 2019. Plant growth promoting rhizobacteria (PGPR) for sustainable agriculture: perspectives and challenges. In: Singh Amit Kishore, Kumar Ajay and Singh Pawan Kumar (eds), *PGPR Amelioration in Sustainable Agriculture*. Woodhead Publishing, Sawston, United Kingdom, pp:129–157.

Raaijmakers, J. M., Vlami, M. and De Souza, J. T. 2002. Antibiotic production by bacterial biocontrol agents. *Antonie van leeuwenhoek* 81(1–4):537.

Rezzonico, F., Zala, M., Keel, C., Duffy, B., Moenne-Loccoz, Y. and Defago, G. 2007. Is the ability of biocontrol fluorescent pseudomonads to produce the antifungal metabolite 2,4-diacetylphloroglucinol really synonymous with higher plant protection? *New Phytol* 173(4):861–872.

Rodriguez, H. and Fraga, R. 1999. Phosphate solubilizing bacteria and their role in plant growth promotion. *Biotechnol Adv* 17(4–5):319–339.

Rodriguez, H., Gonzalez, T., Goire, I. and Bashan, Y. 2004. Gluconic acid production and phosphate solubilization by the plant growth-promoting bacterium *Azospirillum* spp. *Naturwissenschaften* 91(11):552–555.

Safronova, V. I., Piluzza, G., Zinovkina, N. Y., Kimeklis, A. K., Belimov, A. A. and Bullitta, S. 2012. Relationships between pasture legumes, rhizobacteria and nodule bacteria in heavy metal polluted mine waste of SW Sardinia. *Symbiosis* 58(1–3):149–159.

Sahni, S. and Prasad, B. D. 2020. Management of collar rot disease using vermicompost and a PGPR strain *Pseudomonas* sp. and their effect on defense-related enzymes in chickpea. *Indian Phytopathol* 1–11.

Schalk, I. J., Hannauer, M. and Braud, A. 2011. New roles for bacterial siderophores in metal transport and tolerance. *Environ Microbiol* 13(11):2844–2854.

Schwyn, B. and Neilands, J. B. 1987. Universal chemical assay for the detection and determination of siderophores. *Anal Biochem* 160(1):47–56.

Sessitsch, A., Coenye, T., Sturz, A. V., Vandamme, P., Barka, E. A., Salles, J. F., Van Elsas, J. D., Faure, D., Reiter, B., Glick, B. R. and Wang-Pruski, G. 2005. *Burkholderia phytofirmans* sp. nov., a novel plant-associated bacterium with plant-beneficial properties. *Int J SystEvol Microbiol* 55(3):1187–1192.

Sharma, A. and Johri, B. N. 2003. Growth promoting influence of siderophore-producing Pseudomonas strains GRP3A and PRS9 in maize (*Zea mays* L.) under iron limiting conditions. *Microbiol Res* 158 (3):243–248.

Sharma, S. B., Sayyed, R. Z., Trivedi, M. H. and Gobi, T. A. 2013. Phosphate solubilizing microbes: sustainable approach for managing phosphorus deficiency in agricultural soils. *Springer Plus* 2(1):587.

Shedova, E., Lipasova, V., Velikodvorskaya, G., Ovadis, M., Chernin, L. and Khmel, I. 2008. Phytase activity and its regulation in a rhizospheric strain of *Serratia plymuthica. Folia Microbiol (Praha)* 53(2):110.

Shoebitz, M., Ribaudo, C. M., Pardo, M. A., Cantore, M. L., Ciampi, L. and Cura, J. A. 2009. Plant growth promoting properties of a strain of *Enterobacter ludwigii* isolated from *Lolium perenne* rhizosphere. *Soil Biol Biochem* 41 (9):1768–1774.

Siddiqui, Z. A. 2005. PGPR: prospective biocontrol agents of plant pathogens. In: Siddiqui, Zaki A. (ed), *PGPR: Biocontrol and Biofertilization.* Springer, Dordrecht, pp:111–142.

Singh, D., Ghosh, P., Kumar, J. and Kumar, A. 2019. Plant growth-promoting rhizobacteria (PGPRs): functions and benefits. In: Singh Dhananjaya Pratap, Gupta Vijai Kumar and Prabha, Ratna (eds), *Microbial Interventions in Agriculture and Environment.* Springer, Singapore, pp:205–227.

Somers, E., Vanderleyden, J. and Srinivasan, M. 2004. Rhizosphere bacterial signalling: a love parade beneath our feet. *Crit Rev Microbiol* 30(4):205–240.

Spaepen, S. and Vanderleyden, J. 2011. Auxin and plant-microbe interactions. *Cold Spring Harb Perspect Biol* 3(4):a001438.

Spaepen, S., Vanderleyden, J. and Remans, R. 2007. Indole-3-acetic acid in microbial and microorganism-plant signaling. *FEMS Microbiol Rev* 31(4):425–448.

Subiramani, S., Ramalingam, S., Muthu, T., Nile, S. H. and Venkidasamy, B. 2020. Development of abiotic stress tolerance in crops by plant growth-promoting rhizobacteria (PGPR). In: Kumar Manoj, Kumar Vivek and Prasad Ram (eds), *Phyto-Microbiome in Stress Regulation.* Springer, Singapore, pp:125–145.

Sujatha, N. and Ammani, K. 2013. Siderophore production by the isolates of fluorescent Pseudomonads. *Int J Curr Res Rev* 5(20):1.

Tariq, M., Hameed, S., Yasmeen, T., Zahid, M. and Zafar, M. 2014. Molecular characterization and identification of plant growth promoting endophytic bacteria isolated from the root nodules of pea (*Pisumsativum* L.). *World J Microbiol Biotechnol* 30(2):719–725.

Tilak, K. V. B. R., Ranganayaki, N., Pal, K. K., De, R., Saxena, A. K., Nautiyal, C. S., Mittal, S., Tripathi, A. K. and Johri, B. N. 2005. Diversity of plant growth and soil health supporting bacteria. *Curr Sci* 36–150.

Timmusk, S. and Wagner, E. G. H. 1999. The plant-growth-promoting rhizobacterium *Paenibacillus polymyxa* induces changes in *Arabidopsis thaliana* gene expression: a possible connection between biotic and abiotic stress responses. *Mol Plant Microbe Interact* 12(11):951–959.

Tsavkelova, E. A., Klimova, S. Y., Cherdyntseva, T. A. and Netrusov, A. I. 2006. Microbial producers of plant growth stimulators and their practical use: a review. *Appl Biochem Microbiol* 42(2):117–126.

Van Loon, L. C.2007. Plant responses to plant growth-promoting rhizobacteria. In: Bakker P. A. H. M., Raaijmakers J. M., Bloemberg G., Höfte M., Lemanceau P. and Cooke B. M. (eds), *New Perspectives and Approaches in PlantGrowth-Promoting Rhizobacteria Research*. Springer, Dordrecht, pp: 243–254.

Vejan, P., Abdullah, R., Khadiran, T., Ismail, S. and Nasrulhaq Boyce, A. 2016. Role of plant growth promoting rhizobacteria in agricultural sustainability—a review. *Molecules* 21(5):573.

Velusamy, P., Immanuel, J. E., Gnanamanickam, S. S. and Thomashow, L. 2006. Biological control of rice bacterial blight by plant-associated bacteria producing 2, 4-diacetylphloroglucinol. *Can J Microbiol* 52(1):56–65.

Verma, P., Yadav, A. N., Kumar, V., Singh, D. P. and Saxena, A. K. 2017. Beneficial plant-microbes interactions: biodiversity of microbes from diverse extreme environments and its impact for crop improvement. In: Singh Dhananjaya Pratap, Singh Harikesh Bahadur and Prabha Ratna (eds), *Plant-Microbe Interactions in Agro-Ecological Perspectives*. Springer, Singapore, pp:543–580.

Verma, R. K., Sachan, M., Vishwakarma, K., Upadhyay, N., Mishra, R. K., Tripathi, D. K. and Sharma, S. 2018. Role of PGPR in sustainable agriculture: molecular approach toward disease suppression and growth promotion. In: Meena Vijay Singh (eds), *Role of Rhizospheric Microbes in Soil*. Springer, Singapore, pp: 259–290.

Vessey, J. K. 2003. Plant growth promoting rhizobacteria as biofertilizers. *Plant Soil* 255(2):571–586.

Voisard, C., Keel, C., Haas, D. and Defago, G. 1989. Cyanide production by *Pseudomonas fluorescens* helps suppress black root rot of tobacco under gnotobiotic conditions. *EMBO J* 8(2):351–358.

Yadav, A. N., Sachan, S. G., Verma, P. and Saxena, A. K. 2015. Prospecting cold deserts of north western Himalayas for microbial diversity and plant growth promoting attributes. *J Biosci Bioeng* 119 (6):683–693.

Yadav, A. N., Yadav, N., Sachan, S. G. and Saxena, A. K. 2019. Biodiversity of psychrotrophic microbes and their biotechnological applications. *J Appl Biol Biotechnol* 7:99–108.

Yi, Y., Huang, W. and Ge, Y. 2008. Exopolysaccharide: a novel important factor in the microbial dissolution of tricalcium phosphate. *World J Microbiol Biotechnol* 24(7):1059–1065.

Zafar-Ul-Hye, M., Hussain, N. M., Danish, S., Aslam, U. and Zahir, Z. A. 2019. Multi-strain bacterial inoculation of *Enterobacter cloacae*, *Serratia ficaria* and *Burkholderia phytofirmans* with fertilizers for enhancing resistance in wheat against salinity stress. *Pak J Bot* 51(5):1839–1846.

14

Omics Technologies to Unravel Plant-Microbe Interactions

Amrina Shafi[1,2], Insha Zahoor[2,3], and Huma Habib[4]
[1] *Education Department, Government of J&K, Srinagar, India*
[2] *Department of Biotechnology, School of Biological Sciences, University of Kashmir, Srinagar, India*
[3] *Department of Neurology, Henry Ford Hospital, Detroit, Michigan, USA*
[4] *Department of Biochemistry, Islamia College of Science and Commerce, Kashmir, India*

CONTENTS

14.1 Introduction

Diversity in plant and microbe interaction is due to critical relationship they share, as microbes niche below, above the soil and endogenously within the plant cells. (Bulgarelli et al. 2013). Thus plant-microbe (PM) interaction can be both epiphytic and endophytic and can be useful, neutral, and harmful which, directly affects the growth and development of crops (Newton et al. 2010). There is a favorable PM interaction, enhancing agricultural productivity, and various useful metabolites secreted in the rhizosphere via this interaction (Talboys et al. 2014; Zhang et al. 2014), which act as signaling molecules assisting microbe in identifying it's interacting plant system. The kind and structure of these metabolites can modify the soil's microbial diversity and dynamics, which can either favor the growth of beneficial microorganisms for improving plant strength and efficiency or inhibit pathogenic microorganism's growth (Chaparro et al. 2012; Dutta et al. 2013). Phytometabolites collected from the *Arabidopsis* root secretions have helped understand the importance of these exudates in the variation of microbial composition (Badri et al. 2013). While some microorganisms are associated with plant pathogenicity and yield losses, plentiful bacteria and fungi species have a favorable effect on plant growth and productivity (Venkateshwaran et al. 2013). Thus, this beneficial partnership between plants and microbes also increases crop efficiency at minimal expenses. Even though our interpretation of responses occurring during PM interactions has helped in many ways, there are still various facets and hitches to answer several queries about these beneficial PM interactions for generating resistant crops for humans' benefit (Misra and Chaturvedi 2015; Thynne et al. 2015).

We still have limited knowledge about the complete picture of various stress reactions against abiotic and biotic stresses in plants. Therefore, varied biological approaches and their updated improvement are essential to assemble and assimilate the evidence into a comprehensive depiction (Knief 2014). Kyrpides (2009) has reported that with the advancement and developments in DNA sequencing technologies (HTS), many microbial genomes from various families have been sequenced, and several more are approaching accomplishment. The omics tools have delivered enormous information about genome structures, microbe genomes evolution, gene function, and regulatory networks with the help of comparative genomic analyses and annotation of DNA sequences, which has helped us understand microbial metabolism and interaction on the whole (Cardenas and Tiedje 2008). In the recent years, studies of PM interactions have improved with the dawn of novel sequencing methods and several omics approaches (e.g., transcriptomics, proteomics, metabolomics, etc.). Remarkable improvement in the novel omics approaches (e.g., phenomics, epigenomics, lipidomics, ionomics, etc.) has helped better understand defense systems in plants against abiotic and biotic stresses. The new Green Revolution has begun with the advent of these comparatively novel strategies for understanding whole PM interactions (Reid 2011). To completely elucidate the metabolism and interaction of microbes with the plant and environmental factors, it is imperative to comprise accurate quantification and description of gene products (i.e., mRNA, proteins, and metabolites) along at cellular level interaction. Thus, with the advent of current omics technologies, it has become easier to study PM interactions. These omics approaches can be applied to understand the plant genome and its substantial effect on the microbe community structure (the rhizosphere, epiphytic, or endophytic microbes). Various omics approaches have helped elucidate the critical signals between microbes and plants and are promising strategies to know the PM interacti, improving agricultural production and productivity. This chapter will discuss the improvement of various "omics" platforms in discovering the intricate regulatory mechanisms and metabolic systems involved in PM interactions. We have emphasized on the omics platforms used to understand PM interactions, more current findings related to secreted phytomolecules. We have also discussed the latest novel approaches used to understand the PM interactions and their associations.

14.2 Plant-associated Microbiome

In nature, during different stages of the plant growth cycle, plants constantly encounter several microbial species and live together with microbiota (e.g., archaea, protists, bacteria, and fungi) (Bulgarelli et al. 2013). From the previous studies we know that, during the evolutionary process green algae and plants have originated from cyanobacterial endosymbionts, which are now known as chloroplasts or plastids (Keeling 2010). This evolutionary history indicates that the plant and microbes share similar ancestries and are symbiotic. Therefore, the "plant microbiota" and significant part of microbes in PM communications has gained more attention during the past decade (Vorholt 2010; Bulgarelli et al. 2013; Waghunde et al. 2017). The exploitation of this favorable plant microbiota seems to be a better strategy for collaborative development and producing crops in the long run (Reid 2011). The soil microbiome comprises diverse types of organisms, but bacteria, fungi, and archaea are those receiving by far more attention in soil microbiology studies and a less explored diversity of viruses (Berendsen et al. 2012; Agler et al. 2016). The three primary forms of PM associations which have been explored are a symbiosis among mycorrhizae and plants (Denison and Kiers 2011), between plants and rhizobacteria (Hirsch et al. 2013; Vejan et al. 2016), and between plants and pathogenic microbes (de Souza et al. 2016). The soil microbiome composition is designed by intricate multilateral interactions between the abiotic factors, its biotic populaces and varied networking aspects for sustainable and healthy crop production. There are multiple interacting partners of the microbial society with unique diversity and complexity that work together for plant development and growth. The cross-talk in the rhizosphere between interspecies or inter-kingdom through PM or plant-plant communications is indispensable for the environment's farming environment and sustainability (Broz et al. 2010; Pellegrino and Bedini 2014).

Furthermore, the widely explored biotic associations for plant growth and yield are shaped by mycorrhizae fungi, alien plant species, and plant growth-promoting rhizobacteria (PGPRs). PGPRs are beneficial to soil bacteria and represent a wide range of genera recruited by roots, mainly *ludebium*, *Pseudomonas*, *Azospirillus*, and others (Bhattacharyya and Jha 2012). PGPRs are unique in their versatile roles, such as phosphorus solubilization and uptake, synthesizing plant growth hormones, and providing resistance toward environmental stresses (Bhattacharyya and Jha 2012). Shen et al. (2013) reported that PGPRs interaction increases crop growth and yield by enhancing the rhizosphere efficiency in nutrient mobilization and uptake in plants at a high rate (Parmar and Sindhu 2013). Therefore, sustainable agriculture needs beneficial PGPRs to assist nutrient accessibility, enhance plant growth and seed emergence, and favor plant vigor (Mia et al. 2010). Microbiome profiling of plants and root-associated soils has led to the identification of varied and exceedingly active PM interactions that have been discovered from. Vorholt (2012) has reported that various microbial communities have a dynamic influence in shaping our environment (i.e., season, daytime, soil, air) as well as plants (species, developmental stage, and compartment). A unique and robust gradient of microbes abundance in plants is concentrated in bulk soil declines in the roots and is lowermost inside plants. Copeland et al. (2015) have described that various factors affect plant microbiota association during the commencement of the growing season, e.g., soil stimulates microbial communities aboveground. This potential interaction is probably due plants the ability to release and made available some essential metabolites to microbes, which metabolize as nearly 40% of the host carbon assimilated. It indicates the active role of plant channelizing the specific microbial populations (Bais et al. 2006). Thus, a well-designed study pipeline will help classify the functions of various genes, mRNA, metabolites, and proteins that take an active part during PM interactions, which will further highlight the vital processes that strengthen PM associations.

14.3 Exploring the Fundamentals of the Plant-Microbiome Interactions

Several mechanisms are inbuilt in the microbiota that encompasses the plant capacity to adapt to the inconsistent environmental circumstances. Constructive PM interactions comprise promoting plant growth, adaptation and defense toward environmental stresses via induction of various defensive pathways and immune system activation, symbiotic relationships, efficient nutrient translocation, and solubilization of inaccessible nutrients (Hassani et al. 2018). These PM associations are symbiotic (both host plants and microbes gets benefited) and this symbiosis among microbe and plant advances into unique associations, which depends on its influence on overall plant growth and vigor, i.e., symbiotic (favorable to both the partners), neutral (commensalism), or detrimental (pathogenic) (Rodriguez et al. 2019). These PM associations are critical for both the environment and sustainable agriculture. Rosenberg and Zilber-Rosenberg (2018), investigation suggests that the microbe and host plant interactions resulted in co-evolution of various interactions such as plant-plant, microbe-microbe, and plant-microbe. Currently, profiling of these plant-associated microbiomes is an evolving perception to comprehend the PM interactions. According to Carvalhais et al. (2013), the challenge now relies primarily on profiling the processes, which are diverse and highly complicated, involving tangled uncultivable microbial populations. Nevertheless, we have several molecular techniques accessible. We are presently being used to decode the unknown assortment of microbiomes residing in roots and soil niche. These culture-independent techniques are applied in understanding PM interactions at the cellular level. Omics methods (e.g., genomics, metabolomics, transcriptomics, proteomics, and metagenomics) and, very lately, epigenomics have helped understand the PM interactions at the rhizospheric level and the key metabolites, protins, and mechanisms involved (Figure 14.1). Nevertheless, there are still many secrets hidden in these PM interactions to be uncovered. In the current worldwide revolution, these omics tools are also vital to assess the influences of perturbations triggered by numerous stress factors on microbial range and PM associations. The exploration of plant and microbiome genes, which are intricate in the PM interfaces, is indispensable for the forthcoming usage of the plant microbiome. Omics approaches, next-generation sequencing (NGS) technologies and molecular biology, have significantly explored both beneficial and pathogenic microbes (Basu et al. 2018; Jansson and Hofmockel 2018). Barret et al. (2013) based on the functional genomics analysis model, has reviewed the PM association of soil microbes at the metabolomics, genomic, proteomic and transcriptomic levels. Bioinformatics study helps in the construction of richness and dispersal models of functional characters of diverse microbial genomes (e.g., mostly involved in root colonization). Microbial consortia is used to understand PM interactions in sustainable agriculture, whereas another involves a particular genetic change of both bacteria and its host.

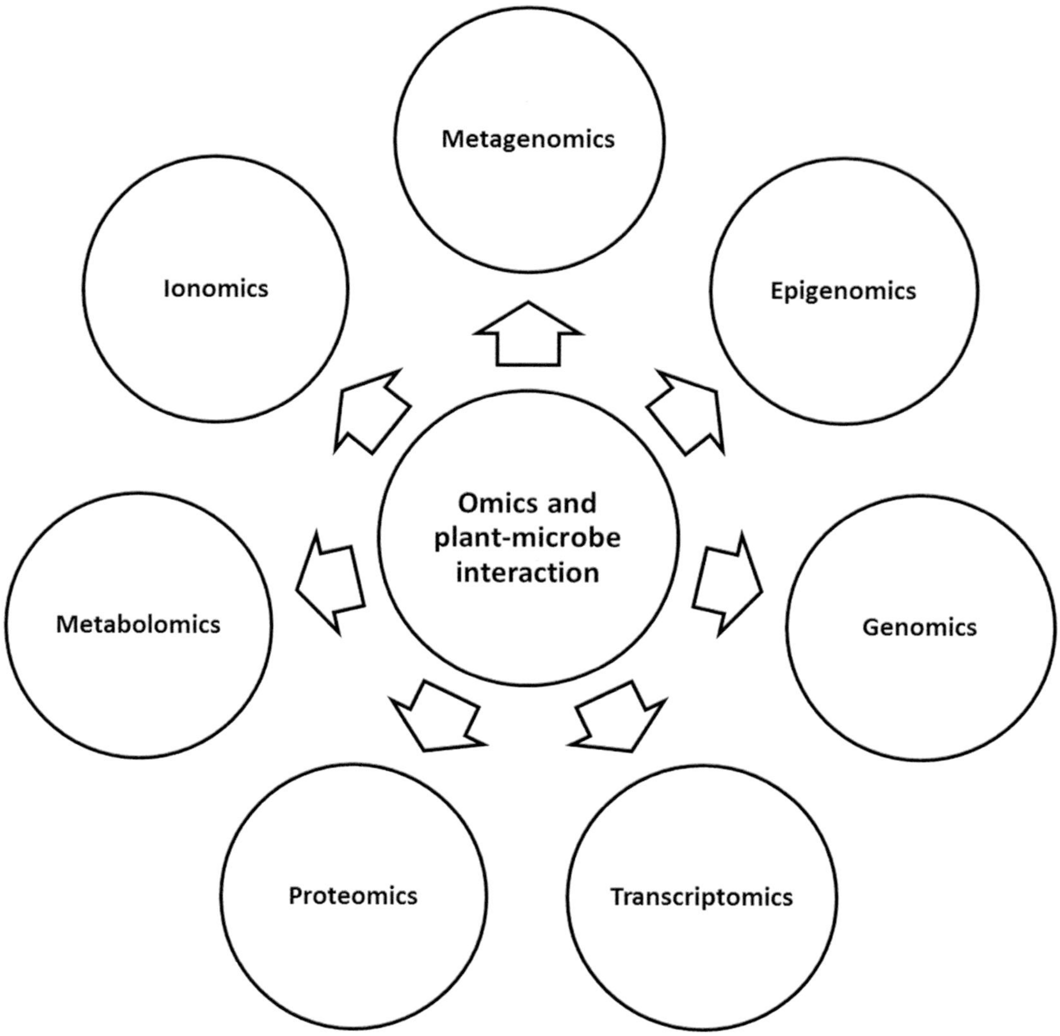

FIGURE 14.1 Omics approaches to the study dynamics of plant-microbe interactions in order to improve plant productivity.

Earlier, gene knockdown and transgenic technology was a promising approach to study gene functions or trait improvement to achieve fast and reliable results. However, the amalgamation of alien genetic entity bounds its prevalence as it is still in its infancy stages (Muñoz et al. 2019). Currently, genome editing (GE) technology which assists in editing the DNA sequences in a more accurate direction, is of completed attentiveness (Knott and Doudna 2018), a double-strand break or nick (DSB) is created with the help of specialized restriction endonucleases (Gaj et al. 2013). Revolution in GE has become with the discovery of CRISPR/Cas systems in eukaryotic GE (Mali et al. 2013) and has appeared as the utmost prevalent GE tool (Sakuma and Yamamoto 2017). Thus, with the latest advancements in these techniques, elucidation of PM interaction, microbial dynamics, and root exudates has expanded exponentially (Dennis et al. 2010). Some scientific work has taken the research to a different horizon and highlighted how we can explore and harness these beneficial microbes' potential and their interactions for improving plant growth and health (Farrar et al. 2014). Plants, as well as microbes, can modify in response to the microenvironment at different omics levels, e.g., fungi modify plant transcriptome by changing expression levels of various genes (Liu et al. 2003) as well as adjust the plant metabolome to build a mutual communication with rhizosphere (Zhao et al. 2013). Similarly, pathogen contact with Arabidopsis modifies the proteomics (De-la-Peña et al. 2008) and epigenomics (De-la-Peña et al. 2012). Consequently, we can utilize all these latest omics tools to increase plant growth and yield and enhance beneficial microbe diversity.

14.4 Epigenomics and Genomic Study in Plant-Microbe Interaction

The invention of numerous NGS technologies has resulted in the completion of overwhelming numbers of plant and their pathogens genome sequencing and re-sequencing. A huge amount of data has been generated from these NGS platforms, which has led to an era of genomics, epigenomics and post-genomics with a challenge to decipher this huge data for plant growth and development through wider tolerance ranges (Knief 2014). To prepare crops against current thwarting climatic conditions and pathogenic bacteria, fungi, and viruses, translational genomics presented a better resolution (Knief 2014). NGS platforms can produce enormous data, which is expected to transform our understanding of epigenomics functional and comparative genomics (Zhang et al. 2011). Epigenomics (latest omics) has likely improved in elucidating the rhizospheric microbes' dynamics and their function in refining plant efficiency. Understanding the complexity of gene regulation for agricultural applications has become easier with the advent of epigenetics (Boyko and Kovalchuk 2013) and different environmental circumstances modify plant epigenetics (Boyko and Kovalchuk 2013), and these regulatory mechanisms to survive such stresses can be inherited. Peng et al. (2004) has reported a finding that could improve plant productivity, they have observed that climate change has resulted in huge crop damage and plants exposed to such temperature extremes have inherited epigenomic variations which has benefited them for better survival and adaptation (Mirouze and Paszkowski 2011). These epigenetic changes could be due to the secretion of certain metabolites or changes in protein secretion into the rhizosphere during plant-pathogen interaction (De-la-Peña and Loyola-Vargas 2012). Plant genotype and environmental conditions affect the plant-microbe ecological communities (Figure 14.2), the variance between genotypes results in changes in biological and immunity which eventually primes to plant-specific bacterial populations. Metagenomics of rhizosphere interactions has resulted in enormous contribution in genomics and this metagenomic approach has identified and characterized many microbes (Mendes et al. 2011). Some outstanding advancement through the metagenomics field has been achieved, as it permits DNA profiling of bacterial communities (environmental sample) and has the great advantage of accessing a much greater diversity of complex microbial communities that are amenable to laboratory cultivation (Kakirde et al. 2010). Mocali and Benedetti (2010) suggested that with the advent of the metagenomics, novel genes, metabolites and species have been discovered that have immense potential in agriculture and various areas. As soil is rich in microbial biodiversity and each of these microbes have definite biological roles in plants, Sugawara et al. (2013) have conducted a metagenomics study on five different *Sinorhizobium* spp. (*Sinorhizobium meliloti*) and has established the fact that each microbe has implemented its unique tactic to associate with various host plants (*Medicago* spp.) and microenvironments. Further, they have reported that the genes associated with the synthesis of polysaccharides, Nod factor biosynthesis (lipochito-oligosaccharides), denitrification, and secretion systems differ within and among species. Localized cloning and insertional mutagenesis techniques have helped understand the rhizobacteria nodulation mechanisms with its host plant (Webb et al. 2000) and provide some evidence about the specific PM interactions. Furthermore, Takahara et al. (2013) characterized and recognized a root factor in *Lotus japonicus* which is involved in nodulation autoregulation with the combination of deletion technology, cloning based on maps, and NGS analysis.

Thus, with the advent of various genomic approaches, present understanding of PM interactions, the environmental significances of the microbiome and host-bacterial metabolic changes have become more comfortable but are still inadequate to clarify the plants and pathogens interactions (Bender 2005). Furthermore, the omics approach helps detect the genes evolved during PM interactions that assist them in acclimatizing to microenvironments. Levy et al. (2018b) have isolated and sequenced 484 genomes of the rhizosphere, particularly from *Brassicaceae*, poplar, and maize and then applied comparative genomic studied to recognize plant-associated gene clusters. They have

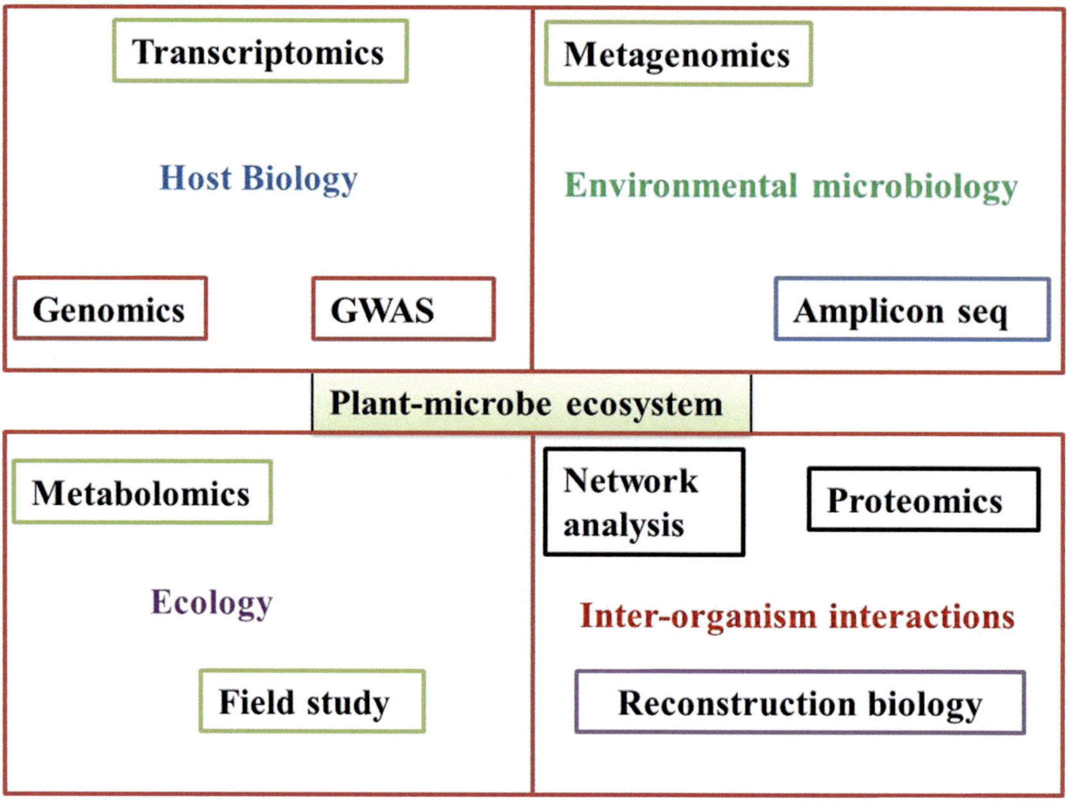

FIGURE 14.2 Genomic tools to understand major units of the plant-microbe ecological system.

reported that genomes of microbiome code more sugar metabolism roles and many plant-associated protein domains that resemble with those domains. This information has extended the genomics-based elucidation of PM interactions and delivers possible leads for sustainable and productive agriculture (Levy et al. 2018b).

14.5 Transcriptomic in Deciphering Plant-Microbiome Interaction

Among the omics approaches, transcriptomics represents a dominant strategy that can generate massive data, which further has to identify the probable genes responsible for crop growth and yield (Figure 14.3). Comparative analysis of transcriptomic data helps find varied expressed genes linked with two naturally different groups under different situations (Bräutigam and Gowik 2010). Transcriptomic technologies have found application in exploring the genome-wide transcriptional activity of various microbial systems, which helps detect genes and regulons responsible for PM interactions (Lakshmanan et al. 2013). RNA-sequencing analysis and microarray technique are the most commonly employed methodologies that can produce transcriptome level information of PM interactions (Wang et al. 2016). Ramachandran et al. (2011) has used ta ranscriptomics approach to perform a comparative analysis of how microbe (*R. leguminosae*) adapt and grow in the roots of a host legume pea, in non-host alfalfa, and a non-legume. They further showed that in the pea rhizosphere, the microbe expressed genes related to Nod factor biosynthesis and metabolism, lignin degradation, and metabolite uptake genes were differentially regulated in alfalfa rhizospheric region. Fan et al. (2012), reported that 8% of the transcriptome of rhizobacterium *Bacillus amyloliquefaciens* was modified due to the root secretion of maize, and the bulk of these genes showed increased expression, especially those which are intricate in microbe movement, nutrient uptake,

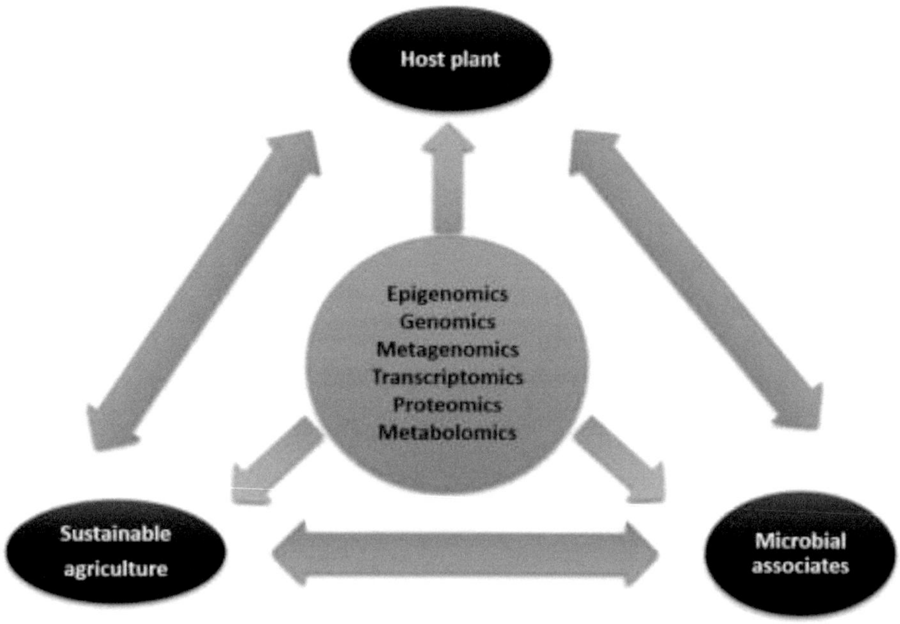

FIGURE 14.3 Omics strategies to the study role of plant-microbe interactions in sustainable agriculture.

motility, and antibacterial molecules. Transcriptomic tools were further explored by Penterman et al. (2014) to discover essential genes e.g., the nodule Cys-rich antimicrobial peptides (NCR) genes (which regulates *S. meliloti bacteroids* growth by controlling the cell cycle). They delivered robust evidence about the molecular mechanism by which NCR peptides switch the *S. meliloti* cell cycle through interaction. RNAseq helps in identifying genes induced in microbe *Sinorhizobium meliloti* (Indole acetic acid, IAA mutant strains) against stress response (Defez et al. 2016); further, they did comparative transcription profiling of two *S. meliloti* strains and found that the genes responsible for stress responses and induction of sigma factor RpoH1 in IAA mutant strain. In another report Alavi et al. (2013), with the transcriptome analysis, have recognized a unique plant growth hormone spermidine synthesized by rapeseed and its associated symbiont against abiotic stress. Further, gene/DNA arrays have been established to identify and assess the differential expression patterns of various gene transcripts along with the particular functional bacterial actions. In this perspective, DNA/gene microarrays are routinely used for comparative analysis of rhizospheric microbe response to various external factors at a transcriptional level. Transcriptome analyses with the help of these fast and robust microarrays have been done for a large number of harmful/pathogenic PM associations (Schenk et al. 2009). The comparative transcriptome in contrasting situations can be used to evaluate the differential gene expression studies based on microarrays. Gene/DNA microarray studies of plant transcriptomics have led to discovery of an essential controlling and resistance genes and complex signal transduction pathways which are activated after elicitor identification and defense signaling plant growth regulators (Livaja et al. 2008; Schwessinger and Zipfel, 2008; Sana et al. 2010). Xu et al. (2011) have employed this RNA-seq and microarray technique for identification of plant genes facilitating defense reactions to the root pathogen.

Meta-transcriptome analysis is another technique used for comparative analysis of microbial transcriptomes obtained from ecological samples and is being used understand the consequences of various environmental factors on microbe transcriptome (Morales and Holben 2011; Simon and Daniel 2011). Meta-transcriptome tools can transform research on PM associations, enabling better exploration of changing transcriptomes of microbes against different environmental factors. It mainly emphases depicting gene expression arrays exhibited by bacterial populations (Morales and Holben 2011). Metatranscriptomics for assessing expressed profiles of microbial functional genes, accountable for important associations, serve as challenging approaches (Carvalhais et al. 2013). Though interactions of crops with mycorrhizae, pathogens, or distinct PGPR have been studied comparatively well, it is expected meta-transcriptomic approach leads to the identification of possibly fascinating and novel PM interactions. Studying the microbiome population has the prospective of finding unknown valuable PM relations, as root growth of wheat seedling has increased when colonized with different *Pseudomonas* spp (Campbell and Greaves 1990). The recent growth in the field of microbial meta-transcriptomics and genomics (Kakirde et al. 2010; Morales and Holben 2011) has observed vast application to further gain insights into multiple interactions of rhizospheric microbes. DNA oligonucleotide microarray or PhyloChip based metagenomics study of the root microbes has led to the identification of more than various archaea and microbes species in the rhizospheric of plants grown-up in sterile soil (Mendes et al. 2011). Earlier, investigation of gene transcript profiling of symbionts and their hosts was more focused on either one partner (Delmotte et al. 2010), and only a rare real-time transcriptomic profiling approach has been reported (Ithal et al. 2007). Current transcriptomic analyses of interactions between pathogenic versus beneficial microbes have led to supportive hypothesis data, which explains how favorable fungi had minimal effect on host gene expression profiles in the deficiency of pathogenic species (Brotman et al. 2011; Molitor et al. 2011). Another transcriptional profiling study by Breuillin et al. (2010) on the interaction between Petunia hybrid host plants with beneficial arbuscular mycorrhiza has discovered a novel role for phosphate (Pi) suppressing vital symbiosis related genes. This discovery has significant inferences in the management of Pi levels to maximize plant development and productivity. Several such works on transcriptomic profiling of plants and beneficial bacteria such as Pseudomonas spp. have been reported (Lim et al. 2010). In interactive transcriptomics study, where concurrent studies of plant and its interacting microbes are done, only minimal transcriptional profiling reports are available. A favorable or opportunistic interactions amongst host and bacteria are strongly interconnected, we need more such studies, A study of legume-rhizobium associations involved in nodule establishment in roots of host plant has been well documented for understanding developmental changes and gene expression (Lohar et al. 2006; Stancey et al. 2006). Symbiosis Chips or customized dual genome chip has been developed (Barnett et al. 2004), where real-time expression analyses of the associated microbes and plant gene are typically delimited to nitrogen-fixing nodules analysis. Also, Brechenmacher et al. (2004) have characterized gene expression of arbuscular mycorrhiza collaborations with the rhizospheric region. Other recent interaction transcriptome analyses were documented among *Oryzae* (rice) and *Magnaporthe oryzae*, which has directed to the elucidation of how plant cells are prepared attachment and invasion of fungal hyphae (Mosquera et al. 2009). The interaction transcriptome analyses should be linked with proper bioinformatics approaches, including databases like Plant-Associated Microbe Gene Ontology Consortium, where various microbial species genomes have been identified and annotated (Torto-Alalibo et al. 2009). Transcriptomics analysis of several species can help in characterization of crucial genes involved in PM interaction which provides new approaches for disease resistance. Data sets from OMICs combined with statistical tools and bioinformatics help in identifying and integrating vital processes, as well as predicting them via modelling. Short-read RNA sequencing data has been used to understand the ectomycorrhizal communications and detect significantly expressed transcripts that can be mapped to specific pathways (Larsen et al. 2011) and used to develop an ectomycorrhizal metabolome model.

14.6 Proteomics

Proteins represent vital biomolecules of all living systems, as they serve in various regulatory and metabolic pathways and several functions at the cell and molecular level (Graham et al. 2007). Further, proteins are vital for maintaining cellular, biochemical and, physiological processes in the living system. Proteomics, as an emerging field for the investigating various proteins, synthesized during PM interactions and also protein-protein interactions studies. Thus, the recent advent in high-throughput proteomics methodologies has made it possible in the identification and quantification on a massive scale. Current progress in the research exhibited that combo of genomics and proteomics can help in discovering of novel molecules particularly in the field and health enhancement (Scherp et al. 2011; Jayaraman et al. 2012). The latest combination of bioinformatics, proteomics, and computational biology approaches, which helps identify and quantify synthesized molecules within the cell at particular time and conditions, it has decoded the information involved in PM associations (Imam et al. 2014). The proteome-level studies provide a deeper understanding of the essential proteins and facilitates definite processes (Kav et al. 2007). PM proteomics represents a difficult task to identify, study and characterize the interactions involving proteins of both partners (Mathesius et al. 2003). Proteomics is considered an important biotechnological tool to identify microbial metabolism, differentiation and relationship with environments (Lacerda and Reardon 2009) and to discover specific proteins intricate in PM associations (Wang et al. 2013), for crop improvement (Eldakak et al. 2013). Antelmann et al. (2006) have used a proteomics approach to study bacterial and fungal secretomes, where they found that *B. subtilis* and fungal partner discharges various proteins that have a vital role in plant surface adhesion and attachment of these microbes (Newey et al. 2007), virulence-related proteins (Song et al. 2009), plant tissue invasive and penetrating proteins (Dong et al. 2011), and symbiotic related proteins (Plett et al. 2011). The understanding of microbial biofilm creation and quorum recognizing signals is critical, as these signals assist microbes in adhesion processes leading to favorable PM associations. Thus, these signals have a deep consequence of plants proteomics and transcriptomics analysis (Mathesius et al. 2003). Budiharjo et al. (2014) identified numerous genes which are involved in promoting plant growth, microbial root and biofilm formation in *B. amyloliquefaciens*. Thus, targeting these any of these signals either quorum sensing or biofilm formation, we can assist more microbial colonization on plant surfaces, which can result in more PM interactions and improvement in plant health, growth and yield (Farrar et al. 2014). Proteomics filed has significant implications in understanding PM interactions, endo or epiphytic microbial communities and plant facilitated signaling responses of associated microbes are intricate (Kosova et al. 2015). Thus, more precise identification of critical proteins involved in adjusting processes of a biological system, stress mitigation and in physiological status will give us more understanding of intricate PM interaction mechanisms (Silva-Sanchez et al. 2015). Various research groups (Kosova et al. 2015; Xu et al. 2015; Wang et al. 2016) have extensively studied a relative analysis of control, stressed and bacteria linked plants in identifying protein targets and networks in plants. Proteomics analysis by using bioinformatics tools, assisted in better understanding the active interchanges in the functional groups of protein, protein-protein interfaces, signaling proteins and adjusting pathways (Kosova et al. 2015). As we know that the cellular life is well-structured via a multifaceted protein-protein system, where we can find numerous proteins involved in regulatory pathways and these proteins tend to assemble in huge protein complexes which are known as interactomics. Identification of these interactomics signifies the vital directions of proteomic providing a better understanding of cellular uptake in bacteria (Singh and Nagaraj 2006). Proteomics field has some demerits especially with proteins of the unknown role and function Choi et al. 2010), with minimal understanding of these unknown proteins. Dam et al. (2014) has reported and highlighted a diverse group of proteins or any of the protein domain which is involved in PM associations in harmful and beneficial microbes (Dam et al. 2014).

14.7 Metabolomics Applications for Understanding Complex Plant-Microbes Interactions

Microbial metabolomics represents an extremely useful technique in interpreting metabolite production in microbes, this information of metabolism will help in connecting the gene-appearance understandings, as it augments variations at the protein level and delivers an improved illustration of the microbiome's appearance (Cascante and Marin 2008). The metabolomics studies involve the depiction of total cell metabolites synthesized by a living system due to the stimulus of ecological situations. Bundy et al. (2005) stated that the metabolomics study helps in correlating different cellular processes and pathways occurring, this information will give us approachability of equivalent information at the genome level. Understanding how plants scrutinize its rhizospheric region for selective microbes by limiting some bacterial movement, this will offer us an innovative prospect to enhance crop growth and yield (Berendsen et al. 2012). Additionally, metabolomics study gives us a broader picture of environmental significance in shaping the association better understanding of the ecological of the interaction between various constituents which assist in the advance of viable management techniques. In recent times, an improved understanding of PM interactions at the cellular and complex signaling processes level which else organize different defense response in the plant system. Even though with the advancement in genomics and post-genomic era, we have got improved information and a vast data accessible and, there is still a large void in the research which aims at the cellular level which may be due to the highly complex molecular system and interconnected components to various interior and exterior situations (Collakova et al. 2012). Further, the presence of diverse bioactive chemicals in plant system (Ketchum et al. 2003) was identified and this statement connects with the other information about the documentation of numerous signaling metabolites synthesized by plants to invite and prompt imperative biological pathways in interacting microbiome population (Micallef et al. 2009). Establishment of PGPR-plant assemblages involves the application of the new technique which has allowed in the identification and understanding the functions of various signaling molecules and signaling pathways. Drogue et al. (2013) have categorized the signaling molecules based on site of production as follows: First, is the PGPR expressed under the influence of host-plant, second where PGPR disturbs plant growth and well-being, and third category, where PGPR influences the expression by another PGPR genes. These PGPR synthesize small-signal molecules (phytohormones) that influence plant development and stress mitigation (Drogue et al. 2013). Interestingly, signaling processes in symbiotic exchanges occur in PM interactions linked to nutrient supply, i.e., nodulation and (Bonfante and Desirò 2015).

Metabolic interactions are one of the essential values of PM associations and numerous novel and sophisticated tools have evolved to study plant metabolomics, such as GC-MS, NMR spectroscopy, LC-MS, HPLC, LC-tandem MS, and capillary electrophoresis-time-of-flight-mass spectrometry, which are used to recognize and identify some vital metabolites (López-Ráez et al. 2008). A huge number of these metabolites have a role in cross-talk signaling in among rhizospheric microbes and are involved in primary recognition and species reliant responses. Using NMR spectroscopy, Vauclare et al. (2013) have analyzed metabolic profiles soybean nodules colonized by the *Bradyrhizobium japonicum* cells. Similarly, Fan et al. (2001) studied the role of root exudates from wheat and barley in the Cd uptake by using NMR spectroscopy. Besides, Tawaraya et al. (2013) have identified metabolites from root metabolites secreted from rice using the novel technique of capillary electrophoresis-spectrometry throughout PM interaction. Growing stage of the plant also affects the metabolites from roots exudates (De-la-Peña et al. 2010) and have varied effects on root microbes (Ziegler et al. 2013). Gas chromatography-mass spectrometry was used by Badri et al. (2013) to characterized phytochemicals of root exudates. They established that some phenolic-related compounds from root exudates can modulate soil microbes. It is estimated that about half of multifaceted carbons in plants are synthesized by plant via roots into the rhizosphere to nurture its microbes (Lynch and Whipps 1990). On the contrary, the microbiome enables solubilization and plant P, N, and Fe uptake (Jacoby et al. 2017). This mineral/nutrient re-localization is an imperative aim of plant re-programming by microbiome via phytohormone and effectors signaling. The untargeted metabolomics tools have an exceptional possibility in enhancing food safety strategies and plant defense, particularly in exploring the general adaption make-up of the host plant with its interacting microbes. The Genome-scale modeling ("GEMing") is an *in silico* metabolic flux prototype obtained from the presently existing genomics information for understanding the metabolism. GEMing is exploited to analyze the phenotype during PM interactions (Collakova et al. 2012). Metabolic modeling of both prokaryotes (microbes) and eukaryotes (higher plants) has routinely been generated (Heavner and Price 2015; Botero et al. 2018). These metabolic modeling of beneficial and harmful microbes are analyzed by comparative networking models, as Mithani et al. (2011) verified the assumption that pathogenicity regime of Pseudomonas *syringae* metabolomics has progressed to be specifically devoted toward its host plant. Lately, an explicit regime of a metabolomic model for the pathogen *Phytophthora infestans* was generated, which envisages various biological processes which occur in different sections of pathogenic cellular spaces and phases (Rodenburg et al. 2018). Thus it would be imperative to limit the application of metabolic models to metabolite level quantification as we can find its application at all levels of molecular processes that can generate a more clear-cut picture of the cellular, molecular and metabolic variations brought in the plant during microbial colonization. Transcriptomic and metabolomic analysis cocktail highlights the vital role of amino acid metabolism and ABA metabolism and signaling pathway for stress tolerance in sesame (You et al. 2019). Chowdhury et al. (2015) have reported that host plants control the metabolites enabled tap cellular signaling in microbes with the help of various volatile compounds. Similarly, Han et al. (2016), has verified this statement in a beneficial microbe (*Acidovorax radices*), where they reported that a lactone (N-Acyl homoserine lactones or AHL) was a vital metabolite in microbe cellular signaling and AHL was also be sensed by host plant (barley). Schenk et al. (2014), has also demonstrated the growth-enhancing and grooming consequences of AHLs on various host plants (Arabidopsis, Medicago, and tomato etc.). Even though the biological influence of these AHLs is well understood, still there are some plants recognition mechanisms and processes of these microbial metabolites persist unidentified (Schikora et al. 2016). There are some lipo-chito-oligosaccharides (e.g., Nod and Myc) association elements, which assist in plant growth and development processes without associating with host plant (Tanaka et al. 2015), this helps us in understanding the fact that the association elements perception and signaling organization are partly autonomous of the inter-dependence competency of the plant. We still require more research in this field to elucidate mechanism and range of microbial signals are released, perceived and inferred by the specific host-plant along with positive or negative effects of these varied metabolites on plant development and confrontation. Collin et al. (2018), has elucidated a term known as stress microbiome, which specifies the elucidation of complex and highly organized interactions among the microbial population and its host counterpart, this shows how a preserved microbe populations organization is linked with higher plants and trees under diverse growing environments. They have applied a bacterial booster approach in studying the effect of changing environmental (heavy metal, light and drought stress) conditions on the microbiome of *Populus deltoids* species. This study helped in identification of a group of differentially regulated microbial genes in response to these stress conditions given to host plant, indicating a significant structure in the PM population and identified probable metabolites of the microbiome responses to stress conditions (Collin et al. 2018). Currently, combined metabolomics and transcriptomic investigation have the potential to serve as an influential technique to understand and figure out the complex association among various components (i.e., genes, transcripts, and functional metabolites) within the cells. Earlier it was reported by ecologists that simultaneous analysis of metabolites-transcripts connection can help us in the identification of functionally essential metabolites (Askenazi et al. 2003) and genes (Urbanczyk-Wochniak et al. 2003). Depuydt et al. (2009) used combined metabolomics and transcriptomic tool to understand the interaction mechanism of Arabidopsis thaliana and *Rhodococcus fascians*, they identified pronounced cytokinin response from transcript results of *R. fascians*.

14.8 CRISPR-mediated Plant-Microbiome Interaction

The PM association information at the gene level is available for various plants and microbial associates, exploring a mechanistic aspect of this association at the genetic level which will prove to be a vital step concerned with the exploitation of microbial genome for sustainable agriculture (Bulgarelli et al. 2015). Keeping this in mind, Knott and Doudna (2018) have developed an innovative stage to understand the basics of the PM associations in an easy and speedy technique known as CRISPR (clustered regularly interspaced short palindromic repeats)-based GE. This GE tool is skillful at tempering with genome bringing specific and precise genetic alterations which can improve crop efficiency and infection tolerance (Zaidi et al. 2018). CRISPR-based GE tool has delivered data about the alignment, assembly, and aspects determining the PM associations; several reports have documented the application of CRISPR-based GE tools in plants predominantly emphasizing on PM interactions (Yin et al. 2017; Chen et al. 2019). As we know that PM symbiosis is dependent on both the microbiome and the host plant genome (Levy et al. 2018a), some plant immune system pathways have evolved to identify signaling molecules synthesized by an associated microbe which aid in eliciting the host defense against specific pathogenic microbes (Netea and van der Meer 2017). Among these signaling molecules that assist PM interactions are plant growth regulators such as jasmonic acid, ethylene, melatonin, and salicylic acid (Pieterse et al. 2014; Han 2019). Bisht et al. (2019) have utilized this GE tools in target gene silencing to understand their role in plants or microbes at the molecular level. Exploration of microbial models will help in understanding the mechanistic facet of genomic factors that are responsible for beneficial or harmful associations with its host plant (Glick 2014; Xin et al. 2018). Recently, these GE tools were used in the rhizospheric bacterium for inducing genetic alterations (Aparicio et al. 2018; Sun et al. 2018).

14.8.1 Interactome Network Analysis

The interaction-based network analysis has been an effective strategy to recognize various factors or elements, pathways, and processes involved molecular PM associations (Marin-de-la-Rosa and Falter-Braun 2015). To elucidate the data involved in interaction mechanisms between plant and microbe, we need a framework of plant ecology which will serve as an obligatory reference protein linkage. Arabidopsis Interactome Mapping Consortium (2011), also known as Arabidopsis Interactome-1 (AI-1), was the pioneer project in developing plant interactome map based on experimental data of Arabidopsis protein-protein domain interactions and this interactome map has offered a combined structural understanding of proteins networking in the plant. Since then matching and more focused interactome maps were generated created and this has further simplified our understanding of complex and specialized pathways. Jones et al. (2014) have developed an interactome map of membrane protein-protein domain interactions with the help of the split-ubiquitin strategy. Similarly, Klopffleisch et al. (2011) have generated a G-protein-protein interaction map, which discovered a novel regulatory role of G-proteins in cell wall alterations involved in defense purpose. An interactome map for the fungus *Phomopsis longicolla* (affecting soybean) was created, which assisted in revealing disease-linked sub-networking pathways (Yu et al. 2004; Li et al. 2018).

14.8.2 Plant-Microbe Interaction Databases

As we know that the microbial population can be both harmful and beneficial to plants and it is a known living entity evolving on earth promptly and continuously. There are some databases which represent the information libraries of various genes, proteins, and metabolites which directly or indirectly affect the host-pathogenic microbe associations. Among these databases is the pathogen-host interaction database (PHI-base) which was recognized in 2005 (Winnenburg et al. 2008). It is an online database comprising data from various pathogenic species (Urban et al. 2015) that assists in the investigational confirmation of effect or genes virulence and pathogenic data. This database also gives significant information about agriculturally important microbes which can be potentially targeted and harvested for their agronomic importance. PHI database is user friendly, allowing users to do a comparative analysis of various microbial system and plant endophytes (Kersey et al. 2014). Another database which is used for studying PM association and interaction is a functional genomics database; it includes libraries of potential genes which have a role in PM associations (Levy et al. 2018b). This is a joint venture of researchers from U.S. Department of Energy (DOE), Joint Genome Institute (JGI), and the Howard Hughes Medical Institute at the University of North Carolina at Chapel Hill (UNC). These researchers have explored, identified, and annotated huge libraries of bacterial genomes, which can help in the more practical application of these candidate genes in acclimatizing plants and improving rhizospheric colonization of beneficial microbes. Further, researcher groups from UNC, Oak Ridge National Lab, and the Max Planck Institute have identified and isolated novel microbes from the rhizospheric region of poplar trees, maize plants, *Arabidopsis* and *Brassicaceae* species. The authors have then integrated, sequenced, assembled, and annotated the data at JGI and made it publicly accessible for using it for comparative genomics analysis. Various investigators have recognized the potential genes that are enhanced in the plant-rhizospheric and endophytic microbe's genome (Levy et al. 2018b). The comprehensive list of novel genes and genomes is made accessible at online web portal known as "Genomic Features of Bacterial Adaptation to Plants" and this database is a valued source for those research groups which are studying PM associations.

14.9 Conclusion and Future Perspective

There are diverse forms of interactions which plants display with several species of bacteria and among these microbes some have not yet been characterized. Thus, the complete investigation of the diverse microbial population which exist in its natural habitat represents a huge task. Understanding of PM interactions requires classical as well as the latest "omics" approaches and quantifiable modeling methods.

These approaches help in generating models which further remove any lacunae or gaps and organize the molecular plant pathology and microbial ecology for elucidating mechanisms involved in the PM interactions. Therefore, omics approaches used for understanding microbiome inhabiting the rhizospheric regions of plants are critical as this locus represents most of the PM associations. This understanding of PM and MM interactions will deliver a helpful future perception in tempering microbial genome for improving plant yield and curtailing disease occurrence. With advances in NGS platforms, which can deliver advanced analysis combined with statistical analyses and repetitive profiling, more precise expression data involved in PM interactions can be generated. This improved decoding of information related to PM interactions (harmful or valuable to plants) will help us in perception and improved handling of the microbiome for various dedications. Despite so much of progress in better understanding and analysis of mechanisms involved in PM interactions with help of these high-throughput omics tools, we are still at the infancy stage of decoding these particularly complex associations. Judging from the present situation, we can predict the forthcoming research advances in giving us a clearer picture that will be valued for making progress in this context of PM associations for future sustainable agriculture. Although it is a massive task, with these analysis tools and strategies it is possible to have an improved understanding of PM interactions which can be implemented for solving various agriculture-related issues.

REFERENCES

Agler, M. T., Ruhe, J., Kroll, S., Morhenn, C., Kim, S. T., Weigel, D. and Kemen, E. M. 2016. Microbial hub taxa link host and abiotic factors to plant microbiome variation. *PLoS Biol* 14:e1002352.

Alavi, P., Starcher, M. R., Zachow, C., Müller, H. and Berg, G. 2013. Root microbe systems: the effect and mode of interaction of stress protecting agent (SPA) *Stenotrophomonas rhizophila* DSM14405T. *Front Plant Sci* 4:141.

Antelmann, H., van Dijk, J. M., Bron, S. and Hecker, M. 2006. Proteomic survey through secretome of *Bacillus subtilis*. In: Humphery-Smith I. and Hecker M (eds), *Microbial Proteomics: Functional Biology of Whole Organisms*. John Wiley & Sons, New York, pp:179–208.

Aparicio, T., de Lorenzo, V. and Martínez-García, E. 2018. CRISPR/Cas9-based counter selection boosts recombineering efficiency in *Pseudomonas putida*. *Biotech J* 13:1–10.

Arabidopsis Interactome Mapping Consortium. 2011. Evidence for network evolution in an Arabidopsis interactome map. *Science* 333:601–607.

Askenazi, M., Driggers, E. M., Holtzman, D. A. and Norman, T. C. et al. 2003. Integrating transcriptional and metabolite profiles to direct the engineering of lovastatin-producing fungal strains. *Nat Biotech* 21:150–156.

Badri, D. V., Chaparro, J. M., Zhang, R., Shen, Q. and Vivanco, J. M. 2013. Application of natural blends of phytochemicals derived from the root exudates of Arabidopsis to the soil reveal that phenolic-related compounds predominantly modulate the soil microbiome. *J Biol Chem* 288:4502–4512.

Bais, H. P., Weir, T. L., Perry, L. G., Gilroy, S. and Vivanco, J. M. 2006. The role of root exudates in rhizosphere interactions with plants and other organisms. *Ann Rev Plant Biol* 57:233–266.

Barnett, M. J., Toman, C. J., Fisher, R. F. and Long, S. R. 2004. A dual-genome symbiosis chip for coordinate study of signal exchange and development in a prokaryote-host interaction. *Proc Natl Acad Sci U S A* 101:16636–16641.

Barret, M., Tan, H., Egan, F., Morrissey, J. P., Reen, J. and O'Gara, F. 2013. Exploiting new systems-based strategies to elucidate plant–bacterial interactions in the rhizosphere. In: de Bruijn F. J. (ed), *Molecular Microbial Ecology of the Rhizosphere*, vol 1. Wiley Blackwell, Hoboken, New Jersey, pp:57–68.

Basu, S., Rabara, R. C., Negi, S. and Shukla, P. 2018. Engineering PGPMOs through gene editing and systems biology: a solution for phytoremediation? *Tren Biotech* 36:499–510.

Bender, C. L. 2005. The post-genomic era: new approaches for studying bacterial diseases of plants. *Aust Plant Pathol* 34(4):471–474.

Berendsen, R. L., Pieterse, C. M. J. and Bakker, P. A. H. M. 2012. The rhizosphere microbiome and plant health. *Tren Plant Sci* 17:478–486.

Bhattacharyya, P. N. and Jha, D. K. 2012. Plant growth-promoting rhizobacteria (PGPR): emergence in agriculture. *World J Microbiol Biotechnol* 28:1327–1350.

Bisht, D. S., Bhatia, V. and Bhattacharya, R. 2019. Improving plant-resistance to insect-pests and pathogens: the new opportunities through targeted genome editing. *Sem Cell Deve Biol* 96:65–76.

Bonfante, P. and Desirò, A. 2015. Arbuscular mycorrhizas: the lives of beneficial fungi and their plant host. In: Lugtenberg B. (ed), *Principles of Plant-Microbe Interactions.* Springer International Publishing Switzerland, Heidelberg, pp:235–245.

Botero, D., Alvarado, C., Bernal, A., Danies, G. and Restrepo, S. 2018. Network analyses in plant pathogens. *Front Microbiol* 9:35.

Boyko, A. and Kovalchuk, I. 2013. Epigenetic modifications in plants under adverse conditions: agricultural applications. In: Tuteja N. and Singh Gill S. (eds), *Plant Acclimation to Environmental Stress.* Springer, New York, pp:233–267.

Bräutigam, A. and Gowik, U. 2010. What can next generation sequencing do for you? Next generation sequencing as a valuable tool in plant research. *Plant Biol* 12:831–841.

Brechenmacher, L., Weidmann, S. and VanTuinen, D. et al. 2004. Expression profiling of up-regulated plant and fungal genes in early and late stages of *Medicago truncatula-Glomus mossea* interactions. *Mycorrhiza* 14:53–262.

Breuillin, F., Schramm, J., Hajirezaei, M., Ahkami, A. and Favre, P. et al. 2010. Phosphate systemically inhibits development of arbuscular mycorrhiza in *Petunia hybrid* and represses genes involved in mycorrhizal functioning. *Plant J* 64:1002–1017.

Brotman, Y., Lisec, J., Meret, M. and Chet, I. et al. 2011. Transcript and metabolite analysis of the *Trichoderma* induced systemic resistance response to *Pseudomonas syringae* in *Arabidopsis thaliana. Microbiologia* 158:139–146.

Broz, A. K., Broeckling, C. D., De-la-Peña, C. and Lewis, M. R. et al. 2010. Plant neighbor identity influences plant biochemistry and physiology related to defense. *BMC Plant Biol* 10:115.

Budiharjo, A., Chowdhury, S. P., Dietel, K., Beator, B., Dolgova, O. and Fan, B. et al. 2014. Transposon mutagenesis of the plant-associated *Bacillus amyloliquefaciens* ssp. Plantarum FZB42 revealed that the nfrA and RBAM17410 genes are involved in plant-microbe-interactions. *PLoS One* 9:5.

Bulgarelli, D., Garrido-Oter, R., Münch, P. C., Weiman, A. and Dröge, J. et al. 2015. Structure and function of the bacterial root microbiota in wild and domesticated barley. *Cell Host Micro* 17:392–403.

Bulgarelli, D., Schlaeppi, K., Spaepen, S., van Themaat, E. V. L. and Schulze-Lefert, P. 2013. Structure and functions of the bacterial microbiota of plants. *Ann Rev Plant Biol* 64:807–838.

Bundy, J. G., Willey, T. L., Castell, R. S., Ellar, D. J. and Brindle, K. M. 2005. Discrimination of pathogenic clinical isolates and laboratory strains of *Bacillus cereus* by NMR-based metabolomic profiling. *FEMS Microbiol Lett* 242:127–136.

Campbell, R. and Greaves, M. P. 1990. Anatomy and community structure of the rhizosphere. In: Lynch J. M. (ed), *The Rhizosphere.* John Wiley & Sons, UK, pp:11–34.

Cardenas, E. and Tiedje, J. M. 2008. New tools for discovering and characterizing microbial diversity. *Curr Opin Biotechnol* 19:544–549.

Carvalhais, L. C., Dennis, P. G., Tyson, G. W. and Schenk, P. M. 2013. Rhizosphere metatranscriptomics: challenges and opportunities. In: de Bruijn F.J. (ed), *Molecular Microbial Ecology of the Rhizosphere*, vol 2. Wiley Blackwell, Hoboken, New Jersey, pp:1137–1144.

Cascante, M. and Marin, S. 2008. Metabolomics and fluxomics approaches. *Essays Biochem* 45:67–82.

Chaparro, J., Sheflin, A., Manter, D. and Vivanco, J. 2012. Manipulating the soil microbiome to increase soil health and plant fertility. *Biol Fert Soils* 48:489–499.

Chen, K., Wang, Y., Zhang, R., Zhang, H. and Gao, C. 2019. CRISPR/Cas genome editing and precision plant breeding in agriculture. *Ann Rev Plant Biol* 70:667–697.

Choi, J., Park, J., Kim, D., Jung, K., Kang, S. and Lee, Y. H. 2010. Fungal secretome database: integrated platform for annotation of fungal secretomes. *BMC Genom* 11:105.

Chowdhury, S. P., Hartmann, A., Gao, X. and Borriss, R. 2015. Biocontrol mechanism by root-associated *Bacillus amyloliquefaciens* FZB42-a review. *Front Microbiol* 6:780.

Collakova, E., Yen, J. Y. and Senger, R. S. 2012. Are we ready for genome-scale modeling in plants. *Plant Sci* 19:53–70.

Collin, M., Timm, A. B., Kelsey, R., Carter, C. and Alyssa, A. et al. 2018. Abiotic stresses shift belowground *Populus*-associated bacteria toward a core stress microbiome. *mSys* 3:e00070–17.

Copeland, J. K., Yuan, L., Layeghifard, M., Wang, P. W. and Guttman, D. S. 2015. Seasonal community succession of the phyllosphere microbiome. *Mol Plant Microbe Interac* 28:274–285.

Da., S., Dyrlund, T. F., Ussatjuk, A., Jochimsen, B. and Nielsen, K. et al. 2014. Proteome reference maps of the *Lotus japonicus* nodule and root. *Proteomics* 14:230–240.

de Souza, E. M., Granada, C. E. and Sperotto, R. A. 2016. Plant pathogens affecting the establishment of plant-symbiont interaction. *Front Plant Sci* 7:15.

Defez, R., Esposito, R., Angelini, C. and Bianco, C. 2016. Overproduction of indole-3-acetic acid in free-living rhizobia induces transcriptional changes resembling those occurring in nodule bacteroids. *Mol Plant Microbe Interact* 29:484–495.

De-la-Peña, C., Badri, D. V., Lei, Z. and Watson, B. S. et al. 2010. Root secretion of defense-related proteins is development-dependent and correlated with flowering time. *J Biol Chem* 285:30654–30665.

De-la-Peña, C., Lei, Z., Watson, B. S., Sumner, L. W. and Vivanco, J. M. 2008. Root microbe communication through protein secretion. *J Biol Chem* 283:25247–25255.

De-la-Peña, C. and Loyola-Vargas, V. M. 2012. The hidden chemical cross-talk between roots and microbes: a proteomic approach. *Curr Prot* 9:103–117.

De-La-Peña, C., Rangel-Cano, A. and Alvarez-Venegas, R. 2012. Regulation of disease-responsive genes mediated by epigenetic factors: interaction of *Arabidopsis-Pseudomonas*. *Mol Plant Pathol* 13:388–398.

Delmotte, N., Ahrens, C. H., Knief, C., Qeli, E. and Koch, M. et al. 2010. An integrated proteomics and transcriptomics reference data set provides new insights into *Bradyrhizobium japonicum* bacteroid metabolism in soybean root nodules. *Proteomics* 10:1391–1400.

Denison, R. F. and Kiers, E. T. 2011. Life histories of symbiotic rhizobia and mycorrhizal fungi. *Current Biol* 21(18):R775–R785.

Dennis, P. G., Miller, A. J. and Hirsch, P. R. 2010. Are root exudates more important than other sources of rhizodeposits in structuring rhizosphere bacterial communities? *FEMS Microbiol Ecol* 72:313–327.

Depuydt, S., Trenkamp, S., Fernie, A. R. and Elftieh, S. et al. 2009. An integrated genomics approach to define niche establishment by *Rhodococcus fascians*. *Plant Physiol* 149:1366–1386.

Dong, S., Yin, W., Kong, G., Yang, X., Qutob, D. and Chen, Q. et al. 2011. *Phytophthora sojaea* virulence effector Avr3b is a secreted NADH and ADP-ribose pyrophosphorylase that modulates plant immunity. *PLoS Pathol* 7:e1002353.

Drogue, B., Combes-Meynet, E., Moënne-Loccoz, Y., Wisniewski-Dyé, F. and Prigent-Combaret, C. 2013. Control of the cooperation between plant growth-promoting rhizobacteria and crops by rhizosphere signals. In: de Bruijn F. J. (ed). *Molecular Microbial Ecology of the Rhizosphere*, vol 1. Wiley Blackwell, Hoboken, New Jersey, pp:281–293.

Dutta, S., Rani, T. S. and Podile, A. R. 2013. Root exudate-induced alterations in *Bacillus cereus* cell wall contribute to root colonization and plant growth promotion. *PLoS One* 8:e78369.

Eldakak, M., Milad, S. I., Nawar, A. I. and Rohila, J. S. 2013. Proteomics: a biotechnology tool for crop improvement. *Front Plant Sci* 4:35.

Plett, J. M., Kemppainen, M., Kale, S. D., Kohler, A., Legué, V. and Brun, A. et al. 2011. A secreted effector protein of *Laccaria bicolor* is required for symbiosis development. *Current Biol* 21:1197–1203.

Fan, B., Carvalhais, L. C., Becker, A., Fedoseyenko, D., von Wirén, N. and Borriss, R. 2012. Transcriptomic profiling of *Bacillus amyloliquefaciens* FZB42 in response to maize root exudates. *BMC Microbiol* 12:116.

Fan, T. W. M., Lane, A. N., Shenker, M., Bartley, J. P., Crowley, D. and Higashi, R. M. 2001. Comprehensive chemical profiling of gramineous plant root exudates using high-resolution NMR and MS. *Phytochem* 57:209–221.

Farrar, K., Bryant, D. and Cope-Selby, N. 2014. Understanding and engineering beneficial plant microbe interactions: plant growth promotion in energy crops. *Plant Biot J* 12:1193–1206.

Gaj, T., Gersbach, C. A. and Barbas, C. F. 2013. ZFN, TALEN and CRISPR/Cas-based methods for genome engineering. *Tren Biot* 31:397–405.

Glick, B. R. 2014. Bacteria with ACC deaminase can promote plant growth and help to feed the world. *Microbiol Res* 169:30–39.

Graham, R. L. J., Graham, C. and McMullan, G. 2007. Microbial proteomics: a mass spectrometry primer for biologists. *Microb Cell Fact* 6:26.

Han, G. Z. 2019. Origin and evolution of the plant immune system. *New Phytol* 222:70–83.

Han, S., Li, D., Trost, E., Mayer, K. F., Vlot, A. C. and Heller, W. et al. 2016. Systemic responses of barley to the 3-hydroxy-decanoyl-homoserine lactone producing plant beneficial endophyte *Acidovorax radicis* N35. *Front Plant Sci* 7:1868.

Hassani, M. A., Durán, P. and Hacquard, S. 2018. Microbial interactions within the plant holobiont. *Microbiom* 6:58.

Heavner, B. D. and Price, N. D. 2015. Comparative analysis of yeast metabolic network models highlights progress, opportunities for metabolic reconstruction. *PLoS Comp Biol* 11:e1004530.

Hirsch, P. R., Miller, A. J. and Dennis, P. G. 2013. Do root exudates exert more influence on rhizosphere bacterial community structure than other rhizodeposits? In: de Bruijn F. J. (ed), *Molecular Microbial Ecology of the Rhizosphere*, vol 1. Wiley Blackwell, Hoboken, New Jersey, pp:229–242.

Imam, J., Nitin, M., Toppo, N. N., Mandal, N. P. and Kumar, Y. et al. 2014. A comprehensive overview on application of bioinformatics and computational statistics in rice genomics towards an amalgamated approach for improving acquaintance base. In Kishor P. B. K., Bandyopadhyay R. and Suravajhala P. (eds), *Agriculture Bioinformatics*. Springer, Berlin.

Ithal, N., Recknor, J., Nettleton, D., Hearne, L. and Maier, T., et al. 2007. Parallel genome-wide expression profiling of host and pathogen during soybean cyst nematode infection of soybean. *Mol Plant Micro Interact* 20, 293–305.

Jacoby, R., Peukert, M., Succurro, A., Koprivova, A. and Kopriva, S. 2017. The role of soil microorganisms in plant mineral nutrition—current knowledge and future directions. *Front Plant Sci 8*:1617.

Jansson, J. K. and Hofmockel, K. S. 2018. The soil microbiome—From metagenomics to metaphenomics. *Curr Opin Microbiol* 43:162–168.

Jayaraman, D., Forshey, K. L. and Grimsrud, P. A. 2012. Leveraging proteomics to understand plant–microbe interactions. *Front Plant Sci* 3:44.

Jones, A. M., Xuan, Y., Xu, M., Wang, R. S. and Ho, C. H. et al. 2014. Border control–a membrane-linked interactome of *Arabidopsis*. *Science* 344:711–716.

Kakirde, K. S., Parsley, L. C. and Liles, M. R. 2010. Size does matter: application-driven approaches for soil metagenomics. *Soil Biol Biochem* 42:1911–1923.

Kav, N. N. V., Srivastava, S., Yajima, W. and Sharma, N. 2007. Application of proteomics to investigate plant-microbe interactions. *Curr Prot* 4:28–43.

Keeling, P. J. 2010. The endosymbiotic origin, diversification and fate of plastids. *Philo Trans Royal Soci B Biol Sci* 365:729–748.

Ketchum, R. E., Rithner, C. D., Qiu, D., Kim, Y. S., Williams, R. M. and Croteau, R. B. 2003. Taxus metabolomics: methyl jasmonate preferentially induces production of taxoids oxygenated at C-13 in Taxus x media cell cultures. *Phytochemistry* 62:901–909.

Kersey, P. J., Allen, J. E., Christensen, M., Davis, P., Falin, L. J. and Grabmueller, C. et al. 2014. Ensembl genomes 2013: scaling up access to genome-wide data. *Nucleic Acids Res* 42:546–552.

Klopffleisch, K., Phan, N., Augustin, K. and Bayne, R. S. et al. 2011. Arabidopsis G-protein interactome reveals connections to cell wall carbohydrates and morphogenesis. *Mol Sys Biol* 7:532.

Knief, C. 2014. Analysis of plant microbe interactions in the era of next generation sequencing technologies. *Front Plant Sci* 5:216.

Knott, G. J. and Doudna, J. A. 2018. CRISPR-Cas guides the future of genetic engineering. *Science* 361:866–869.

Kosova, K., Vitamvas, P., Urban, M. O., Klima, M., Roy, A. and Prasil, I. T. 2015. Biological networks underlying abiotic stress tolerance in temperate crops—a proteomic perspective. *Int J MolSci* 16:20913–20942.

Kyrpides, N. C. 2009. Fifteen years of microbial genomics: meeting the challenges and fulfilling the dream. *Nat Biotech* 27:627–632.

Lacerda, C. M. and Reardon, K. F. 2009. Environmental proteomics: applications of proteome profiling in environmental microbiology and biotechnology. *Brief Func Genom Prot* 8:75–87.

Lakshmanan, V., Castaneda, R., Rudrappa, T. and Bais, H. P. 2013. Root transcriptome analysis of *Arabidopsis thaliana* exposed to beneficial *Bacillus subtilis* FB17 rhizobacteria revealed genes for bacterial recruitment and plant defense independent of malate efflux. *Planta* 238:657–668.

Larsen, P. E., Sreedasyam, A., Trivedi, G. and Podila, G. K. et al. 2011. Using next generation transcriptome sequencing to predict an ectomycorrhizal metabolome. *BMC Sys Biol* 5:70.

Levy, A., Conway, J. M., Dangl, J. L. and Woyke, T. 2018a. Elucidating bacterial gene functions in the plant microbiome. *Cell Host Micro* 24:475–485.

Levy, A., Salas-Gonzalez, I. and Mittelviefhaus, M. et al. 2018b. Genomic features of bacterial adaptation to plants. *Nat Gen* 50:138–150.

Li, S., Musungu, B., Lightfoot, D. and Ji, P. 2018. The interactomic analysis reveals pathogenic protein networks in *Phomopsis longicolla* underlying seed decay of soybean. *Front Gen* 9:104.
Lim, K. A., Shamsuddin, Z. H. and Ho, C. L. 2010. Transcriptomic changes in the root of oil palm (*Elaeis guineensis* jacq.) upon inoculation with *Bacillus sphaericus* UPMB10. *Tree Genet Geno* 6(5):793–800.
Liu, J., Blaylock, L. A., Endre, G., Cho, J., Town, C. D., Vanden Bosch, K. A. and Harrison, M. J. 2003. Transcript profiling coupled with spatial expression analyses reveals genes involved in distinct developmental stages of an arbuscular mycorrhizal symbiosis. *Plant Cell* 15:2106–2123.
Livaja, M., Zeidler, D., Von-Rad, U. and Durner, J. 2008. Transcriptional responses of *Arabidopsis thaliana* to the bacteria derived PAMPs harpin and lipopolysaccharide. *Immunobiology* 213:161–171.
Lohar, D. P., Sharopova, N., Endre, G., Peñuela, S. and Samac, D. et al. 2006. Transcript analysis of early nodulation events in *Medicago truncatula. Plant Physiol* 140:221–234.
López-Ráez, J. A., Charnikhova, T. and Gómez-Roldán, V. et al. 2008. Tomato strigolactones are derived from carotenoids and their biosynthesis is promoted by phosphate starvation. *New Phytol* 178:863–874.
Lynch, J. M. and Whipps, J. M. 1990. Substrate flow in the rhizosphere. *Plant Soil* 129:1–10.
Mali, P., Yang, L., Esvelt, K. M., Aach., J., Güell, M., Dicarlo, J. E., Norville, J. E. and Church, G. M. 2013. RNA-guided human genome engineering via Cas9. *Science* 339:823–826.
Marin-de-la-Rosa, N. and Falter-Braun, P. 2015. Primer on protein-protein interaction maps. *Amer Can Socie* LS Vol 1, 1–9.
Mathesius, U., Mulders, S., Gao, M. and Teplitski, M. et al. 2003. Extensive and specific responses of a eukaryote to bacterial quorum-sensing signals. *Proc Natl Acad Sci U S A* 100:1444–1449.
Mendes, R., Kruijt, M., de Bruijn, I., Dekkers, E. and van der Voort, M. et al. 2011. Deciphering the rhizosphere microbiome for disease-suppressive bacteria. *Science* 332(6033):1097–1100.
Mia, M. B., Shamsuddin, Z., Wahab, Z. and Marziah, M. 2010. Effect of plant growth promoting rhizobacterial (PGPR) inoculation on growth and nitrogen incorporation of tissue cultured Musa plantlets under nitrogen free hydroponics condition. *Aust J Crop Sci* 4:85–90.
Micallef, S. A., Channer, S., Shiaris, M. P. and Colón-Carmona, A. 2009. Plant age and genotype impact the progression of bacterial community succession in the *Arabidopsis* rhizosphere. *Plant Signal Behav* 4:777–780.
Mirouze, M. and Paszkowski, J. 2011. Epigenetic contribution to stress adaptation in plants. *Curr Opin Plant Biol* 14:267–274.
Misra, B. B. and Chaturvedi, R. 2015. When plants braces for the emerging pathogens. *Physiol Mol Plant Pathol* 92:181–185.
Mithani, A., Hein, J. and Preston, G. M. 2011. Comparative analysis of metabolic networks provides insight into the evolution of plant pathogenic and nonpathogenic lifestyles in *Pseudomonas. Mol Biol Evo* 28:483–499.
Mocali, S. and Benedetti, A. 2010. Exploring research frontiers in microbiology: the challenge of metagenomics in soil microbiology. *Res Microbiol* 161:497–505.
Molitor, A., Zajic, D., Voll, L. M. and Hnemann, J. P. K. et al. 2011. Barley leaf transcriptome and metabolite analysis reveals new aspects of compatibility and *Piriformospora indica*-mediated systemic induced resistance to powdery mildew. *Mol Plant Microbe Interac* 24(12):1427–1439.
Morales, S. E. and Holben, W. E. 2011. Linking bacterial identities and ecosystem processes: 369 can "omic" analyses be more than the sum of their parts? *FEMS Microbiol Ecol* 75:2–16.
Mosquera, G., Giraldo, M. C., Khang, C. H., Coughlan, S. and Valent, B. 2009. Interaction transcriptome analysis identifies *Magnaporthe oryzae* BAS1-4 as biotrophy-associated secreted proteins in rice blast disease. *Plant Cell* 21:1273–1290.
Muñoz, I. V., Sarrocco, S., Malfatti, L., Baroncelli, R. and Vannacci, G. 2019. CRISPR-Cas for fungal genome editing: a new tool for the management of plant diseases. *Front Plant Sci* 10:1–5.
Netea, M. G. and van der Meer, J. W. 2017. Trained immunity: an ancient way of remembering. *Cell Host Micro* 21:297–300.
Newey, L. J., Caten, C. E. and Green, J. R. 2007. Rapid adhesion of *Stagonospora nodorum* spores to a hydrophobic surface requires pre-formed cell surface glycoproteins. *Myco Res* 111:1255–1267.
Newton, A. C., Fitt, B. D. L., Atkins, S. D., Walters, D. R. and Daniell, T. J. 2010. Pathogenesis, parasitism and mutualism in the trophic space of microbe-plant interactions. *Tren Microbiol* 18:365–373.
Parmar, P. and Sindhu, S. 2013. Potassium solubilization by rhizosphere bacteria: influence of nutritional and environmental conditions. *J Microbiol Res* 3:25–31.

Pellegrino, E. and Bedini, S. 2014. Enhancing ecosystem services in sustainable agriculture: biofertilization and biofortification of chickpea (*Cicer arietinum*) by arbuscular mycorrhizal fungi. *Soil Biol Biochem* 68:429–439.

Peng, S., Huang, J., Sheehy, J. E., Laza, R. C. and Visperas, R. M. et al. 2004. Rice yields decline with higher night temperature from global warming. *Proc Natl Acad Sci U S A* 101:9971–9975.

Penterman, J., Abo, R. P., De Nisco, N. J., Arnold, M. F. F., Longhi, R., Zanda, M. and Walker, G. C. 2014. Host plant peptides elicit a transcriptional response to control the *Sinorhizobium meliloti* cell cycle during symbiosis. *Proc Natl Acad Sci U S A* 111:3561–3566.

Pieterse, C. M., Zamioudis, C., Berendsen, R. L., Weller, D. M., VanWees, S. C. and Bakker, P. A. 2014. Induced systemic resistance by beneficial microbes. *Ann Rev Phytopathol* 52:347–375.

Ramachandran, V. K., East, A. K., Karunakaran, R., Downie, J. A. and Poole, P. S. 2011. Adaptation of *Rhizobium leguminosarum* to pea, alfalfa and sugar beet rhizospheres investigated by comparative transcriptomics. *Genome Biol* 12:R106.

Reid, A. 2011. Microbes helping to improve crop productivity. *Microbe* 6:435–439.

Rodenburg, S. Y. A., Seidl, M. F., de Ridder, D. and Govers, F. 2018. Genome-wide characterization of *Phytophthora infestans* metabolism: a systems biology approach. *Mol Plant Pathol* 19:1403–1413.

Rodriguez, P. A., Rothballer, M., Chowdhury, S. P., Nussbaumer, T., Gutjahr, C. and Falter-Braun, P. 2019. Systems biology of plant-microbiome interactions. *Mol Plant* 12:804–821.

Rosenberg, E. and Zilber-Rosenberg, I. 2018. The hologenome concept of evolution after 10 years. *Microbiom* 6:78.

Sakuma, T. and Yamamoto, T. 2017. Magic wands of CRISPR—lots of choices for gene knock-in. *Cell Biol Toxi* 33:501–505.

Sana, T. R., Fischer, S., Wohlgemuth, G. and Katrekar, A. et al. 2010. Metabolomic and transcriptomic analysis of the rice response to the bacterial blight pathogen Xanthomonas oryzae. *Metabolomics* 6:451–465.

Schenk, S. T., Hernandez-Reyes, C., Samans, B. and Stein, E. et al. 2014. N-acyl-homoserine lactone primes plants for cell wall reinforcement and induces resistance to bacterial pathogens via the salicylic acid/oxylipin pathway. *Plant Cell* 26:2708–2723.

Schenk, S. T., Peer, M. P., Choo, J. H. and Wong, C. L. 2009. Microarray analyses to study plant defence and rhizosphere-microbe interactions. *CAB Rev* 4:1–14.

Scherp, P., Ku, G., Coleman, L. and Kheterpal, I. 2011. Gel-based and gel-free proteomic technologies. *Methods Mol Biol* 702:163–190.

Schikora, A., Schenk, S. T. and Hartmann, A. 2016. Beneficial effects of bacteria-plant communication based on quorum sensing molecules of the N-acyl homoserine lactone group. *Plant Mol Biol* 90:605–612.

Schwessinger, B. and Zipfel, C. 2008. News from the frontline: recent insights into PAMP triggered immunity in plants. *Curr Opini Plant Biol* 11:389–395.

Shen, J., Li, C., Mi, G., Li, L., Yuan, L., Jiang, R. and Zhang, F. 2013. Maximizing root/rhizosphere efficiency to improve crop productivity and nutrient use efficiency in intensive agriculture of China. *J Exp Bot* 64:1181–1192.

Silva-Sanchez, C., Li, H. and Chen, S. 2015. Recent advances and challenge in plant phosphoproteomics. *Proteomics* 15:1127–1141.

Simon, C. and Daniel, R. 2011. Metagenomic analyses: past and future trends. *App Environ Microbiol* 77:1153–1161.

Singh, O. V. and Nagaraj, N. S. 2006. Transcriptomics, proteomics and interactomics: unique approaches to track the insights of bioremediation. *Brie Func Genom Prot* 4:355–362.

Song, J., Win, J., Tian, M., Schornack, S. and Kaschani, F. et al. 2009. Apoplastic effectors secreted by two unrelated eukaryotic plant pathogens target the tomato defense protease Rcr3. *PNASUSA* 106:1654–1659.

Stancey, G., Libault, M., Brechenmacher, L., Wan, J. and May, G. D. 2006. Genetics and functional genomics of legume nodulation. *Curr Opin Plant Biol* 9:110–121.

Sugawara, M., Epstein, B., Badgley, B. D., Unno, T. and Xu, L. et al. 2013. Comparative genomics of the core and accessory genomes of 48 *Sinorhizobium* strains comprising five geno species. *Genome Biol* 14:R17.

Sun, J., Wang, Q., Jiang, Y., Wen, Z., Yang, L., Wu, J. and Yang, S. 2018. Genome editing and transcriptional repression in *Pseudomonas putida* KT2440 via the type II CRISPR system. *Microb Cell Fact* 17:41.

Takahara, M., Magori, S., Soyano, T. and Okamoto, S. et al. 2013. Too much love, a novel kelch repeat-containing F-box protein, functions in the long distance regulation of the legume-rhizobium symbiosis. *Plant Cell Physiol* 54:433–447.

Talboys, P. J., Owen, D. W., Healey, J. R., Withers, P. J. and Jones, D. L. 2014. Auxin secretion by *Bacillus amyloliquefaciens* FZB42 both stimulates root exudation and limits phosphorus uptake in *Triticum aestivum*. *BMC Plant Biol* 14:51.

Tanaka, K., Cho, S. H., Lee, H., Pham, A. Q. and Batek, J. M. et al. 2015. Effect of lipo-chitooligosaccharide on early growth of C4 grass seedlings. *J Exp Bot* 66:5727–5738.

Tawaraya, K., Horie, R., Saito, A., Shinano, T., Wagatsuma, T., Saito, K. and Oikawa, A. 2013. Metabolite profiling of shoot extracts, root extracts, and root exudates of rice plant under phosphorus deficiency. *J Plant Nutri* 36:1138–1159.

Thynne, E., McDonald, M. C. and Solomon, P. S. 2015. Phytopathogen emergence in the genomics era. *Tren Plant Sci* 20:246–255.

Torto-Alalibo, T., Collmer, C. W. and Gwinn-Giglio, M. 2009. The Plant-Associated Microbe Gene Ontology (PAMGO) Consortium: community development of new gene ontology terms describing biological processes involved in microbe-host interactions. *BMC Microbiol* 9:S7.

Urban, M., Pant, R., Raghunath, A., Irvine, A. G., Pedro, H. and Hammond-Kosack, K. E. 2015. The pathogen-host interactions database (PHI-base): additions and future developments. *Nucleic Acids Res* 43:645–655.

Urbanczyk-Wochniak, E., Luedemann, A., Kopka, J. and Selbig, J. et al. 2003. Parallel analysis of transcript and metabolic profiles: a new approach in systems biology. *EMBO Rep* 4:989–993.

Vauclare, P., Bligny, R., Gout, E. and Widmer, F. 2013. An overview of the metabolic differences between *Bradyrhizobium japonicum* 110 bacteria and differentiated bacteroids from soybean (*Glycine max*) root nodules: an in vitro 13C- and 31P-nuclear magnetic resonance spectroscopy study. *FEMS Microbiol Lett* 343:49–56.

Vejan, P., Abdullah, R., Khadiran, T., Ismail, S. and Nasrulhaq-Boyce, A. 2016. Role of plant growth promoting rhizobacteria in agricultural sustainability—a review. *Molecules* 21(5):573.

Venkateshwaran, M., Volkening, J. D., Sussman, M. R. and Ané, J. M. 2013. Symbiosis and the social network of higher plants. *Curr Opin Plant Biol* 16:118–127.

Vorholt, J. A. 2012. Microbial life in the phyllosphere. *Nature Rev Microbiol* 10(12):828–840.

Waghunde, R. R., Shelake, R. M., Shinde, M. S. and Hayashi, H. 2017. Endophyte microbes: a weapon for plant health management. In: *Microorganisms for Green Revolution*. Springer, Singapore, pp:303–325.

Wang, Y., Hu, B., Du, S., Gao, S., Chen, X. and Chen, D. 2016. Proteomic analyses reveal the mechanism of *Dunaliella salina* Ds-26-16 gene enhancing salt tolerance in *Escherichia coli*. *PLoS ONE* 11:e0153640.

Wang, Y., Kim, S. G., Wu, J., Huh, H. H., Lee, S. J. and Rakwal, R. et al. 2013. Secretome analysis of the rice bacterium *Xanthomonas oryzae* (Xoo) using in vitro and in planta systems. *Proteomics* 13:1901–1912.

Webb, K. J., Skøt, L., Nicholson, M. N., Jørgensen, B. and Mizen, S. 2000. *Mesorhizobium loti* increases root-specific expression of a calcium-binding protein homologue identified by promoter tagging in *Lotus japonicus*. *Mol Plant Microbe Intera* 13:606–616.

Winnenburg, R., Urban, M., Beacham, A., Baldwin, T. K., Holland, S. and Lindeberg, M. et al. 2008. PHI-base update: additions to the pathogen host interaction database. *Nucleic Acids Res* 36:572–576.

Xin, X. F., Kvitko, B. and He, S. Y. 2018. *Pseudomonas syringae*: what it takes to be a pathogen. *Nat Rev Genet* 16:316–328.

Xu, Z., Jiang, Y. and Zhou, G. 2015. Response and adaptation of photosynthesis, respiration, and antioxidant systems to elevated CO_2 with environmental stress in plants. *Front Plant Sci* 6:701.

Xu, L., Zhu, L., Tu, L., Liu, L. and Yuan, D.et al. 2011. Lignin metabolism has a central role in the resistance of cotton to the wilt fungus *Verticillium dahliae* as revealed by RNA-seq-dependent transcriptional analysis and histochemistry. *J Exp Bot* 62(15):5607–5621.

Yin, K., Gao, C. and Qiu, J. L. 2017. Progress and prospects in plant genome editing. *Nat Plants* 3:17107.

You, J., Zhang, Y., Liu, A., Li, D. and Wang, X. et al. 2019. Transcriptomic and metabolomics profiling of drought-tolerant and susceptible sesame genotypes in response to drought stress. *BMC Plant Biol* 19:267.

Yu, H., Luscombe, N. M., Lu, H. X., Zhu, X., Xia, Y. and Han, J. D. et al. 2004. Annotation transfer between genomes: protein-protein interologs and protein-DNA regulogs. *Genome Res* 14:1107–1118.

Zaidi, S. S. E. A., Mukhtar, M. S. and Mansoor, S. 2018. Genome editing: targeting susceptibility genes for plant disease resistance. *Tren Biotech* 36:898–906.

Zhang, J., Chiodini, R., Badr, A. and Zhang, G. 2011. The impact of next-generation sequencing on genomics. *J Genet Genom* 38:95–109.

Zhang, N., Wang, D., Liu, Y., Li, S., Shen, Q. and Zhang, R. 2014. Effects of different plant root exudates and their organic acid components on chemotaxis, biofilm formation and colonization by beneficial rhizosphere-associated bacterial strains. *Plant Soil* 374:689–700.

Zhao, N., Wang, G., Norris, A., Chen, X. and Chen, F. 2013. Studying plant secondary metabolism in the age of genomics. *Crit Rev Plant Sci* 32:369–382.

Ziegler, M., Engel, M., Welzl, G. and Schloter, M. 2013. Development of a simple root model to study the effects of single exudates on the development of bacterial community structure. *J Microbiol Meth* 94:30–36.

15

Cas9 Technology: An Innovative Approach to Enhance Phytoremediation

Deepu Pandita
Government Department of School Education Jammu, Jammu and Kashmir, India

CONTENTS

15.1 Introduction

Heavy metals (HMs) are natural compounds having high atomic numbers, atomic weights, and atomic densities and constitute approximately 5% of earth's crust. These occur as rock-loving lithophiles or ore-loving chalcophiles (Lide 2004), usually at trace levels beneath 1 g/kg (Ahmed 2014). HMs are classified as precious metals or essential elements, e.g., platinum (Pt), silver (Ag), palladium (Pd), gold (Au), ruthenium (Ru); nonessential elements or toxic metals, e.g., aluminum (Al), cadmium (Cd), lead (Pb), zinc (Zn), cobalt (Co), arsenic (As), chromium (Cr), nickel (Ni), copper (Cu), iron (Fe), molybdenum (Mb); and radionuclides, e.g., uranium (U), thorium (Th), radium (Ra), and americium (Am) (Hossain and Komatsu 2013; Ahmed 2014; Noman and Aqeel 2017; Noman et al. 2018). Besides natural sources, various major anthropogenic practices, for instance, excavating and casting of metalliferous ores, metal processing, fossil fuel (hydrocarbon deposit) burning, sludges, and agriculture are also cradles of HM accretion in natural ecosystem (Gopinath et al. 2020). The haphazard migration and release of toxic and mutagenic HMs cause substantial increase in ecological pollution in the environs. HMs such as Zn, Ni, Cu, and Cr are essential micronutrients for biological organisms at lower concentrations, whereas at higher concentrations they act as key toxicants to all living beings. However, these are critical in cell defense, signal transduction, regulation of genes, and their deficiency impedes plant growth, reproduction, and tolerance to various plant stresses (Broadley et al. 2012). Excessive concentration of biologically nonessential HMs, for instance Cd, Pb, Cr, Al, Cu, Hg, Zn, and Ni, causes austere extensive interactions like mutagenicity, plant toxicity, cytotoxicity, and carcinogenicity (Agwu et al. 2018; Kanwal et al. 2018) and mostly impacts and interferes with mitochondrial functioning by generating reactive oxygen species (ROS); interfering in lipids biomembranes and DNA damage repair, cell organelles and cell components, cellular functions and essential cell metabolic functions, plant photosynthesis and respiration rates; causing low biomass accumulation, stunted growth and development, reduced harvest, and even plant death (Oves et al. 2013; Karthik et al. 2016; Malar et al. 2016; Wu et al. 2016; Khalid et al. 2017, 2018c). After toxic HMs enter into the agricultural and ecological units, these pose health jeopardies to other biological organisms and impend security and safety of food by transfer to the upper levels of food chain. HM accretions in microbes initiate DNA synthesis inhibition, instabilities in cell enzymatic activities, interference in plant-microbe interactions, and attacks by pathogens (Wu et al. 2016). HMs cause health disorders like cognitive impairment, cancer, cardiovascular diseases, anemia, damage of kidneys, neurological diseases, and skin allergic reactions and irritations in humans (Gopinath et al. 2020). Owing to mutagenicity and carcinogenicity, HMs are well thought-out "priority pollutant" by United States Environmental Protection Agency (USEPA 1996) and so, HM remediation is an indispensable gait for sustainable environment. The strategy for HM remediation comprises innumerable approaches of physical, chemical, and biological nature. Physiochemical remediation techniques are rapid but are expensive, difficult to apply, produce secondary pollutants, and change physiochemistry and biology of soil. Therefore, adaptation of an alternate, cost-effective and sustainable bio-strategy is the need of the hour (Ullah et al. 2015; Senthil and Gunasundari 2017). In this scenario, bioremediation approach of HMs is promising, most effective, conservational, and low cost strategy and it does not cause any alteration in soil physiochemical properties and has other benefits over conventional physical and chemical processes (Jacob et al. 2018; Pratush et al. 2018). The biological remediation is defined as the process in which HM pollutants are biologically diminished under controlled state of events to a safe level (Pratush et al. 2018). The biological approaches include phytoremediation, land farming, windrows, bioventing, biosparging, bioreactor, bioslurping, biopile, bioaugmentation, permeable reactive barrier, bioleaching, and composting. Among these bioremediation approaches, phytoremediation is the most potent and widely acknowledged because of its cost-effectiveness and sustainability (Shahabaldin et al. 2016; Jacob et al. 2018). Plants have developed regulator homeostatic networks for HM uptake, HM tracking, HM stowage, and essential and nonessential metal detoxification. Plants under HM stress evolve two complex mechanisms to escape, tolerate, or defend HMs up to certain limit. The utmost general approach is HM exclusion, in which HM accrual gets restricted to underground plant parts and its uptake and translocation from root xylem to shoots remains under regulation to reserve less quantity in shoots. Antagonistically plants, e.g., metal (hyper) accumulators possess trait to accrue HM concentrations in shoots and low concentrations in roots and connected with the enhanced ability for detoxification of elevated metal levels in aboveground plant parts (Krämer 2010). In response to HM toxicity, plants decrease or avoid their uptake by limiting them to root cell wall apoplasts and if HM ions enter into the cell cytoplasm, at that point plants undergo detoxification mechanism or storage mechanisms (sequestration into vacuoles, chelation, trafficking or transport) to overcome toxic effects of HMs (Manara 2012). But, factors like reduced biomass at higher metal concentrations, sensitivity to HMs and meagre superficial root system restricted efficacy of phytoremediation and the improvement in the phytoremediation proficiency and efficacy is developing approach (Ullah et al. 2015). Omics approaches which include DNA genomics, RNA transcriptomics, proteomics and metabolomics assist the system biology investigations of microorganisms in evaluating regulation at genetic level for bioremediation (DeSousa et al. 2018). The advanced research approaches of editing genes and systemic biology finds application in bioremediation of HMs, persistent organic pollutants (POPs), drainage of acids, petroleum and xenobiotics (Basu et al. 2018; Dai et al. 2018; Gaur et al. 2018). CRISPR (Clustered Regularly Interspaced Short Palindromic Repeats)-associated protein 9 genetic modification has high target efficiency in various species (Gaj et al. 2013). CRISPR-Cas9-based systems derived from naturally occurring genome editing immune system in bacteria have revolutionized plant science and are widely used, programmable, efficient and precise, target-specific genome editing tools for various biological applications (Kang et al. 2020; Liu et al. 2020; Ma et al. 2020). CRISPR-Cas9 contains Cas9 protein and single guide RNA (sgRNA) and introduces precise double-strand breaks (DSB) in DNA. SpCas9 requires NGG Protospacer Adjacent Motif for genome targeting (Jinek et al. 2012). CRISPR-Cas9 generated mutations are insertions and deletions (Hsu et al. 2013). Genome editing by CRISPR-Cas9 phytoremediation is novel, emerging and proficient method to enhance plant remediation ability. Genomes of model phytoremediator hyper-accumulators for instance *Thlaspi caerulescens* (Cadmium, Ni, and Zn), *Arabidopsis halleri* (Cadmium and Zinc), *Pteris vittata* (Arsenic), *Hirschfeldia incana* (Pb), *Brassica juncea*, and a small number of energy crops have been fully or partially sequenced (Xie et al. 2009; Mandáková et al. 2015; Auguy et al. 2016; Yang et al. 2016; Estrela and Cate 2016; Briskine et al. 2017) and modifying their genomes by CRISPR-Cas9 for enhanced HMs tolerance may provide manifold advantages. Their genomic sequence information of different phytoremediators may identify and characterize key gene networks involved in the phytoremediation processes which can be exploited by CRISPR/Cas 9 based target editing for enhanced phytoremediation.

15.2 Phytoremediation

Phytoremediation (phyto = plant; remediation = cleaning) term denotes utilization of plants for eradication of HM contaminants from polluted environments. Phytoremediation has certain benefits as primarily based on autotrophic system needs low amount of nutrients, modest to cope and commonly approved by society due to its attractiveness and conservational sustainability (Mahajan and Kaushal 2018). Several phytoremediation plants accumulate efficiently toxic HMs as inorganic ions into their somatic and sexual parts and are usually native species having rapid growth, good biomass yield and tolerance, extended root system, adapt, absorb and accumulate the HM pollutants or ionic compounds even at lower concentrations through their cationic transporters (Khan et al. 2009). HMs and metalloid accumulation and tolerance properties categorized plants into excluders (HM sensitive plants), indicators (less HMs uptake and translocation) and accumulators (notable HMs uptake and accumulation) (Khan et al. 2009). Hyper-accumulators tolerate high amount of HM concentrations and allocated HMs in aboveground tissues at increased levels (Ma et al. 2016). The hyper-accumulator efficiency depends on several soil and edaphic factors like bioavailability of HMs, cation exchange capacity (CEC), temperature, organic material, redox potential, pH, and amount of water which change or limit plant growth, plant biomass and plant HMs uptake (Andrea et al. 2019). Biotic factors also affect soil nutrient status, thus decline phytoremediation efficiency by limiting plant growth and HMs uptake (Jacob et al. 2018) and about 400 known metal hyper-accumulator plants occur worldwide (Reeves and Baker 2000). Phytoremediators should have rapid growth, high biomass of aerial organs and tolerance to HMs stress, strong root architecture, ability to stop and/or consumption of conspicuous quantities of toxic metals as in hyper-accumulators. Hyper-accumulators have high metal accumulation of 1–3 scale over other plants of that area (Chaney et al. 2000; Lasat 2002; McGrath et al. 2002; Padmavathiamma and Li 2007; Pollard et al. 2014). Various hyper-accumulators oozes out some metal chelators like citric acid, malic acid, malonic acid and oxalic acids (Ma et al. 2001) which decrease the rhizospheric pH and make metal ions unavailable to the plant (Ross 1994). These organic acids enable metal ingestion or immobilize metals by complex formation with roots (Huang et al. 1988; Murphy et al. 1999; Pineros and Kochian 2001). *Populus* and *Salix* spp. store comparatively higher foliar contents of Cd and Zn HMs in metal-contaminated lands (Mertens et al. 2007; Dickinson et al. 2009). Various plants show highest phytoremediation efficiency in low HMs polluted areas while in higher HM concentration areas, same plants lesser growth and biomass yield (Zhaoyu et al. 2017). Rhizospheric microorganisms support plants with tolerance to elevated HMs toxicity and in phytoremediation process (Karthik and Arulselvi 2017). Over-expression of NAS1 gene in *Arabidopsis* and *Nicotiana* increased tolerance towards Cadmium, Copper, Iron, Nickle and Zinc and over-expression of MTA1, MT1 and MT2 metallothione in encoding-genes increased endurance and accumulation of Cadmium, Copper and Zinc (Xia et al. 2012; Sebastian et al. 2019). Pollutant removal enzymes should be targeted by CRISPR-Cas tools (Pandey and Singh 2019).

15.2.1 Strategies for Phytoremediation

On the basis of direction of HMs uptake and translocation, phytoremediation can be categorized as the phytoextraction (HMs uptake from different environments into plant biomass), phytostabilization (declining the HMs movement by plants), phytofiltration (adsorption and/or absorption of HMs by aquatic plant parts) and phytovolatilization (conversion and volatilization of HM contaminants into air by plant leaves) (DalCorso et al. 2019), phytoaccumulation (hyper-accumulation in cell vacuoles of hyper-accumulators due to hyper tolerance), phytostimulation (stimulation of microbes to degrade HMs), phytoassimilation (transportation and metabolization of HM contaminants within plant tissues), and phytodegradation (degradation of HM contaminants within plant tissues) (Chaudhry et al. 1998). The phytoextraction is second mechanism of phytoremediation in which plants uptake, transport, accumulates and concentrates HMs into aboveground tissues (Fasani 2012) through vascular system and transporter proteins. HM uptake in plants varies on the basis of species and HM type (Ma et al. 2016; Qin et al. 2017). Metal extraction process begins by metal ingestion from soil solution to the exterior layer of the roots and then undergo selective permeable recognition or substrate specificity of HMs by ligation of HM inorganic ions to the cells of root system and then enters into the cells through cation transporters and finally transported HMs to the above ground parts (Manara, 2012; Mahajan and Kaushal, 2018; Suman et al. 2018). In aquatic environment, phytofiltration of hyper-accumulator plants takes place by passive metal transport/direct absorption from contaminated wastewater and accumulates HMs in aboveground parts (Shahabaldin et al. 2016). Microalgae eradicate bromide (Br), cadmium (Cd), mercury (Hg), and lead (Pb) from plant effluents, waste waters, and agricultural wastes (Abdel-Raoufa et al. 2012). In phytovolatilization Hg, arsenic (As), and selenium (Se) undergo root absorption, translocation, and volatilization into less toxic volatile compounds that finally discharge into the atmosphere. Inorganic Se (chemically analogous to sulfur) gets converted to less toxic dimethyl selenide by the enzymes of plant sulfur metabolism that assimilates and volatilizes it into vaporous state and finally released into air (D´Souza et al. 2000). Soil microorganisms formarsenite consumed by plants and crops later on. Rice mobilizes arsenite by the silicon uptake pathway which causes arsenite poisoning in consumers (Zhu et al. 2008). Leaf absorbs methyl mercury (MeHg) by atmosphere, as Hg translocation via vascular tissues is rare even water-soluble Hg quickly adsorbs and transfer to shoots in rice (Meng et al. 2014). In phytostabilization or rhizostabilization categories of phytoremediation, plants stabilize metal ions in chemical form within the soil vadose zone which are then absorbed and accumulated into the plant roots and precipitate to rhizospheric zone, regulate leach out, mobility, bioavailability and hazardous effects. Plants accumulate HMs in underground parts. The metal ions with positive charge bind to cell walls pectins and negative charged metal ions bind to plasma membranes (Berti and Cunningham, 2000; Mench et al. 2006; Brunner et al. 2008). The different natural phytoremediation and phytotechniques used for decreasing HMs pollution in the environment are listed in Figure 15.1.

15.2.2 Molecular Mechanisms of Phytoremediation

The HM stress triggers a network of signaling pathways which includes calcium signaling, hormones signaling and MAPK signaling resulting in synthesis of HMs binding and transporter proteins (Peng and Gong 2014). Plant genomes are composed of different genes and gene families which play key roles in specific expression array son HMs uptake, metal tracking, translocation and accumulation. Appropriate modules of homeostatic networks in HMs tolerance and detoxification comprise ion transporters, metallo-chaperons, and ligands worked in coordination manner for HMs uptake, translocation to various tissues and transport into individual cells (Kramer et al. 2007; Pilon et al. 2009; Maestri et al. 2010; Gustin et al. 2011; White and Broadley 2011). The physical homology of HMs to essential metal nutrients, disguises their transfer to root system via nutrient translocation network (import, tracking, sequestration and export) wherein function of several transporters is known. Most of the HMs transported through extremely evolved co-transporters and metal pumps (a group of membrane proteins) localized on plasma membrane, vacuole tonoplast and additional sub-cellular membrane enveloped organelles, organs, tissues or at the entire plant system. Membrane proteins transfer various HMs diagonally through cellular membranes and perform key function in all influx-eflux phases of translocation route from underground root system to the aboveground shoot system (Kramer et al. 2007; Pilon et al. 2009; Maestri et al. 2010; Gustin et al. 2011; White and Broadley 2011). Intracellular transport of toxic HMs takes place via a widespread variety of transporters belonging to zinc-iron permease (ZIP) protein family, multidrug and toxin efflux proteins (MATE), cation diffusion facilitator proteins (CDF), natural resistance-associated macrophage proteins (NRAMP)and P-type P1B-ATPases (Axelsen and Palmgren 1998a; b; Guerinot

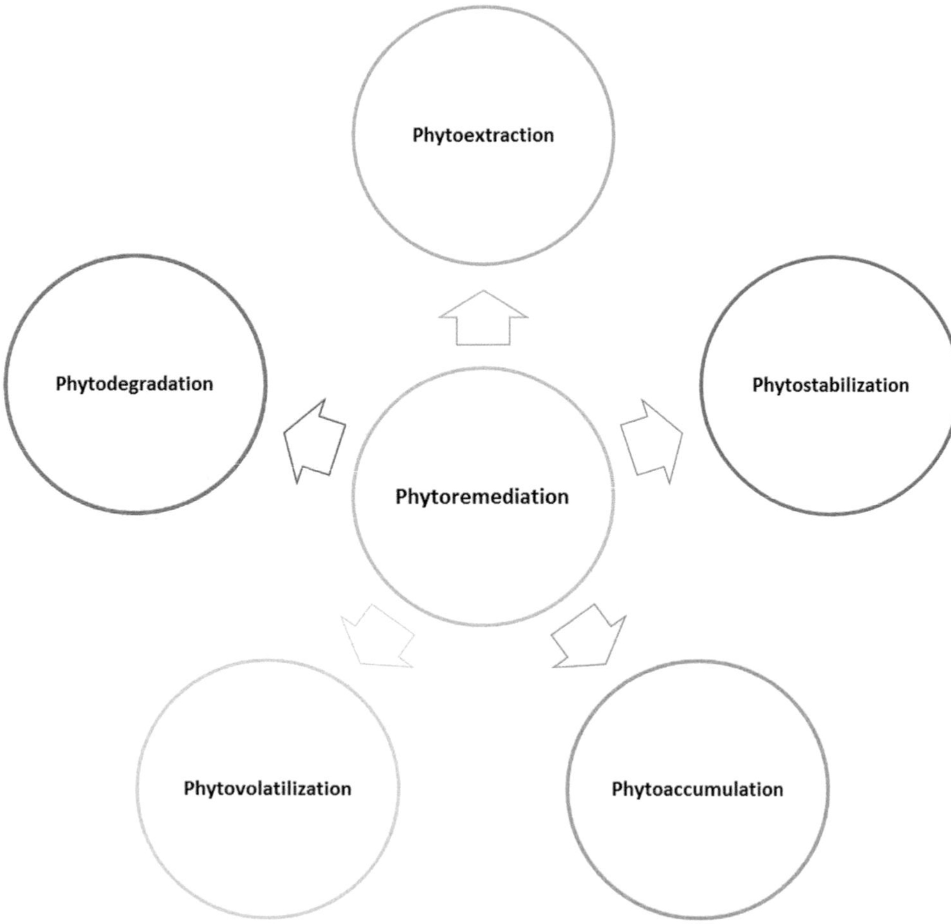

FIGURE 15.1 Different mechanisms of phytoremediation process for heavy metal contaminants.

2000; Williams et al. 2000; Takahashi et al. 2012). Metal transporter gene families are classified on the basis of sequence homology include ZIP family transporter genes (ZRT and IRT like protein) encoding cell membrane cation transporters that involved in consumption and translocation of metals from root system to shoot system (Guerinot 2000; Milner et al. 2013) and also regulate the high Zn cytoplasmic concentration and membrane transport in plants. On the basis of the Zn concentration and plant tissue type, the ZIP family undergoes gene expression (Chen et al. 2008). Iron-regulated transporter (IRT like protein) and NRAMPs assist in uptake of HMs to root cells from the environment (Larsen et al. 2005; Sasaki et al. 2012; Takahashi et al. 2012). *T. caerulescens* which is a zinc hyper-accumulator plant expresses genes encoding Zn transporters continuously irrespective of zinc levels in the soil and accumulates high zinc quantity inside the plant (Pence et al. 2000). Overexpression of NcZNT1 in roots of *Noccae caerulescens* indicates its association with root-to-shoot transport by retaining influx of zinc and cadmium for xylem loading (Lin et al. 2016). IRT1 overexpression in *Arabidopsis thaliana* roots has proven its role in the uptake of Fe in Fe deficient soils (Vert et al. 2002). CDF family members translocate HMs into internal cellular compartments and extracellular spaces (Gustin et al. 2011) whereas numerous metal tolerance proteins (MTPs) characterized in plants include vacuolar MTP1 transporters localized in tonoplast. MTP1 is a Zn^{2+}/H^{+} anti-porter and transports PC-Cd (phytochelatin-cadmium) complexes and play significant roles in excess Zn accumulation in vacuoles and Zn tolerance (Ortiz et al. 1995; Dräger et al. 2004; Desbrosses-Fonrouge et al. 2005; Hammond et al. 2006; Kuriakose and Prasad 2008; Gustin et al. 2009). HM ATPase (HMA) gene families and AtALS3 gene enables HM absorption, resistance and metal transport from the root into shoot system via xylem (Takahashi et al. 2012). P1B-type ATPase (HMA proteins) pumps cations outside of cell cytoplasm through ATP hydrolysis. On the basis of metal substrate specificity, HMAs are categorized into Zn/Co/Cd/Pb P1B-ATPase group and Cu/Ag P1B-ATPase group (Axelsen and Palmgren 1998a; b) however, in *Arabidopsis halleri*, *T. caerulescens*, *Triticum aestivum*, *Hordeum vulgare* and *A. thaliana* (8 members), P1B-type ATPases have been characterized (Williams and Mills 2005; Deng et al. 2013). HMA1 localized in chloroplast membrane helps in Zn detoxification in plastids under high concentration of Zn (Kim et al. 2009) while as, HMA3 (HM ATPase 3) detoxifies zinc, cadmium, cobalt, and Pb by sequestration into cell vacuoles. Over-expression of AtHMA3 induced tolerance to zinc, cadmium, cobalt and Pb in hyper-accumulators which store HMs in cell vacuole (Morel et al. 2009). HM ATPase 4 transporters which are localized in plasma membrane, transports Zn metal into the cytoplasm (Williams and Mills, 2005; Hanikenne and Baurain 2014). OsHMA2 (HM ATPase 2) is a dual transporter of Zn and Cd from root to shoot in rice via xylem (Takahashi et al. 2011; Satoh-Nagasawa et al. 2012; Takahashi et al. 2012). OsHMA2 transporter gene mutant shows reduced Cd amount in leaves and grains and lower growth and yield in rice plant (Yamaji et al. 2013). HMA3 over-expression increased accumulation of Cd in root and shoot system (Morel et al. 2009) and OsHMA3 sequesters Cd in the cell vacuoles of root (Ueno et al. 2010; 2011). AtHMA4 plays an imperative function in hyper accumulation and translocation of Cd and Zn in hyper-accumulator plant *Anemone halleri* (Hanikenne et al. 2008). NcHMA3 and NcHMA4 expression in *N. acaerulescens* ecotypes have been reported to impart tolerance to Cd (Ueno et al. 2011; Craciun et al. 2012). *Aeluropus littoralis* phytoremediates Pb and Hg HMs by the regulation of H^+-ATPase gene of plasma membrane (Jam et al. 2014). HMA genes showed differential gene expression in various tissues against copper, cadmium, zinc, cobalt, Ag, Pb, and Mn toxic effect in *Populus trichocarpa*. HMA1 and HMA4 genes showed tissue-specific up regulation in leaves while HMA5and HMA8 were highly expressed in roots (Li et al. 2015a). FDR3 gene over-expression encodes for MATE protein family transporters were also reported in roots of *Arabidopsis halleri* and *T. caerulescens* and have function in HMs translocation (Talke et al. 2006; van de Mortel et al. 2006; Durrett et al. 2007; Krämer et al. 2007). Yellow stripe1 (Ys1) gene encodes Fe(III)-PS transporter of *Zea mays* and transports Mn, Ni, Zn, Fe(III)-NA or Fe(II), Cu, and Cd (Schaaf et al. 2004). NRAMPs, another type of tonoplast localized and membrane-bound metal transporters of root and shoot cells, transports HMs like Ni, Cd, Co, Zn, and Mn (Nevo and Nelson 2006). NRAMP1 localized in plasma membrane of *A. thaliana* transported Fe and shows high-affinity for Mn uptake from the soil (Cailliatte et al. 2010). OsNRAMP1 transports Fe via endodermis and pericycle of rice cells and its over-expression enhanced tolerance of Cd and As in root-shoot system of *A. thaliana* (Tiwari et al. 2014). Vacuolar Iron Transporter (VIT1) exports Fe expressed into vacuoles during seed development (Kim et al. 2006) whereas and NRAMP3 and NRAMP4 transporters localized in vacuole envelope export sequestered iron out of vacuole during seed germination in *A. thaliana* (Bastow et al. 2018). ABC transporter, Low-affinity Cation Transporters (LCT), Cation proton Exchanger (CAX) and Calcium Cation Exchangers (CCX) play significant roles in HMs translocation and accumulation (Nakanishi et al. 2006; Sasaki et al. 2012; Takahashi et al. 2012). OsLCT1 which is a cell membrane localized Cd transporter gene expresses strongly during grain-ripening stage in rice and through xylem-mediated root-shoot and phloem-mediated Cd transport, transfers Cd into grains (Uraguchi et al. 2011). Many Cd transporters have been identified and characterized which include OsIRT1, OsIRT2 and two NRAMP family members, however, OsNRAMP5 and OsNRAMP1 facilitate the uptake of Cd by roots (Nakanishi et al. 2006; Senoura et al. 2011; Ishikawa et al. 2012; Sasaki et al. 2012). A higher exposure of Cd toxicity efficiently induces expression of genes of phytochelatin synthase (PCS) protein and F-box protein (Jebara et al. 2018) and the F-box protein upgrades growth of plant and chelation and also translocation of Cd ions (Brunetti et al. 2011). In detoxification and sequestration of HMs, intracellular transporters ABC {Multidrug Resistance associated Proteins (MRP) and Pleiotropic Drug Resistance proteins (PDR) families}, CDF (translocates Cadmium, Nickle, Iron, Mn, Cobalt, and Zinc metal cation from cytosol into apoplast, vacuole and endoplasmic reticulum), CaCA, NRAMP, and HMA sequester and transported chelated HMs and xenobiotics into cell vacuoles (Kramer et al. 2007; Montanini et al. 2007; Peiter et al. 2007). Plant responded toward HMs stress by alterations in expression of genes and methylation profiling in the metabolic and biochemical modifications along with toxic HM ions cavenging and improving stress symptoms. The citrate, malate, nicotinamine, methionine, and histidine chelators translocated HMs by binding and making complexes with the HM ions (Callahan et al. 2006; Anwer et al. 2017) whereas, proline and histidine induced tolerance against HMs by chelation of HM ions inside cells and xylem sap (Rai 2002). These organic acids, amino acids, along with jasmonic acid, ethylene, abscisic acid hormones, phytochelatins, and metallothioneins are induced or produced for chelation of HM ions (Manara 2012; Dalvi and Bhalerao 2013). The phytochelatin induction takes place due to elevated HMs levels; however Cd is the best effective stimulator (Howden et al. 1995). Metallothioneins are gene-encoded polypeptides important for intracellular HM ion homeostasis and sequestration (Kohler et al. 2004; Guo et al. 2013). Besides, chelating compounds are subscribed to HM tolerance by eliminating toxic ions from contaminated sites by sequestration and vacuolar compartmentalization. Cells triggered production of ROS and activated ROS-scavenging mechanisms viz., enzymatic (superoxide dismutases [SOD], peroxidises, and catalases) and non-enzymatic (flavonoids, glutathione, and carotenoids) to resist stress conditions (Huang et al. 2012).

15.2.2.1 Phytoremediation through Plant Growth-Promoting Rhizobacteria, Plant Growth-Promoting Endophytes, and Genetically Engineered PGPR Strains

The metalliferous soils are characterized by the soil particles bound by the metals that remain unavailable for consumption by plants. PGPR and PGPE (e.g., *Arthrobacter, Bacillus, Microbacterium, Pseudomonas, Kocuria*) solubilizes water undissolved Zn, Ni and Cu HMs by local soil acidification via acetate, tartrate, lactate, oxalate, succinate, gluconate, citrate, ketogluconate, and glycolate secretions (Becerra-Castro et al. 2011). The interaction between plants with plant growth-promoting microorganisms (PGPMOs) has successfully affected quest for Phytoremediation of U, Pb, and Zn (Viktorova et al. 2016; Ahsan et al. 2017). Zn activating bacteria, sequestered from serpentine terrains stimulate Zn, Cu, and Ni accumulations in *Ricinus communis* (Rajkumar and Freitas 2008), while *Rahnella* sp. JN6, identified from *Polygonum pubescens* stimulates plant growth and uptake of Cd, Zn, and Pb HMs in *Brassica napus* (He et al. 2013). The bacteria found in rhizosphere of *Betula celtiberica* pseudometallophyte, proliferate storage of As in its leaves and roots in arsenic contaminated soils. *Ensifer adhaerens* rhizobacterium strain 91R promotes *in vitro* plant growth (Mesa et al. 2017). Some PGPR and PGPE of hyper-accumulator plants stimulate growth rate and phytoextraction of various plant species cultivated in single and multiple HM polluted territories (Ma et al. 2009; 2011) and PGPRs discharge biosurfactants responsible for the property of HM bioavailability in the specific locations (Gamalero and Glick 2012). In Fe-deficient soils, PGPRs discharged low molecular weight siderophores that act as iron chelators of ferric iron (Fe^{3+}), thus augmenting iron accessibility to microorganisms as well as to plants and also has the ability to chelate other HMs for instance, Cr (III), Cu, Cd, Mg, Mn, Zn, Ni, Pb, and As (Schalk et al. 2011). *B. juncea* on inoculation with mutant SD1 of phosphate-solubilizing *Enterobacter sp.* NBRI K28, increased siderophore formation which in turn increased plant biomass and bioavailability of Zn, Cr and Ni (Kumar et al. 2008). PGPRs also secrete bacterial auxin IAA, which triggers physiological fluctuations in cell metabolism of the plant and stimulates growth, cell multiplication in root system, lateral root initiation process and production of surface root hairs on metal-contaminated sites (Glick 2010). PGPR and PGPE also synthesized ACC deaminase, which inhibits plant ethylene production by degrading ACC into 2-oxobutanoate and ammonia. Inoculation of hyper-accumulator plants with ACC deaminase secreting PGPR leads to wide spread proliferation of roots and efficient phytoremediation (Arshad et al. 2007). Genomics approach enabled PAHs (Polycyclic Aromatic Hydrocarbons), para-nitrophenol, organophosphate and phenanthrene pesticide compounds in *Rhodococcus wratislaviensis* strain 9, Mycobacterium (Birolli et al. 2018; Li et al. 2018; Subashchandrabose et al. 2018). The Ohb (ortho-dechlorination gene) of *Pseudomonas aeruginosa* when expressed in *Comamonas testosterone strain* VP44 produces enzymes that degrade chlorobenzoic acids (Aken et al. 2010). The ArsB/ArsB/ArsC genes of *Staphylococcus aureus* when expressed in *E. coli* confers As resistance (Abbas et al. 2015). The mtL (laccase gene) gene of *Myceliophthora thermophile* when expressed in *Saccharomyces cerevisiae* degrades lignin and polyaromatic hydrocarbons (Yang et al. 2017). Cytochrome P450CAM gene of *Pseudomonas putida* oxidizes hexane and 3 methylpentane (Kelly and Kelly 2013). The dszA/B/C of *Rhodococcus erythropolis* and *Chelatococcus* sp. when expressed in *Pseudomonas* strains desulfurized DBT and removes S (Khairy et al. 2015). The pnp operon (transforming p-nitrophenol into b-ketoadipate) of *Delftia acidovorans* when expressed in *Pseudomonas putida* degrades paraoxon (Kim et al. 2011). Engineered *Pseudomonas putida* 06909 bioinoculant reduced phytotoxic effect of Cd and enhanced Cd uptake by 40% (Wu et al. 2006). The bi-functional glutathione synthase gene (gcsgs) inserted in *Enterobacter sp.* CBSB1 improves efficiency of HM elimination in *B. juncea* (Qiu et al. 2014). *Rhodopseudomonas palustris* strain with MerP and MerT proteins showed simultaneous expression of Hg transport system and metallothione in for Hg2þ removal (Xu and Pei 2011). Engineered PGPR strains accumulated HMs and defend their hosts from HMs toxicity and improves rate of seed germination and plant physiological activities (Yong et al. 2014).

15.2.2.2 Phytoremediation via Genetically Engineered Plants

The phytoremediation efficiency can be enhanced by introgression of HM genes of accumulation and tolerance of wild hyper-accumulators into species with higher biomass yield (Dushenkov et al. 2002). Scientists enhanced phytoremediation in several plant species via genetic engineering (Cherian and Oliveira 2005). With somatic hybridizations, HM-resistant capability of *Noccaeac aerulescens* was introduced in *B. juncea* by protoplast fusion resulting in *B. juncea* hybrids with high Zn and Ni accumulation and biomass yield (Brewer et al. 1999). T-DNA insertion mutant of OsNRAMP5 showed reduced Cd (ranging from 0.01–0.03 mg/kg) and Mn accumulation in grains; however, the grain yield was only 11% of the wild rice (Ishikawa et al. 2012; Sasaki et al. 2012). Radionuclide hyper-accumulator plants accumulate toxic HMs and show HM phytoremediation of polluted sites. *Pteris vittata* which is As hyper-accumulator showed accumulation of Al, Cu, Fe, Ni, Zn, Se, Mn, U, and V near radionuclide and HM-contaminated uranium mine tailing ponds in eastern India (Laxman Singh et al. 2015). OsHMA3 overexpression considerably reduces sequestration of Cd accumulation in vacuole of leaf straw and seed grain, although OsHMA3 mutant promotes translocation of Cd from root-shoot (Miyadate et al. 2010; Ueno et al. 2010; Sasaki et al. 2014; Shao et al. 2017). OsLCD knockout (with low Cd) declines Cd accumulation in rice grains by 43–55%, nevertheless its purpose as Cd transporter is unclear (Shimo et al. 2011). Plasma membrane channel protein (NtCBP4)(identified with transcriptomics) over-expression in *Nicotiana* elevated cadmium (Cd) or lead (Pb2+) accumulation and plant's sensitivity to metal (Arazi et al. 1999; Sunkar et al. 2000). MerC gene expression in *A. thaliana* and *Nicotiana* via genetic engineering approach elevated Hg HM accumulation by 2-folds compared to their wild plants however, plants become hypersensitive to Hg HM (Sasaki et al. 2006). Transgenic *Nicotiana* or *Brassica* plants with human MT-2 gene increased tolerance of Cd (Misra and Gedamu 1989) and *Noccaeac aerulescens* shoots accumulated highest Zn concentrations and are appropriate for soil remediation (Barrutia et al. 2011). *In vitro* and green house conditions enabled *Daphne tangutica* to accumulate high Pb contents and biosynthesis of chlorophyll and carotenoid pigments was greater than *Daphne jasminea* (Wiszniewska et al. 2015). Some of the genetically engineered plants utilized for phytoremediation process are discussed in Table 15.1.

15.2.2.3 Emerging Area of Phytoremediation

MicroRNAs (miRNA) are micro and single stranded and composed of 18–25 conserved nucleotides. The non-coding RNAs regulate expression of genes by posttranscriptional gene silencing and these non-coding genes formerly designated as "JUNK" (Pandita and Pandita 2016) which include miRNAs are documented to play pivotal functions in abiotic stress responses both in model crops and wild plants (Pandita 2018; 2019; Pandita and Wani 2019). The microRNAs assist molecular mechanisms by providing plants tolerance against HMs (Fang et al. 2013). Presently, there are a limited number of investigations about functions of key HMs transporter and regulatory genes and miRNAs based post transcriptional gene regulation in plants in HMs stress conditions (Gao et al. 2019; Wang et al. 2019). ABC or NRAMP transporters having role in metal tolerance pathways are targeted by miRNA (Noman and Aqeel 2017) and a complex response in *indica* and *japonica* rice species, miRNome expression profiling indicated 16 different miRNAs (Srivastava et al. 2012), 18 diverse microRNAs in rice (Liu and Zhang 2012) and 69 novel microRNAs in *B. juncea* under aluminum stress (Lima et al. 2011). MicroRNA156, microRNA171 and microRNA396a show down regulation in *Brassica napus* under Cd stress (Zhou et al. 2012a; b). In *Brassica vulgaris*, 13 different miRNAs are reported to regulate HMs stress (Li et al. 2015a; b) besides, Medicago under Cd stress reported down regulation of miR166 and up regulation of miR393, miR171, miR319 and miR529 respectively (Zhou et al. 2008). However in *M. truncatula* under Hg stress, microRNA171, microRNA 319, microRNA 393 and microRNA529 were up regulated (Zhou et al. 2008).

TABLE 15.1

Genetically Engineered Plants for Phytoremediation

Transgenic Organisms Generated	Genes Manipulated	Effects	References
Arabidopsis thaliana	Transfer of bacterial genes, *XplA*, and *XplB*	Efficiency for removal and detoxification of RDX by means of cytochrome P450-reductase complex	Rylott et al. 2011
Arabidopsis thaliana	Transformation with glutathione-S-transferase gene from tobacco	Tolerance to Al, Cu, and Na	Ezaki et al. 2000
Arabidopsis thaliana	Transformation and collected expression of SRSIp/ArsC and ACT 2p/γ-ECS	Enhanced accumulation and tolerance to As, more As/gram of tissue than wild plants or transgenics with only γ-ECS or ArsC expression	Dhankher et al. 2002; Mello-Farias and Chaves 2008
Arabidopsis thaliana	Transformation with MTA gene from pea	Accumulates Cu	Evans et al. 1992
Arabidopsis thaliana	Overexpression of At MTPI	Accumulates Zn	Van der Zaal et al. 1999
Arabidopsis thaliana	Transformation with Znt A-heavy metal transporters from *E. coli*	Cd and Pb resistant	Lee et al. 2003
Arabidopsis thaliana	Transformation with selenocysteine methyl transferase from *A. bisculatus*	Selenite resistant	Ellis et al. 2004
Arabidopsis thaliana	Transformation with yeast cadmium factor 1 (YCF1)	Cd and Pb tolerant	Song et al. 2003
Arabidopsis thaliana, Nicotiana tabacum	Overexpression of NAS1 gene encoding NA synthase	Enhanced tolerance toward Ni, Cd, Mn, Cu, Fe, and Zn and enhanced Mn and Ni uptake	Kim et al. 2005; Pianelli et al. 2005
Arabidopsis thaliana, Oryza sativa	Expressing the genes of naphthalene dioxygenase system	Tolerance to and metabolization of naphthalene and phenanthrene	Peng et al. 2014
Arabidopsis thaliana, Nicotiana tabacum	Transformation with *merA* gene from bacterial host	Resistance to Hg	Bizily et al. 2000; Rugh et al. 2000
Brassica juncea	Integration of *APS* and *SMT* genes, encoding ATP sulfurylase and selenocysteine methyl transferase	Enriched tolerance and accretion of Se	LeDuc et al. 2006
Brassica juncea	Investigation of rate-limiting factors for the production of glutathione and phytochelatin. gshl gene from *E. coli* was introduced to generate γ-ECS transgenics	Escalated tolerance to cadmium. The greater concentration of γ-GluCys, glutathione, phytochelatins, and nonproteinthiols	Ow 1996; Fulekar et al. 2008
Brassica juncea	Transformation with a bacterial glutathione reductase in cytoplasm and chloroplast	Tolerance to Se	D´Souza et al. 2000
Brassica juncea	Arsenate reductase g-glutamyl cysteine synthetase from bacteria	As tolerant	Dhankher et al. 2002

(Continued)

TABLE 15.1 (*Continued*)

Genetically Engineered Plants for Phytoremediation

Transgenic Organisms Generated	Genes Manipulated	Effects	References
Brassica juncea	ATP sulfurylase CAPS	Tolerance to Se	Van Huysen et al. 2004
Brassica juncea	Cystathionine-gamma synthase (CGS)	Se volatilization	Van Huysen et al. 2004
Brassica oleracea	Transformation with CUP-1 gene from Yeast	Accumulates Cd	Hasegawa et al. 1997
Hirschfeldia incana	Over expression of MT2b metallothionein gene	Enhanced tolerance and accumulation of Pb	Auguy et al. 2016
Medicago sativa	Expression of the gene, Bph C. B	Enhanced PCBs and 2,4-dichlorophenol (2,4-DCP) tolerance, dissipation and removal of 2,4-DCP and PCBs	Wang et al. 2015
Nicotiana glauca	Transformation with phytochelatin synthase (Ta PCS) from wheat	Accumulation of Pb	Gisbert et al. 2003
Nicotiana tabacum	Transformation with gene vacuolar transporter *CAX-2* gene from *A. thaliana*	Amassing of Ca, Cd, and Mn	Hirschi et al. 2000
Nicotiana tabacum	Transformation with gene Nt CBP4	Tolerance to Ni and accumulation of Pb	Arazi et al. 1999
Nicotiana tabacum	Transformation with gene Ferretin from soybean	Enhanced iron accumulation	Goto et al. 1998
Nicotiana tabacum	Transformation with *MT-1* gene from mouse	Cd tolerant	Pan et al. 1994
Nicotiana tabacum, Brassica napus subsp. napus	Transformation with *MT2* gene from human	Tolerance to Cd	Misra and Gedamu 1989
Nicotiana tabacum, Populus angustifolia, or *Silenecucubalis*	Overexpression of glutamyl cysteine syntlietase	Enhanced HMs accumulation	Fulekar et al. 2008
Oryza sativa	with gene Ferretin from soybean	Increased Fe accumulation	Goto et al. 1998; 1999
Populus angustifolia, Nicotiana tabacum, Arabidopsis thaliana	Overexpression of MTA1, MT1 and MT2 metallothionein encoding-genes	Enhanced endurance and accumulates Cu, Zn and Cd	Turchi et al. 2012; Lv et al. 2013; Xia et al. 2012

The plants under Hg or Cd stress conditions and deficiency of nutrients exhibited down regulation of miR398 (Kuo and Chiou 2011; Yang and Chen 2013). Under As stress, exogenous application of JA and IAA modify expression of microRNA 167, microRNA 319 and microRNA854, sequentially enhancing growth of plants (Gupta et al. 2014). The stress responsible microRNAs were identified in *Cicer Arietinum* L. under inoculation of *Pseudomonas putida* RA PGPR (Jatan et al. 2019).

15.3 CRISPR-Cas9 Genome Editing Tools

CRISPR-Cas9 (Clustered Regularly Interspaced Short Palindromic Repeats and Cas9 associated proteins) initiates from immune system response of bacteria and archaea against foreign viral and plasmid DNA.A bacterium endures initial bacteriophage infection by inserting phage genome segments into the bacterial genome. CRISPR is an array of 30–40 bp direct repeat sequences separated by foreign spacer sequences and transcribes CRISPR RNA crRNA (Zhang et al. 2018) and when the same bacterium is infected second time, from already adapted phage genome segment (CRISPR) and directs CRISPR/Cas endonuclease for degradation of the phage DNA (Johnson et al. 2015). Cas9is RNA guided DNA endonuclease which cleaves foreign invader DNA (Mahas and Mahfouz 2018) and the single guide RNA acquired by CRISPR (Listgarten et al. 2018) specifically binds to target DNA site (Shah et al. 2018). CRISPR RNA and Cas endonucleases form CRISPR-Cas crRNP (Ribonucleoprotein) complex degrades the foreign DNA/RNA (Majumdar et al. 2017) and CRISPR-Cas systems have six main types; Type I (endonucleases Cas3, Cas8a, Cas8b, Cas8c, Cas10d, Cse1(Cas8e), Cas8f (Csy1), Cas5f1 (Csy2), Cas7f1 (Csy3), Cas6f, Cas5f2, Cas7f2), Type II (endonucleases Csn2, Cas9) and Type III (endonucleases Csm2, Cmr5, Cas10 (Csx11), gRAMP) (Chylinski et al. 2014; Zhu et al. 2018), with specific Cas with respect to organism (Cooper et al. 2018) Type IV (endonucleasesCas5 and Cas7), V (Cas12a (Cpf1), Cas12b (C2c1), Cas12c (C2c3), Cas12d (CasY), Cas12e (CasX), Cas12f, Cas12g, Cas12h, Cas12i, Cas12k endonucleases) and VI (Cas13a (C2C2), Cas13b1, Csx28, Cas13b2, Csx27, Cas13c (C2C7), Cas13dendonucleases) systems (Makarova et al. 2015). In CRISPR/Cas9 system, mode of action depends on the Cas9 protein (CRISPR-associated protein 9) and single guide RNA (sgRNA) (Liu et al. 2017). The Cas9 endonuclease mainly occur in bacterial species like *Brevibacillus laterosporus* (Karvelis et al. 2015), *Staphylococcus aureus* (Ran et al. 2015), and other *Streptococcus* species (Steinert et al. 2015). Cas9 endonuclease proteins contains HNH-like nuclease (nuclease-associated proteins) domains cleave DNA strand complementary to sequence of target strand or guide RNA and RuvC-like nuclease domain cleaves non-target DNA strand (Liu et al. 2017). Synthetic single guide RNA possesses length of <100 bp with a 20 bp sequence coupled to 5′ end and guides target sequence identification via PAM-Protospacer Adjacent Motifs (Liu et al. 2017) and the single guide RNA 3′end is looped for precise linkage to target DNA sequence. Cas9 endonuclease introduced dsDNA cleavage as double-stranded breaks (DSBs) (Jacobs et al. 2015) at the target site where the gene of interest (GOI) gets inserted (Shapiro et al. 2018). CRISPR-Cas9 is governed by RNA-DNA recognition to generate DSBs and activated the host cell's repair machinery against damaged DNA mostly by error-prone Non-Homologous End Joining (NHEJ) or in contrast by Homology Directed Repair (HDR) method in which novel donor DNA fragment flanked by two homologous arms that can be integrated at the DSB site and becomes the basis for precise genome engineering(Mojica et al. 2005; Lieber 2010; Xie and Yang 2013; Jacobs et al. 2015; Liu et al. 2017). An overview of CRISPR-Cas 9 mediated genome editing system is shown diagrammatically in Figure 15.2.

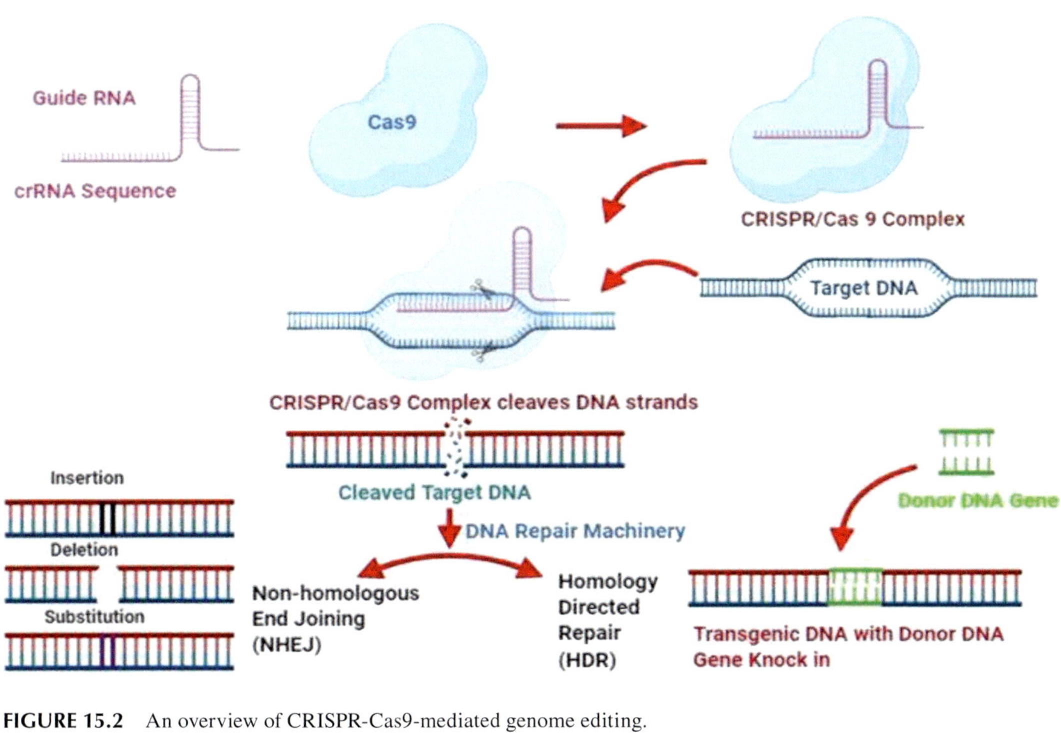

FIGURE 15.2 An overview of CRISPR-Cas9-mediated genome editing.

15.4 Applications of CRISPR-Cas9 in Phytoremediation Approach

Discovery of CRISPR draws considerable attention by scientific community worldwide due to user-friendliness, simple target design, high Cas9 protein efficiency and guide RNA, production of transgenics with less sensitivity of vector designing, and multiplex gene editing cause synchronized targeted mutations in multiple genes (Mali et al. 2013; Ma et al. 2015; Malzahn et al. 2017). CRISPR-Cas9 toolbox allows efficient, precise and transgene-free plant gene editing by transformation of the protoplasts or direct bombardment of guide RNA and Cas9 into cells (Liang et al. 2018), transient expression at the callus stage (Zhang et al. 2016), CRISPR-Cas9based cytidine base editor technology (Veillet et al. 2019) and transgene killer CRISPR (TKC) Technology (He et al. 2019)and has found use in many plant species, as well as in rice (Ishizaki 2016; Jung et al. 2017). Virus-based gRNA delivery system (VIGE) for CRISPR-Cas9 mediated editing of plant genomes is an efficient tool (Yin et al. 2015) as RNA virus (Tobacco rattle virus replicates in cytoplasm) (Ali et al. 2015) and DNA virus (Cabbage Leaf Curl virus replicates in nucleus) (Yin et al. 2015) delivering Cas9 gRNAs into transgenic plants. Phytoremediation with biotechnological strategies includes CRISPR-Cas 9 mediated gene expression profiling to escalate synthesis of HM ligands (for instance metallothioneins and phytochelatins), HM transport proteins (CDF, MATE, HMA, YSL, and ZIP), growth hormones (CKs, AUXs, and GAs) and low molecular weight organic acid root exudates and siderophores (Basharat et al. 2018). Genes of the plants and bacteria transferred to target plants resulted in beneficial effects for phytoremediation which may be improved by CRISPR technology (Fasani et al. 2018). Genetic engineering promoted and enhanced metal tolerance and accumulation by introgression and expression of gene/genes involved in metal ingestion, transportation, accumulation and detoxification and against oxidative stress or escalation in production of target hyper-accumulator plant biomass (Fasani et al. 2018). Genes encoding HM ion transporters are currently extensively utilized for the advancement in the potential of phytoremediation approaches (Fasani et al. 2018). CRISPR-Cas9 systems find application in genome editing of phytoremediator species with phytoremediation potential like in poplar and maize (Miglani 2017; Fan et al. 2015). *Populus* is a good candidate for phytoremediation process because of high production of the plant biomass, rapid growth rate, deep and widespread root architecture, distinctive adaptation capability to varied conditions, tolerance to pollutants and outstanding capability in vegetative reproduction which accelerates its propagation (Baldantoni et al. 2014). In addition to this, *Z. mays* also has fast-growth, high biomass production with a distinct ability of HM accretion and has offered constructive consequences in phytoremediation and phytomining processes (Anderson et al. 2005; Ali et al. 2013). Genome editing breakthroughs in *Z. mays* by the CRISPR-Cas9 system are previously documented (Agarwal et al. 2018) and poplar and *Z. mays* were progressively employed to combine their phytoremediation abilities for the generation of the bio-energy signifying the appropriateness of CRISPR for progressing the process efficiency and uphold sustainability (Meerset al. 2010; Pandey et al. 2016). The gRNA-guided dCas9 CRISPR research holds great potential in phytoremediation for modulation of gene expression (Liu and Stewart 2015; Vazquez-Vilar et al. 2016). Transcription factors joined with dCas9 lead to RNA transcription repression or enhancement sequentially, up regulation or down regulation expression of gene or genes of interest (Miglani, 2017). The CRISPRi (CRISPR-interference) and CRISPRa (CRISPR-activation) modify expression of genes above 1000-fold range (Gilbert et al. 2014) and CRISPR-Cas9 genome editing has been efficiently employed in wheat (Liang et al. 2018), apple and grapevine (Osakabe et al. 2018), *Oryza sativa* (Khanday et al. 2019), *Arabidopsis* (Endo et al. 2019), maize (Svitashev et al. 2016), *Camelina sativa* (Jiang et al. 2017), tomato (Li et al. 2018) and *Brassica napus* (Okuzaki et al. 2018). CRISPR-Cas9 based OsLCT1 and OsNRAMP5 mutants exhibited low-Cd-accumulated rice genotypes however, in their knockdown mutants, one target site was selected for OsLCT1 Cd transporter gene at exon 1 and two sites (exon 7 and exon 9) for OsNRAMP5 Cd transporter genes for generating mutants (Liu et al. 2019). CRISPR-Cas9 vectors had CYP81A6-hpRNAi element (Lu et al. 2017), adapted from pHun4c12 (Xu et al. 2014). OsLCT1 mutant had reduced Cd accumulation in grains of *O. sativa* and the plant growth was unaffected (Uraguchi et al. 2011). An ion beam-mutant and CRISPR-Cas9 designed mutant lines of nramp5 considerably reduce Cd accumulation devoid of substantial effects on agronomic traits (Ishikawa et al. 2012; Tang et al. 2017). With CRISPR-Cas9 system, Cd resistant *O. sativa* indica

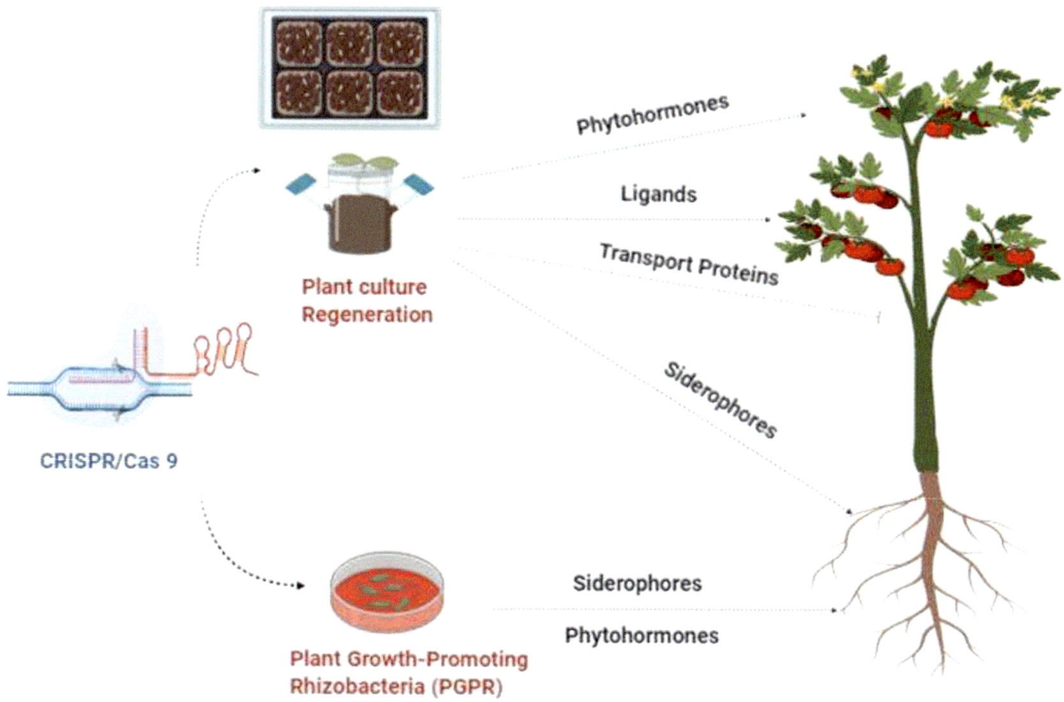

FIGURE 15.3 Prospective applications of CRISPR-Cas9 for phytoremediation. (Adapted from Basharat et al. 2018.)

cultivars with decreased Cd accumulation in shoots and roots without any significant effect on plant yield were generated by knockdown ofOsNRAMP5exon IX (Tang et al. 2017). Phytoremediation can be manipulated by CRISPR-Cas9 tools by producing efficient PGPR (Mosa et al. 2016). The genes of phytohormone biosynthesis (Boivin et al. 2016), nitrogenase complex (Chinnaswamy et al. 2018) and siderophore formation (Thode et al. 2018) are involved in PGPR-plant interaction. Hence, editing PGPR genes may benefit the plant's ability of growth and pollutant grip by enhancing nitrogen (N) fixation, phosphate solubilization, FE sequestration, phytohormone production and biocontrol (Novo et al. 2018). Next generation phytoremediators may be developed from the plant HM-responsive genes, miRNAs and PGPR CRISPR manipulated versions, permitting the efficacious phytoremediation of soil and aquatic environments are possible in limited time frames. The prospective applications of CRISPR-Cas9 for different strategies of phytoremediation are described in Figure 15.3.

15.5 Limitations of CRISPR-Cas 9 System in Phytoremediation

The key limiting factor of CRISPR-Cas9 technology is the probability to produce 'off target mutations' due to incorrect pairings of single guide RNAs on undesirable DNA locations which is a backbone in CRISPR-Cas system improvement programs, mainly for transgene free crops (Hahn and Nekrasov 2019). Research investigations on genomes of *A. thaliana* (Zhang et al. 2018a; b), *O. sativa* (Baysal et al. 2016), citrus (Jia et al. 2017) and *Solanum lycopersicum* (Li et al. 2017) modified with CRISPR-Cas9 system has confirmed that CRISPR-Cas9 system is very specific in plants and "off-target mutations" occur rarely. In *O. sativa* "off-target mutations" were identified (Xie and Yang 2013) and complete mutant genome sequencing confirms "off-target mutations" or only sequencing of potential off-target locations predicted bioinformatically can fulfill the requirement (Wolt et al. 2016; Puchta 2017). Plant and animal "off-target mutations" are not that challenging but stratagems to reduce their occurrence involving SpCas9 alternatives with precise nature because of longer Protospacer Adjacent Motif sequence (Kangquan et al. 2017;

Schindele et al. 2018) and insertion of CRISPR-Cas system as Ribo-Nucleo-Protein are accessible (Hahn and Nekrasov 2019). Target site recognition by the single guide RNA PAM sequence declined plasticity of probable targets sites and single guide RNA pattern (Zhou et al. 2018b). CRISPR-Cas9 has limitations in RNA and DNA viruses (Mehta et al. 2019). Protection and commercialization concerns, as genome edited transgenic crops are not permitted in all countries (Abbott 2018). CRISPR edited transgene-free oilseed products are allowed in United States and other edited plants ready for commercialization include mushroom, *O. sativa*, *Z. mays*, *Glycine max*, and bristle grass (Waltz 2018). Nonetheless, it can be expected that CRISPR-Cas9 tools would enable the application probability of Genetically Modified Organisms for phytoremediation.

15.6 Conclusion

CRISPR-Cas-mediated genome editing though in infancy in phytoremediation area, heralds great prospective for editing and engineering plant genomes (Peng et al. 2016; Yin et al. 2017) for enhancing phytoremediation prospective to its zenith in both plants and microorganisms. But, breakthroughs in this biotechnological domain of CRISPR-Cas technology will be achieved after appreciating the cross-talk of molecular mechanisms of various process of phytoremediation to bio-accumulate the HMs thoroughly. Modifying genes of interest (GOI) are metallothioneins and phytochelatins; CDF, HMA, MATE, YSL, and ZIP transport proteins; genes coding for AUXs, CKs, and GAs growth hormones; and LMWOA (low-molecular-weight organic acid) and siderophores (Basharat et al. 2018). Other genes are listed in the Table 15.1. Stress responsible microRNAs, PGPR, PGPE, and genetically engineered PGPR strains, whole metabolic pathways, and HMs contaminant homeostasis networks of hyper-accumulation, degradation, or tolerance can prove to be ground-breaking as has been obtained in CRISPR-Cas-mediated mutants of OsLCT1 and OsNRAMP5 showing low-Cd-accumulation in rice plants.

REFERENCES

Abbas, S. Z., Riaz, M., Ramzan, N., Zahid, M. T., Shakoori, F. R. and Rafatullah, M. 2015. Isolation and characterization of arsenic resistant bacteria from wastewater. *Braz J Microbiol* 45(4):1309–1315.

Abbott, A. 2018. European court suggests relaxed gene-editing rules. *Nature* [(accessed on 19 January 2018)]; https://www.nature.com/articles/d41586-018-01013-5.

Abdel-Raoufa, N., Al-Homaidan, A. A. and Ibraheem, I. B. M. 2012. Microalgae and waste water treatment. *Saudi J Biol Sci* 19:257–275.

Agarwal, A., Yadava, P., Kumar, K., Singh, I., Kaul, T., Pattanayak, A. and Agrawal, P. K. 2018. Insights into maize genome editing via CRISPR/Cas9. *Physiol Mol Biol Plants* 24:175–183.

Agwu, K. K., Okoye, C. M. I., Okeji, M. C. and Clifford, E. O. 2018. Potential health impacts of heavy metal concentrations in fresh and marine water fishes consumed in southeast, Nigeria. *Pakistan j Nutr* 17:647–653.

Ahmed, M. 2014. Phosphate-solubilizing bacteria-assisted phytoremediation of metalliferous soils: a review. *3 Biotech* 5:111–121.

Ahsan, M. T., Najam-Ul-Haq, M., Idrees, M., Ullah, I. and Afzal, M. 2017. Bacterial endophytes enhance phytostabilization in soils contaminated with uranium and lead. *Int J Phytoremediat* 19(10):937–946.

Aken, B. V., Correa, P. A. and Schnoor, J. L. 2010. Phytoremediation of polychlorinated biphenyls: new trends and promises. *Environ Sci Technol* 44(8):2767–2776.

Ali, H., Khan, E. and Sajad, M. A. 2013. Phytoremediation of heavy metals- concepts and applications. *Chemosphere* 91:869–881.

Ali, Z., Abul-Faraj, A., Li, L., Ghosh, N., Piatek, M., Mahjoub, A., Aouida, M., Piatek, A., Baltes, N. J., Voytas D. F., Dinesh-Kumar, S. and Mahfouz M. M. 2015. Efficient virus-mediated genome editing in plants using the CRISPR/Cas9system. *Mol Plant* 8(8):1288–1291.

Anderson, C., Moreno, F. and Meech, J. 2005. A field demonstration of gold phytoextraction technology. *Miner Eng* 18:385–392.

Andrea, M. M. E., Carolina, T. E. A., Jose, C. B. T., Luis, M. N. J. and Carlos, G. M. L. 2019. Evaluation of contaminants in agricultural soils in an Irrigation District in Colombia. *Heliyon* 5:e02217.

Anwer, S., Khan, S., Ashraf, M. Y., Noman, A., Baloch, S. U., Zafar, S. and Fahad, S. 2017. Impact of chelator-induced phytoextraction of cadmium on yield and ionic uptake of maize. *Int J Phytoremediat* 19(6):505–513.

Arazi, T., Sunkar, R., Kaplan, B. and Fromm, H. A. 1999. A tobacco plasma membrane calmodulin-binding transporter confers Ni^{2+} tolerance and Pb^{2+} hypersensitivity in transgenic plants. *Plant J* 20:171–182.

Arshad, M., Saleem, M. and Hussain, S. 2007. Perspectives of bacterial ACC deaminase in phytoremediation. *Trends Biotechnol* 25:356–362.

Auguy, F., Fahr, M., Moulin, P., El Mzibri, M., Smouni, A., Filali-Maltouf, A., Béna, G. and Doumas, P. 2016. Transcriptome changes in *Hirschfeldia incana* in response to lead exposure. *Front Plant Sci* 6:1–13.

Axelsen, K. B. and Palmgren, M. G. 1998a. Evolution of substrate specificities in the P-type ATPase superfamily. *J Mol Evol* 46:84–101.

Axelsen, K. B. and Palmgren, M. G. 1998b. Inventory of the superfamily of P-type ion pumps in Arabidopsis. *Plant Physiol* 126:696–706.

Baldantoni, D., Cicatelli, A., Bellino, A. and Castiglione, S. 2014. Different behaviours in phytoremediation capacity of two heavy metal tolerant poplar clones in relation to iron and other trace elements. *J Environ Manag* 146:94–99.

Barrutia, O., Artetxe, U., Hernández, A., Olano, J. M., García-Plazaola, J. I., Garbisu, C. and Becerril, J. M. 2011. Native plant communities in an abandoned Pb-Zn mining area of northern Spain: implications for phytoremediation and germplasm preservation. *Int J Phytoremediat* 13:256–270.

Basharat, Z., Novo, L. A. B. and Yasmin, A. 2018. Genome editing weds CRISPR: what is in it for phytoremediation? *Plants* 7:51.

Bastow, E. L., dela Torre, V. S. G., Maclean, A. E., Green, R. T., Merlot, S., Thomine, S. and Balk, J. 2018. Vacuolar iron stores gated by NRAMP3 and NRAMP4 are the primary source of iron in germinating seeds. *Plant Physiol* 177:1267–1276.

Basu, S., Rabara, R. C., Negi, S. and Shukla, P. 2018. Engineering PGPMOs through gene editing and systems biology: a solution for phytoremediation? *Trends Biotechnol* 36:499–510.

Baysal, C., Bortesi, L., Zhu, C., Farré, G., Schillberg, S. and Christou, P. 2016. CRISPR/Cas9 activity in the rice OsBEIIb gene does not induce off-target effects in the closely related paralog OsBEIIa. *Mol Breed* 36:108.

Becerra-Castro, C., Prieto-Fernández, Á, Álvarez-López, V., Cabello-Conejo, M. I., Acea, M. J. and Kidd, P. S. 2011. Nickel solubilizing capacity and characterization of rhizobacteria isolated from hyperaccumulating and non-hyperaccumulating subspecies of Alyssum serpyllifolium. *Int J Phytoremed* 13:229–244.

Berti, W. R. and Cunningham, S. D. 2000. Phytostabilization of metals. In: Raskin I. and Ensley B. D. (eds), *Phytoremediation of Toxic Metals: Using Plants to Clean up the Environment*. Wiley, New York, pp:71–88.

Birolli, W. G., Santos, D. D. A., Alvarenga, N., Garcia, A. C., Romão, L. P. and Porto, A. L. 2018. Biodegradation of anthracene and several PAHs by the marine-derived fungus *Cladosporium* sp. CBMAI 1237. *Mar Pollut Bull* 129:525–533.

Bizily, S. P., Rugh, C. L. and Meagher, R. B. 2000. Phytodetoxification of hazardous organomercurials by genetically engineered plants. *Nature Biotechnol* 18:213–217.

Boivin, S., Fonouni-Farde, C. and Frugier, F. 2016. How auxin and cytokinin phytohormones modulate root microbe interactions. *Front Plant Sci* 7:1–12.

Brewer, E. P., Saunders, J. A., Scott, A. J., Chaney, R. L. and McIntosh, M. S. 1999. Somatic hybridization between the zinc accumulator *Thlaspi caerulescens* and *Brassica napus*. *Theor Appl Genet* 99:761–771.

Briskine, R. V., Paape, T., Shimizu-Inatsugi, R., Nishiyama, T., Akama, S., Sese, J. and Shimizu, K. K. 2017. Genome assembly and annotation of *Arabidopsis halleri*, a model for heavy metal hyperaccumulation and evolutionary ecology. *Mol Ecol Resour* 17:1025–1036.

Broadley, M., Brown, P., Cakmak, I., Rengel, Z. and Zhao, F. 2012. Function of nutrients: micronutrients. In: Marschner P. (ed), *Mineral Nutrition of Higher Plants*. Academic Press Inc., San Diego, pp:191–248.

Brunetti, P., Zanella, L., Proi, A., Paoli, A., Falasca, G., Altamura, M. M., Sanita di Toppi, L., Costantino, P. and Cardarelli, M. 2011. Cadmium tolerance and phytochelatin content of Arabidopsis seedlings overexpressing the phytochelatin synthase gene AtPCS1. *J Exp Bot* 62:5509–5519.

Brunner, I., Luster, J., Günthardt-Goerg, M. S. and Frey, B. 2008. Heavy metal accumulation and phytostabilisation potential of tree fine roots in a contaminated soil. *Environ Pollut* 152:559–568.

Cailliatte, R., Schikora, A., Briat, J. F., Mari, S. and Curie, C. 2010. High-affinity manganese uptake by the metal transporter NRAMP1 is essential for Arabidopsis growth in low manganese conditions. *Plant Cell* 22:904–917.

Callahan, D. L., Baker, A. J. M., Kolev, S. D. and Wedd, A. G. 2006. Metal ion ligands in hyperaccumulating plants. *J Biol Inorg Chem* 11:2–12.

Chaney, R. L., Li, Y. M., Brown, S. L., Homer, F. A., Malik, M. and Angle, J. S. 2000. Improving metal hyperaccumulator wild plants to develop phytoextraction systems: approaches and progress. In: Terry N. and Banuelos G. (eds), *Phytoremediation of Contaminated Soil and Water.* CRC Press, Boca Raton, FL, USA pp:129–158.

Chaudhry, T. M., Hayes, W. J., Khan, A. G. and Khoo, C. S. 1998. Phytoremediation—focusing on accumulator plants that remediate metal contaminated soils. *Australasian J Ecotoxicol* 4:37–51.

Chen, W. R., Feng, Y. and Chao, Y. E. 2008. Genomic analysis and expression pattern of OsZIP1, OsZIP3, and OsZIP4 in two rice (*Oryza sativa* L.) genotypes with different zinc efficiency. *Russ J Plant Physiol* 55:400–409.

Cherian, S. and Oliveira, M. M. 2005. Transgenic plants in phytoremediation: recent advances and new possibilities. *Environ Sci Technol* 39:9377–9390.

Chinnaswamy, A., Coba de la Peña, T., Stoll, A., de la Peña Rojo, D., Bravo, J., Rincón, A., Lucas, M. M. and Pueyo, J. J. 2018. A nodule endophytic *Bacillus megaterium* strain isolated from *Medicago polymorpha* enhances growth, promotes nodulation by *Ensifer medicae* and alleviates salt stress in alfalfa plants. *Ann Appl Biol* 172:295–308.

Chylinski, K., Makarova, K. S., Charpentier, E. et al. 2014. Classification and evolution of type II CRISPR-Cas systems. *Nucleic Acids Res* 42:6091–6105.

Cooper, L. A., Stringer, A. M. and Wade, J. T. 2018. Determining the specificity of cascade binding, interference, and primed adaptation in vivo in the *Escherichia coli* type IE CRISPR-cas system. *mBio* 9:e02100–17.

Craciun, A. R., Meyer, C. L., Chen, J., Roosens, N., Groodt, R. D., Hilson, P. and Verbruggen, C. 2012. Variation in HMA4 gene copy number and expression among *Noccaea caerulescens* populations presenting different levels of Cd tolerance and accumulation. *J Exp Bot* 63:4179–4189.

D´Souza, M. P., Pilon-Smits, E. A. H. and Terry, N. 2000. The physiology and biochemistry of selenium volatilization by plants. In: Raskin I. and Ensley B. D. (ed.), *Phytoremediation of Toxic Metals: Using Plants to Clean up the Environment*, Wiley and Sons, New York, pp:171–188.

Dai, Z., Zhang, S., Yang, Q., Zhang, W., Qian, X., Dong, W., Jiyang, M. and Xin, F. 2018. Genetic tool development and systemic regulation in biosynthetic technology. *Biotechnol Biofuels* 11:152.

DalCorso, G., Fasani, E., Manara, A., Visioli, G. and Furini, A. 2019. Heavy metal pollutions: state of the art and innovation in phytoremediation. *Int J Mol Sci* 20:3412.

Dalvi, A. A. and Bhalerao, S. A. 2013. Response of plants towards heavy metal toxicity: an overview of avoidance, tolerance and uptake mechanism. *Ann Plant Sci* 2:362–368.

Deng, F., Yamaji, N., Xia, J. and Ma, J. F. 2013. A member of the heavy metal P-type ATPase OsHMA5 is involved in xylem loading of copper in rice. *Plant Physiol* 163:1353–1362.

Desbrosses-Fonrouge, A. G., Voigt, K., Schröder, A., Arrivault, S., Thomine, S. and Krämer, U. 2005. *Arabidopsis thaliana* MTP1 is a Zn transporter in the vacuolar membrane which mediates Zn detoxification and drives leaf Zn accumulation. *FEBS Lett* 579:4165–4174.

DeSousa, C. S., Hassan, S. S., Pinto, A. C., Silva, W. M., De Almeida, S. S., Soares, S. D. C., Azevedo, M. S. P., Rocha, C. S., Barh, D. and Azevedo, V. 2018. Microbialomics: applications in biotechnology. In: Barh D. and Azevedo V. (eds), *Omics Technologies and Bio-Engineering.* Academic Press, Cambridge, pp:3–20.

Dhankher, O. P., Li, Y., Rosen, B. P., Shi, J., Salt, D., Senecoff, J. F., Sashti, N. A. and Meagher, R. B. 2002. Engineering tolerance and hyperaccumulation of arsenic in plants by combining arsenate reductase and g-glutamylcysteine synthetase expression. *Nature Biotechnol* 20:1140–1145.

Dickinson, N. M., Baker, A. J., Doronila, A., Laidlaw, S. and Reeves, R. D. 2009. Phytoremediation of inorganics: realism and synergies. *Int J Phytoremediat* 11:97–114.

Dräger, B. D., Desbrosses-Fonrouge, A. G., Krach, C., Chardonnens, A. N., Meyer, R. C., Saumitou-Laprade, P. and Krämer, U. 2004. Two genes encoding *Arabidopsis halleri* MTP1 metal transport proteins cosegregate with zinc tolerance and account for high MTP1 transcript levels. *Plant J* 39:425–43.

Durrett, T. P., Gassmann, W. and Rogers, E. E. 2007. The FRD3-mediated efflux of citrate into the root vasculature is necessary for efficient iron translocation. *Plant Physiol* 144:197–205.

Dushenkov, S., Skarzhinskaya, M., Glimelius, K., Gleba, D. and Raskin, I. 2002. Bioengineering of a phytoremediation plant by means of somatic hybridization. *Int J Phytoremediat* 4:117–126.

Ellis, D. R., Sors, T. G., Brunk, D. G., Albrecht, C., Orser, C., Lahner, B., Wood, K. V., Harris, H. H., Pickering, I. J. and Salt, D. E. 2004. Production of S methyl selenocysteine in transgenic plants expressing selenocysteine methyltransferase. *BMC Plant Biol* 28 (4):1.

Endo, M., Mikami, M., Endo, A., Kaya, H., Itoh, T., Nishimasu, H., Nureki, O. and Toki, S. 2019. Genome editing in plants by engineered CRISPR–Cas9 recognizing NG PAM. *Nature Plants* 5:14–17.

Estrela, R. and Cate, J. H. D. 2016. Energy biotechnology in the CRISPR-Cas9 era. *Curr Opin Biotechnol* 38:79–84.

Evans, K. M., Gatehouse, J. A., Lindsay, W. P., Shi, J., Tommey, A. M. and Robinson, N. J. 1992. Expression of the pea metallothionein like gene Ps MTA in Escherichia coli and *Arabidopsis thaliana* and analysis of trace metal ion accumulation: implications of Ps MTA function. *Plant Mol Biol* 20:1019–1028.

Ezaki, B., Gardner, R. C., Ezaki, Y. and Matsumoto, H. 2000. Expression of aluminium induced genes in transgenic Arabidopsis plants can ameliorate aluminium stress and/or oxidative stress. *Plant Physiol* 122:657–665.

Fan, D., Liu, T., Li, C., Jiao, B., Li, S., Hou, Y. and Luo, K. 2015. Efficient CRISPR/Cas9-mediated targeted mutagenesis in Populus in the first generation. *Sci Rep* 5:12217.

Fang, X., Zhao, Y., Ma, Q., Huang, Y., Wang, P., Zhang, J., Nian, H. and Yang, C. 2013. Identification and comparative analysis of cadmium tolerance associated miRNAs and their targets in two soybean genotypes. *PLoS One* 8:81471.

Fasani, E. 2012. Plants that hyperaccumulate heavy metals. In: Furini A. (ed), *Plants and Heavy Metals*. Springer, Cham, Switzerland, pp:55–74.

Fasani, E., Manara, A., Martini, F., Furini, A. and DalCorso, G. 2018. The potential of genetic engineering of plants for the remediation of soils contaminated with heavy metals. *Plant Cell Environ* 41:1201–1232.

Fulekar, M. H., Singh, A. and Bhaduri, A. M. 2008. Genetic engineering strategies for enhancing phytoremediation of heavy metals. *African J Biotechnol* 8(4):529–535.

Gaj, T., Gersbach, C. A. and Barbas, C. F. 2013. ZFN, TALEN, and CRISPR/Cas-based methods for genome engineering. *Trends Biotechnol* 31(7):397–405.

Gamalero, E. and Glick, B. R. 2012. Plant growth-promoting bacteria and metals phytoremediation. In: Anjum N. A., Pereira M. E., Ahmad I., Duarte A. C., Umar S. and Khan N. A. (eds), *Phytotechnologies: Remediation of Environmental Contaminants*. CRC Press, Boca Raton, pp:361–376.

Gao, J., Luo, M., Peng, H., Chen, F. and Li, W. 2019. Characterization of cadmium-responsive MicroRNAs and their target genes in maize (*Zea mays*) roots. *BMC Mol Biol* 20:14.

Gaur, N., Narasimhulu, K. and Pydisetty, Y. 2018. Recent advances in the bioremediation of persistent organic pollutants and its effect on environment. *J Clean Prod* 198:1602–1631.

Gilbert, L. A., Horlbeck, M. A., Adamson, B., Villalta, J. E., Chen, Y., Whitehead, E. H., Guimaraes, C., Panning, B., Ploegh, H. L., Bassik, M. C., Qi, L. S., Kampmann, M. and Weissman, J. S. 2014. Genome-scale CRISPR-mediated control of Gene repression and activation. *Cell* 159:647–661.

Gisbert, C., Ros, R., De Haro, A., Walker, D. J., Pilar Bernal, M., Serrano, R. and Avino, J. N. 2003. A plant genetically modified that accumulates Pb is especially promising for phytoremediation. *Biochem Biophys Res Commun* 303:440–445.

Glick, B. R. 2010. Using soil bacteria to facilitate phytoremediation. *Biotechnol Adv* 28:367–374.

Gopinath, A., Krishna, K. and Karthik, C. 2020. Adsorptive removal and recovery of heavy metal ions from aqueous solution/effluents using conventional and non- conventional materials. In: *Modern Age Waste Water Problems*. Springer Nature Switzerland AG, pp:309–328.

Goto, F., Yoshihara, T. and Saiki, H. 1998. Iron accumulation in tobacco plants expressing soybean ferritin gene. *Transgenic Res* 7:173–180.

Goto, F., Yoshihara, T., Shigemoto, N., Toki, S. and Takaiwa, F. 1999. Iron accumulation in rice seed by soybean ferritin gene. *Nature Biotechnol* 17:282–286.

Guerinot, M. L. 2000. The ZIP, family of metal transporters. *Biochim Biophys Acta* 1465:190–198.

Guo, J., Xu, L., Su, Y., Wang, H., Gao, S., Xu, J. and Que, Y. 2013. ScMT2-1-3, a metallothionein gene of sugarcane, plays an important role in the regulation of heavy metal tolerance/accumulation. *Bio Med Res Int* 904769:12.

Gupta, O., Sharma, P., Gupta, R. and Sharma, I. 2014. MicroRNA mediated regulation of metal toxicity in plants: present status and future perspectives. *Plant Mol Biol* 84:1–18.

Gustin, J. L., Loureiro, M. E., Kim, D., Na, G., Tikhonova, M. and Salt, D. E. 2009. MTP1-dependent Zn sequestration into shoot vacuoles suggests dual roles in Zn tolerance and accumulation in Zn hyperaccumulating plants. *Plant J* 57:1116–1127.

Gustin, J. L., Zanis, M. J. and Salt, D. E. 2011. Structure and evolution of the plant cation diffusion facilitator family of ion transporters. *BMC Evol Biol* 11:76–89.

Hahn, F. and Nekrasov, V. 2019. CRISPR/Cas precision: do we need to worry about off-targeting in plants? *Plant Cell Rep* 38(4):437–441.

Hammond, J. P.,, Bowen, H. C., White, P. J., Mills, V., Pyke, K. A., Baker, A. J., Whiting, S. N., May, S. T. and Broadley, M. R. 2006. A comparison of the Thlaspi caerulescens and Thlaspi arvense shoot transcriptomes. *New Phytol* 170:239–260.

Hanikenne, M. and Baurain, D. 2014. Origin and evolution of metal P-type ATPases in Plantae (Archaeplastida). *Front Plant Sci* 4:544.

Hanikenne, M., Talke, I. N., Haydon, M. J., Lanz, C., Nolte, A., Motte, P., Kroymann, J., Weigel, D. and Krämer, U. 2008. Evolution of metal hyperaccumulation required cis regulatory changes and triplication of HMA4. *Nature* 453:391–395.

Hasegawa, I., Terada, E., Sunairi, M., Wakita, H., Shinmachi, F., Noguchi, A., Nakajim, M. and Yazaki, J. 1997. Genetic improvement of heavy metal tolerance in plants by transfer of the yeast metallothionein gene (CUPI). *Plant Soil* 196:277–281.

He, H., Ye, Z., Yang, D., Yan, J., Xiao, L., Zhong, T., Yuan, M., Cai, X., Fang, Z. and Jing, Y. 2013. Characterization of endophytic *Rahnella* sp. JN6 from *Polygonum pubescens* and its potential in promoting growth and Cd, Pb, Zn uptake by *Brassica napus*. *Chemosphere* 90:1960–1965.

He, Y., Zhu, M., Wang, L., Wu, J., Wang, Q., Wang, R. and Zhao, Y. 2019. Improvements of TKC technology accelerate isolation of transgene-free CRISPR/Cas9-edited rice plants. *Rice Sci* 26(2):109–117.

Hirschi, K. D., Korenkov, V. D., Wilganowski, N. L. and Wagner, G. J. 2000. Expression of Arabidopsis CAX2 in tobacco altered metal accumulation and increased manganese tolerance. *Plant Physiol* 124:125–133.

Hossain, Z. and Komatsu, S. 2013. Contribution of proteomic studies towards understanding plant heavy metal stress response. *Front Plant Sci* 3:310.

Howden, R., Goldsbrough, P. B., Andersen, C. R. and Cobbett, C. S. 1995. Cadmium-sensitive, cad1 mutants of Arabidopsis thaliana are phytochelatin deficient. *Plant Physiol* 107:1059–1066.

Hsu, P. D., Scott, D. A., Weinstein, J. A., Ran, F. A., Konermann, S., Agarwala, V., Li, Y., Fine, E. J., Wu, X., Shalem, O., Cradick., T. J., Marraffini, L. A., Bao, G. and Zhang, F. 2013. DNA targeting specificity of RNA-guided Cas9 nucleases. *Nat Biotechnol* 31(9):827–832.

Huang, H., Gupta, D. K., Tian, S., Yang, X. and Li, T. 2012. Lead tolerance and physiological adaptation mechanism in roots of accumulating and non-accumulating ecotypes of Sedum alfredii. *Environ Sci Pollut Res* 19:1640–1651.

Huang, J. W., Blaylock, M. J., Kapulnik, Y. and Ensley, B. D. 1988. Phytoremediation of uranium contaminated soils: role of organic acids in triggering uranium hyperaccumulation in plants. *Environ Sci Technol* 32:2004–2008.

Ishikawa, S., Ishimaru, Y., Igura, M., Kuramata, M., Abe, T., Senoura, T., Hase, Y., Arao, T., Nishizawa, N. K. and Nakanishi, H. 2012. Ion-beam irradiation, gene identification, and marker-assisted breeding in the development of low-cadmium rice. *Proc Natl Acad Sci USA* 109:19166–19171.

Ishizaki, T. 2016. CRISPR/Cas9 in rice can induce new mutations in later generations, leading to chimerism and unpredicted segregation of the targeted mutation. *Mol Breed* 36:165.

Jacob, J. M., Karthik, C. S. R. G., Kumar, S. S., Prabakar, D., Kadirvelu, K. and Pugazhendhi, A. 2018. Biological approaches to tackle heavy metal pollution: a survey of literature. *J Environ Manag* 217:56–70.

Jacobs, T. B., La Fayette, P. R., Schmitz, R. J. and Parrott, W. A. 2015. Targeted genome modifications in soybean with CRISPR/Cas9. *BMC Biotechnol* 15:16.

Jam, M., Alemzadeh, A., Tale, A. M. and Esmaeili-Tazangi, S. 2014.Heavy metal regulation of plasma membrane H+-ATPase gene expression in halophyte *Aeluropus littoralis*. *Mol Biol Res Commun* 3(2):129–139.

Jatan, R., Tiwari, S., Asif, M. H. and Lata, C. 2019. Genome-wide profiling reveals extensive alterations in *Pseudomonas putida* mediated miRNAs expression during drought stress in chickpea (*Cicer arietinum* L.). *Environ Exp Bot* 157:217–227.

Jebara, S. H., Chiboub, M. and Jebara, M. 2018. Antioxidant responses and gene level expressions of *Sulla coronaria* inoculated by heavy metals resistant plant growth promoting bacteria under cadmium stress. In: Kallel A., et al. (eds), *Recent Advances in Environmental Science from the Euro-Mediterranean and Surrounding Regions, Advances in Science, Technology & Innovation*. Springer International Publishing AG, New York. https://doi.org/10.1007/978-3-319-70548-4_106.

Jia, H., Zhang, Y., Orbović, V., Xu, J., White, F. F., Jones, J. B. and Wang, N. 2017. Genome editing of the disease susceptibility gene CsLOB1 in citrus confers resistance to citrus canker. *Plant Biotechnol J* 15:817–823.

Jiang, W. Z., Henry., I. M., Lynagh, P. G., Comai, L., Cahoon, E. B. and Weeks, D. P. 2017. Significant enhancement of fatty acid composition in seeds of the allohexaploid, *Camelina sativa*, using CRISPR/Cas9 gene editing. *Plant Biotechnol J* 15:648–657.

Jinek, M., Chylinski, K., Fonfara, I., Hauer, M., Doudna, J. A. and Charpentier, E. 2012. A programmable dual-RNA-guided DNA endonuclease in adaptive bacterial immunity. *Science* 337(6096):816–821.

Johnson, R. A., Gurevich, V., Filler, S. et al. 2015. Comparative assessments of CRISPR-Cas nucleases' cleavage efficiency in planta. *Plant Mol Biol* 87:143–156.

Jung, C., Capistrano-Gossmann, G., Braatz, J., Sashidhar, N. and Melzer, S. 2017. Recent developments in genome editing and applications in plant breeding. *Plant Breed* 137(1):1–9.

Kang, S., Lee, W., An, J. H., Lee, J. H., Kim, Y. H., Kim, H., Oh, Y., Park, Y. H., Jin, Y. B., Jun, B. H., Hur, J. K., Kim, S. U. and Lee, S. H. 2020. Prediction-based highly sensitive CRISPR off-target validation using target-specific DNA enrichment. *Nat Commun* 11:3596.

Kangquan, Y., Caixia, G., Jin-Long, Q. 2017. Progress and prospects in plant genome editing. *Nat Plants*. https :// doi.org/10.1038/nplan ts.2017.107.

Kanwal, R., Fiza, F., Iqra, W. and Muhammed, S. A. A. 2018. Prevalence of exposure of heavy metals and their impact on health consequences. *J Cell Biochem* 119:157–184.

Karthik, C. and Arulselvi, P. I. 2017. Biotoxic effect of chromium (VI) on plant growth- promoting traits of novel *Cellulosimicrobium funkei* strain AR8 isolated from *Phaseolus vulgaris* rhizosphere. *Geomicrobiol J* 34:434–442.

Karthik, C., Oves, M., Thangabalu, R., Sharma, R., Santhosh, S. B. and Arulselvi, P. I. 2016. *Cellulosi microbium* funkei-like enhances the growth of *Phaseolus vulgaris* by modulating oxidative damage under chromium (VI) toxicity. *J Adv Res* 7:839–850.

Karvelis, T., Gasiunas, G., Young, J., Bigelyte, G., Silanskas, A., Cigan, M. and VSiksnys, V. 2015. Rapid characterization of CRISPR-Cas9 protospacers adjacent motif sequence elements. *Genome Biol* 16:253.

Kelly, S. L. and Kelly, D. E. 2013. Microbial cytochromes P450: biodiversity and biotechnology. Where do cytochromes P450 come from, what do they do and what can they do for us? Philos. *Trans R Soc Lond B Biol Sci* 368:20120476.

Khairy, H., Wübbeler, J. H. and Steinbüchel, A. 2015. Biodegradation of the organic disulfide 4,40 -dithiodibutyric acid by *Rhodococcus* spp. *Appl Environ Microbiol* 81:8294–8306.

Khalid, N., Hussain, M., Hameed, M. and Ahmad, R. 2017. Physiological, biochemical and defense system responses of *Parthenium hysterophorus* to vehicular exhaust pollution. *Pak J Bot* 49(1):67–75.

Khalid, N., Noman, A., Sanaullah, T., Akram, M. A. and Aqeel, A. 2018. Vehicle pollution toxicity induced changes in physiology, defence system and biochemical characteristics of *Calotropis procera* L. *Chem Ecol* 34(6):565–568.

Khan, M. S., Zaidi, A., Ahmad, W. P. and Oves, M. 2009. Role of plant growth promoting rhizobacteria in the remediation of metal contaminated soils. *Environ Chem Lett* 7:1–19.

Khanday, I., Skinner, D., Yang, B., Mercier, R. and Sundaresan, V. 2019. A male-expressed rice embryogenic trigger redirected for asexual propagation through seeds. *Nature* 565:91–95.

Kim, K., Tsay, O. G., Atwood, D. A. and Churchill, D. G. 2011. Destruction and detection of chemical warfare agents. *Chem Rev* 111(9):5345–5403.

Kim, S., Takahashi, M., Higuchi, K., Tsunoda, K., Nakanishi, H., Yoshimura, E., Mori, S. and Nishizawa, N. K. 2005. Increased nicotianamine biosynthesis confers enhanced tolerance of High levels of metals, in particular nickel, to plants. *Plant Cell Physiol* 46:1809–1818.

Kim, S. A., Punshon, T., Lanzirotti, A., Li, L., Alonso, J. M., Ecker, J. R., Kaplan, J. and Guerinot, M. L. 2006. Localization of iron in Arabidopsis seed requires the vacuolar membrane transporter VIT1. *Science* 314:1295–1298.

Kim, Y. Y., Choi, H., Segami, S., Cho, H. T., Martinoia, E., Maeshima, M. and Lee, Y. 2009. AtHMA1contributes to the detoxification of excess Zn(II) in Arabidopsis. *Plant J* 58:737–753.

Kohler, A., Blaudez, D., Chalot, M. and Martin, F. 2004. Cloning and expression of multiple metallothioneins from hybrid poplar. *New Phytol* 164:83–93.

Krämer, U. 2010. Metal hyperaccumulation in plants. *Annu Rev Plant Biol* 61:517–534.

Krämer, U., Talke, I. N. and Hanikenne, M. 2007. Transition metal transport. *FEBS Lett* 581:2263–2272.

Kumar, K. V., Singh, N., Behl, H. M. and Srivastava, S. 2008. Influence of plant growth promoting bacteria and its mutant on heavy metal toxicity in *Brassica juncea* grown in fly ash amended soil. *Chemosphere* 72:678–683.

Kuo, H. F. and Chiou, T. J. 2011. The role of microRNAs in phosphorus deficiency signaling. *Plant Physiol* 156:1016–1024.

Kuriakose, S. V. and Prasad, M. N. V. 2008. Cadmium stress affects seed germination and seedling growth in sorghum bicolor (L.) Moench by changing the activities of hydrolysing enzymes. *Plant Growth Regul* 54:143–156.

Larsen, P. B., Geisler, M. J., Jones, C. A., Williams, K. M. and Cancel, J. D. 2005. ALS3 encodes a phloem-localized ABC transporter-like protein that is required for aluminum tolerance in Arabidopsis. *Plant J* 41:353–363.

Lasat, M. M. 2002. Phytoextraction of toxic metals: a review of biological mechanisms. *J Environ Qual* 31:109–20.

Laxman Singh, K., Sudhakar, G., Swaminathan, S. K. and Muralidhar, R. C. 2015. Identification of elite native plants species for phytoaccumulation and remediation of major contaminants in uranium tailing ponds and its affected area. *Environ Dev Sustain* 17:57–81.

LeDuc, D. L., AbdelSamie, M., Móntes-Bayon, M., Wu, C. P., Reisinger, S. J. and Terry, N. 2006. Overexpressing both ATP sulfurylase and selenocysteine methyltransferase enhances selenium phytoremediation traits in Indian mustard. *Environ Pollut* 144:70–76.

Lee, J., Bae, H., Jeong, J., Lee, J. Y., Yang, Y. Y., Hwang, I., Martinoia, E. and Lee, Y. 2003. Functional expression of a bacterial heavy metal transporter in Arabidopsis enhances resistance to and decreases uptake of heavy metals. *Plant Physiol* 133:589–596.

Li, D., Xu, X., Hu, X., Liu, Q., Wang, Z., Zhang, H., Wang, H., Wei, M., Wang, H., Liu, H. and Li, C. 2015a. Genome-wide analysis and heavy metal-induced expression profiling of the HMA gene family in *Populus trichocarpa*. *Front Plant Sci* 6:1149.

Li, J. L., Cui, J. and Cheng, D. Y. 2015b. Computational identification and characterization of conserved miRNAs and their target genes in beet (*Beta vulgaris*). *Genet Mol Res* 14:9103–9108.

Li, R., Fu, D., Zhu, B., Luo, Y. and Zhu, H. 2018. CRISPR/Cas9-mediated mutagenesis of lncRNA1459 alters tomato fruit ripening. *Plant J* 94:513–24.

Li, R., Li, R., Li, X., Fu, D., Zhu, B., Tian, H., Luo, Y. and Zhu, H. 2018. Multiplexed CRISPR/Cas9-mediated metabolic engineering of γ-aminobutyric acid levels in *Solanum lycopersicum*. *Plant Biotechnol J* 16:415–427.

Li, R., Li, R., Li, X. et al. 2017. Multiplexed CRISPR/Cas9-mediated metabolic engineering of gamma-aminobutyric acid levels in Solanum lycopersicum. *Plant Biotechnol J* 16:415–427.

Liang, Z., Chen, K., Zhang, Y., Liu, J., Yin, K., Qiu, J. L. and Gao, C. 2018. Genome editing of bread wheat using biolistic delivery of CRISPR/Cas9 in vitro transcripts or ribonucleoproteins. *Nat Protoc* 13(3):413–430.

Lide, D. R. 2004. *CRC Handbook of Chemistry and Physics 2004–2005: a Ready-Reference Book of Chemical and Physical Data*, 85th ed. CRC Press, Boca Raton, pp:4–26.

Lieber, M. R. 2010. The mechanism of double-strand DNA break repair by the nonhomologous DNA end joining pathway. *Ann Review Biochem* 79:181–211.

Lima, J. C., Arenhart, R. A., Margis-Pinheiro, M. and Margis, R. 2011. Aluminum triggers broad changes in microRNA expression in rice roots. *Genet Mol Res* 10:2817–2832.

Lin, Y. F., Hassan, Z., Talukdar, S., Schat, H. and Aarts, M. G. 2016. Expression of the ZNT1 zinc transporter from the metal hyperaccumulator *Noccaea caerulescens* confers enhanced zinc and cadmium tolerance and accumulation to *Arabidopsis thaliana*. *PLoS One* 11(3):e0149750.

Listgarten, J., Weinstein, M., Kleinstiver, B. P., Sousa, A. A., Joung, J. K., Crawford, J., Gao, K., Hoang, L., Elibol, M., Doench, J. G. and Fusi, N. 2018. Prediction of off-target activities for the end-to-end design of CRISPR guide RNAs. *Nat Biomed Eng* 2:38.

Liu, M. S., Gong, S., Yu, H. H., Jung, K., Johnson, K. A. and Taylor, D. W. 2020. Engineered CRISPR/Cas9 enzymes improve discrimination by slowing DNA cleavage to allow release of off-target DNA. *Nat Commun* 11:3576.

Liu, Q. and Zhang, H. 2012. Molecular identification and analysis of arsenite stress responsive miRNAs in rice. *J Agric Food Chem* 60:6524–6536.

Liu, S., Jiang, J., Liu, Y., Meng, J., Xu, S., Tan, Y., Li, Y., Shu, Q. and Huang, J. 2019. Characterization and evaluation of OsLCT1 and OsNramp5 mutants generated through CRISPR/Cas9-mediated mutagenesis for breeding low Cd rice. *Rice Sci* 26(2):88–97.

Liu, W. and Stewart, C. N. 2015. Plant synthetic biology. *Trends Plant Sci* 20:309–317.

Liu, X., Wu, S., Xu, J., Sui, C. and Wei, J. 2017. Applications of CRISPR/Cas9 in plant biology. *Acta Pharmaceutica Sinica B* 7:292–302.

Lu, H. P., Liu, S. M., Xu, S. L., Chen, W. Y., Zhou, X., Tan, Y. Y., Huang, J. Z. and Shu, Q. Y. 2017. CRISPR-S: an active interference element for a rapid and inexpensive selection of genome-edited, transgene-free rice plants. *Plant Biotechnol J* 15(11):1371–1373.

Lv, Y., Deng, X., Quan, L., Xia, Y. and Shen, Z. 2013. Metallothioneins BcMT1 and BcMT2 from Brassica campestris enhance tolerance to cadmium and copper and decrease production of reactive oxygen species in Arabidopsis thaliana. *Plant Soil* 367:507–519.

Ma, J. F., Ryan, P. R. and Delhaize, E. 2001. Aluminium tolerance in plants and the complexing role of organic acids. *Trends Plant Sci* 6:273–278.

Ma, X., Zhang, Q., Zhu, Q., Liu, W., Chen, Y., Qiu, R., Wang, B., Yang, Z., Li, H., Lin, Y., Xie, Y., Shen, R., Chen, S., Wang, Z., Chen, Y., Guo, J., Chen, L., Zhao, X., Dong, Z. and Liu, Y. G. 2015. A robust CRISPR/Cas9 system for convenient, High-efficiency multiplex genome editing in monocot and dicot plants. *Mol Plant* 8(8):1274–1284.

Ma, X., Zhang, X., Liu, H. and Li, Z. 2020. Highly efficient DNA-free plant genome editing using virally delivered CRISPR–Cas9. *Nat Plants* 6:773–779.

Ma, Y., Oliveira, R. S., Freitas, H. and Zhang, C. 2016. Biochemical and molecular mechanisms of plant-microbe-metal interactions: relevance for phytoremediation. *Front Plant Sci* 7:918.

Ma, Y., Rajkumar, M. and Freitas, H. 2009. Isolation and characterization of Nimobilizing PGPB from serpentine soils and their potential in promoting plant growth and Ni accumulation by Brassica spp. *Chemosphere* 75:719–725.

Maestri, E., Marmiroli, M., Visioli, G. and Marmiroli, N. 2010. Metal tolerance and hyperaccumulation: costs and trade-offs between traits and environment. *Environ Exp Bot* 68:1–13.

Mahajan, P. and Kaushal, J. 2018. Role of phytoremediation in reducing cadmium toxicity in soil and water. *J Toxicol* 2018:4864365.

Mahas, A. and Mahfouz, M. 2018. Engineering virus resistance via CRISPR–Cas systems. *Curr Opin Virol* 32:1–8.

Majumdar, S., Ligon, M., Skinner, W. C., Terns, R. M. and Terns, M. P. 2017. Target DNA recognition and cleavage by a reconstituted type IG CRISPR-Cas immune electorcomplex. *Extremophiles* 21:95–107.

Makarova, K. S., Wolf, Y. I., Alkhnbashi, O. S., Costa, F., Shah, S. A., Saunders, S. J., Barrangou, R., Brouns, S. J., Charpentier, E., Haft, D. H., Horvath, P., Moineau, S., Mojica, F. J., Terns, R. M., Terns, M. P., White, M. F., Yakunin, A. F., Garrett, R. A., van der Oost, J., Backofen, R. and Koonin, E. V. 2015. An updated evolutionary classification of CRISPR-Cas systems. *Nat Rev Microbio* 113(11):722–736.

Malar, S., Vikram, S. S., Favas, P. J. and Perumal, V. 2016. Lead heavy metal toxicity induced changes on growth and antioxidative enzymes level in water hyacinths [Eichhornia crassipes (Mart.)]. *Bot Stud* 55(1):54.

Mali, P., Yang, L., Esvelt, K. M., Aach, J., Guell, M., DiCarlo, J. E., Norville, J. E. and Church, G. M. 2013. RNA-guided human genome engineering via Cas9. *Science* 339(6121):823–826.

Malzahn, A., Lowder, L. and Qi, Y. 2017. Plant genome editing with TALEN and CRISPR. *Cell Biosci* 7(1):21.

Manara, A. 2012. Plant responses to heavy metal toxicity. In: Furini A (ed), *Plants and Heavy Metals*. Springer, Cham Switzerland, pp:27–53.

Mandáková, T., Singh, V., Krämer, U. and Lysak, M. A. 2015. Genome structure of the heavy metal hyperaccumulator Noccaea caerulescens and its stability on metalliferous and Non-metalliferous soils. *Plant Physiol* 169:674–689.

McGrath, S. P., Zhao, F. J. and Lombi, E. 2002. Phytoremediation of metals, metalloids and radionuclides. *Adv Agron* 75:1–56.

Mehta, D., Stürchler, A., Anjanappa, R. B., Zaidi, S. S., Hirsch-Hoffmann, M., Gruissem, W. and Vanderschuren, H. 2019. Linking CRISPR-Cas9 interference in cassava to the evolution of editing-resistant geminiviruses. *Genome Biol* 20(1):80.

Mello-Farias, P. C. and Chaves, A. L. S. 2008. Biochemical and molecular aspects of toxic metals phytoremediation using transgenic plants. In: Tiznado-Hernandez M. E.. Troncoso-Rojas R. and Rivera Domínguez M. A. (eds), *Transgenic Approach in Plant Biochemistry and Physiology.* Research Signpost, Kerala, India, pp:253–266.

Mench, M., Vangronsveld, J., Bleeker, P., Ruttens, A., Geebelen, W. and Lepp, N. 2006. Phytostabilisation of Metal-Contaminated Sites. In: Morel J. L., Echevarria G. and Goncharova N., (eds) *Phytoremediation of Metal-Contaminated Soils.* Springer, Dordrecht. pp:109–190.

Meng, M., Li, B., Shao, J. J., Wang, T., He, B., Shi, J. B., Ye, Z. H. and Jiang, G. B. 2014. Accumulation of total mercury and methyl mercury in rice plants collected from different mining areas in China. *Environ Pollut* 184:179–186.

Mertens, J., VanNevel, L., DeSchrijver, A., Piesschaert, F., Oosterbaan, A., Tack, F. M. and Verheyen, K. 2007. Tree species effect on the redistribution of soil metals. *Environ Pollut* 149:173–181.

Mesa, V., Navazas, A., González-Gil, R., González, A., Weyens, N., Lauga, B., Gallego, J. L. R., Sánchez, J. and Peláez, A. 2017. Use of endophytic and rhizosphere bacteria to improve phytoremediation of arsenic-contaminated industrial soils by autochthonous Betula celtiberica. *Appl Environ Microbiol* 183:e03411–16.

Miglani, G. S. 2017. Genome editing in crop improvement: Present scenario and future prospects. *J Crop Improv* 31:453–559.

Milner, M. J., Seamon, J., Craft, E. and Kochian, L. V. 2013. Transport properties of members of the ZIP family in plants and their role in Zn and Mn homeostasis. *J Exp Bot* 64:369–381.

Misra, S. and Gedamu, L. 1989. Heavy metal tolerant transgenic *Brassica napus* L. and *Nicotiana tabacum* L. Plants. *Theor Appl Genet* 78:161–168.

Miyadate, H., Adachi, S., Hiraizumi, A., Tezuka, K., Nakazawa, N., Kawamoto, T., Katou, K., Kodama, I., Sakurai, K., Takahashi, H., Satoh-Nagasawa, N., Watanabe, A., Fujimura, T. and Akagi, H. 2010. OsHMA3, a P1B-type of ATPase affects root-to-shoot cadmium translocation in rice by mediating efflux into vacuoles. *New Phytol* 189(1):190–199.

Mojica, F., Garcia-Martinez, J. and Soria, E. 2005. Intervening sequences of regularly spaced prokaryotic repeats derive from foreign genetic elements. *J Mol Evol* 60:174–182.

Montanini, B., Blaudez, D., Jeandroz, S., Sanders, D. and Chalot, M. 2007. Phylogenetic and functional analysis of the cation diffusion facilitator (CDF) family: improved signature and prediction of substrate specificity. *BMC Genomics* 8:107.

Morel, M., Crouzet, J., Gravot, A., Auroy, P., Leonhardt, N., Vavasseur, A. and Richaud, P. 2009. AtHMA3, a P1B- ATPase allowing Cd/Zn/Co/Pb vacuolar storage in Arabidopsis. *Plant Physiol* 149:894–904.

Mosa, K. A., Saadoun, I., Kumar, K., Helmy, M. and Dhankher, O. P. 2016. Potential biotechnological strategies for the cleanup of heavy metals and metalloids. *Front Plant Sci* 7:303.

Murphy, A. S., Eisenger, W. R., Shaff, J. E., Kuchian, L. V. and Taiz, L. 1999. Early copper induced leakage of K+ from Arabidopsis seedlings is mediated by ion channels and coupled to citrate efflux. *Plant Physiol* 121:1375–1382.

Nakanishi, H., Ogawa, I., Ishimaru, Y., Mori, S. and Nishizawa, N. K. 2006. Iron deficiency enhances cadmium uptake and translocation mediated by the Fe2þ transporters OsIRT1 and OsIRT2 in rice. *Soil Sci Plant Nutr* 52:464–469.

Nevo, Y. and Nelson, N. 2006. The NRAMP family of metal-ion transporters. *Biochim Biophys Acta* 1763:609–620.

Noman, A. and Aqeel, M. 2017. miRNA-based heavy metal homeostasis and plant growth. *Environ Sci Pollut Res* 24:10068–10082.

Noman, A., Ali, Q., Maqsood, J., Iqbal, N., Javed, M. T., Rasool, N. and Naseem, J. 2018. Deciphering physio-biochemical, yield, and nutritional quality attributes of water-stressed radish (*Raphanus sativus* L.) plants grown from Zn-Lys primed seeds. *Chemosphere* 195:175–189.

Novo, L. A. B., Castro, P. M. L., Alvarenga, P. and da Silva, E. F. 2018. Plant Growth–Promoting rhizobacteria-assisted phytoremediation of mine soils. In: Prasad M. N.V., Favas P. J. C., Maiti S. K. (eds), *Bio-Geotechnologies for Mine Site Rehabilitation*, pp. 281–295. Elsevier. https://doi.org/10.1016/B978-0-12-812986-9.00016-6.

Okuzaki, A., Ogawa, T., Koizuka, C., Kaneko, K., Inaba, M., Imamura, J. and Koizuka, N. 2018. CRISPR/Cas9-mediated genome editing of the fatty acid desaturase 2 gene in *Brassica napus*. *Plant Physiol Biochem* 131:63–69.

Ortiz, D. F., Ruscitti, T., McCue, K. F. and Ow, D. W. 1995. Transport of metal-binding peptides by HMT1, a fission yeast ABC-type vacuolar membrane protein. *J Biol Chem* 270:4721–4728.

Osakabe, Y., Liang, Z., Ren, C., Nishitani, C., Osakabe, K., Wada, M., Komori, S., Malnoy, M., Velasco, R., Poli, M., Jung, M. H., Koo, O. J., Viola, R. and Kanchiswamy, C. N. 2018. CRISPR–Cas9-mediated genome editing in apple and grapevine. *Nat Protoc* 13:2844–2863.

Oves, M., Khan, M. S. and Zaidi, A. 2013. Chromium reducing and plant growth promoting novel strain pseudomonas aeruginosa OSG41 enhance chickpea growth in chromium amended soils. *Eur J Soil Biol* 56:72–83.

Ow, D. W. 1996. Heavy metal tolerance genes-prospective tools for bioremediation. *Resour Conserv Recycl* 18:135–149.

Padmavathiamma, P. K. and Li, L. Y. 2007. Phytoremediation technology: hyper-accumulation metals in plants. *Water Air Soil Pollut* 184:105–126.

Pan, A., Yang, M., Tie, F., Li, L., Chen, Z. and Ru, B. 1994. Expression of mouse metallothionein-1-gene confers cadmium resistance in transgenic tobacco plants. *Plant Mol Biol* 24:341–351.

Pandey, V. C. and Singh, V. 2019. Exploring the potential and opportunities of current tools for removal of hazardous materials from environments. In: Pandey V. C. and Bauddh K. (eds), *Phyto-management of Polluted Sites*. Elsevier, Amsterdam, Netherlands, pp:501–516.

Pandey, V. C., Bajpai, O. and Singh, N. 2016. Energy crops in sustainable phytoremediation. *Renew Sustain Energy Rev* 54:58–73.

Pandita, D. 2018. Plant miRNAs: micro structure and macro character. *Res Rev J Agric Allied Sci* 7:83–84.

Pandita, D. 2019. Plant MIRnome: miRNA biogenesis and abiotic stress response. In: Hasanuzzaman M., Hakeem K., Nahar K. and Alharby H. (eds), *Plant Abiotic Stress Tolerance*. Springer, Cham. https://doi.org/10.1007/978-3-030-06118-0_18.

Pandita, D. and Wani, S. H. 2019. MicroRNA as a tool for mitigating abiotic stress in rice (*oryza sativa* L.). In: Wani S. (ed), *Recent Approaches in Omics for Plant Resilience to Climate Change*. Springer, Cham. https://doi.org/10.1007/978-3-030-21687-0_6.

Pandita, D. and Pandita, A. 2016. Jumping genes-"The other half of the human genome" and the missing heritability conundrum of human genetic disorders. *British Biotechnol J* 11(3):1–18.

Peiter, E., Montanini, B., Gobert, A., Pedas, P., Husted, S., Maathuis, F., Blaudez, M., Chalot, D. and Sanders, M. 2007. A secretory pathway-localized cation diffusion facilitator confers plant manganese tolerance. *Proc Natl Acad Sci USA* 104:8532–8537.

Pence, N. S., Larsen, P. B., Ebbs, S. D., Letham, D. L., Lasat, M. M., Garvin, D. F., Eide, D. and Kochian, L. V. 2000. The molecular physiology of heavy metal transport in the Zn/Cd hyperaccumulator *Thlaspi caerulescens*. *Proc Natl Acad Sci USA* 97(9):4956–4960.

Peng, J. S. and Gong, J. M. 2014. Vacuolar sequestration capacity and long-distance metal transport in plants. *Front Plant Sci* 5:19.

Peng, R., Lin, G. and Li, J. 2016. Potential pitfalls of CRISPR/Cas9-mediated genome editing. *FEBS J* 283:1218–1231.

Peng, R. H., Fu, X. Y., Zhao, W., Tian, Y. S., Zhu, B., Han, H. J., Xu, J. and Yao, Q. H. 2014. Phytoremediation of phenanthrene by transgenic plants transformed with a naphthalene dioxygenase system from pseudomonas. *Environ Sci Technol* 48:12824–12832.

Pianelli, K., Mari, S., Marquès, L., Lebrun, M. and Czernic, P. 2005. Nicotianamine over-accumulation confers resistance to nickel in Arabidopsis thaliana. *Transgenic Res* 14:739–748.

Pilon, M., Cohu, C. M., Ravet, K., Abdel-Ghany, S. E. and Gaymard, F. 2009. Essential transition metal homeostasis in plants. *Curr Opin Plant Biol* 12:347–357.

Pineros, M. A. and Kochian, L. V. 2001. A patch clamp study on the physiology of aluminium toxicity and aluminium tolerance in maize identification and characterization of Al3+ induced anion channels. *Plant Physiol* 125:292–305.

Pollard, A. J., Reeves, R. D. and Baker, A. J. M. 2014. Facultative hyperaccumulation of heavy metals and metalloids. *Plant Sci* 217(218):8–17.

Pratush, A., Kumar, A. and Hu, Z. 2018. Adverse effect of heavy metals (As, Pb, Hg, and Cr) on health and their bioremediation strategies: a review. *Int Microbiol* 21:97–106.

Puchta, H. 2017. Applying CRISPR/Cas for genome engineering in plants: the best is yet to come. *Curr Opin Plant Biol* 36:1–8.

Qin, D., Pei, X. X. and Zhao, L. W. 2017. Differential cadmium distribution and translocation in roots and shoots related to hyper-tolerance between tall fescue and Kentucky bluegrass. *Front Plant Sci* 8:113.

Qiu, Z., Tan, H., Zhou, S. and Cao, L. 2014. Enhanced phytoremediation of toxic metals by inoculating endophytic enterobacter sp. CBSB1 expressing bifunctional glutathione synthase. *J Hazard Mater* 267:17–20.

Rai, V. K. 2002. Role of amino acids in plant responses to stress. *Biol Plant* 45:481–487.

Rajkumar, M. and Freitas, H. 2008. Influence of metal resistant-plant growth promoting bacteria on the growth of *Ricinus communis* in soil contaminated with heavy metals. *Chemosphere* 71:834–842.

Ran, F. A., Cong, L., Yan, W. X., Scott, D. A., Gootenberg, J. S., Kriz, A. J., Zetsche, B., Shalem, O., Wu, X., Makarova, K. S., Koonin, E. V., Sharp, P. A. and Zhang, F. 2015. In vivo genome editing using *Staphylococcus aureus* Cas9. *Nature* 520:186–191.

Reeves, R. D. and Baker, A. J. H. 2000. Metal accumulating plants. In: Raskin I and Ensley (eds), *Phytoremediation of Toxic Metals: Using Plants to Clean up the Environment*. Wiley, New York, pp:193–229.

Ross, S. M. (ed.) 1994. Retention, transformation and mobility of toxic metals in Soils. In: Ross, S. M. (ed.), *Toxic Metals in Soil-Plant Systems*. John Wiley & Sons, Chichester, pp:63–152.

Rugh, C. L., Bizily, S. P. and Meagher, R. B. 2000. Phytoremediation of environmental mercury pollution. In: Raskin, I. and Ensley, B. D. (eds), *Phytoremediation of Toxic Metals Using Plants to Clean up the Environment*. Wiley and Sons, New York, pp:151–171.

Rylott, E. L., Lorenz, A. and Bruce, N. C. 2011. Biodegradation and biotransformation of explosives. *Curr Opin Biotechnol* 22:434–440.

Sasaki, A., Yamaji, N. and Ma, J. F. 2014. Overexpression of OsHMA3 enhances Cd tolerance and expression of Zn transporter genes in rice. *J Exp Bot* 65(20):6013–6021.

Sasaki, A., Yamaji, N., Yokosho, K. and Ma, J. F. 2012. Nramp5 is a major transporter responsible for manganese and cadmium uptake in rice. *Plant Cell* 24:2155–2167.

Sasaki, Y., Hayakawa, T., Inoue, C., Miyazaki, A., Silver, S. and Kusano, T. 2006. Generation of mercury-hyperaccumulating plants through transgenic expression of the bacterial mercury membrane transport protein MerC. *Transgenic Res* 15:615–625.

Satoh-Nagasawa, N., Mori, M., Nakazawa, N., Kawamoto, T., Nagato, Y., Sakurai, K., Takahashi, H., Watanabe, A. and Akagi, H. 2012. Mutations in rice (Oryza sativa) heavy metal ATPase 2 (OsHMA2) restrict the translocation of zinc and cadmium. *Plant Cell Physiol* 53(1):213–224.

Schaaf, G., Ludewig, U., Erenoglu, B. E., Mori, S., Kitahara, T. and von Wirén, N. 2004. ZmYS1 functions as a proton-coupled symporter for phytosiderophore- and nicotianamine-chelated metals. *J Biol Chem* 279:9091–9096.

Schalk, I. J., Hannauer, M. and Braud, A. 2011. New roles for bacterial siderophores in metal transport and tolerance. *Environ Microbiol* 13:2844–2854.

Schindele, P., Wolter, F. and Puchta, H. 2018. Transforming plant biology and breeding with CRISPR/Cas9, Cas12 and Cas13. *FEBS Lett* 592:1954–1967.

Sebastian, A., Shukla, P., Nangia, A. K. and Prasad, M. N. V. 2019. Transgenics in Phytoremediation of Metals and Metalloids: from Laboratory to Field. In: Prasad M. N. V. (ed), *Transgenic Plant Technology for Remediation of Toxic Metals and Metalloids*. Academic Press USA, pp:3–22.

Senoura, T., Shimo, H., Ishikawa, S. and Nishizawa, N. K. 2011. The OsNRAMP1 iron transporter is involved in Cd accumulation in rice. *J Exp Bot* 62:4843–4850.

Senthil, P. K. and Gunasundari, E. 2017. Bioremediation of heavy metals. In: Varjani S., Agarwal A., Gnansounou E. and Gurunathan B. (eds), *Bioremediation: Applications for Environmental Protection and Management. Energy, Environment, and Sustainability*. Springer, Singapore. pp:165–195. https://doi.org/10.1007/978-981-10-7485-1_9.

Shah, T., Andleeb, T., Lateef, S. and Noor, M. A. 2018. Genome editing in plants: advancing crop transformation and overview of tools. *Plant Physiol Biochem* 131:12–21.

Shahabaldin, R., Shazwin, M. T., Mohd, F. M. D., Farrah, A. D. and Hesam, K. 2016. Comprehensive review on phytotechnology: heavy metals removal by diverse aquatic plants species from wastewater: review. *J Hazard Mater* 318:587–599.

Shao, J. F., Fujii-Kashino, M., Yamaji, N., Fukuoka, S., Shen, R. F. and Ma, J. F. 2017. Isolation and characterization of a rice line with high Cd accumulation for potential use in phytoremediation. *Plant Soil* 410:357–368.

Shapiro, R. S., Chavez, A. and Collins, J. J. 2018. CRISPR-based genomic tools for the manipulation of genetically intractable microorganisms. *Nat Rev Microbiol* 16:1.

Shimo, H., Ishimaru, Y., An, G., Yamakawa, T., Nakanishi, H. and Nishizawa, N. K. 2011. Low cadmium (LCD), a novel gene related to cadmium tolerance and accumulation in rice. *J Exp Bot* 62(15):5727–5734.

Song, W. Y., Sohn, E. J., Martinoia, E., Lee, Y. J., Yang, Y. Y., Jasinski, M., Forestier, C., Hwang, I. and Lee, Y. 2003. Engineering tolerance and accumulation of lead and cadmium in transgenic plants. *Nat Biotechnol* 21(8):914–919.

Srivastava, S., Suprasanna, P. S. and D'Souza, S. F. 2012. Mechanisms of arsenic tolerance and detoxification in plants and their application in transgenic technology: a critical appraisal. *Int J Phytoremediat* 14:506–517.

Steinert, J., Schiml, S., Fauser, F. and Puchta, H. 2015. Highly efficient heritable plant genome engineering using Cas9 orthologues from *Streptococcus thermophilus* and *Staphylococcus aureus*. *Plant J* 84:1295–1305.

Subashchandrabose, S. R., Venkateswarlu, K., Krishnan, K., Naidu, R., Lockington, R. and Megharaj, M. 2018. Rhodococcus wratislaviensis strain 9: an efficient p-nitrophenol degrader with a great potential for bioremediation. *J Hazard Mater* 347:176–83.

Suman, J., Uhlik, O., Viktorova, J. and Macek, T. 2018. Phytoextraction of heavy metals: a promising tool for clean-up of polluted environment. *Front Plant Sci* 9:1476.

Sunkar, R., Kaplan, B., Bouché, N., Arazi, T., Dolev, D., Talke, I. N., Maathuis, F. J., Sanders, D., Bouchez, D. and Fromm, H. 2000. Expression of a truncated tobacco NtCBP4 channel in transgenic plants and disruption of the homologous Arabidopsis CNGC1 gene confer Pb2+ tolerance. *Plant J* 24(4):533–542.

Svitashev, S., Schwartz, C., Lenderts, B., Young, J. K. and Mark Cigan, A. 2016. Genome editing in maize directed by CRISPR-Cas9 ribonucleoprotein complexes. *Nat Commun* 7:13274.

Takahashi, R., Bashir, K., Ishimaru, Y., Nishizawa, N. K. and Nakanishi, H. 2012. The role of heavy-metal ATPases, HMAs, in zinc and cadmium transport in rice. *Plant Signal Behav* 7:1799–1801.

Takahashi, R., Ishimaru, Y., Senoura, T., Shimo, H., Ishikawa, S., Arao, T., Nakanishi, H. and Nishizawa, N. K. 2011. The OsNRAMP1 iron transporter is involved in Cd accumulation in rice. *J Exp Bot* 62(14):4843–4850.

Takahashi, R., Shimaru, Y., Shimo, H., Ogo, Y., Senoura, T., Nishizawa, N. K. and Nakanishi, H. 2012. The OsHMA2 transporter is involved in root-to-shoot translocation of Zn and Cd in rice plant. *Cell Environ* 35:1948–1957.

Talke, I. N., Hanikenne, M. and Krämer, U. 2006. Zinc-dependent global transcriptional control, transcriptional deregulation, and higher gene copy number for genes in metal homeostasis of the hyperaccumulator *Arabidopsis halleri*. *Plant Physiol* 142:148–167.

Tang, L., Mao, B. G., Li, Y. K., Lv, Q. M., Zhang, L. P., Chen, C. Y., He, H. J., Wang, W. P., Zeng, X. F., Shao, Y., Pan, Y. L., Hu, Y. Y., Peng, Y., Fu, X. Q., Li, H. Q., Xia, S. T. and Zhao, B. R. 2017. Knockout of OsNramp5 using the CRISPR/Cas9 system produces low Cd-accumulating indica rice without compromising yield. *Sci Rep* 7(1): 14438.

Thode, S. K., Rojek, E., Kozlowski, M., Ahmad, R. and Haugen, P. 2018. Distribution of siderophore gene systems on a vibrionaceae phylogeny: database searches, phylogenetic analyses and evolutionary perspectives. *PLoS One* 13:1–20.

Tiwari, M., Sharma, D., Dwivedi, S., Singh, M., Tripathi, R. D. and Trivedi, P. K. 2014. Expression in *Arabidopsis* and cellular localization reveal involvement of rice NRAMP, OsNRAMP1, in arsenic transport and tolerance. *Plant Cell Environ* 37:140–152.

Turchi, A., Tamantini, I., Camussi, A. M. and Racchi, M. L. 2012. Expression of a metallothionein A1 gene of pisum sativum in white poplar enhances tolerance and accumulation of zinc and copper. *Plant Sci 1*83:50–56.

Ueno, D., Milner, M. J., Yamaji, N., Yokosho, K., Koyama, E., Zambrano, C. M., Kaskie, M., Ebbs, S., Kochian, L. V. and Ma, J. F. 2011. Elevated expression of TcHMA3 plays a key role in the extreme Cd tolerance in a Cd-hyperaccumulating ecotype of *Thlaspi caerulescens*. *Plant J* 66:852–862.

Ueno, D., Yamaji, N., Kono, I., Huang, C. F., Ando, T., Yano, M. and Ma, J. F. 2010. Gene limiting cadmium accumulation in rice. *Proc Natl Acad Sci* 107:16500–16505.

Ullah, A., Heng, S., Munis, M. F. H., Fahad, S. and Yang, X. 2015. Phytoremediation of heavy metals assisted by plant growth promoting (PGP) bacteria: a review. *Environ Exp Bot* 117:28–40.

Uraguchi, S., Kamiya, T., Sakamoto, T., Kasai, K., Sato, Y., Nagamura, Y., Yoshida, A., Kyozuka, J., Ishikawa, S. and Fujiwara, T. 2011. Low-affinity cation transporter (OsLCT1) regulates cadmium transport into rice grains. *Proc Natl Acad Sci* 108:20959–20964.

USEPA. 1996. *The Metals Translator: Guidance for Converting a Total Recoverable Permit Limit from a Dissolved Criterion*, vol. 823. Office of Water EPA, USA (96-007).

Van de Mortel, J. E., Villanueva, L. A., Schat, H., Kwekkeboom, J., Coughlan, S., Moerland, P. D., Themaat, E. V. L., Koornneef, M. and Aarts, M. G. M. 2006. Large expression differences in genes for iron and zinc homeostasis, stress response and lignin biosynthesis distinguish roots of *Arabidopsis thaliana* and the related metal hyperaccumulator *Thlaspi caerulescens*. *Plant Physiol* 142:1127–1147.

Van der Zaal, B. J., Neuteboom, L. W., Pinas, J. E., Chardonnen, A. N., Schat, H., Verkleij, J. A. C. and Hooykaas, P. J. J. 1999. Overexpression of a novel Arabidopsis gene related to putative zinc transporter genes from animals can lead to enhanced zinc resistance and accumulation. *Plant Physiol* 119:1047–55.

Van Huysen, T., Terry, N. and Pilon-Smits, E. A. 2004. Exploring the selenium phytoremediation potential of transgenic Indian mustard overexpressing ATP sulfurylase or cystathione gamma synthase. *Int J Phytoremediat* 6:111–118.

Vazquez-Vilar, M., Bernabé-Orts, J. M., Fernandez-del-Carmen, A., Ziarsolo, P., Blanca, J., Granell, A. and Orzaez, D. 2016. A modular toolbox for gRNA–Cas9 genome engineering in plants based on the Golden braid standard. *Plant Methods* 12:10.

Veillet, F., Perrot, L., Chauvin, L., Kermarrec, M. P., Guyon-Debast, A., Chauvin, J. E., Nogué, F. and Mazier, M. 2019. Transgene-free genome editing in tomato and potato plants using agrobacterium mediated delivery of a CRISPR/Cas9 cytidine base editor. *Int J Mol Sci* 20(2):402.

Vert, G., Grotz, N., Dédaldéchamp, F., Gaymard, F., Guerinot, M. L., Briat, J. F. and Curie, C. 2002. IRT1, an Arabidopsis transporter essential for iron uptake from the soil and for plant growth. *Plant Cell* 14:1223–1233.

Viktorova, J., Jandova, Z., Madlenakova, M., Prouzova, P., Bartunek, V., Vrchotova, B., Lovecka, P., Musilova, L. and Macek, T. 2016. Native phytoremediation potential of Urtica dioica for removal of PCBs and heavy metals can be improved by genetic manipulations using constitutive CaMV 35S promoter. *PLoS One* 11(12):e0167927.

Waltz, E. 2018. With a free pass, CRISPR-edited plants reach market in record time. *Nat Biotechnol* 36(1):6–7.

Wang, B., Cheng, D., Chen, Z., Zhang, M., Zhang, G., Jiang, M. and Tan, M. 2019. Bioinformatic exploration of the targets of xylem sap miRNAs in maize under cadmium stress. *Int J Mol Sci* 20:1474.

Wang, Y., Ren, H., Pan, H., Liu, J. and Zhang, L. 2015. Enhanced tolerance and remediation to mixed contaminates of PCBs and 2, 4-DCP by transgenic alfalfa plants expressing the 2, 3-dihydroxybiphenyl-1,2-dioxygenase. *J Hazard Mater* 286:269–275.

White, P. J. and Broadley, M. R. 2011. Physiological limits to zinc biofortification of edible crops. *Front Plant Sci* 2:80.

Williams, L. E. and Mills, R. F. 2005. P1B-ATPases-an ancient family of transition metal pumps with diverse functions in plants. *Trends Plant Sci* 10:491–502.

Williams, L. E., Pittman, J. K. and Hall, J. L. 2000. Emerging mechanisms for heavy metal transport in plants. *Biochim Biophys Acta* 1465:104–126.

Wiszniewska, A., Hanus-Fajerska, E., Smolen´, S. and Muszyn´ska, E. 2015. In vitro selection for lead tolerance in shoot culture of daphne species. *Acta Sci Pol Hortorum Cultus* 14:129–142.

Wolt, J. D., Wang, K., Sashital, D. and Lawrence-Dill, C. J. 2016. Achieving plant CRISPR targeting that limits off-target effects. *Plant Genome* 9(3):1–8.

Wu, S. C., Cheung, K. C., Luo, Y. M. and Wong, M. H. 2006. Effects of inoculation of plant growth-promoting rhizobacteria on metal uptake by *Brassica juncea*. *Environ Pollut* 140:124–135.

Wu, X., Cobbina, S. J., Mao, G., Xu, H., Zhang, Z. and Yang, L. 2016. A review of toxicity and mechanisms of individual and mixtures of heavy metals in the environment. *Environ Sci Pollut Res* 23:8244–8259.

Xia, Y., Qi, Y., Yuan, Y., Wang, G., Cui, J., Chen, Y., Zhang, H. and Shen, Z. 2012. Overexpression of Elsholtzia haichowensis metallothionein 1 (EhMT1) in tobacco plants enhances copper tolerance and accumulation in root cytoplasm and decreases hydrogen peroxide production. *J Hazard Mater* 233:65–71.

Xie, K. B. and Yang, Y. N. 2013. RNA-guided genome editing in plants using a CRISPR-Cas system. *Mole Plant* 6:1975–1983.

Xie, Q. E., Yan, X. L., Liao, X. Y. and Li, X. 2009. The arsenic hyperaccumulator fern *Pteris vittata* L. *Environ Sci Technol* 43:8488–8495.

Xu, D. and Pei, J. 2011. Construction and characterization of a photosynthetic bacterium genetically engineered for Hg2þ uptake. *Bioresour Technol* 102:3083–3088.

Xu, R. F., Li, H., Qin, R. Y., Wang, L., Li, L., Wei, P. C. and Yang, J. B. 2014. Gene targeting using the *Agrobacterium tumefaciens*-mediated CRISPR-Cas system in rice. *Rice* 7(1):5.

Yamaji, N., Xia, J. X., Mitani-Ueno, N., Yokosho, K. and Ma, J. F. 2013. Preferential delivery of zinc to developing tissues in rice is mediated by P-type heavy metal ATPase OsHMA2. *Plant Physiol* 162(2):927–939.

Yang, J., Li, W., Ng, T. B., Deng, X. and Lin, J. 2017. Laccases: production, expression regulation, and applications in pharmaceutical biodegradation. *Front Microbiol* 8:832.

Yang, J., Liu, D., Wang, X., Ji, C., Cheng, F., Liu, B., Hu, Z., Chen, S., Pental, D., Ju, Y., Yao, P., Li, X., Xie, K., Zhang, J., Wang, J., Liu, F., Ma, W., Shopan, J., Zheng, H., Mackenzie, S. A. and Zhang, M. 2016. The genome sequence of allopolyploid *Brassica juncea* and analysis of differential homoeolog gene expression influencing selection. *Nat Genet* 48:1225–1232.

Yang, Z. M. and Chen, J. 2013. A potential role of microRNAs in plant response to metal toxicity. *Metallomics* 5:1184–1190.

Yin, K., Han, T., Liu, G., Chen, T., Wang, Y., Yu, A. Y. L. and Liu, Y. 2015. A geminivirus-based guide RNA delivery system for CRISPR/Cas9 mediated plant genome editing. *Sci Rep* 5:14926.

Yong, X., Chen, Y., Liu, W., Xu, L., Zhou, J., Wang, S., Chen, P., Ouyang, P. and Zheng, T. 2014. Enhanced cadmium resistance and accumulation in *Pseudomonas putida* KT2440 expressing the phytochelatin synthase gene of Schizo saccharomyces pombe. *Lett Appl Microbiol* 58:255–261.

Zhang, C., Konermann, S., Brideau, N. J., Lotfy, P., Novick, S. J., Strutzenberg, T., Griffin, P. R., Hsu, P. D. and Lyumkis, D. 2018. Structural basis for the RNA-guided ribonuclease activity of CRISPR-Cas13d. *Cell* 175:212–223.

Zhang, Q., Xing, H. L., Wang, Z. P., Zhang, H. Y., Yang, F., Wang, X. C. and Chen, Q. J. 2018a. Potential high-frequency off-target mutagenesis induced by CRISPR/Cas9 in Arabidopsis and its prevention. *Plant Mol Biol* 96:445–456.

Zhang, X., Jin, M., Tadesse, N., Dang, J., Zhou, T., Zhang, H., Wang, S., Guo, Z. and Ito, Y. 2018. Dioscorea zingiberensis C. H. Wright: an overview on its traditional use, phytochemistry, pharmacology, clinical applications, quality control, and toxicity. *J Ethnopharmacol* 220:283–293.

Zhang, Y., Liang, Z., Zong, Y., Wang, Y., Liu, J., Chen, K., Qui, J. and Gao, C. 2016. Efficient and transgene-free genome editing in wheat through transient expression of CRISPR/Cas9 DNA or RNA. *Nat Commun* 7:12617.

Zhaoyu, K., Zhenshan, D., Bernard, R. G., Gehong, W. and Minxia, C. 2017. A nodule endophytic plant growth-promoting Pseudomonas and its effects on growth, nodulation and metal uptake in *Medicago lupulina* under copper stress. *Ann Microbiol* 67:49–58.

Zhou, H., Liu, Q., Li, J., Jiang, D., Zhou, L., Wu, P., Lu, S., Li, F., Zhu, L. and Liu, Z. 2012a. Photoperiod and thermosensitive genic male sterility in rice are caused by a point mutation in a novel noncoding RNA that produces a small RNA. *Cell Res* 22:649–660.

Zhou, J., Jiao, F., Wu, Z., Li, Y., Wang, X., He, X., Zhong, W. and Wu, P. 2008. OsPHR2 is involved in phosphate- starvation signaling and excessive phosphate accumulation in shoots of plants. *Plant Physiol* 146:1673–1686.

Zhou, L., Peng, R., Zhang, R. and Li, J. 2018. The applications of CRISPR/Cas system in molecular detection. *J Cell Mol Med* 22:5807–5815.

Zhou, Z. S., Song, J. B. and Yang, Z. M. 2012b. Genome wide identification of Brassica napus microRNAs and their targets in response to cadmium. *J Exp Bot* 63:4597–461.

Zhu, Y. G., Williams, P. N. and Meharg, A. A. 2008. Exposure to inorganic arsenic from rice: a global health issue? *Environ Pollut* 154:169–171.

Index

H

I

J

K

L

M

N

O

P

Q

R

S

T